MANUAL FOR PROCESS ENGINEERING CALCULATIONS

MANUAL FOR PROCESS ENGINEERING CALCULATIONS

LOYAL CLARKE

Chief Process Engineer, Organics Division Engineering
Olin Mathieson Chemical Corporation
New Haven, Conn.

ROBERT L. DAVIDSON

Editor, Petro/Chem Engineer, Dallas

SECOND EDITION

McGRAW-HILL BOOK COMPANY

New York San Francisco Toronto London Sydney

MANUAL FOR PROCESS ENGINEERING CALCULATIONS

 Library of Congress Catalog Card Number: 61-14353

11249

3 4 5 6 - MP - 10 9

PREFACE

This book is intended as a convenience for chemical engineers and others who may be required to present quick answers without adequate time for recourse to the many comprehensive texts and reference books. It is not intended as a substitute for either. An accurate and complete answer to many problems will require more information than is contained here. Industry is full of equipment that has been installed because someone timidly ventured, "It seems about the right size." To increase the accuracy in this kind of guesswork every process engineer has his own file of useful data charts and whatnot. These he makes as brief as possible for ready reference. This is essentially such a set of notes; it is only a start toward a handy set of working data for rapid-process calculations. The hope is that it may be found useful.

The second edition enlarges coverage of fluid flow, heat transfer, mass transfer, and many other subjects. The chapter on heat transfer has been entirely rewritten with more emphasis on fundamental data and theory, and it features use of the "transfer unit" as a design tool. The chapter on absorption, stripping, and distillation has been replaced by two chapters: one covering phase equilibria; the other covering absorption, stripping, distillation, and liquid-liquid extraction under the heading of Countercurrent Separations. The latter chapter is organized around a consolidated treatment of the features common to the four techniques.

The presentation of data includes discussion of methods of application to process design calculations and gives a partial exposition of the basic underlying theories. The novice is cautioned about the danger in the use of data beyond the range of validity; he is, moreover, advised of the desirability of making estimates by more than one method, using basic data from different sources.

This book includes compilations of data from numerous sources. Indebtedness is acknowledged to the various individuals who have given assistance. In particular, M. H. Douthitt and the late J. L. Schlitt made valuable contributions to the first edition. Dr. Leland M. Reed offered suggestions that have strengthened the chapter on heat transfer. E. H. Lebeis reviewed the chapter on countercurrent separations and contributed constructive suggestions.

Loyal Clarke
Robert L. Davidson

CONTENTS

CHAPTER A

PERSPECTIVE

This chapter indulges in a discussion of the nature and philosophy of process engineering, presents some general references, and outlines the nomenclature used.

1. Philosophical Notes

a. Process Engineers and Definitions.—Philosophy and definitions are hopelessly intertwined. Tell me your definitions and I will tell you your philosophy. One ordinarily thinks of an engineer as one who deals with precisely defined terms suited for detailed analyses of problems. This is true, but it should not be overemphasized. As "go-betweens," dealing with assistant operators and vice-presidents in charge of research, all engineers must put up with vaguely defined terms since the pipe fitter does not understand the precise terminology of engineering, and the plant superintendent—who does—refuses to be bothered. It is appropriate that this should be so. Very specific terms are essential for getting down to details; general terms are needed, if consideration of the broad aspects of a problem is not to be clouded by useless detail. To put it another way, engineering problems require flexible thinking, which is favored by the inclusion of flexible terms in the vocabulary.

The division of science and engineering into various fields is necessary because a lifetime is insufficient for an individual to become conversant with more than a small fraction of the sum total of knowledge. It is also important that the division be a liberal one. An attempt to group into definite compartments would fail because many gaps would result. Moreover, the scientists and engineers would not tolerate such a restriction of their activities. It is better to define the centers of interest and allow each individual to expand in as many directions as his ability permits.

The process engineer's principal duty is to translate the findings of the laboratory into a basic commercial plant design ready for the designers and estimators to complete the final mechanical design and cost estimates. His chief weapons are the slide rule, graph sheets, sketch pads, reference books, a sharp pencil, and his background of experience, education, and common sense. A good education in chemical engineering is essential, and some additional mathematics, physics, chemistry, and mechanical engineering thrown in do no harm. Equally essential is common sense; process engineering is an art as well as a science. There are usually so many ways of manufacturing anything, that the engineer must confine his attention to only the most promising, in order to keep within the allotted time.

b. Operating against Investment Costs.—Modern methods have progressed to the point where a plant may easily be overmechanized. The question of balance between investment and operating costs requires careful consideration.

In public utilities, rates are usually set by commissions who base sales prices on just rates on capital invested. This, together with the stability of demand, favors high investments. In other industries the vagaries of market prices of raw materials and products often do not permit investment on a long-term basis so that a plant that cannot "pay out" in a short time is considered a doubtful venture. The sum of annual interest, depreciation, maintenance, taxes, and overhead charges made by the cost accountant is usually at 20 per cent of the gross capital outlay. In making very rough estimates, the costs of major items may be added together, but the sum will represent only a small fraction of the total outlay, which must include piping, wiring, sewers, foundations, yard structures, various construction overheads, etc. The inclusion of these items may *more than triple* the total.

c. The Case for Inefficiency.—The process engineer, whether trained originally as a physicist, mechanical engineer, chemist, or chemical engineer, is a scientist and is inclined to make a fetish of efficiency. Frequently, therefore, he has the natural tendency always to choose apparatus and processes that give the highest engineering performance. The results achieved for each dollar expended are the true measures of industrial utility. By the choice of less efficient processes or equipment, the investment cost may often be lowered with no alarming increase in operating cost.

Consequently, the process engineer should avoid specifying heat exchangers with too close approach, steam turbines of very high efficiencies, etc. On the other hand, the use of added utilities also represents capital outlay in the power plant, water plant, and piping.

d. Sizing of Equipment.—If plant equipment is oversized, the investment is unnecessarily large. If equipment is too small, production suffers. Since it is often impossible to state capacities exactly, reasonable factors of safety for different types of equipment must be adopted. These factors depend on the relative uncertainties involved and on the costs of purchasing oversize equipment. They should, when possible, be adopted as a result of discussion between design engineers, process engineers, estimators, and those charged with operation. Only one thing will be stressed here: regardless of the method used in choosing the factors of safety, a summary of them for each piece of equipment should be recorded in one place. Hidden factors of safety introduced by one member of the staff may be overlooked by others and thereby compounded.

e. Process Control.—The process engineer can usually compute suitable flows for good operation. The whole process from there rises or falls on the ability to operate the plant at these flows. Consequently, simplicity in the control system and other factors favoring ease of operations are important considerations in choosing between competitive proposals.

f. The Case for Efficiency.—The preceding comment stresses simplicity and low cost as factors of first importance. There is no intent to disparage efficiency. The modern trend has been toward ever-increasing performance factors. An inefficient plant is not likely to be competitive in present-day industry. Usually the management would be glad to spend 7 or 8 per cent more to secure a machine of 5 per cent greater efficiency but would balk at double the cost.

To be most successful, the plant must combine efficiency and simplicity. This sounds like a difficult assignment, and it is. Usually, solutions are available and will be found, provided the various engineers on the project have imagination and are ever conscious of the need of securing the best performance with economy.

g. Process Calculations.—The first consideration is the over-all process: determining the quantity of products to be made and the raw materials required. This may be divided into a number of separate units. Finally, the process is broken down to *unit operations*. For many of these, standard equipment is employed, and methods of estimation based on theory and experience are available. These procedures are frequently sufficiently simple and standardized so that they may be used without much knowledge of theory.

When new types of equipment are encountered, recourse must be made to the basic chemical, physical, and engineering principles before it can be stated whether any of the standard methods are applicable. A surprising number of problems may be wholly or partly solved by (1) material balances, (2) energy balances, (3) consideration of equilibrium conditions at key points, and (4) equipment performance factors. Material and energy balances are applicable to the whole process, any unit, any single piece of equipment, or any minute operation. A consideration of the mechanism is necessary to determine where close approaches to equilibrium are involved. Performance factors are certain only when experimentally determined, though analogies to other types of equipment may afford a guide. The twins, trial and error, are a great nuisance, but without them process engineers would be helpless in cases where recycles between units make everything dependent on everything else. Frequently something must be assumed in order to get started.

The process engineer should not complain because all roads are not paved. Rather he should rejoice that there are still frontiers affording the opportunity to use his ingenuity to the fullest extent.

2. General References

The beginning of each chapter gives a list of references pertaining to the particular subject. Here are listed some comprehensive texts which cover a number of fields of interest to process engineers. Only treatises in the English language are included.

a. General Source Material:

Dyson, "A Short Guide to Chemical Literature," Longmans, Green & Co., Ltd., London, 1951.
Crane, Patterson, and Marr, "A Guide to the Literature of Chemistry," John Wiley & Sons, Inc., New York, 1957.
"Chemical Abstracts," published by the American Chemical Society, Washington, D.C.
"British Chemical Abstracts," published by the Chemical Society, London, England.
"Engineering Index," published by Engineering Index, Inc., New York.
Kirk and Othmer, "Encyclopedia of Chemical Technology," Interscience Publishers, Inc., New York.

b. Chemistry and Physics:

"International Critical Tables," McGraw-Hill Book Company, Inc., New York.
Lange, "Handbook of Chemistry," 10th ed., McGraw-Hill Book Company, Inc., New York, 1961.
Taylor and Glasstone, "Treatise of Physical Chemistry," 3d ed., D. Van Nostrand Company, Inc., Princeton, N.J., 1942.
Mellor, "A Comprehensive Treatise of Inorganic and Theoretical Chemistry," Longmans, Green & Co., Inc., New York.
Whitmore, "Organic Chemistry," 2d ed., D. Van Nostrand Company, Inc., Princeton, N.J., 1951.
Timmermanns, "Physico-chemical Constants of Pure Organic Compounds," Elsevier Publishing Company, Amsterdam, 1950.
Reid and Sherwood, "The Properties of Gases and Liquids," McGraw-Hill Book Company, Inc., New York, 1958.

c. Industrial Chemistry:

Shreve, "The Chemical Process Industries," 2d ed., McGraw-Hill Book Company, Inc., New York, 1956.

Groggins, "Unit Processes in Organic Synthesis," 5th ed., McGraw-Hill Book Company, Inc., New York, 1958.

Hougen, Watson, and Ragatz, "Chemical Process Principles: Part I—Materials and Energy Balances; Part II—Thermodynamics," John Wiley & Sons, Inc., New York, 1954 and 1959.

Hougen and Watson, "Chemical Process Principles: Part III—Kinetics and Catalysis," John Wiley & Sons, Inc., New York, 1947.

Mantell, "Electrochemical Engineering," 4th ed. of "Industrial Electrochemistry," McGraw-Hill Book Company, Inc., New York, 1960.

d. Chemical Engineering:

Perry, "Chemical Engineers' Handbook," 3d ed., McGraw-Hill Book Company, Inc., New York, 1950.

McCabe and Smith, "Unit Operations of Chemical Engineering," McGraw-Hill Book Company, Inc., New York, 1956.

Badger and Banchero, "Introduction to Chemical Engineering," McGraw-Hill Book Company, Inc., 1955.

Foust, Wenzel, Clump, Maus, and Anderson, "Principles of Unit Operations," John Wiley & Sons, Inc., New York, 1960.

Vilbrandt and Dryden, "Chemical Engineering Plant Design," 4th ed., McGraw-Hill Book Company, Inc., New York, 1959.

Riegel, "Chemical Machinery," 5th ed., Reinhold Publishing Corporation, New York, 1949.

Hesse and Rushton, "Chemical Equipment Design," D. Van Nostrand Company, Princeton, N.J., 1944.

Nelson, "Petroleum Refinery Engineering," 4th ed., McGraw-Hill Book Company, Inc., New York, 1958.

e. Other Engineering Fields:

Knowlton, "Standard Handbook for Electrical Engineers," 9th ed., McGraw-Hill Book Company, Inc., 1957.

Marks, "Mechanical Engineers' Handbook," 6th ed., McGraw-Hill Book Company, Inc., New York, 1958.

Urquhart, "Civil Engineering Handbook," 4th ed., McGraw-Hill Book Company, Inc., New York, 1959.

f. Legal:

Buckles, "Ideas, Inventions, and Patents," John Wiley & Sons, Inc., New York, 1957.

Canfield and Bowman, "Business, Legal and Ethical Phases of Engineering," 2d ed., McGraw-Hill Book Company, Inc., 1954.

g. Economics:

Perry, "Chemical Business Handbook," McGraw-Hill Book Company, Inc., New York, 1954.

Happel, "Chemical Process Economics," John Wiley & Sons, Inc., New York, 1958.
Peters, "Plant Design and Economics for Chemical Engineers," McGraw-Hill Book Company, Inc., New York, 1958.

h. Manufacturers:

"Thomas' Register," published annually by the Thomas Publishing Company, New York.
"Chemical Engineering Catalog," published annually by Reinhold Publishing Corporation, New York.

i. Miscellaneous:

Comings, "High Pressure Technology," McGraw-Hill Book Company, Inc., New York, 1956.
Mickley, Sherwood, and Reed, "Applied Mathematics in Chemical Engineering," 2d ed., McGraw-Hill Book Company, Inc., New York, 1957.
Pierce, "A Short Table of Integrals," 3d ed., Ginn & Company, Boston.
Stibitz and Larrivee, "Mathematics and Computers," McGraw-Hill Book Company, Inc., New York, 1956.
Considine, "Process Instruments and Controls Handbook," McGraw-Hill Book Company, Inc., New York, 1957.

3. Nomenclature

An effort has been made throughout the book to define the symbols used in each equation on the same or at least on an adjacent page. Below are listed the more frequent symbols used throughout the book. The standards of the American Institute of Chemical Engineers[1] have been used as the general basis. However, several changes were found necessary to prevent confusion.

A Work content (Helmholz free energy); absorption factor
a Activity; relative volatility
α Ostwald coefficient of (gas) solubility
β Bunsen coefficient of (gas) solubility; ratio of orifice to pipe diameter
C Coefficient
C_p Heat capacity [Btu/(lb mole)(°F)] at constant pressure
C_v Heat capacity [Btu/(lb mole)(°F)] at constant volume
c_p Specific heat [Btu/(lb)(°F)] at constant pressure
c_v Specific heat [Btu/(lb)(°F)] at constant volume
D Diameter (larger of two); diffusivity
d Diameter (smaller of two); differential operator
Δ Prefixed to indicate that value is change accompanying some process (*i.e.*, Δp would be a pressure drop)

[1] 1946 version; new designation of dimensionless numbers has not been adopted.

E	Efficiency
e	Base of natural logarithms = 2.718
ϵ	Emissivity; roughness; efficiency parameter; fraction of voids
F	Free energy (Lewis); enrichment factor
f	Friction factor; fugacity; functional relationship
G	Mass (or mole) velocity; mole flow rate
g	Acceleration due to gravity
γ	Ratio of specific heats at constant pressure, volume
H	Heat content (enthalpy); Henry's-law constant in pressure units
h	Fluid head; heat content of liquid when gas = H
I	Liquid irrigation rate in packed columns
J	Mechanical equivalent of heat
j	j factor
K	Equilibrium constant for chemical reactions; coefficient of discharge from orifice including velocity of approach
k	Thermal conductivity; vaporization equilibrium constant = y/x
L	Latent heat of vaporization; length of pipe; mass (or mole) flow of liquid corresponding to G for gas in same system
l	Length or thickness
ln	Denotes "natural logarithm of . . . "
M	Flow of gas, standard cfm
m	Flow, gpm; molecular weight
μ	Viscosity; Joule-Thomson coefficient
N	Rpm; mole fraction (liquid or gas)
n	Number (as number of theoretical plates)
ν	Kinematic viscosity
P	Percentage
Pr	Prandtl number
p	Pressure
π	Ratio of circle circumference to diameter = 3.1416
Ψ	Shape parameter for annuli
Q	Heat absorbed
R	Gas-law constant
Re	Reynolds number
r	Radius; compression ratio
ρ	Density
S	Entropy; specific gravity of gases (air = 1); stripping factor; length of stroke
St	Stanton number
s	Specific gravity of liquids (water = 1); entropy of liquid when gas = S

- T Absolute temperature (°R or °K)
- t Relative temperature (°F or °C)
- θ Time
- U Intrinsic (internal) energy; over-all coefficient of heat transfer
- u Velocity; mean velocity
- V Volume
- W Weight flow
- w Work (done by system); weight per cent
- X Mole ratio, liquids
- x Mole fraction in liquid
- Y Expansion factor (for gases through orifices); mole ratio, gases
- y Mole fraction in gas
- Z Height above datum plane; height of transfer unit; mole ratio in second liquid phase
- z Compressibility factor in reduced equation of state; fluid head; mole fraction in second liquid phase

CHAPTER B

NUMERICAL AND MATHEMATICAL DATA

1. Logarithms

Table B-1 gives four-place logarithms; Table B-2 gives four-place antilogarithms. Natural logarithms may be obtained by multiplying the common logarithms (to base 10) by 2.3026.

$$\ln x = (\ln 10) \cdot \log x$$
$$= 2.3026 \log x$$

The table of antilogs may be used to evaluate exponentials as follows:

$$10^x = \text{antilog } x$$
$$e^x = \text{antilog}\left(\frac{x}{2.3026}\right)$$

Powers of numbers can be computed as follows:

$$a \log x = \log (x^a)$$

i.e.,

$$x^a = \text{antilog } (a \log x)$$

Logarithmic tables are occasionally used for multiplication and division as follows:

$$a \times b \div c = \text{antilog } (\log a + b - \log c)$$

It should be noted that the tables cover only the base logarithms, i.e., mantissas, of numbers from 1 to 9.99. The complete logarithms of larger or smaller numbers are obtained by adding or subtracting 1 for every factor of ten, i.e.:

$$\log 2 = 0.3010$$
$$\log 20{,}000 = 0.3010 + 4 \times 1 = 4.3010$$
$$\log 0.02 = 0.3010 - 2 = -1.6990$$

TABLE B-1.—FOUR-PLACE LOGARITHMS

N.	0	1	2	3	4	5	6	7	8	9	Proportional parts				
											1	2	3	4	5
10	0000	0043	0086	0128	0170	0212	0253	0294	0334	0374	4	8	12	17	21
11	0414	0453	0492	0531	0569	0607	0645	0682	0719	0755	4	8	11	15	19
12	0792	0828	0864	0899	0934	0969	1004	1038	1072	1106	3	7	10	14	17
13	1139	1173	1206	1239	1271	1303	1335	1367	1399	1430	3	6	10	13	16
14	1461	1492	1523	1553	1584	1614	1644	1673	1703	1732	3	6	9	12	15
15	1761	1790	1818	1847	1875	1903	1931	1959	1987	2014	3	6	8	11	14
16	2041	2068	2095	2122	2148	2175	2201	2227	2253	2279	3	5	8	11	13
17	2304	2330	2355	2380	2405	2430	2455	2480	2504	2529	2	5	7	10	12
18	2553	2577	2601	2625	2648	2672	2695	2718	2742	2765	2	5	7	9	12
19	2788	2810	2833	2856	2878	2900	2923	2945	2967	2989	2	4	7	9	11
20	3010	3032	3054	3075	3096	3118	3139	3160	3181	3201	2	4	6	8	11
21	3222	3243	3263	3284	3304	3324	3345	3365	3385	3404	2	4	6	8	10
22	3424	3444	3464	3483	3502	3522	3541	3560	3579	3598	2	4	6	8	10
23	3617	3636	3655	3674	3692	3711	3729	3747	3766	3784	2	4	5	7	9
24	3802	3820	3838	3856	3874	3892	3909	3927	3945	3962	2	4	5	7	9
25	3979	3997	4014	4031	4048	4065	4082	4099	4116	4133	2	3	5	7	9
26	4150	4166	4183	4200	4216	4232	4249	4265	4281	4298	2	3	5	7	8
27	4314	4330	4346	4362	4378	4393	4409	4425	4440	4456	2	3	5	6	8
28	4472	4487	4502	4518	4533	4548	4564	4579	4594	4609	2	3	5	6	8
29	4624	4639	4654	4669	4683	4698	4713	4728	4742	4757	1	3	4	6	7
30	4771	4786	4800	4814	4829	4843	4857	4871	4886	4900	1	3	4	6	7
31	4914	4928	4942	4955	4969	4983	4997	5011	5024	5038	1	3	4	6	7
32	5051	5065	5079	5092	5105	5119	5132	5145	5159	5172	1	3	4	5	7
33	5185	5198	5211	5224	5237	5250	5263	5276	5289	5302	1	3	4	5	6
34	5315	5328	5340	5353	5366	5378	5391	5403	5416	5428	1	3	4	5	6
35	5441	5453	5465	5478	5490	5502	5514	5527	5539	5551	1	2	4	5	6
36	5563	5575	5587	5599	5611	5623	5635	5647	5658	5670	1	2	4	5	6
37	5682	5694	5705	5717	5729	5740	5752	5763	5775	5786	1	2	3	5	6
38	5798	5809	5821	5832	5843	5855	5866	5877	5888	5899	1	2	3	5	6
39	5911	5922	5933	5944	5955	5966	5977	5988	5999	6010	1	2	3	4	5
40	6021	6031	6042	6053	6064	6075	6085	6096	6107	6117	1	2	3	4	5
41	6128	6138	6149	6160	6170	6180	6191	6201	6212	6222	1	2	3	4	5
42	6232	6243	6253	6263	6274	6284	6294	6304	6314	6325	1	2	3	4	5
43	6335	6345	6355	6365	6375	6385	6395	6405	6415	6425	1	2	3	4	5
44	6435	6444	6454	6464	6474	6484	6493	6503	6513	6522	1	2	3	4	5
45	6532	6542	6551	6561	6571	6580	6590	6599	6609	6618	1	2	3	4	5
46	6628	6637	6646	6656	6665	6675	6684	6693	6702	6712	1	2	3	4	5
47	6721	6730	6739	6749	6758	6767	6776	6785	6794	6803	1	2	3	4	5
48	6812	6821	6830	6839	6848	6857	6866	6875	6884	6893	1	2	3	4	4
49	6902	6911	6920	6928	6937	6946	6955	6964	6972	6981	1	2	3	4	4
50	6990	6998	7007	7016	7024	7033	7042	7050	7059	7067	1	2	3	3	4
51	7076	7084	7093	7101	7110	7118	7126	7135	7143	7152	1	2	3	3	4
52	7160	7168	7177	7185	7193	7202	7210	7218	7226	7235	1	2	2	3	4
53	7243	7251	7259	7267	7275	7284	7292	7300	7308	7316	1	2	2	3	4
54	7324	7332	7340	7348	7356	7364	7372	7380	7388	7396	1	2	2	3	4
N.	0	1	2	3	4	5	6	7	8	9	1	2	3	4	5

TABLE B-1.—FOUR-PLACE LOGARITHMS (*Concluded*)

N.	0	1	2	3	4	5	6	7	8	9	Proportional parts 1	2	3	4	5
55	7404	7412	7419	7427	7435	7443	7451	7459	7466	7474	1	2	2	3	4
56	7482	7490	7497	7505	7513	7520	7528	7536	7543	7551	1	2	2	3	4
57	7559	7566	7574	7582	7589	7597	7604	7612	7619	7627	1	2	2	3	4
58	7634	7642	7649	7657	7664	7672	7679	7686	7694	7701	1	1	2	3	4
59	7709	7716	7723	7731	7738	7745	7752	7760	7767	7774	1	1	2	3	4
60	7782	7789	7796	7803	7810	7818	7825	7832	7839	7846	1	1	2	3	4
61	7853	7860	7868	7875	7882	7889	7896	7903	7910	7917	1	1	2	3	4
62	7924	7931	7938	7945	7952	7959	7966	7973	7980	7987	1	1	2	3	3
63	7993	8000	8007	8014	8021	8028	8035	8041	8048	8055	1	1	2	3	3
64	8062	8069	8075	8082	8089	8096	8102	8109	8116	8122	1	1	2	3	3
65	8129	8136	8142	8149	8156	8162	8169	8176	8182	8189	1	1	2	3	3
66	8195	8202	8209	8215	8222	8228	8235	8241	8248	8254	1	1	2	3	3
67	8261	8267	8274	8280	8287	8293	8299	8306	8312	8319	1	1	2	3	3
68	8325	8331	8338	8344	8351	8357	8363	8370	8376	8382	1	1	2	3	3
69	8388	8395	8401	8407	8414	8420	8426	8432	8439	8445	1	1	2	3	3
70	8451	8457	8463	8470	8476	8482	8488	8494	8500	8506	1	1	2	2	3
71	8513	8519	8525	8531	8537	8543	8549	8555	8561	8567	1	1	2	2	3
72	8573	8579	8585	8591	8597	8603	8609	8615	8621	8627	1	1	2	2	3
73	8633	8639	8645	8651	8657	8663	8669	8675	8681	8686	1	1	2	2	3
74	8692	8698	8704	8710	8716	8722	8727	8733	8739	8745	1	1	2	2	3
75	8751	8756	8762	8768	8774	8779	8785	8791	8797	8802	1	1	2	2	3
76	8808	8814	8820	8825	8831	8837	8842	8848	8854	8859	1	1	2	2	3
77	8865	8871	8876	8882	8887	8893	8899	8904	8910	8915	1	1	2	2	3
78	8921	8927	8932	8938	8943	8949	8954	8960	8965	8971	1	1	2	2	3
79	8976	8982	8987	8993	8998	9004	9009	9015	9020	9025	1	1	2	2	3
80	9031	9036	9042	9047	9053	9058	9063	9069	9074	9079	1	1	2	2	3
81	9085	9090	9096	9101	9106	9112	9117	9122	9128	9133	1	1	2	2	3
82	9138	9143	9149	9154	9159	9165	9170	9175	9180	9186	1	1	2	2	3
83	9191	9196	9201	9206	9212	9217	9222	9227	9232	9238	1	1	2	2	3
84	9243	9248	9253	9258	9263	9269	9274	9279	9284	9289	1	1	2	2	3
85	9294	9299	9304	9309	9315	9320	9325	9330	9335	9340	1	1	2	2	3
86	9345	9350	9355	9360	9365	9370	9375	9380	9385	9390	1	1	2	2	3
87	9395	9400	9405	9410	9415	9420	9425	9430	9435	9440	0	1	1	2	2
88	9445	9450	9455	9460	9465	9469	9474	9479	9484	9489	0	1	1	2	2
89	9494	9499	9504	9509	9513	9518	9523	9528	9533	9538	0	1	1	2	2
90	9542	9547	9552	9557	9562	9566	9571	9576	9581	9586	0	1	1	2	2
91	9590	9595	9600	9605	9609	9614	9619	9624	9628	9633	0	1	1	2	2
92	9638	9643	9647	9652	9657	9661	9666	9671	9675	9680	0	1	1	2	2
93	9685	9689	9694	9699	9703	9708	9713	9717	9722	9727	0	1	1	2	2
94	9731	9736	9741	9745	9750	9754	9759	9763	9768	9773	0	1	1	2	2
95	9777	9782	9786	9791	9795	9800	9805	9809	9814	9818	0	1	1	2	2
96	9823	9827	9832	9836	9841	9845	9850	9854	9859	9863	0	1	1	2	2
97	9868	9872	9877	9881	9886	9890	9894	9899	9903	9908	0	1	1	2	2
98	9912	9917	9921	9926	9930	9934	9939	9943	9948	9952	0	1	1	2	2
99	9956	9961	9965	9969	9974	9978	9983	9987	9991	9996	0	1	1	2	2
N.	0	1	2	3	4	5	6	7	8	9	1	2	3	4	5

TABLE B-2.—ANTILOGARITHMS

	0	1	2	3	4	5	6	7	8	9	Proportional parts				
											1	2	3	4	5
.00	1000	1002	1005	1007	1009	1012	1014	1016	1019	1021	0	0	1	1	1
.01	1023	1026	1028	1030	1033	1035	1038	1040	1042	1045	0	0	1	1	1
.02	1047	1050	1052	1054	1057	1059	1062	1064	1067	1069	0	0	1	1	1
.03	1072	1074	1076	1079	1081	1084	1086	1089	1091	1094	0	0	1	1	1
.04	1096	1099	1102	1104	1107	1109	1112	1114	1117	1119	0	1	1	1	1
.05	1122	1125	1127	1130	1132	1135	1138	1140	1143	1146	0	1	1	1	1
.06	1148	1151	1153	1156	1159	1161	1164	1167	1169	1172	0	1	1	1	1
.07	1175	1178	1180	1183	1186	1189	1191	1194	1197	1199	0	1	1	1	1
.08	1202	1205	1208	1211	1213	1216	1219	1222	1225	1227	0	1	1	1	1
.09	1230	1233	1236	1239	1242	1245	1247	1250	1253	1256	0	1	1	1	1
.10	1259	1262	1265	1268	1271	1274	1276	1279	1282	1285	0	1	1	1	1
.11	1288	1291	1294	1297	1300	1303	1306	1309	1312	1315	0	1	1	1	2
.12	1318	1321	1324	1327	1330	1334	1337	1340	1343	1346	0	1	1	1	2
.13	1349	1352	1355	1358	1361	1365	1368	1371	1374	1377	0	1	1	1	2
.14	1380	1384	1387	1390	1393	1396	1400	1403	1406	1409	0	1	1	1	2
.15	1413	1416	1419	1422	1426	1429	1432	1435	1439	1442	0	1	1	1	2
.16	1445	1449	1452	1455	1459	1462	1466	1469	1472	1476	0	1	1	1	2
.17	1479	1483	1486	1489	1493	1496	1500	1503	1507	1510	0	1	1	1	2
.18	1514	1517	1521	1524	1528	1531	1535	1538	1542	1545	0	1	1	1	2
.19	1549	1552	1556	1560	1563	1567	1570	1574	1578	1581	0	1	1	1	2
.20	1585	1589	1592	1596	1600	1603	1607	1611	1614	1618	0	1	1	1	2
.21	1622	1626	1629	1633	1637	1641	1644	1648	1652	1656	0	1	1	2	2
.22	1660	1663	1667	1671	1675	1679	1683	1687	1690	1694	0	1	1	2	2
.23	1698	1702	1706	1710	1714	1718	1722	1726	1730	1734	0	1	1	2	2
.24	1738	1742	1746	1750	1754	1758	1762	1766	1770	1774	0	1	1	2	2
.25	1778	1782	1786	1791	1795	1799	1803	1807	1811	1816	0	1	1	2	2
.26	1820	1824	1828	1832	1837	1841	1845	1849	1854	1858	0	1	1	2	2
.27	1862	1866	1871	1875	1879	1884	1888	1892	1897	1901	0	1	1	2	2
.28	1905	1910	1914	1919	1923	1928	1932	1936	1941	1945	0	1	1	2	2
.29	1950	1954	1959	1963	1968	1972	1977	1982	1986	1991	0	1	1	2	2
.30	1995	2000	2004	2009	2014	2018	2023	2028	2032	2037	0	1	1	2	2
.31	2042	2046	2051	2056	2061	2065	2070	2075	2080	2084	0	1	1	2	2
.32	2089	2094	2099	2104	2109	2113	2118	2123	2128	2133	0	1	1	2	2
.33	2138	2143	2148	2153	2158	2163	2168	2173	2178	2183	0	1	1	2	2
.34	2188	2193	2198	2203	2208	2213	2218	2223	2228	2234	1	1	2	2	3
.35	2239	2244	2249	2254	2259	2265	2270	2275	2280	2286	1	1	2	2	3
.36	2291	2296	2301	2307	2312	2317	2323	2328	2333	2339	1	1	2	2	3
.37	2344	2350	2355	2360	2366	2371	2377	2382	2388	2393	1	1	2	2	3
.38	2399	2404	2410	2415	2421	2427	2432	2438	2443	2449	1	1	2	2	3
.39	2455	2460	2466	2472	2477	2483	2489	2495	2500	2506	1	1	2	2	3
.40	2512	2518	2523	2529	2535	2541	2547	2553	2559	2564	1	1	2	2	3
.41	2570	2576	2582	2588	2594	2600	2606	2612	2618	2624	1	1	2	2	3
.42	2630	2636	2642	2649	2655	2661	2667	2673	2679	2685	1	1	2	2	3
.43	2692	2698	2704	2710	2716	2723	2729	2735	2742	2748	1	1	2	3	3
.44	2754	2761	2767	2773	2780	2786	2793	2799	2805	2812	1	1	2	3	3
.45	2818	2825	2831	2838	2844	2851	2858	2864	2871	2877	1	1	2	3	3
.46	2884	2891	2897	2904	2911	2917	2924	2931	2938	2944	1	1	2	3	3
.47	2951	2958	2965	2972	2979	2985	2992	2999	3006	3013	1	1	2	3	3
.48	3020	3027	3034	3041	3048	3055	3062	3069	3076	3083	1	1	2	3	4
.49	3090	3097	3105	3112	3119	3126	3133	3141	3148	3155	1	1	2	3	4
	0	1	2	3	4	5	6	7	8	9	1	2	3	4	5

TABLE B-2.—ANTILOGARITHMS (*Concluded*)

	0	1	2	3	4	5	6	7	8	9	Proportional parts				
											1	2	3	4	5
.50	3162	3170	3177	3184	3192	3199	3206	3214	3221	3228	1	1	2	3	4
.51	3236	3243	3251	3258	3266	3273	3281	3289	3296	3304	1	2	2	3	4
.52	3311	3319	3327	3334	3342	3350	3357	3365	3373	3381	1	2	2	3	4
.53	3388	3396	3404	3412	3420	3428	3436	3443	3451	3459	1	2	2	3	4
.54	3467	3475	3483	3491	3499	3508	3516	3524	3532	3540	1	2	2	3	4
.55	3548	3556	3565	3573	3581	3589	3597	3606	3614	3622	1	2	2	3	4
.56	3631	3639	3648	3656	3664	3673	3681	3690	3698	3707	1	2	3	3	4
.57	3715	3724	3733	3741	3750	3758	3767	3776	3784	3793	1	2	3	3	4
.58	3802	3811	3819	3828	3837	3846	3855	3864	3873	3882	1	2	3	4	4
.59	3890	3899	3908	3917	3926	3936	3945	3954	3963	3972	1	2	3	4	5
.60	3981	3990	3999	4009	4018	4027	4036	4046	4055	4064	1	2	3	4	5
.61	4074	4083	4093	4102	4111	4121	4130	4140	4150	4159	1	2	3	4	5
.62	4169	4178	4188	4198	4207	4217	4227	4236	4246	4256	1	2	3	4	5
.63	4266	4276	4285	4295	4305	4315	4325	4335	4345	4355	1	2	3	4	5
.64	4365	4375	4385	4395	4406	4416	4426	4436	4446	4457	1	2	3	4	5
.65	4467	4477	4487	4498	4508	4519	4529	4539	4550	4560	1	2	3	4	5
.66	4571	4581	4592	4603	4613	4624	4634	4645	4656	4667	1	2	3	4	5
.67	4677	4688	4699	4710	4721	4732	4742	4753	4764	4775	1	2	3	4	5
.68	4786	4797	4808	4819	4831	4842	4853	4864	4875	4887	1	2	3	4	6
.69	4898	4909	4920	4932	4943	4955	4966	4977	4989	5000	1	2	3	5	6
.70	5012	5023	5035	5047	5058	5070	5082	5093	5105	5117	1	2	4	5	6
.71	5129	5140	5152	5164	5176	5188	5200	5212	5224	5236	1	2	4	5	6
.72	5248	5260	5272	5284	5297	5309	5321	5333	5346	5358	1	2	4	5	6
.73	5370	5383	5395	5408	5420	5433	5445	5458	5470	5483	1	3	4	5	6
.74	5495	5508	5521	5534	5546	5559	5572	5585	5598	5610	1	3	4	5	6
.75	5623	5636	5649	5662	5675	5689	5702	5715	5728	5741	1	3	4	5	7
.76	5754	5768	5781	5794	5808	5821	5834	5848	5861	5875	1	3	4	5	7
.77	5888	5902	5916	5929	5943	5957	5970	5984	5998	6012	1	3	4	5	7
.78	6026	6039	6053	6067	6081	6095	6109	6124	6138	6152	1	3	4	6	7
.79	6166	6180	6194	6209	6223	6237	6252	6266	6281	6295	1	3	4	6	7
.80	6310	6324	6339	6353	6368	6383	6397	6412	6427	6442	1	3	4	6	7
.81	6457	6471	6486	6501	6516	6531	6546	6561	6577	6592	2	3	5	6	8
.82	6607	6622	6637	6653	6668	6683	6699	6714	6730	6745	2	3	5	6	8
.83	6761	6776	6792	6808	6823	6839	6855	6871	6887	6902	2	3	5	6	8
.84	6918	6934	6950	6966	6982	6998	7015	7031	7047	7063	2	3	5	6	8
.85	7079	7096	7112	7129	7145	7161	7178	7194	7211	7228	2	3	5	7	8
.86	7244	7261	7278	7295	7311	7328	7345	7362	7379	7396	2	3	5	7	8
.87	7413	7430	7447	7464	7482	7499	7516	7534	7551	7568	2	3	5	7	9
.88	7586	7603	7621	7638	7656	7674	7691	7709	7727	7745	2	4	5	7	9
.89	7762	7780	7798	7816	7834	7852	7870	7889	7907	7925	2	4	5	7	9
.90	7943	7962	7980	7998	8017	8035	8054	8072	8091	8110	2	4	6	7	9
.91	8128	8147	8166	8185	8204	8222	8241	8260	8279	8299	2	4	6	8	9
.92	8318	8337	8356	8375	8395	8414	8433	8453	8472	8492	2	4	6	8	10
.93	8511	8531	8551	8570	8590	8610	8630	8650	8670	8690	2	4	6	8	10
.94	8710	8730	8750	8770	8790	8810	8831	8851	8872	8892	2	4	6	8	10
.95	8913	8933	8954	8974	8995	9016	9036	9057	9078	9099	2	4	6	8	10
.96	9120	9141	9162	9183	9204	9226	9247	9268	9290	9311	2	4	6	8	11
.97	9333	9354	9376	9397	9419	9441	9462	9484	9506	9528	2	4	6	9	11
.98	9550	9572	9594	9616	9638	9661	9683	9705	9727	9750	2	4	7	9	11
.99	9772	9795	9817	9840	9863	9886	9908	9931	9954	9977	2	5	7	9	11
	0	1	2	3	4	5	6	7	8	9	1	2	3	4	5

2. Powers of Numbers

Figure B-1 permits the approximate estimation of powers and roots. Because of the large number of operations that can be performed on this single chart, it is likely to be confusing at first. Familiarity with it, however, will disclose its true simplicity.

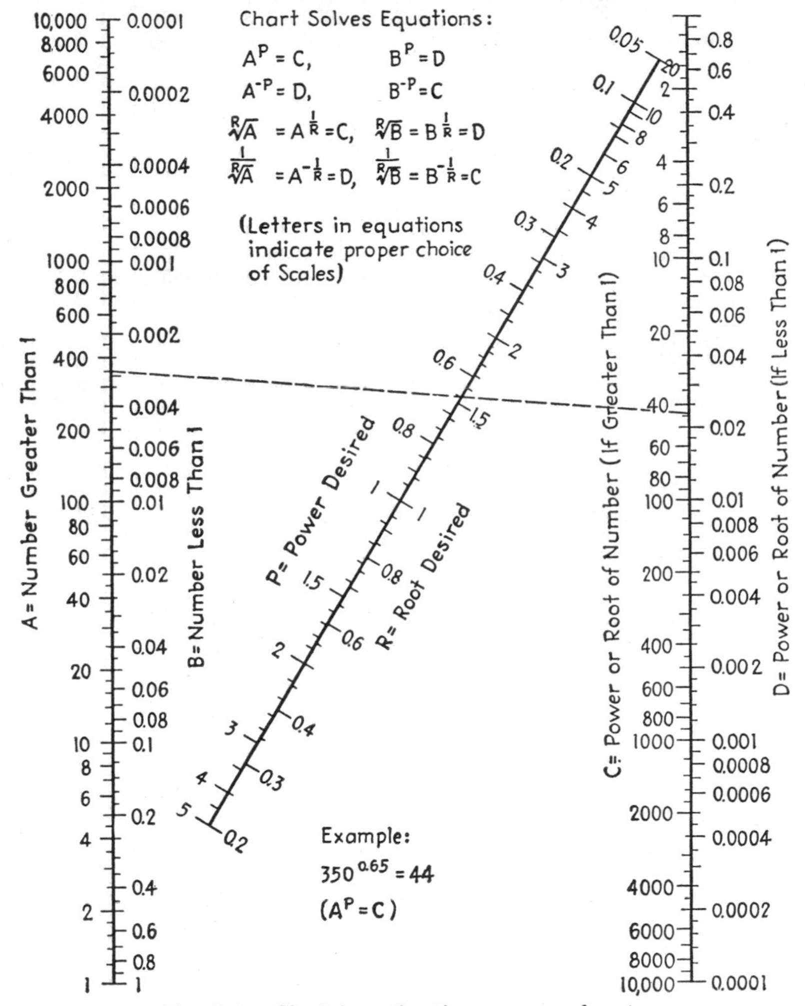

FIG. B-1.—Chart for estimating powers and roots.

3. Formulas for Area

Circle: $\frac{\pi}{4} \times (\text{diameter})^2 = 0.7854\ (\text{diameter})^2$

$$\pi \times (\text{radius})^2 = 3.1416\ (\text{radius})^2$$

Rectangle: Width × length.
Triangles:
Right angle = ½ product of two sides adjacent to right angle
Equilateral = 0.4330 × (side)^2
Surface of sphere: 3.1416 (diameter)^2

4. Formulas for Volume

Sphere: $\frac{\pi}{6}$ (diameter)^3 = 0.5236 (diameter)^3

Regular prism: Area of base × altitude

Cone: $\frac{\pi}{12}$ (diameter)^2 × altitude

Cylindrical tanks: See Tables B-3, B-4, and Fig. B-2.

TABLE B-3.—GALLONS PER INCH OF DEPTH IN VERTICAL CYLINDRICAL TANKS

Diameter												
Ft	In.											
	0	1	2	3	4	5	6	7	8	9	10	11
	Gallons per inch											
0	.....	.....	.....	.....	.....	.....	0.12	0.17	0.22	0.28	0.34	0.41
1	0.49	0.58	0.67	0.77	0.87	0.98	1.10	1.23	1.36	1.50	1.65	1.80
2	1.96	2.12	2.30	2.48	2.67	2.86	3.06	3.27	3.48	3.70	3.93	4.16
3	4.41	4.66	4.91	5.17	5.44	5.72	6.00	6.29	6.58	6.88	7.20	7.51
4	7.85	8.16	8.50	8.84	9.19	9.55	9.91	10.28	10.66	11.05	11.44	11.84
5	12.24	12.65	13.07	13.50	13.93	14.36	14.81	15.26	15.72	16.19	16.66	17.14
6	17.63	18.12	18.62	19.12	19.64	20.16	20.69	21.22	21.76	22.31	22.86	23.42
7	24.0	24.6	25.2	25.7	26.3	26.9	27.5	28.3	28.1	29.4	30.0	30.7
8	31.3	32.0	32.6	33.3	34.0	34.7	35.4	36.1	36.8	37.5	38.2	38.9
9	39.6	40.4	41.1	41.9	42.6	43.4	44.2	45.0	45.8	46.5	47.4	48.2
10	49.0	49.8	50.6	51.4	52.3	53.1	54.0	54.8	55.7	56.6	57.5	58.4
11	59.2	60.1	61.0	62.0	62.9	63.8	64.8	65.7	66.6	67.6	68.6	69.5
12	70.3	71.5	72.5	73.5	74.5	75.5	76.5	77.5	78.6	79.6	80.6	81.7
13	82.7	83.8	84.9	86.0	87.0	88.1	89.2	90.3	91.4	92.6	93.7	94.8
14	96.0	97.1	98.3	99.4	100.6	101.8	102.9	104.1	105.3	106.5	107.7	108.9

TABLE B-4.—CAPACITIES OF HORIZONTAL CYLINDRICAL TANKS IN U.S. GALLONS
(Contents given for 1 ft of length, flat ends)
(Perry, "Chemical Engineers' Handbook," 3d ed., McGraw-Hill Book Company, Inc., New York, 1950)

Diameter of tank, in.	Depth of liquid, in.																
	3	6	9	12	15	18	21	24	27	30	33	36	39	42	45	48	51
12	1.15	2.94	4.73	5.88													
18	1.45	3.86	6.61	9.36	11.77	13.22											
24	1.70	4.60	8.05	11.75	15.45	18.90	21.80	23.50									
30	1.91	5.23	9.27	13.72	18.36	23.00	27.45	31.49	34.81	36.72							
36	2.12	5.79	10.34	15.43	20.85	26.44	32.03	37.45	42.54	47.09	50.76	52.88					
42	2.28	6.31	11.31	16.97	23.07	29.46	35.99	42.51	48.90	55.00	60.66	65.66	69.69	71.97			
48	2.45	6.78	12.20	18.38	25.10	32.20	39.54	47.01	54.47	61.81	68.91	75.63	81.81	87.23	91.56	94.01	
54	2.60	7.22	13.04	19.68	26.97	34.72	42.80	51.08	59.49	67.90	76.18	84.26	92.01	99.30	105.94	111.76	116.38
60	2.75	7.64	13.82	20.91	28.72	37.06	45.82	54.88	64.11	73.45	82.78	92.01	101.07	109.83	118.17	125.98	133.07
66	2.89	8.04	14.56	22.07	30.37	39.28	48.65	58.39	68.42	78.59	88.87	99.14	109.31	119.34	129.08	138.45	147.36
72	3.02	8.42	15.26	23.17	31.92	41.36	51.32	61.71	72.45	83.41	94.54	105.76	116.98	128.11	139.07	149.81	160.20
78	3.15	8.78	15.94	24.21	33.41	43.34	53.86	64.87	76.27	87.97	99.90	111.97	124.12	136.27	148.34	160.27	171.97
84	3.26	9.12	16.57	25.24	34.85	45.24	56.29	67.87	79.91	92.30	104.98	117.85	130.87	143.95	157.03	170.05	182.92
90	3.43	9.46	17.20	26.20	36.21	47.05	58.61	70.75	83.39	96.43	109.81	123.46	137.28	151.22	165.25	179.27	193.21
96	3.50	9.79	17.80	27.13	37.52	48.81	60.84	73.52	86.73	100.39	114.44	128.79	143.40	158.17	173.07	188.01	202.95
102	3.61	10.10	18.37	28.01	39.00	50.49	62.99	76.18	89.94	104.20	118.89	133.92	149.25	164.81	180.53	196.36	212.25
108	3.71	10.39	18.94	28.90	40.03	52.14	65.09	78.74	93.04	107.87	123.17	138.87	154.89	171.19	187.71	204.37	221.15

TABLE B-4—CAPACITIES OF HORIZONTAL CYLINDRICAL TANKS IN U.S. GALLONS *(Concluded)*

Diameter of tank, in.	Depth of liquid, in.																
	54	57	60	63	66	69	72	75	78	81	84	87	90	93	96	102	108
12																	
18																	
24																	
30																	
36																	
42																	
48																	
54	118.98																
60	139.25	144.14	146.89														
66	155.66	163.17	169.69	174.84	177.73												
72	170.16	179.60	188.35	196.26	203.10	208.50	211.52										
78	183.37	194.38	204.90	214.83	224.03	232.30	239.46	245.09	248.24								
84	195.60	207.99	220.03	231.61	242.66	253.05	262.66	271.33	278.78	284.64	287.90						
90	207.03	220.68	234.06	247.10	259.74	271.88	283.44	294.28	304.29	313.29	321.03	327.06	330.49				
96	217.85	232.62	247.23	261.58	275.63	289.29	302.50	315.18	327.21	338.50	348.89	358.22	366.23	372.52	376.02		
102	228.12	243.95	259.67	275.23	290.56	305.59	320.28	334.54	348.30	361.49	373.99	385.48	396.47	406.11	414.38	424.48	
108	238.05	254.75	271.53	288.19	304.71	321.01	337.03	352.73	368.03	382.86	397.16	410.81	423.76	435.87	447.00	465.51	475.90

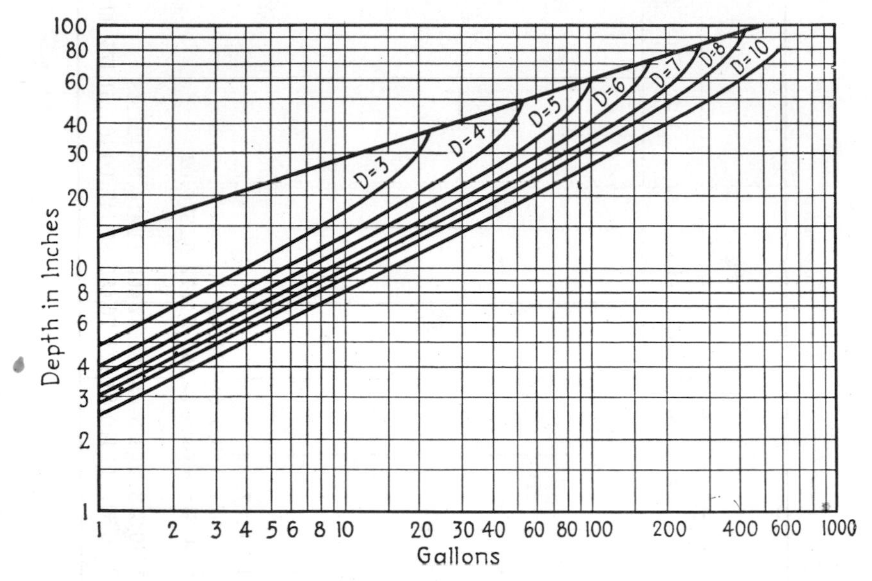

Constructed from Doolittle Formula, *Ind. Eng. Chem.*, March, 1928.

$$V \text{ (both ends)} = 0.0009328h^2(3r - h), \text{ gal}$$

h = depth, in.

r = radius of tank, in.

In curves D = diameter of tank, ft

FIG. B-2.—Volumes of (two) bulged or dished ends of horizontal cylindrical tanks. (*Perry, "Chemical Engineers' Handbook," 3d ed., McGraw-Hill Book Company, Inc., New York,* 1950.) This chart is to be used in conjunction with Table B-4 to secure volumes of horizontal tanks with dished ends.

5. Approximations

$$(1 \pm a)^n = 1 \pm na \qquad \text{(when } a \text{ is small)}$$

$$(1 \pm a)^m(1 \pm b)^n = 1 \pm ma \pm nb \qquad \text{(when } a \text{ and } b \text{ are small)}$$

$$\sqrt{mn} = \frac{n + m}{2} \qquad \text{(when } m - n \text{ is small)}$$

For small angles expressed in radians:

$$\theta = \sin \theta = \tan \theta$$

6. Integrals and Differentials

TABLE B-5.—DIFFERENTIALS AND INTEGRALS

Function, f	Derivative, $\frac{df}{dx}$	Integral, $\int f\,dx$
ax	a	$\frac{a}{2}x^2$
ax^n	anx^{n-1}	$\frac{ax^{n+1}}{n+1}$ (unless $n = -1$)
uv	$u\frac{dv}{dx} + v\frac{du}{dx}$	
$\frac{u}{v}$	$\frac{v\frac{du}{dx} - u\frac{dv}{dx}}{v^2}$	
$\frac{a}{bx+c}$	$-\frac{ab}{(bx+c)^2}$	$\frac{a}{b}\ln(bx+c)$
$\frac{1}{a+bx+cx^2}$	$-\frac{b+2cx}{(a+bx+cx^2)^2}$	$\frac{2}{\sqrt{4ac-b^2}}\tan^{-1}\left(\frac{2cx+b}{\sqrt{4ac-b^2}}\right)$ or, $\frac{2}{\sqrt{b^2-4ac}}\ln\left(\frac{2cx+b-\sqrt{b^2-4ac}}{2cx+b+\sqrt{b^2-4ac}}\right)$
e^{ax}	ae^{ax}	$\frac{1}{a}e^{ax}$
a^{bx}	$ba^{bx}\ln a$	$\frac{a^{bx}}{b\ln a}$
$\ln ax$	$\frac{a}{x}$	$x\ln ax - x$
$\log ax$	$\frac{a}{2.30259x}$	$\frac{x}{2.30259}\log ax - \frac{x}{2.30259}$

NOTE: $\ln x$ = logarithm of x to the base e
$\log x$ = logarithm of x to the base 10
$\tan^{-1} x$ = *principal* angle, in radians, whose tangent is x

CHAPTER C

CONVERSION TABLES

The engineer constantly uses a wide variety of units in his calculations. This chapter gives data for the interconversion of units frequently used by American chemists and engineers. No effort has been made to include units in common usage in other countries or to comment on small differences between different versions of the same unit. For *very precise* work it is, of course, necessary to specify the particular variety of calorie, Btu, watt, volt, etc., that is used.

The tables and figures presented in this chapter are listed below:

TABLE C-1.—FUNDAMENTAL CONSTANTS

Perfect gas-law constant, R*....	1.987 Btu/(lb mole)(°R)
	1,543 ft-lb/(lb mole)(°R)
Faraday constant..............	96,500 int. coulombs/g equivalent
	12,160 amp-hr/lb equivalent
Mechanical equivalent of heat, J †	778.1 ft-lb/Btu
Standard atmosphere‡..........	14.696 psi
Temperature of melting ice.....	491.7°R
Acceleration due to gravity, g:	
Standard value..............	32.17 ft/(sec)^2
0° latitude...............	32.09
20° latitude...............	32.11
40° latitude...............	32.16
60° latitude...............	32.22
80° latitude...............	32.25

[Correction for altitude = −0.003 ft/(sec)^2 for each 1,000 ft above sea level]

* See Table E-1.

† See Table C-9.

‡ See Fig. C-1.

TABLE C-2.—ATOMIC WEIGHTS OF ELEMENTS
(Adapted from "Chart of the Nuclides," 5th ed., revised to April, 1956, prepared at the Atomic Energy Commission's Knolls Atomic Power Laboratory operated by General Electric Co.)

Name	Symbol	Atomic weight	Name	Symbol	Atomic weight
Actinium	Ac	227	Mendelevium	Mv	256*
Aluminum	Al	26.98	Mercury	Hg	200.61
Americium	Am	241*	Molybdenum	Mo	95.95
Antimony	Sb	121.76	Neodymium	Nd	144.27
Argon	A	39.944	Neon	Ne	20.183
Arsenic	As	74.91	Neptunium	Np	237
Astatine	At	211*	Nickel	Ni	58.71
Barium	Ba	137.36	Niobium	Nb	92.91
Berkelium	Bk	244*	Nitrogen	N	14.008
Beryllium	Be	9.013	Osmium	Os	190.2
Bismuth	Bi	209.00	Oxygen	O	16.000
Boron	B	10.82	Palladium	Pd	106.4
Bromine	Br	79.916	Phosphorus	P	30.975
Cadmium	Cd	112.41	Platinum	Pt	195.09
Calcium	Ca	40.08	Plutonium	Pu	239*
Californium	Cf	249*	Polonium	Po	210
Carbon	C	12.011	Potassium	K	39.100
Cerium	Ce	140.13	Praseodymium	Pr	140.92
Cesium	Cs	132.91	Promethium	Pm	147*
Chlorine	Cl	35.457	Protactinium	Pa	231
Chromium	Cr	52.01	Radium	Ra	226.05
Cobalt	Co	58.94	Radon (also emanation)	Rn (also Em)	222
Columbium (now niobium)	Cb (now Nb)	92.91	Rhenium	Re	186.22
Copper	Cu	63.54	Rhodium	Rh	102.91
Curium	Cm	242*	Rubidium	Rb	85.48
Dysprosium	Dy	162.51	Ruthenium	Ru	101.1
Einsteinium	E	253*	Samarium	Sm	150.35
Emanation (also radon)	Em (also Rn)	222	Scandium	Sc	44.96
Erbium	Er	167.27	Selenium	Se	78.96
Europium	Eu	152.0	Silicon	Si	28.09
Fermium	Fm	255*	Silver	Ag	107.880
Fluorine	F	19.00	Sodium	Na	22.991
Francium	Fr	223*	Strontium	Sr	87.63
Gadolinium	Gd	157.26	Sulphur	S	32.066
Gallium	Ga	69.72	Tantalum	Ta	180.95
Germanium	Ge	72.60	Technetium	Tc	99*
Gold	Au	197.0	Tellurium	Te	127.61
Hafnium	Hf	178.50	Terbium	Tb	158.93
Helium	He	4.003	Thallium	Tl	204.39
Holmium	Ho	164.94	Thorium	Th	232.05
Hydrogen	H	1.0080	Thulium	Tm	168.94
Indium	In	114.82	Tin	Sn	118.70
Iodine	I	126.91	Titanium	Ti	47.90
Iridium	Ir	192.2	Tungsten (also wolfram)	W	183.86
Iron	Fe	55.85	Uranium	U	238.07
Krypton	Kr	83.80	Vanadium	V	50.95
Lanthanum	La	138.92	Xenon	Xe	131.30
Lead	Pb	207.21	Ytterbium	Yb	173.04
Lithium	Li	6.940	Yttrium	Y	88.92
Lutetium	Lu	174.99	Zinc	Zn	65.38
Magnesium	Mg	24.32	Zirconium	Zr	91.22
Manganese	Mn	54.94			

* Artificial radioactive elements which do not exist in nature. As each of these elements exists in many isotopic forms, the most commonly accepted isotope for each element has been used in the table.

TABLE C-3.—LENGTH EQUIVALENTS

Inches	Feet	Yards	Meters
1	0.08333	0.02778	0.0254
12	1	0.3333	0.3048
36	3	1	0.9144
39.37	3.281	1.0936	1

1 chain = 66 ft   1 mile = 5,280 ft   1 rod = 16.5 ft
1 nautical mile = 6,080.2 ft   1 fathom = 6 ft

Example: 1 m = 39.37 in.

TABLE C-4.—VOLUME EQUIVALENTS

Cu in.	Cu ft	U.S. apothecary oz	U.S. gal	Liters
1,000	0.5787	554	4.329	16.39
1,728	1	957.5	7.481	28.32
1,805	1.044	1,000	7.812	29.57
231	0.1337	128	1	3.785
61.03	0.03531	33.81	0.2642	1

1 gal = 4 qt = 8 pt
1 petroleum bbl = 42 gal
1 l = 1,000.027 cc
1 imperial gal = 1.2 U.S. gal

Example: 1 oz $= \frac{1{,}805}{1{,}000} = 1.805$ cu in.

TABLE C-5.—WEIGHT EQUIVALENTS

1 lb avoirdupois = 16 oz (avdp) = 453.6 g = 7,000 grains
1 short ton = 2,000 lb = 0.9072 metric ton
1 long ton = 2,240 lb = 1.016 metric tons
1 metric ton = 2,205 lb = 1,000 kg
1 lb troy or apothecary = 12 oz troy or apothecary
= 0.8229 lb avoirdupois
= 373.2 g

TABLE C-6.—CONVERSION TABLE: DEGREES CENTIGRADE TO DEGREES FAHRENHEIT

$°F = 1.8 \times °C + 32$

°C	0	10	20	30	40	50	60	70	80	90
	F	F	F	F	F	F	F	F	F	F
−200	−328	−346	−364	−382	−400	−418	−436	−454	......	......
−100	−148	−166	−184	−202	−220	−238	−256	−274	−292	−310
−0	+32	+14	−4	−22	−40	−58	−76	−94	−112	−130
0	32	50	68	86	104	122	140	158	176	194
100	212	230	248	266	284	302	320	338	356	374
200	392	410	428	446	464	482	500	518	536	554
300	572	590	608	626	644	662	680	698	716	734
400	752	770	788	806	824	842	860	878	896	914
500	932	950	968	986	1004	1022	1040	1058	1076	1094
600	1112	1130	1148	1166	1184	1202	1220	1238	1256	1274
700	1292	1310	1328	1346	1364	1382	1400	1418	1436	1454
800	1472	1490	1508	1526	1544	1562	1580	1598	1616	1634
900	1652	1670	1688	1706	1724	1742	1760	1778	1796	1814
1000	1832	1850	1868	1886	1904	1922	1940	1958	1976	1994
1100	2012	2030	2048	2066	2084	2102	2120	2138	2156	2174
1200	2192	2210	2228	2246	2264	2282	2300	2318	2336	2354
1300	2372	2390	2408	2426	2444	2462	2480	2498	2516	2534
1400	2552	2570	2588	2606	2624	2642	2660	2678	2696	2714
1500	2732	2750	2768	2786	2804	2822	2840	2858	2876	2894
1600	2912	2930	2948	2966	2984	3002	3020	3038	3056	3074
1700	3092	3110	3128	3146	3164	3182	3200	3218	3236	3254
1800	3272	3290	3308	3326	3344	3362	3380	3398	3416	3434
1900	3452	3470	3488	3506	3524	3542	3560	3578	3596	3614
2000	3632	3650	3668	3686	3704	3722	3740	3758	3776	3794
2100	3812	3830	3848	3866	3884	3902	3920	3938	3956	3974
2200	3992	4010	4028	4046	4064	4082	4100	4118	4136	4154
2300	4172	4190	4208	4226	4244	4262	4280	4298	4316	4334
2400	4352	4370	4388	4406	4424	4442	4460	4478	4496	4514
2500	4532	4550	4568	4586	4604	4622	4640	4658	4676	4694
2600	4712	4730	4748	4766	4784	4802	4820	4838	4856	4874
2700	4892	4910	4928	4946	4964	4982	5000	5018	5036	5054
2800	5072	5090	5108	5126	5144	5162	5180	5198	5216	5234
2900	5252	5270	5288	5306	5324	5342	5360	5378	5396	5414
3000	5432	5450	5468	5486	5504	5522	5540	5558	5576	5594
3100	5612	5630	5648	5666	5684	5702	5720	5738	5756	5774
3200	5792	5810	5828	5846	5864	5882	5900	5918	5936	5954
3300	5972	5990	6008	6026	6044	6062	6080	6098	6116	6134
3400	6152	6170	6188	6206	6224	6242	6260	6278	6296	6314
3500	6332	6350	6368	6386	6404	6422	6440	6458	6476	6494
3600	6512	6530	6548	6566	6584	6602	6620	6638	6656	6674
3700	6692	6710	6728	6746	6764	6782	6800	6818	6836	6854
3800	6872	6890	6908	6926	6944	6962	6980	6998	7016	7034
3900	7052	7070	7088	7106	7124	7142	7160	7178	7196	7214
°C	0	10	20	30	40	50	60	70	80	90

°C	°F
1	1.8
2	3.6
3	5.4
4	7.2
5	9.0
6	10.8
7	12.6
8	14.4
9	16.2
10	18.0

°F	°C
1	0.56
2	1.11
3	1.67
4	2.22
5	2.78
6	3.33
7	3.89
8	4.44
9	5.00
10	5.56
11	6.11
12	6.67
13	7.22
14	7 78
15	8.33
16	8.89
17	9.44
18	10.00

Examples: 1347°C = 2444°F + 12.6°F = 2456.6°F
3367°F = 1850°C + 2.78°C = 1852.78°C

TABLE C-7.—CONVERSION TABLE: DEGREES FAHRENHEIT TO DEGREES CENTIGRADE

$$°C = \frac{°F - 32}{1.8}$$

°F	0	10	20	30	40	50	60	70	80	90
	C	C	C	C	C	C	C	C	C	C
−400	−240.0	−245.5	−251.1	−256.6	−262.2	−267.7	....	....	....	....
−300	−184.4	−190.0	−195.5	−201.1	−206.6	−212.2	−217.7	−223.3	−228.8	−234.4
−200	−128.8	−134.4	−140.0	−145.5	−151.1	−156.6	−162.2	−167.7	−173.3	−178.8
−100	−73.3	−78.8	−84.4	−90.0	−95.5	−101.1	−106.6	−112.2	−117.7	−123.3
−0	−17.7	−23.3	−28.8	−34.4	−40.0	−45.5	−51.1	−56.6	−62.2	−67.7
0	−17.7	−12.2	−6.6	−1.1	+4.4	+10.0	+15.5	+21.1	+26.6	+32.2
100	37.7	43.3	48.8	54.4	60.0	65.5	71.1	76.6	82.2	87.7
200	93.3	98.8	104.4	110.0	115.5	121.1	126.6	132.2	137.7	143.3
300	148.8	154.4	160.0	165.5	171.1	176.6	182.2	187.7	193.3	198.8
400	204.4	210.0	215.5	221.1	226.6	232.2	237.7	243.3	248.8	254.4
500	260.0	265.5	271.1	276.6	282.2	287.7	293.3	298.8	304.4	310.0
600	315.5	321.1	326.6	332.2	337.7	343.3	348.8	354.4	360.0	365.5
700	371.1	376.6	382.2	387.7	393.3	398.8	404.4	410.0	415.5	421.1
800	426.6	432.2	437.7	443.3	448.8	454.4	460.0	465.5	471.1	476.6
900	482.2	487.7	493.3	498.8	504.4	510.0	515.5	521.1	526.6	532.2
1000	537.7	543.3	548.8	554.4	560.0	565.5	571.1	576.6	582.2	587.7
1100	593.3	598.8	604.4	610.0	615.5	621.1	626.6	632.2	637.7	643.3
1200	648.8	654.4	660.0	665.5	671.1	676.6	682.2	687.7	693.3	698.8
1300	704.4	710.0	715.5	721.1	726.6	732.2	737.7	743.3	748.8	754.4
1400	760.0	765.5	771.1	776.6	782.2	787.7	793.3	798.8	804.4	810.0
1500	815.5	821.1	826.6	832.2	837.7	843.3	848.8	854.4	860.0	865.5
1600	871.1	876.6	882.2	887.7	893.3	898.8	904.4	910.0	915.5	921.1
1700	926.6	932.2	937.7	943.3	948.8	954.4	960.0	965.5	971.1	976.6
1800	982.2	987.7	993.3	998.8	1004.4	1010.0	1015.5	1021.1	1026.6	1032.2
1900	1037.7	1043.3	1048.8	1054.4	1060.0	1065.5	1071.1	1076.6	1082.2	1087.7
2000	1093.3	1098.8	1104.4	1110.0	1115.5	1121.1	1126.6	1132.2	1137.7	1143.3
2100	1148.8	1154.4	1160.0	1165.5	1171.1	1176.6	1182.2	1187.7	1193.3	1198.8
2200	1204.4	1210.0	1215.5	1221.1	1226.6	1232.2	1237.7	1243.3	1248.8	1254.4
2300	1260.0	1265.5	1271.1	1276.6	1282.2	1287.7	1293.3	1298.8	1304.4	1310.0
2400	1315.5	1321.1	1326.6	1332.2	1337.7	1343.3	1348.8	1354.4	1360.0	1365.5
2500	1371.1	1376.6	1382.2	1387.7	1393.3	1398.8	1404.4	1410.0	1415.5	1421.1
2600	1426.6	1432.2	1437.7	1443.3	1448.8	1454.4	1460.0	1465.5	1471.1	1476.6
2700	1482.2	1487.7	1493.3	1498.8	1504.4	1510.0	1515.5	1521.1	1526.6	1532.2
2800	1537.7	1543.3	1548.8	1554.4	1560.0	1565.5	1571.1	1576.6	1582.2	1587.7
2900	1593.3	1598.8	1604.4	1610.0	1615.5	1621.1	1626.6	1632.2	1637.7	1643.3
3000	1648.8	1654.4	1660.0	1665.5	1671.1	1676.6	1682.2	1687.7	1693.3	1698.8
3100	1704.4	1710.0	1715.5	1721.1	1726.6	1732.2	1737.7	1743.3	1748.8	1754.4
3200	1760.0	1765.5	1771.1	1776.6	1782.2	1787.7	1793.3	1798.8	1804.4	1810.0
3300	1815.5	1821.1	1826.6	1832.2	1837.7	1843.3	1848.8	1854.4	1860.0	1865.5
3400	1871.1	1876.6	1882.2	1887.7	1893.3	1898.8	1904.4	1910.0	1915.5	1921.1
3500	1926.6	1932.2	1937.7	1943.3	1948.8	1954.4	1960.0	1965.5	1971.1	1976.6
3600	1982.2	1987.7	1993.3	1998.8	2004.4	2010.0	2015.5	2021.1	2026.6	2032.2
°F	0	10	20	30	40	50	60	70	80	90

°F	°C
1	0.5
2	1.1
3	1.6
4	2.2
5	2.7
6	3.3
7	3.8
8	4.4
9	5.0

Examples: −246.0°F = −151.11°C − 3.33°C = −154.44°C

2423.5°F = 1326.66°C + 1.66°C + 0.27°C = 1328.61°C

TABLE C-8.—CONVERSION OF DEGREES KELVIN TO DEGREES RANKINE AND DEGREES FAHRENHEIT

°R = 1.8 × °K = 459.7 + °F

Kelvin, °K	Rankine, °R	Fahrenheit, °F
100	180	−280
150	270	−190
200	360	−100
250	450	−10
273.16	491.7	32
298.16	536.7	77
300	540	80
350	630	170
373.16	671.7	212
400	720	260
450	810	350
500	900	440
550	990	530
600	1080	620
650	1170	710
700	1260	800
750	1350	890
800	1440	980
900	1620	1160
1000	1800	1340
1100	1980	1520
1200	2160	1700
1250	2250	1790
1300	2340	1880
1400	2520	2060
1500	2700	2240
1600	2880	2420
1700	3060	2600
1750	3150	2690
1800	3240	2780
1900	3420	2960
2000	3600	3140
2100	3780	3320
2200	3960	3500
2250	4050	3590
2300	4140	3680
2400	4320	3860
2500	4500	4040
2750	4950	4490
3000	5400	4940

NOTE: Fahrenheit temperatures given are rounded to nearest degree.

TABLE C-9.—ENERGY EQUIVALENTS

Ft-lb	Hp-hr	Kwhr	Kg-m	Joules (10^7 ergs)	Btu	Kg-cal
1,000,000	0.5051	0.3766	138,300	1,356,000	1286	324.1
1,980,000	*1*	0.7457	273,750	2,684,500	2545	641.6
2,655,000	1.341	*1*	367,100	3,600,000	3412.8	860.6
723,300	0.3653	0.2724	*100,000*	980,665	929.6	234.4
737,600	0.3725	0.2773	101,970	*1,000,000*	947.8	239
778,160	0.3930	0.2930	107,580	1,055,000	*1000*	252
308,600	0.1558	0.1162	42,690	418,400	396.8	*100*

1 watt = 1 joule/sec = 0.7376 ft-lb/sec
1 metric hp = 0.9863 hp
1 therm = 100,000 Btu

$$\textit{Example: } 1 \text{ joule} = \frac{737{,}600}{1{,}000{,}000} = 0.7376 \text{ ft-lb}$$

TABLE C-10.—SPECIFIC HEAT AND ENTROPY EQUIVALENTS

Units of		Equivalents	
Specific heat	Entropy		
Btu/(lb)(°F)	Btu/(lb)(°R)	1	0.2390
Cal/(g)(°C)	Cal/(g)(°K)	1	0.2390
Joule/(g)(°C)	Joule/(g)(°K)	4.183	1

TABLE C-11.—PRESSURE EQUIVALENTS

Atm	Psi	Kg/sq cm	Mega-bars	Columns of mercury at 32°F		Columns of water at 60°F	
				M	In.	In.	Ft
1	14.696	1.0332	1.0133	0.760	29.92	407.14	33.93
0.06804	1	0.07031	0.06895	0.05171	2.036	27.70	2.309
0.9678	14.22	1	0.9807	0.7355	28.96	394.05	32.84
0.9869	14.50	1.0197	1	0.750	29.53	401.8	33.48
1.316	19.34	1.3596	1.3333	1	39.37	535.7	44.64
0.03342	0.4912	0.03453	0.03386	0.0254	1	13.61	1.134
0.002456	0.0361	0.002538	0.00249	0.00187	0.0735	1	0.0833
0.02947	0.4332	0.03045	0.02986	0.0224	0.8819	12	1

1 psi = 144 psf = 2304 oz/sq ft

Example: 1 kg/sq cm = 0.968 atm

NOTE: Absolute pressure = gauge pressure + atmospheric pressure (see Fig. C-1).

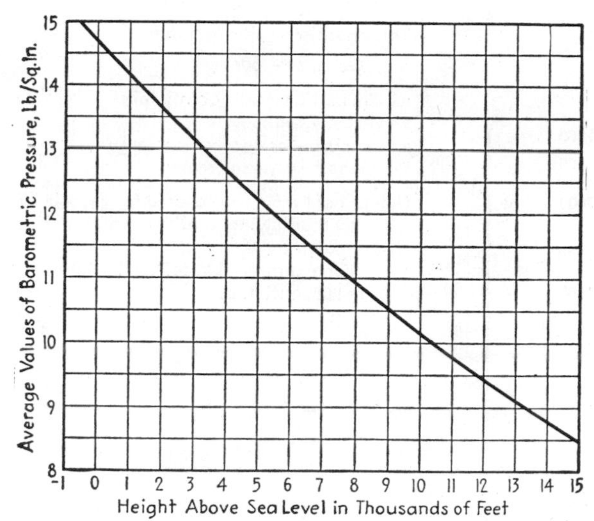

FIG. C-1.—Barometric pressures at different altitudes.

TABLE C-12.—VISCOSITY EQUIVALENTS

Poises = $\frac{\text{dyne-sec}}{\text{sq cm}}$	$\frac{\text{Kg-sec}}{\text{sq m}}$	$\frac{\text{Lb}}{\text{ft-sec}}$	$\frac{\text{Lb-sec}}{\text{sq ft}}$
1	0.0102	0.0672	0.00209
98	1	6.59	0.205
14.88	0.152	1	0.0311
4.78	4.88	32.2	1

1 lb/ft-sec = 3,600 lb/ft-hr   1 lb-hr/sq ft = 3,600 lb-sec/sq ft

1 poise = 100 centipoises = 0.1 kg/m-sec

Examples: 1 centipoise = $\frac{1}{100}$ poise = $\frac{0.0672}{100}$, or 0.000672 lb/ft-sec

1 lb/ft-hr = $\frac{1}{3{,}600}$ lb/ft-sec = $\frac{14.88}{3{,}600}$, or 0.00413 poise

TABLE C-13.—KINEMATIC VISCOSITY SCALES

Scale	Equation (approx)
Saybolt Universal*	$\nu = \frac{\mu}{\rho} = 0.22t - \frac{180}{t}$
Saybolt Furol*	$\nu = \frac{\mu}{\rho} = 2.2t - \frac{1{,}800}{t}$
Redwood	$\nu = \frac{\mu}{\rho} = 0.26t - \frac{171.5}{t}$
Redwood Admiralty	$\nu = \frac{\mu}{\rho} = 0.27t - \frac{2{,}000}{t}$
Engler	$\nu = \frac{\mu}{\rho} = 0.147t - \frac{374}{t}$

NOTE: ν = kinematic viscosity, centistokes
μ = viscosity, centipoises
ρ = density, g/cc
t = time of standard flow, sec.

* See Fig. C-2

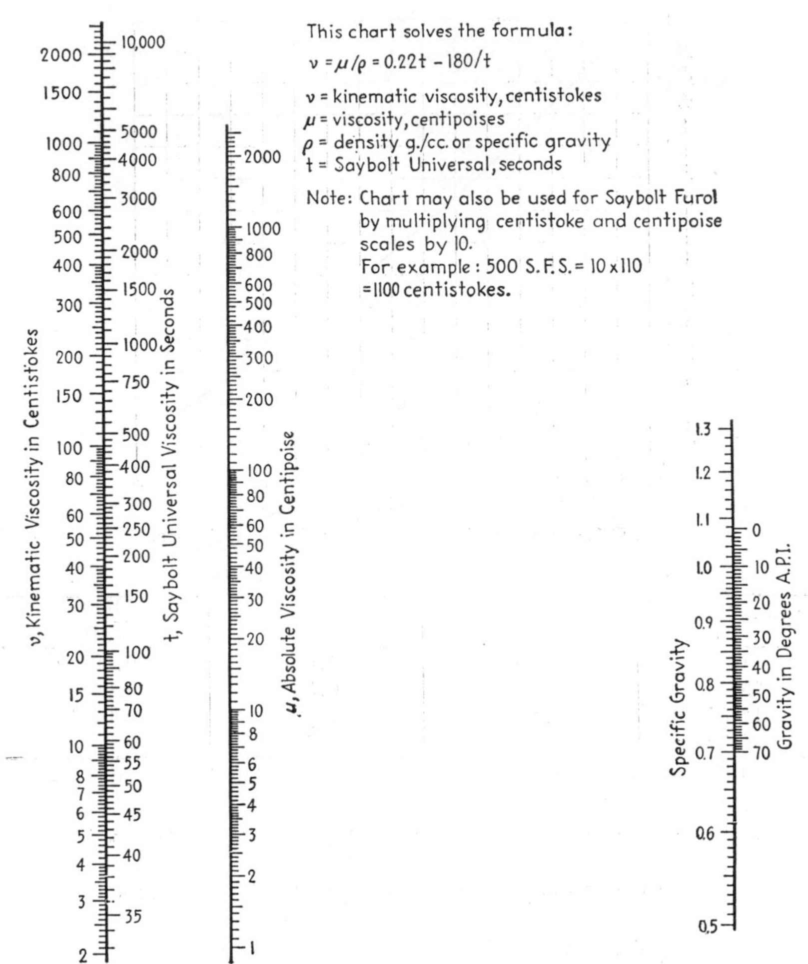

FIG. C-2.—Viscosity conversion chart. *(Reproduced by permission of Product Engineering.)*

TABLE C-14.—THERMAL CONDUCTIVITY EQUIVALENTS

Btu / (hr)(sq ft)(°F/ft)	Btu / (hr)(sq ft)(°F/in.)	Cal / (sec)(sq cm)(°C/cm)	Watts / (sq cm)(°C/cm)
1	12	0.004134	0.01728
0.0833	1	0.0003445	0.001440
241.9	2903	1	4.183
57.83	694.0	0.2391	1

Example: 1 watt/(sq cm) (°C/cm) = 57.83 Btu/(hr)(sq ft)(°F/ft).

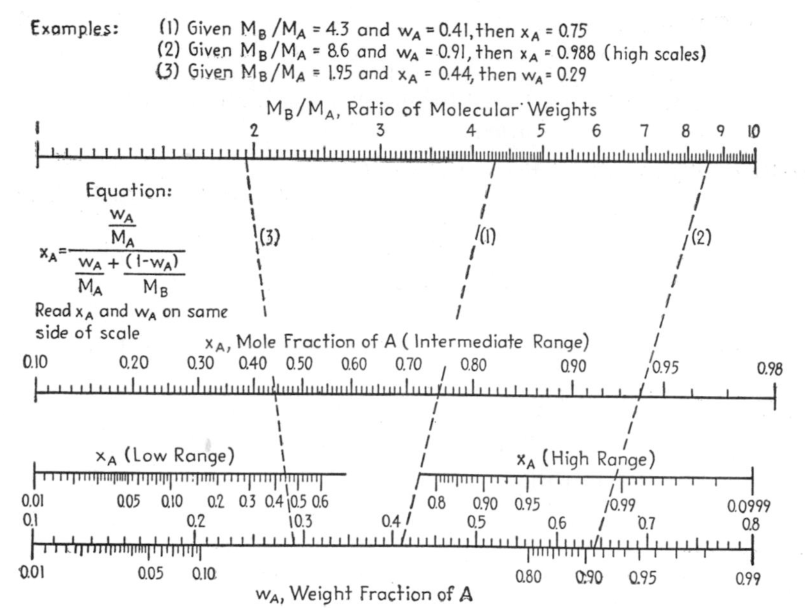

FIG. C-3.—Conversion of weight fractions to mole fractions. (*Bridger, Chem. and Met. Eng., August,* 1937.)

TABLE C-15.—DENSITY EQUIVALENTS

G/cu cm	Lb/cu in.	Lb/cu ft	Short tons/cu yd	Lb/U.S. gal
1	0.03613	62.43	0.8428	8.345
27.68	1	1,728	23.33	231
1.602	0.05787	100	1.350	13.37
1.186	0.04286	74.07	1	9.902
1.198	0.04329	74.81	1.010	10

Example: 1 lb/gal $= \dfrac{74.81}{10} = 7.481$ lb/cu ft.

TABLE C-16.—CONVERSION TABLE: SPECIFIC GRAVITY TO DEGREES BAUMÉ, DEGREES AMERICAN PETROLEUM INSTITUTE, DEGREES TWADDELL, POUNDS PER GALLON, AND POUNDS PER CUBIC FOOT
(Perry, "Chemical Engineers' Handbook," 3d ed., McGraw-Hill Book Company, Inc., New York, 1950)

Specific gravity at 60°F relative to water at 60°F

$$= \text{sp gr } (60/60) = 62.4 \text{ lb/cu ft} = 8.34 \text{ lb/U.S. gal}$$

$$\text{Degrees Baumé} = °\text{Bé} = 145 - \frac{145}{\text{sp gr}} \text{ (heavier than water)}$$

$$= \frac{140}{\text{sp gr}} - 130 \text{ (lighter than water)}$$

$$\text{Degrees American Petroleum Institute} = °\text{API} = \frac{141.5}{\text{sp gr}} - 131.5$$

$$\text{Degrees Twaddell} = °\text{Tw} = 200\,[\text{sp gr } (60/60) - 1]$$

Sp gr (60/60)	°Bé	°A.P.I.	Lb/gal @ 60°F, wt in air	Lb/cu ft @ 60°F, wt in air	Sp gr (60/60)	°Bé	°A.P.I.	Lb/gal @ 60°F, wt in air	Lb/cu ft @ 60°F, wt in air
0.600	103.33	104.33	4.9929	37.350	0.725	63.10	63.67	6.0352	45.146
0.605	101.40	102.38	5.0346	37.662	0.730	61.78	62.34	6.0769	45.458
0.610	99.51	100.47	5.0763	37.973	0.735	60.48	61.02	6.1186	45.770
0.615	97.64	98.58	5.1180	38.285	0.740	59.19	59.72	6.1603	46.082
0.620	95.81	96.73	5.1597	38.597	0.745	57.92	58.43	6.2020	46.394
0.625	94.00	94.90	5.2014	39.910	0.750	56.67	57.17	6.2437	46.706
0.630	92.22	93.10	5.2431	39.222	0.755	55.43	55.92	6.2854	47.018
0.635	90.47	91.33	5.2848	39.534	0.760	54.21	54.68	6.3271	47.330
0.640	88.75	89.59	5.3265	39.845	0.765	53.01	53.47	6.3688	47.642
0.645	87.05	87.88	5.3682	40.157	0.770	51.82	52.27	6.4104	47.953
0.650	85.38	86.19	5.4098	40.468	0.775	50.65	51.08	6.4521	47.265
0.655	83.74	84.53	5.4515	40.780	0.780	49.49	49.91	6.4938	48.577
0.660	82.12	82.89	5.4932	41.092	0.785	48.34	48.75	6.5355	48.889
0.665	80.53	81.28	5.5349	41.404	0.790	47.22	47.61	6.5772	49.201
0.670	78.96	79.69	5.5766	41.716	0.795	46.10	46.49	6.6189	49.513
0.675	77.41	78.13	5.6183	42.028	0.800	45.00	45.38	6.6606	49.825
0.680	75.88	76.59	5.6600	42.340	0.805	43.91	44.28	6.7023	50.137
0.685	74.38	75.07	5.7017	42.652	0.810	42.84	43.19	6.7440	50.448
0.690	72.90	73.57	5.7434	42.963	0.815	41.78	42.12	6.7857	50.760
0.695	71.44	72.10	5.7851	43.275	0.820	40.73	41.06	6.8274	51.072
0.700	70.00	70.64	5.8268	43.587	0.825	39.70	40.02	6.8691	51.384
0.705	68.58	69.21	5.8685	43.899	0.830	38.67	38.98	6.9108	51.696
0.710	67.18	67.80	5.9101	44.211	0.835	37.66	37.96	6.9525	52.008
0.715	65.80	66.40	5.9518	44.523	0.840	36.67	36.95	6.9941	52.320
0.720	64.44	65.03	5.9935	44.834	0.845	35.68	35.96	7.0358	52.632
					0.850	34.71	34.97	7.0775	52.943

TABLE C-16.—CONVERSION TABLE: SPECIFIC GRAVITY TO DEGREES BAUMÉ, DEGREES AMERICAN PETROLEUM INSTITUTE, DEGREES TWADDELL, POUNDS PER GALLON, AND POUNDS PER CUBIC FOOT (*Continued*)

Sp gr (60/60)	°Bé	°A.P.I.	Lb/gal @ 60°F, wt in air	Lb/cu ft @ 60°F, wt in air	Sp gr (60/60)	°Bé	°Tw	Lb/gal @ 60°F, wt in air	Lb/cu ft @ 60°F, wt in air
0.855	33.74	34.00	7.1192	53.255	1.030	4.22	6	8.5784	64.171
0.860	32.79	33.03	7.1609	53.567	1.035	4.90	7	8.6201	64.483
0.865	31.85	32.08	7.2026	53.879	1.040	5.58	8	8.6618	64.795
0.870	30.92	31.14	7.2443	54.191	1.045	6.24	9	8.7035	65.107
0.875	30.00	30.21	7.2860	54.503	1.050	6.91	10	8.7452	65.419
0.880	29.09	29.30	7.3277	54.815	1.055	7.56	11	8.7869	65.731
0.885	28.19	28.39	7.3694	55.127	1.060	8.21	12	8.8286	66.042
0.890	27.30	27.49	7.4111	55.438	1.065	8.85	13	8.8703	66.354
0.895	26.42	26.60	7.4528	55.750	1.070	9.49	14	8.9120	66.666
0.900	25.56	25.72	7.4944	56.062	1.075	10.12	15	8.9537	66.978
0.905	24.70	24.85	7.5361	56.374	1.080	10.74	16	8.9954	67.290
0.910	23.85	23.99	7.5777	56.685	1.085	11.36	17	9.0371	67.602
0.915	23.01	23.14	7.6194	56.997	1.090	11.97	18	9.0787	67.914
0.920	22.17	22.30	7.6612	57.310	1.095	12.58	19	9.1204	68.226
0.925	21.35	21.47	7.7029	57.622	1.100	13.18	20	9.1621	68.537
0.930	20.54	20.65	7.7446	57.934	1.105	13.78	21	9.2038	68.849
0.935	19.73	19.84	7.7863	58.246	1.110	14.37	22	9.2455	69.161
0.940	18.94	19.03	7.8280	58.557	1.115	14.96	23	9.2872	69.473
0.945	18.15	18.24	7.8697	58.869	1.120	15.54	24	9.3289	69.785
0.950	17.37	17.45	7.9114	59.181	1.125	16.11	25	9.3706	70.097
0.955	16.60	16.67	7.9531	59.493	1.130	16.68	26	9.4123	70.409
0.960	15.83	15.90	7.9947	59.805	1.135	17.25	27	9.4540	70.721
0.965	15.08	15.13	8.0364	60.117	1.140	17.81	28	9.4957	71.032
0.970	14.33	14.38	8.0780	60.428	1.145	18.36	29	9.5374	71.344
0.975	13.59	13.63	8.1197	60.740	1.150	18.91	30	9.5790	71.656
0.980	12.86	12.89	8.1615	61.052	1.155	19.46	31	9.6207	71.968
0.985	12.13	12.15	8.2032	61.364	1.160	20.00	32	9.6624	72.280
0.990	11.41	11.43	8.2449	61.676	1.165	20.54	33	9.7041	72.592
0.995	10.70	10.71	8.2866	61.988	1.170	21.07	34	9.7458	72.904
1.000	10.00	10.00	8.3283	62.300	1.175	21.60	35	9.7875	73.216
		°Tw							
1.005	0.72	1	8.3700	62.612	1.180	22.12	36	9.8292	73.528
1.010	1.44	2	8.4117	62.924	1.185	22.64	37	9.8709	73.840
1.015	2.14	3	8.4534	63.236	1.190	23.15	38	9.9126	74.151
1.020	2.84	4	8.4950	63.547	1.195	23.66	39	9.9543	74.463
1.025	3.54	5	8.5367	63.859	1.200	24.17	40	9.9960	74.775

TABLE C-16.—CONVERSION TABLE: SPECIFIC GRAVITY TO DEGREES BAUMÉ, DEGREES AMERICAN PETROLEUM INSTITUTE, DEGREES TWADDELL, POUNDS PER GALLON, AND POUNDS PER CUBIC FOOT (*Continued*)

Sp gr (60/60)	°Bé	°Tw	Lb/gal @ 60°F, wt in air	Lb/cu ft @ 60°F, wt in air	Sp gr (60/60)	°Bé	°Tw	Lb/gal @ 60°F, wt in air	Lb/cu ft @ 60°F, wt in air
1.205	24.67	41	10.0377	75.087	1.380	39.93	76	11.4969	86.003
1.210	25.17	42	10.0793	75.399	1.385	40.31	77	11.5386	86.315
1.215	25.66	43	10.1210	75.711	1.390	40.68	78	11.5803	86.626
1.220	26.15	44	10.1627	76.022	1.395	41.06	79	11.6220	86.938
1.225	26.63	45	10.2044	76.334	1.400	41.43	80	11.6637	87.250
1.230	27.11	46	10.2461	76.646	1.405	41.80	81	11.7054	87.562
1.235	27.59	47	10.2878	76.958	1.410	42.16	82	11.7471	87.874
1.240	28.06	48	10.3295	77.270	1.415	42.53	83	11.7888	88.186
1.245	28.53	49	10.3712	77.582	1.420	42.89	84	11.8304	88.498
1.250	29.00	50	10.4129	77.894	1.425	43.25	85	11.8721	88.810
1.255	29.46	51	10.4546	78.206	1.430	43.60	86	11.9138	89.121
1.260	29.92	52	10.4963	78.518	1.435	43.95	87	11.9555	89.433
1.265	30.38	53	10.5380	78.830	1.440	44.31	88	11.9972	89.745
1.270	30.83	54	10.5797	79.141	1.445	44.65	89	12.0389	90.057
1.275	31.27	55	10.6214	79.453	1.450	45.00	90	12.0806	90.369
1.280	31.72	56	10.6630	79.765	1.455	45.34	91	12.1223	90.681
1.285	32.16	57	10.7047	80.077	1.460	45.68	92	12.1640	90.993
1.290	32.60	58	10.7464	80.389	1.465	46.02	93	12.2057	91.305
1.295	33.03	59	10.7881	80.701	1.470	46.36	94	12.2473	91.616
1.300	33.46	60	10.8298	81.013	1.475	46.69	95	12.2890	91.928
1.305	33.89	61	10.8715	81.325	1.480	47.03	96	12.3307	92.240
1.310	34.31	62	10.9132	81.636	1.485	47.36	97	12.3724	92.552
1.315	34.73	63	10.9549	81.948	1.490	47.68	98	12.4141	92.864
1.320	35.15	64	10.9966	82.260	1.495	48.01	99	12.4558	93.176
1.325	35.57	65	11.0383	82.572	1.500	48.33	100	12.4975	93.488
1.330	35.98	66	11.0800	82.884	1.505	48.65	101	12.5392	93.800
1.335	36.39	67	11.1217	83.196	1.510	48.97	102	12.5809	94.112
1.340	36.79	68	11.1634	83.508	1.515	49.29	103	12.6226	94.424
1.345	37.19	69	11.2051	83.820	1.520	49.61	104	12.6643	94.735
1.350	37.59	70	11.2467	84.131	1.525	49.92	105	12.7060	95.047
1.355	37.99	71	11.2884	84.443	1.530	50.23	106	12.7477	95.359
1.360	38.38	72	11.3301	84.755	1.535	50.54	107	12.7894	95.671
1.365	38.77	73	11.3718	85.067	1.540	50.84	108	12.8310	95.983
1.370	39.16	74	11.4135	85.379	1.545	51.15	109	12.8727	96.295
1.375	39.55	75	11.4552	85.691	1.550	51.45	110	12.9144	96.606

TABLE C-16.—CONVERSION TABLE: SPECIFIC GRAVITY TO DEGREES BAUMÉ, DEGREES AMERICAN PETROLEUM INSTITUTE, DEGREES TWADDELL, POUNDS PER GALLON, AND POUNDS PER CUBIC FOOT (*Continued*)

Sp gr (60/60)	°Bé	°Tw	Lb/gal @ 60°F, wt in air	Lb/cu ft @ 60°F, wt in air	Sp gr (60/60)	°Bé	°Tw	Lb/gal @ 60°F, wt in air	Lb/cu ft @ 60°F, wt in air
1.555	51.75	111	12.9561	96.918	1.730	61.18	146	14.4153	107.834
1.560	52.05	112	12.9978	97.230	1.735	61.34	147	14.4570	108.146
1.565	52.35	113	13.0395	97.542	1.740	61.67	148	14.4987	108.458
1.570	52.64	114	13.0812	97.854	1.745	61.91	149	14.5404	108.770
1.575	52.94	115	13.1229	98.166	1.750	62.14	150	14.5821	109.082
1.580	53.23	116	13.1646	98.478	1.755	62.38	151	14.6238	109.394
1.585	53.52	117	13.2063	98.790	1.760	62.61	152	14.6655	109.705
1.590	53.81	118	13.2480	99.102	1.765	62.85	153	14.7072	110.017
1.595	54.09	119	13.2897	99.414	1.770	63.08	154	14.7489	110.329
1.600	54.38	120	13.3313	99.725	1.775	63.31	155	14.7906	110.641
1.605	54.66	121	13.3730	100.037	1.780	63.54	156	14.8323	110.953
1.610	54.94	122	13.4147	100.349	1.785	63.77	157	14.8740	111.265
1.615	55.22	123	13.4564	100.661	1.790	63.99	158	14.9157	111.577
1.620	55.49	124	13.4981	100.973	1.795	64.22	159	14.9574	111.889
1.625	55.77	125	13.5398	101.285	1.800	64.44	160	14.9990	112.200
1.630	56.04	126	13.5815	101.597	1.805	64.67	161	15.0407	112.512
1.635	56.32	127	13.6232	101.909	1.810	64.89	162	15.0824	112.824
1.640	56.59	128	13.6649	102.220	1.815	65.11	163	15.1241	113.136
1.645	56.85	129	13.7066	102.532	1.820	65.33	164	15.1658	113.448
1.650	57.12	130	13.7483	102.844	1.825	65.55	165	15.2075	113.760
1.655	57.39	131	13.7900	103.156	1.830	65.77	166	15.2492	114.072
1.660	57.65	132	13.8317	103.468	1.835	65.98	167	15.2909	114.384
1.665	57.91	133	13.8734	103.780	1.840	66.20	168	15.3326	114.696
1.670	58.17	134	13.9150	104.092	1.845	66.41	169	15.3743	115.007
1.675	58.43	135	13.9567	104.404	1.850	66.62	170	15.4160	115.318
1.680	58.69	136	13.9984	104.715	1.855	66.83	171	15.4577	115.630
1.685	58.95	137	14.0401	105.027	1.860	67.04	172	15.4993	115.943
1.690	59.20	138	14.0818	105.339	1.865	67.25	173	15.5410	116.255
1.695	59.45	139	14.1235	105.651	1.870	67.46	174	15.5827	116.567
1.700	59.71	140	14.1652	105.963	1.875	67.67	175	15.6244	116.879
1.705	59.96	141	14.2069	106.275	1.880	67.87	176	15.6661	117.191
1.710	60.20	142	14.2486	106.587	1.885	68.08	177	15.7078	117.503
1.715	60.45	143	14.2903	106.899	1.890	68.28	178	15.7495	117.814
1.720	60.70	144	14.3320	107.210	1.895	68.48	179	15.7912	118.126
1.725	60.94	145	14.3737	107.522	1.900	68.68	180	15.8329	118.438

TABLE C-16.—CONVERSION TABLE: SPECIFIC GRAVITY TO DEGREES BAUMÉ, DEGREES AMERICAN PETROLEUM INSTITUTE, DEGREES TWADDELL, POUNDS PER GALLON, AND POUNDS PER CUBIC FOOT (*Concluded*)

Sp gr (60/60)	°Bé	°Tw	Lb/gal @ 60°F, wt in air	Lb/cu ft @ 60°F, wt in air	Sp gr (60/60)	°Bé	°Tw	Lb/gal @ 60°F, wt in air	Lb/cu ft @ 60°F, wt in air
1.905	68.88	181	15.8746	118.740	1.955	70.83	191	16.2915	121.869
1.910	69.08	182	15.9163	119.062	1.960	71.02	192	16.3332	122.181
1.915	69.28	183	15.9580	119.374	1.965	71.21	193	16.3749	122.493
1.920	69.48	184	15.9996	119.686	1.970	71.40	194	16.4166	122.804
1.925	69.68	185	16.0413	119.998	1.975	71.58	195	16.4583	123.116
1.930	69.87	186	16.0830	120.309	1.980	71.77	196	16.5000	123.428
1.935	70.06	187	16.1247	120.621	1.985	71.95	197	16.5417	123.740
1.940	70.26	188	16.1664	120.933	1.990	72.14	198	16.5833	124.052
1.945	70.45	189	16.2081	121.245	1.995	72.32	199	16.6250	124.364
1.950	70.64	190	16.2498	121.557	2.000	72.50	200	16.6667	124.676

CHAPTER D

PHYSICAL AND MECHANICAL PROPERTIES

The start of all engineering work rests on a knowledge of the properties of the materials used. Consequently, various properties are discussed in almost every chapter of this book. This chapter gives liquid densities and references to physical data presented in other chapters; a few mechanical properties of selected construction materials; and finally, a brief treatment of corrosion. Because of the extensive amount of information an engineer is likely to require, no one source book can be entirely adequate. Below are listed a number of useful compilations and reference books.

"International Critical Tables," 7 vols., McGraw-Hill Book Company, Inc., New York, 1926–1930.

Hodgman, "Handbook of Chemistry and Physics," published annually by the Chemical Rubber Publishing Co., Cleveland, Ohio.

Lange, "Handbook of Chemistry," 10th ed., McGraw-Hill Book Company, Inc., New York, 1961.

Perry, "Chemical Engineers' Handbook," 3d ed., McGraw-Hill Book Company, Inc., New York, 1950.

Marks, "Mechanical Engineers' Handbook," 5th ed., McGraw-Hill Book Company, Inc., New York, 1951.

Brady, "Materials Handbook," 8th ed., McGraw-Hill Book Company, Inc., New York, 1956.

"The Metals Handbook," 1948, with supplements in 1954 and 1955, American Society of Metals, Cleveland, Ohio.

Moore and Moore, "Textbook of the Materials of Engineering," 8th ed., McGraw-Hill Book Company, Inc., New York, 1953.

Hoyt, "Metals Data," 2d ed., Reinhold Publishing Corporation, New York, 1952.

Speller, "Corrosion: Causes and Prevention," 3d ed., McGraw-Hill Book Company, Inc., New York, 1951.

Sauveur, "Metallography and Heat Treatment of Iron and Steel," 4th ed., McGraw-Hill Book Company, Inc., New York, 1935.

Burns and Bradley, "Protective Coatings for Metals," Reinhold Publishing Corporation, New York, 1955.

Mantell, "Engineering Materials Handbook," 1st ed., McGraw-Hill Book Company, Inc., New York, 1958.

Uhlig, "Corrosion Handbook," John Wiley & Sons, Inc., New York, 1948.

Reid and Sherwood, "The Properties of Gases and Liquids," McGraw-Hill Book Company, Inc., New York, 1958.

Norton, "Refractories," 3d ed., McGraw-Hill Book Company, Inc., New York, 1949.
DuMond, "Engineering Materials Manual," Reinhold Publishing Corporation, New York, 1951.
Blanks and Kennedy, "The Technology of Cement and Concrete," Vol. I, John Wiley & Sons, Inc., New York, 1955.
Hool and Kinney, "Reinforced Concrete and Masonry Structures," 2d ed., McGraw-Hill Book Company, Inc., New York, 1944.
Brown, Panshin, and Forsaith, "Textbook of Wood Technology," Vol. I, 1949; Vol. II, 1952, McGraw-Hill Book Company, Inc., New York.
"Plastics Progress 1957," Philosophical Library, Inc., New York, 1957.

1. Physical Properties of Process Materials

a. Densities of Liquids.—Values are given in chart form for a number of materials. A list of the figures follows:

Specific gravity or volume of:

Also Fig. E-4 in the next chapter permits the estimation of liquid densities at various temperatures from a single value. Tables E-13 to E-18 give specific volumes of water, carbon dioxide, propane, and ammonia. (The density in pounds per cubic foot is the reciprocal of the specific volume.) Tables H-7 and H-12 give density data on liquid fuels.

Densities of liquid mixtures may, according to the laws of perfect solution, be computed by assuming no volume change on mixing; that is, additive volumes:

$$v = \frac{P_1}{100} v_1 + \frac{P_2}{100} v_2 + \frac{P_3}{100} v_3 + \cdots$$

$$\frac{1}{\rho} = \frac{P_1}{100\rho_1} + \frac{P_2}{100\rho_2} + \frac{P_3}{100\rho_3} + \cdots$$

where v = specific volume of mixture
v_1, v_2, v_3 = specific volume of components 1, 2, 3
ρ = density of mixture
ρ_1, ρ_2, ρ_3 = density of components 1, 2, 3
P_1, P_2, P_3 = weight per cent of components 1, 2, 3

For hydrocarbon mixtures at moderate pressures this rule works fairly well. Similarly, it is likely to work for mixtures of closely allied liquids. However, many solutions do not follow the rule, which, therefore, should be used with caution, particularly in the case of water solutions.

b. Other Properties.—Data are scattered throughout the book in connection with subjects for which the particular properties are most important. However, as no physical datum is the exclusive property of any one subject, the principal tables and figures are listed in the accompanying tabulation.

Physical property	Tables	Figures
Compressibility factors (reduced equation of state)	....	E-1 to E-3
Liquid density, estimation of	....	E-4
Specific heats:		
Gases	E-3	E-5 to E-9
Solids	E-5, E-11	
Liquids	E-6	E-15
Solutions	E-8 to E-10	
Sensible heats:		
Gases	....	E-10 to E-14
Solids	E-4	
Petroleum liquids	E-7	
Vapor pressures	H-12	E-16 to E-18, H-5
Heats of vaporization	E-12	E-19
Thermodynamic properties:		
Steam	E-13 to E-15	
Propane	E-16	E-20
Ammonia	E-17	E-21
Carbon dioxide	E-18	
Air	N-1	
Heats of formation	E-19	
Heats of combustion:		
Gases	H-9 to H-11	
Solids	H-2 to H-4	H-1
Liquids	H-7, H-12	H-4
Viscosities:		
Gases	F-2, F-3	F-2 to F-4
Liquids	F-1	F-1, F-4, H-3
Heat conductivity	D-2, G-4, G-5	
Atomic weights	C-2	
Critical temperatures and pressures	E-2	

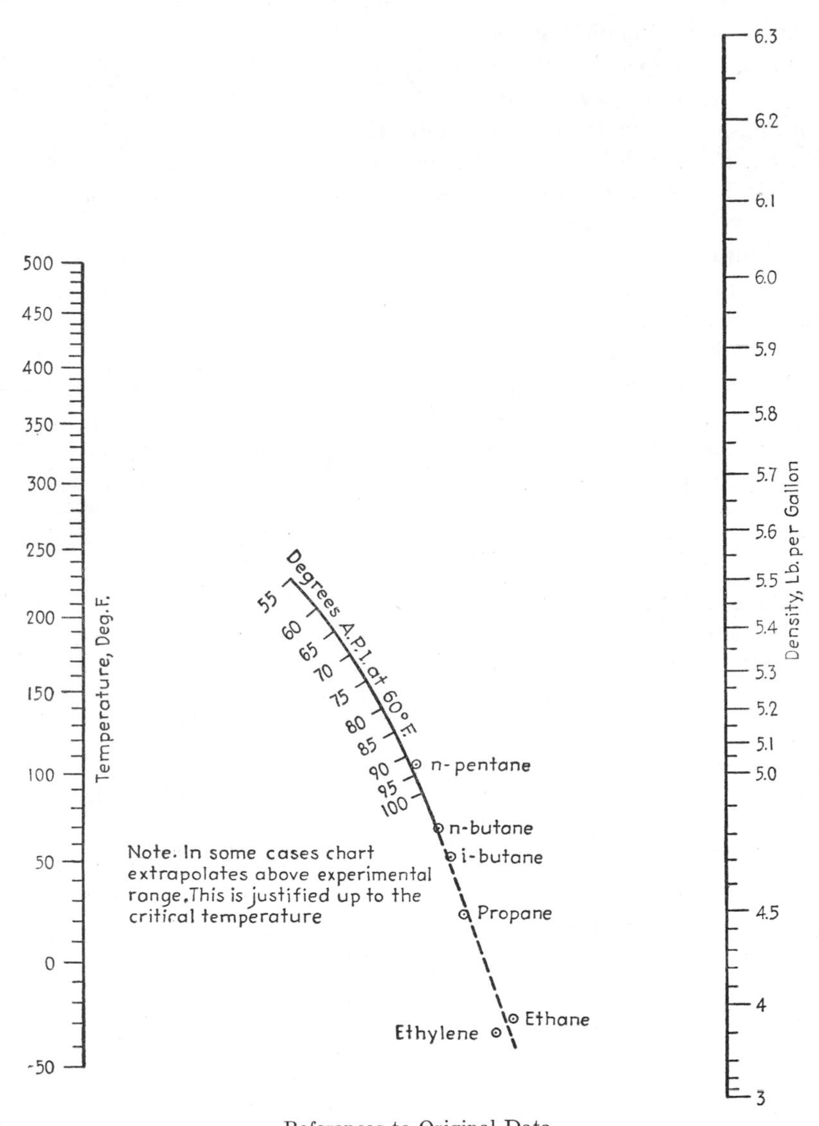

References to Original Data

Natl. Bur. Stands. Circ. 410 (1936). (Oils 65–100° A.P.I. 0° to 90°F, oils 55–65° A.P.I. 0° to 125°F.)

Sage and Lacey, *Ind. Eng. Chem.* **34,** 730 (1942). (*n*-Pentane, 100° to 387°F.)

Stearns and George, *Ind. Eng. Chem.* **35,** 602 (1943). (Propane, −80° to 200°F.)

Edwards, *Refrigerating Eng.*, August, 1944. (Ethane up to 90°F.)

York and White, *Trans. Am. Inst. Chem. Engrs.* **40,** 227 (1944). (Ethylene up to 48°F.)

Carney, *Proc.*, 21st Annual Convention of the Natural Gasoline Association of America, 1942. (Propane, *i*-Butane, *n*-Butane, *i*-Pentane, *n*-Pentane, three mixtures and three natural gasolines. Temperature range, −50 to 140°F.)

FIG. D-1.—Densities of light hydrocarbon liquids.

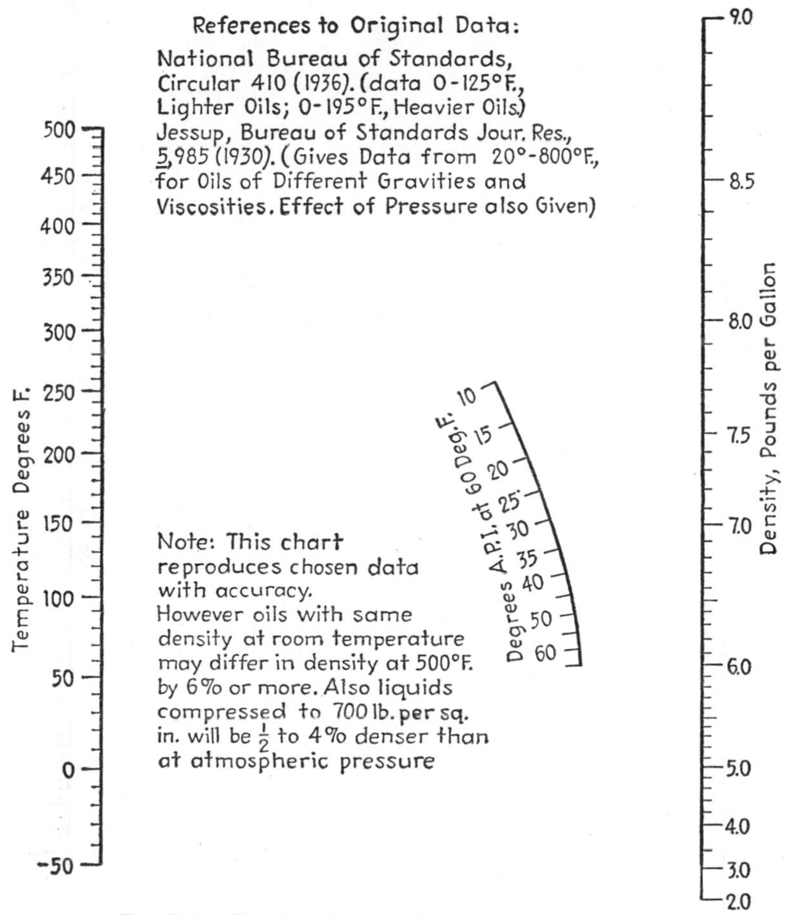

FIG. D-2.—Density of typical liquid petroleum products.

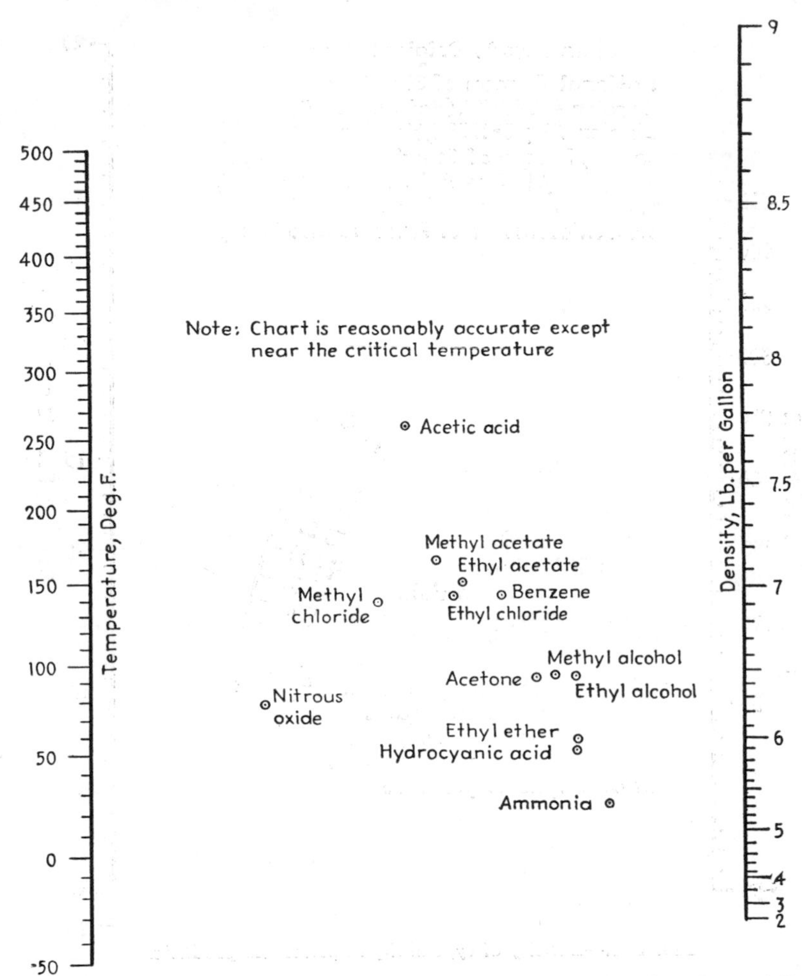

FIG. D-3.—Densities of miscellaneous liquids. (*Based on data from Perry, "Chemical Engineers' Handbook," 2d ed., McGraw-Hill Book Company, Inc., New York,* 1941; *Lange, "Handbook of Chemistry," 3d ed., McGraw-Hill Book Company, Inc., New York,* 1939.)

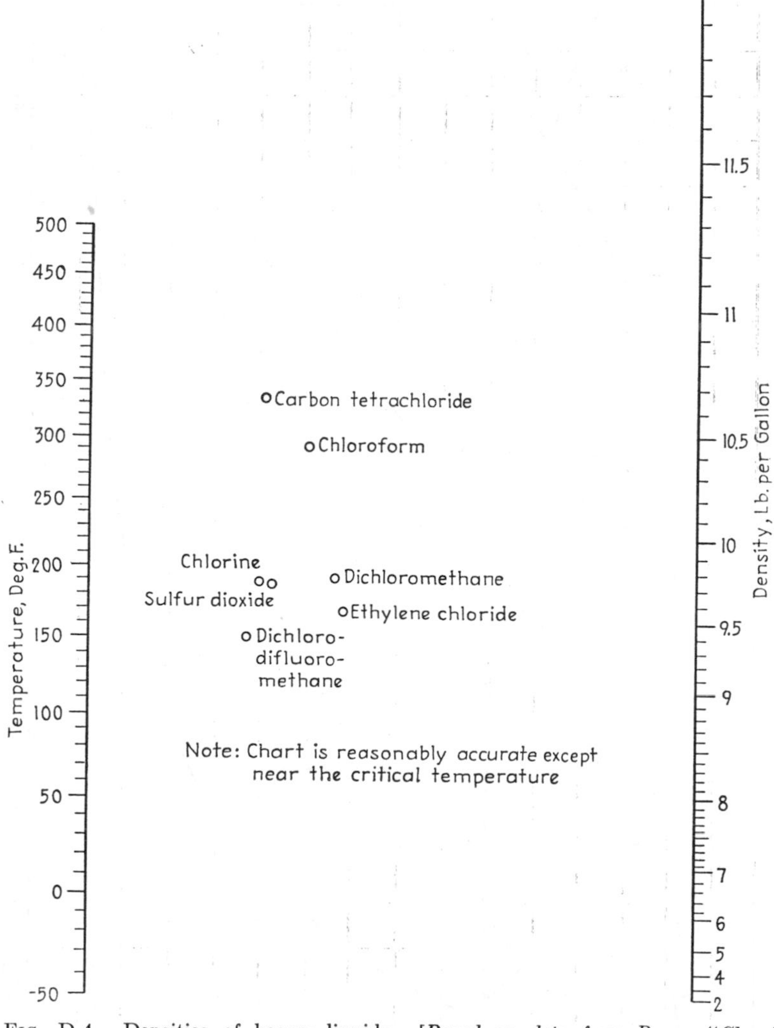

FIG. D-4.—Densities of heavy liquids. [*Based on data from Perry, "Chemical Engineers' Handbook," 2d ed., McGraw-Hill Book Company, Inc., New York,* 1941; *McGovern, Ind. Eng. Chem.*, **35,** 1230 (1943).]

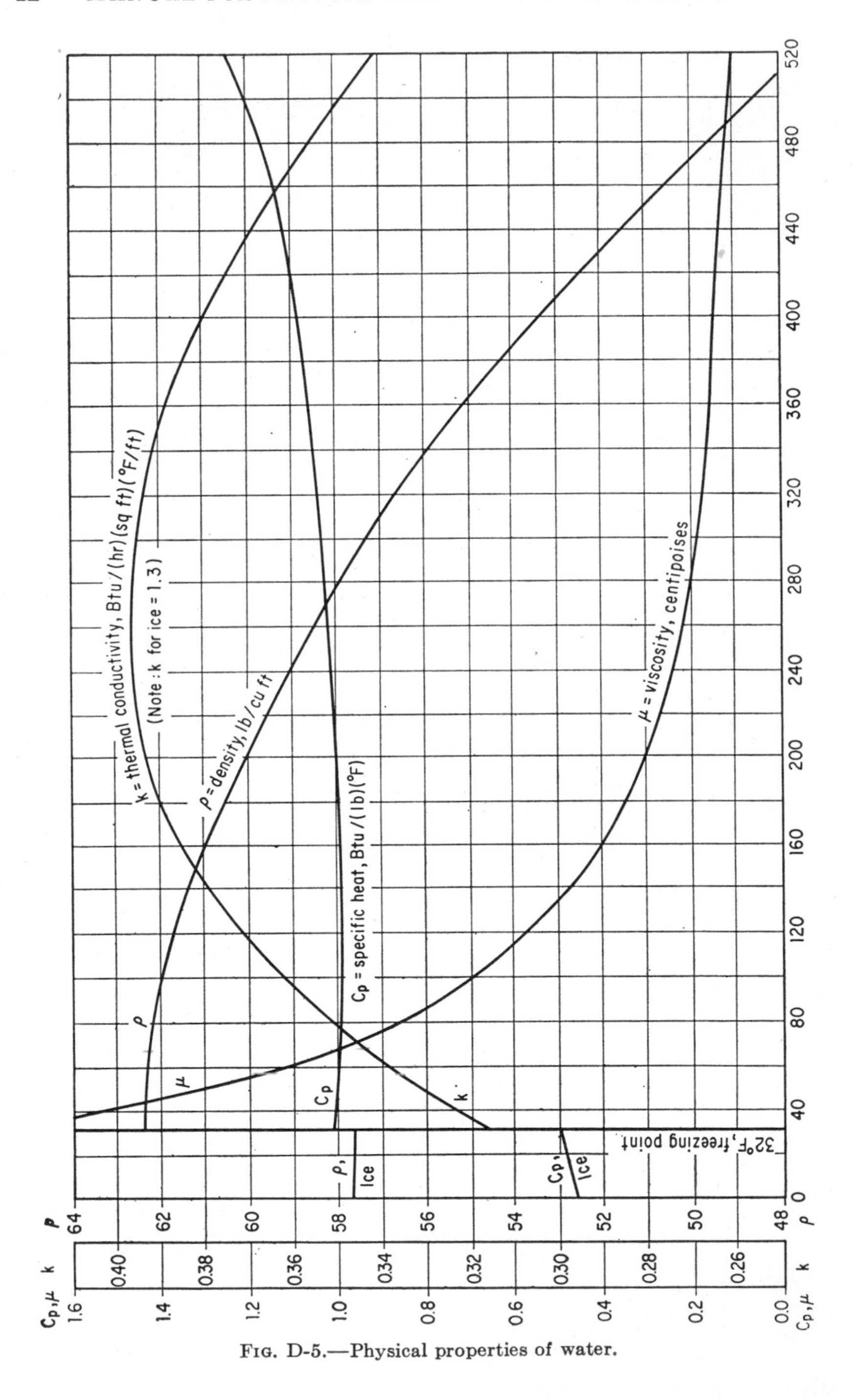

FIG. D-5.—Physical properties of water.

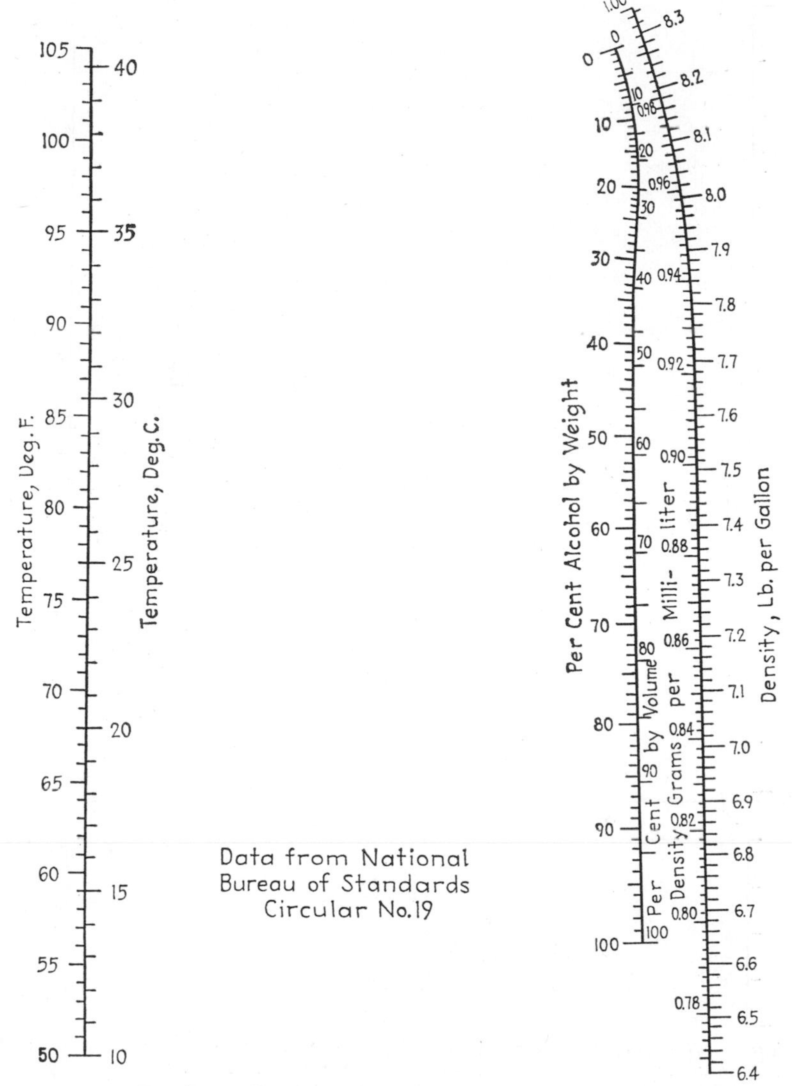

FIG. D-6.—Densities of ethyl alcohol–water mixtures.

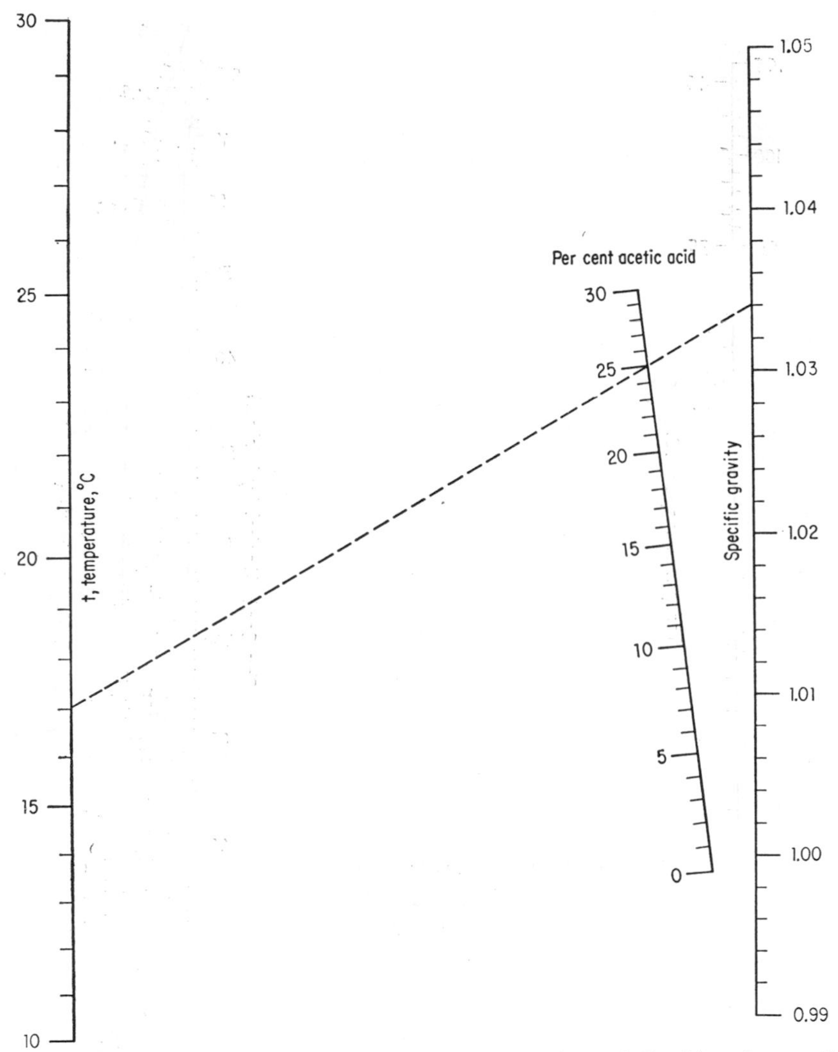

FIG. D-7.—Temperature–specific-gravity–concentration relationships for acetic acid, 0 to 30 per cent. (*Reprinted from Davis, "Chemical Engineering Nomographs," McGraw-Hill Book Company, Inc., New York, 1944.*)

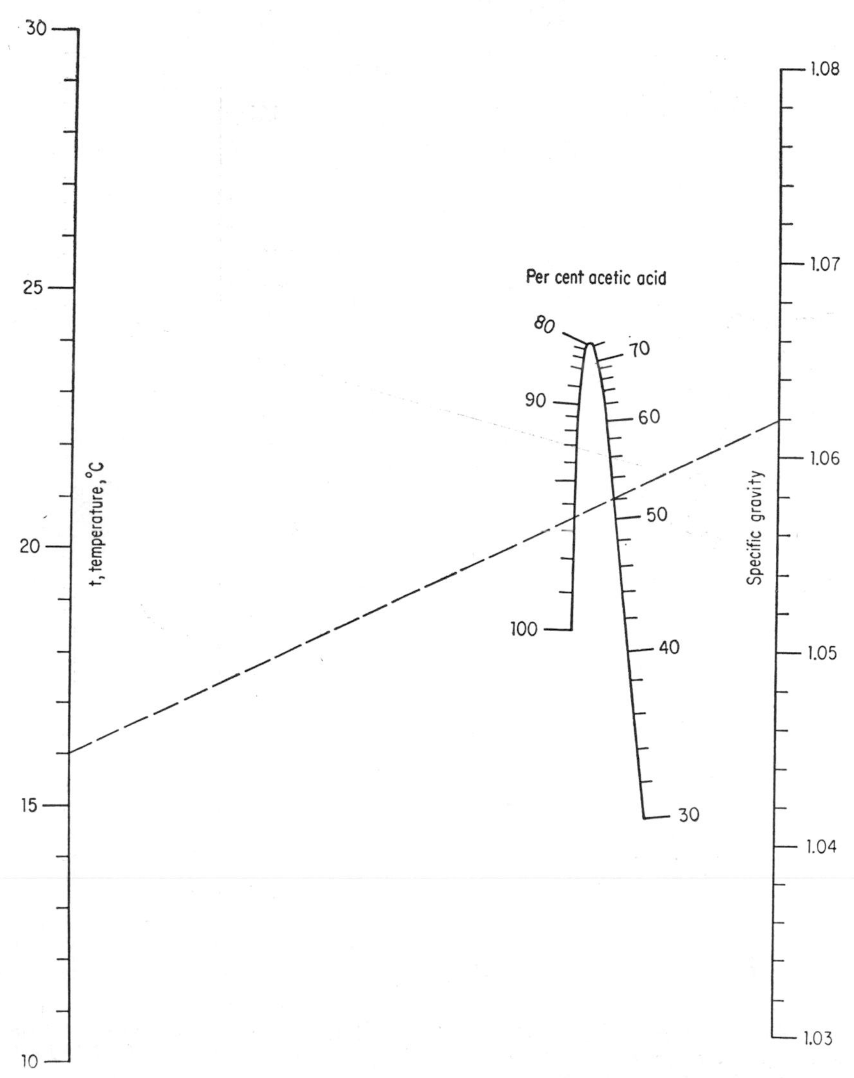

FIG. D-8.—Temperature–specific-gravity–concentration relationships for acetic acid, 30 to 100 per cent. (*Reprinted from Davis, "Chemical Engineering Nomographs," McGraw-Hill Book Company, Inc., New York, 1944.*)

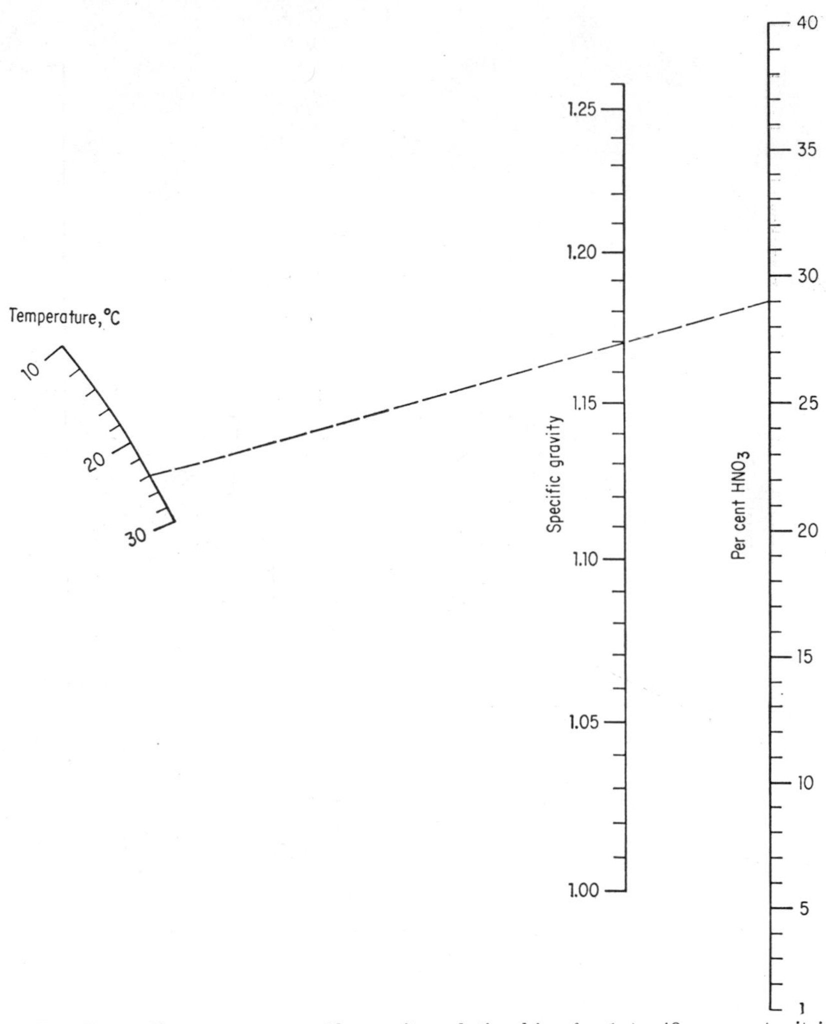

Fig. D-9.—Temperature–specific-gravity relationships for 1 to 40 per cent nitric acid. (*Reprinted from Davis, "Chemical Engineering Nomographs," McGraw-Hill Book Company, Inc., New York,* 1944.)

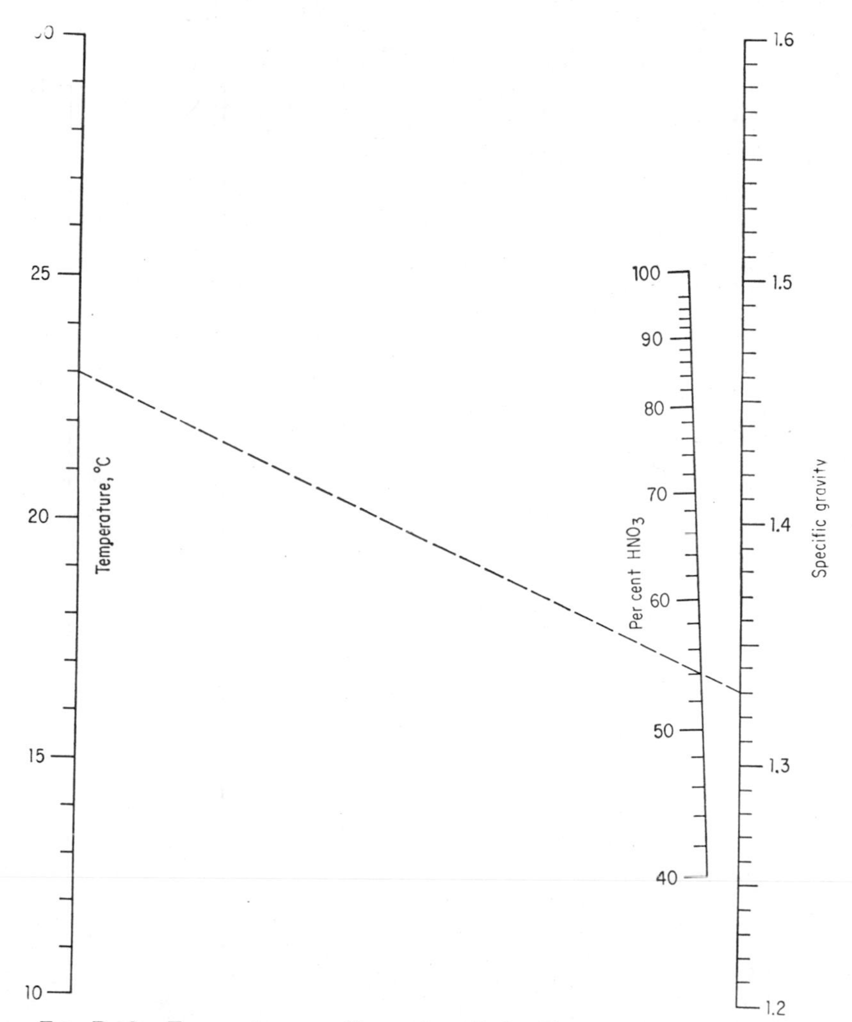

FIG. D-10.—Temperature–specific-gravity relationships from 40 to 100 per cent nitric acid. (*Reprinted from Davis, "Chemical Engineering Nomographs," McGraw-Hill Book Company, Inc., New York,* 1944.)

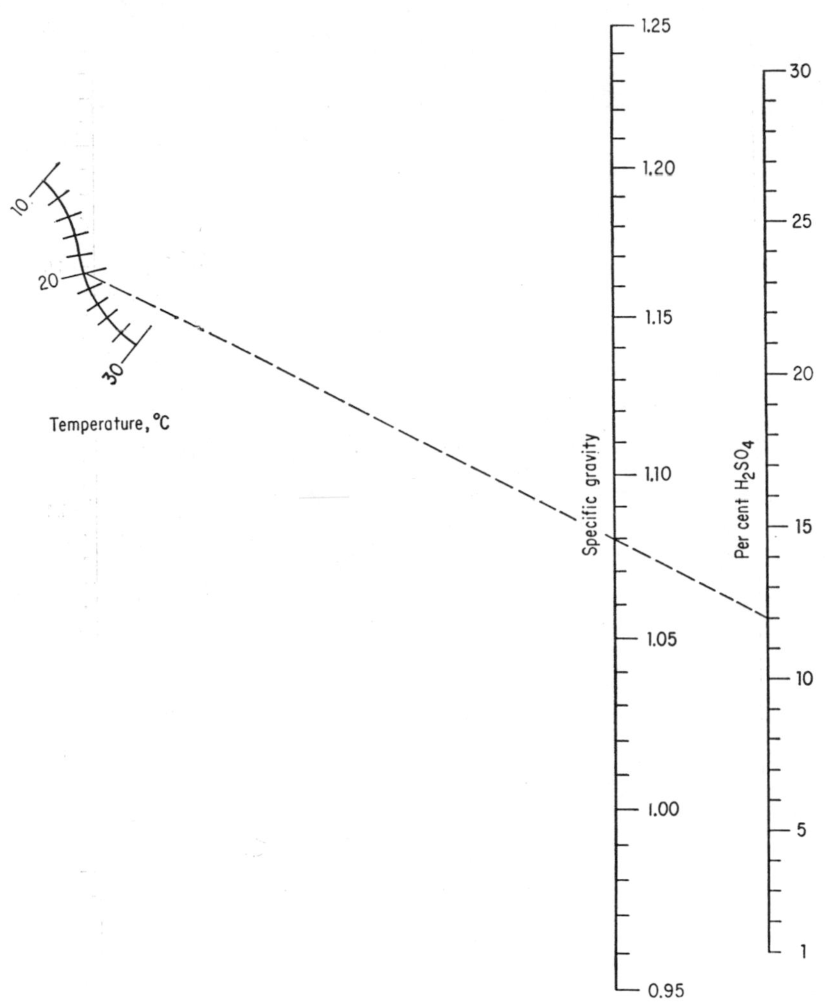

FIG. D-11.—Temperature-specific-gravity relationships for 1 to 30 per cent sulphuric acid. (*Reprinted from Davis, "Chemical Engineering Nomographs," McGraw-Hill Book Company, Inc., New York,* 1944.)

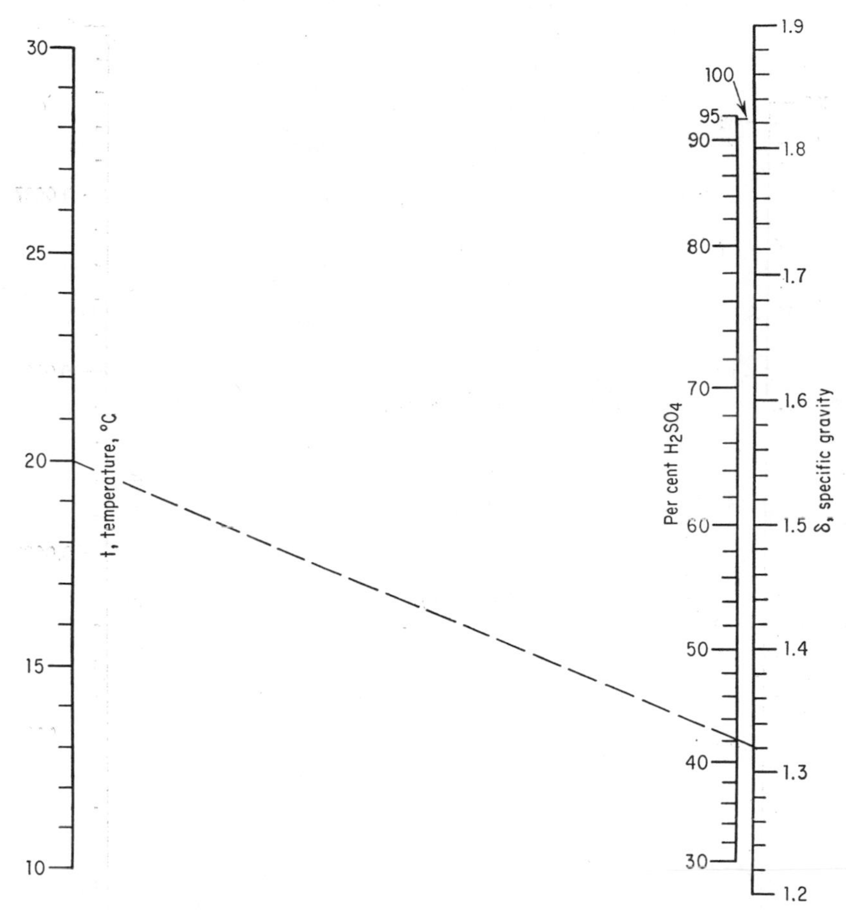

FIG. D-12.—Temperature–specific-gravity relationships for 30 to 100 per cent sulphuric acid. (*Reprinted from Davis, "Chemical Engineering Nomographs," McGraw-Hill Book Company, Inc., New York, 1944.*)

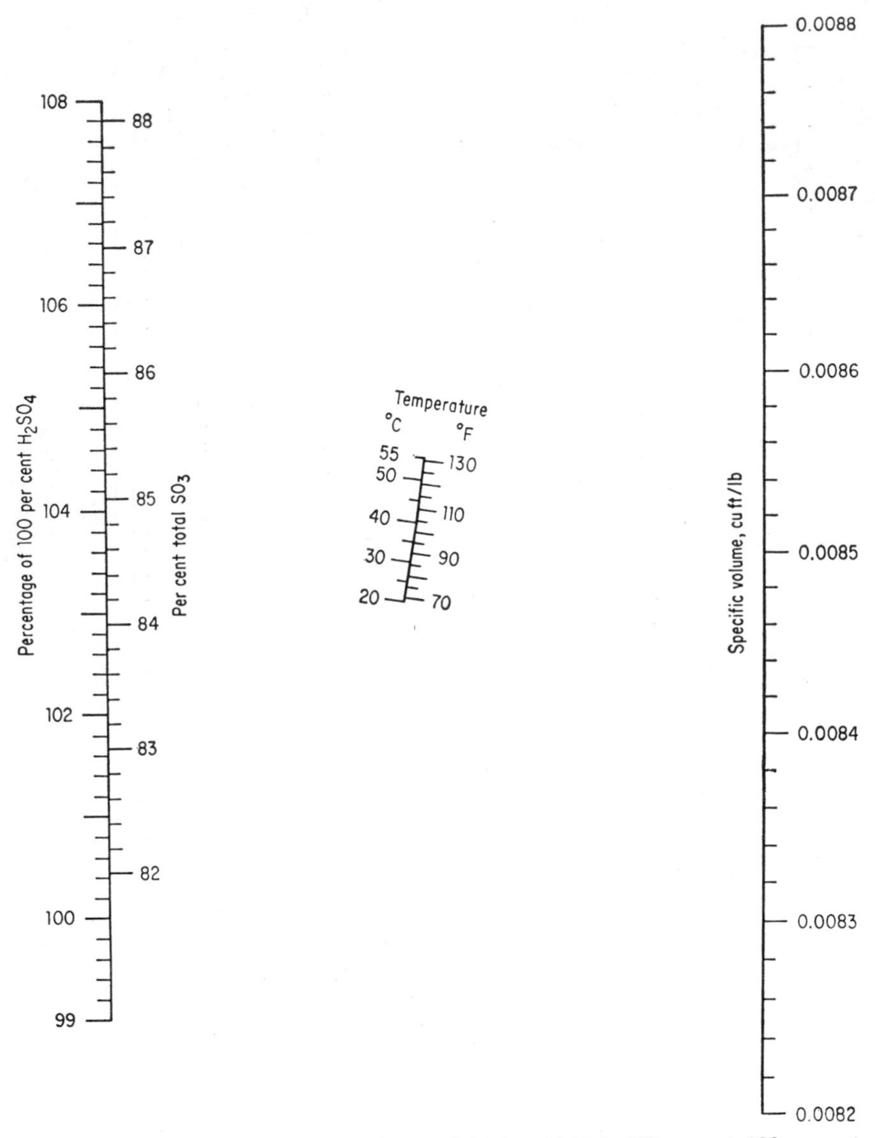

FIG. D-13.—Specific volume of fuming sulphuric acid, 99 to 108 per cent, 100 per cent sulphuric acid, 20 to 55°C. (*Reprinted from Davis, "Chemical Engineering Nomographs," McGraw-Hill Book Company, Inc., New York,* 1944.)

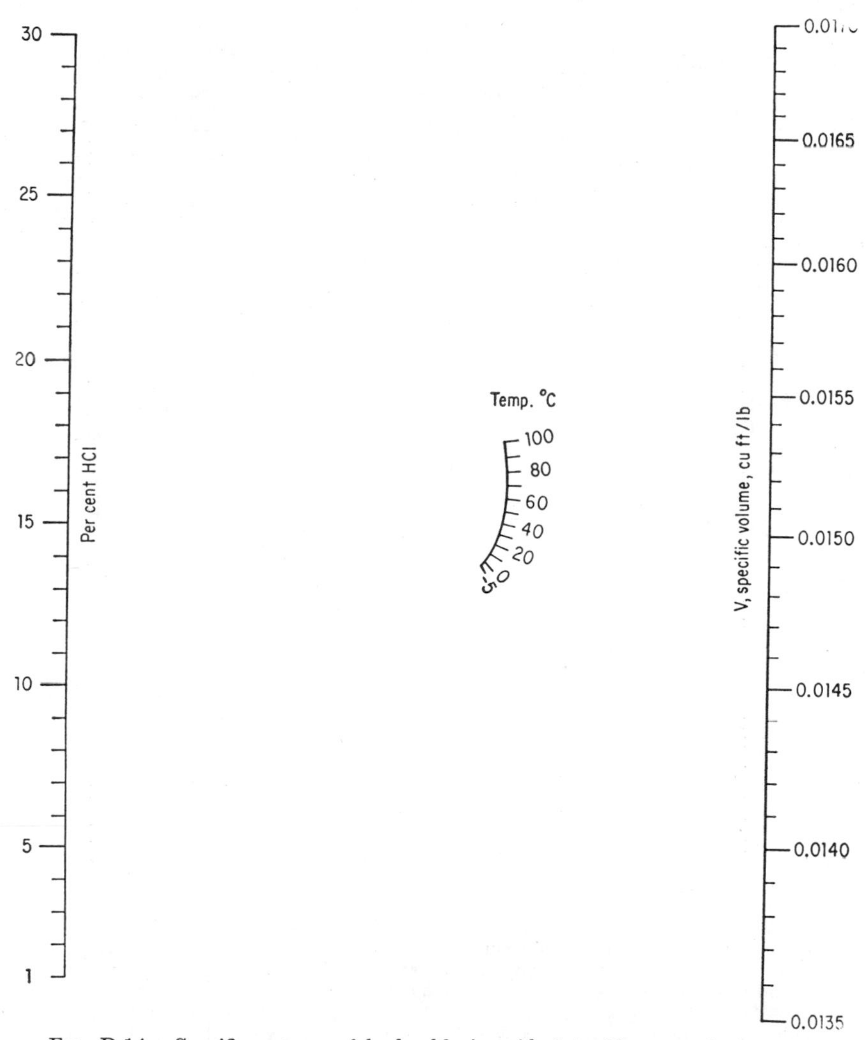

FIG. D-14.—Specific volume of hydrochloric acid, 1 to 30 per cent; temperature range, −5 to 100°C. (*Reprinted from Davis, "Chemical Engineering Nomographs," McGraw-Hill Book Company, Inc., New York,* 1944.)

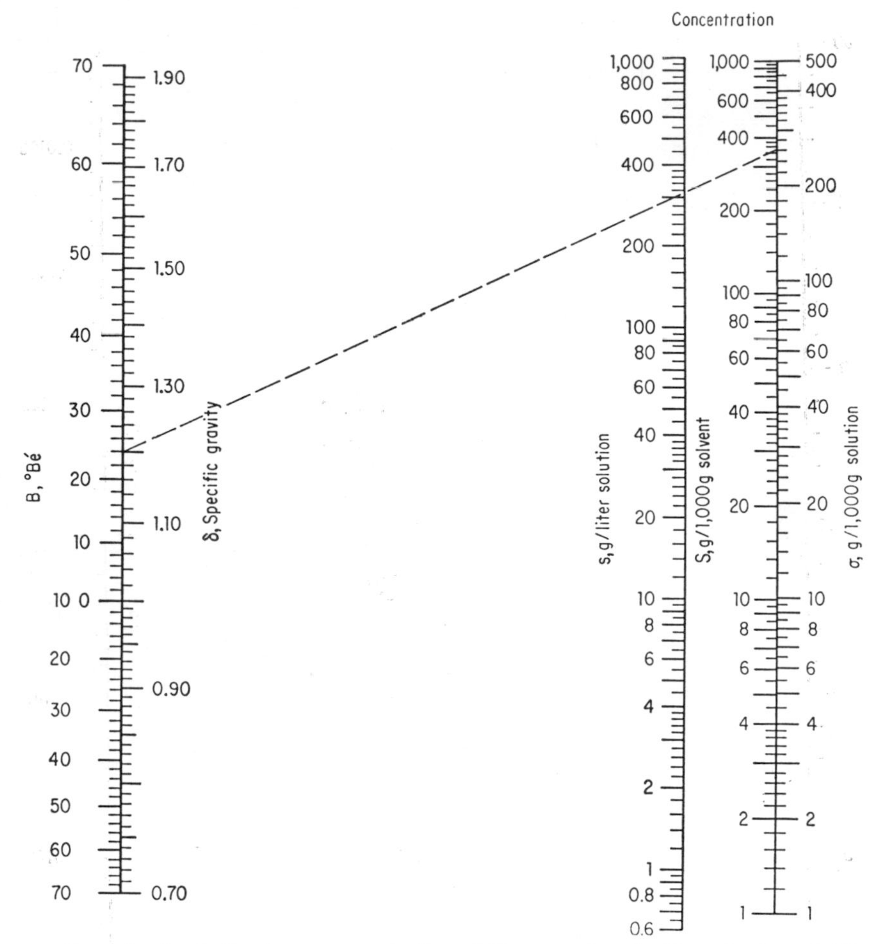

Example: A solution has a concentration of 26 per cent and a Baumé value of 24°. What is the concentration in grams per 1,000 g of solution, grams per 1,000 g of solvent, and in grams per liter of solution? What is the specific gravity of the solution? Twenty-six per cent, or 26 g per 100 g of solution, is the same as σ = 260 g per 1,000 g of solution. Opposite σ = 260, read S = 351 g per 1,000 g of solvent. Connect 260 on the σ axis with 24 on the B scale, and read s as 312 g/liter on the s scale. Opposite 24 on the B scale, read δ as 1.20.

FIG. D-15.—Concentration conversions. (*Reprinted from Davis, "Chemical Engineering Nomographs," McGraw-Hill Book Company, Inc., New York,* 1944.)

2. Metals

a. Mechanical Tests.—For ductile, resilient material such as structural steel the tensile stress-strain test is the most important. This test consists in applying a force (stress) to a standard specimen and measuring the elongation (strain). The force is gradually increased until failure of the specimen occurs. The results of such a test are shown in Fig. D-16. The slope of the line OA, *i.e.*, the ratio of stress to strain below the proportional limit, is *Young's modulus of elasticity*. The *proportional limit* is the highest stress at which the strain is proportional to the stress. (By common consent, stresses are almost universally expressed as pounds per square inch of original cross section.) At stresses above this limit the specimen is permanently deformed. The definitions of *yield point, ultimate tensile strength,* and *breaking strength* are indicated by Fig. D-16. Many materials do not exhibit yield points. In these cases the *yield strength* or stress to produce a standard permanent deformation (per cent set) is usually specified. A method commonly used to determine the yield strength is the offset method, illustrated by Fig. D-17. Very common practice is to base the yield strength on 0.2 per cent set, though for some materials or for special uses the design engineer may prefer yield strengths based on smaller or larger sets.

Compressive strength tests are made in much the same way as tensile tests. For iron and steel products, the yield strengths in tension and compression are nearly the same, though the ultimate strengths differ. Metals are also tested by applying stress in other ways, notably in shear.

The rate of applying stress to a metal may be of importance. For example, brittle materials may have high tensile strengths but may fracture readily under sharp blows. Impact tests such as the Izod and several varieties of Charpy are, therefore, of importance in design where brittle materials must be employed. It should perhaps be noted that some materials, such as iron and lead which are ductile at room temperature, become brittle at low temperatures.

Soft materials such as lead exhibit a slow deformation at comparatively low stress. This phenomenon is known as "creep" and is exhibited by all metals at high temperatures. The rate of creep usually exhibits three stages: The initial rate is comparatively rapid and drops slowly to a steady rate in the second stage. Eventually the rate increases in the third stage which ends in failure. Results are usually presented as the stress that produces a specified creep rate in the second stage.

Failure of metals (even though creep is not measurable) may occur

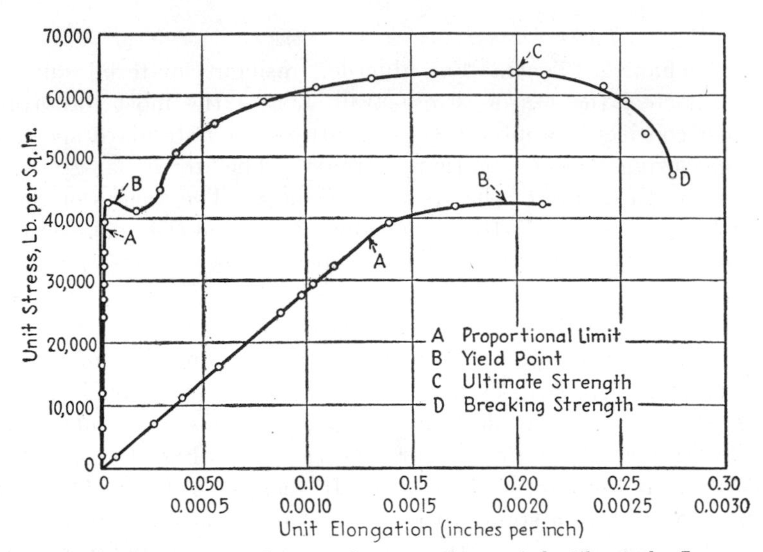

FIG. D-16.—Stress-strain diagram for a specimen of ductile steel. Lower curve is part of upper curve drawn to larger horizontal scale. (*Reprinted by permission from "Mechanics of Materials," 3d ed., by P. G. Laurson and W. J. Cox, published by John Wiley & Sons, Inc., New York,* 1954.)

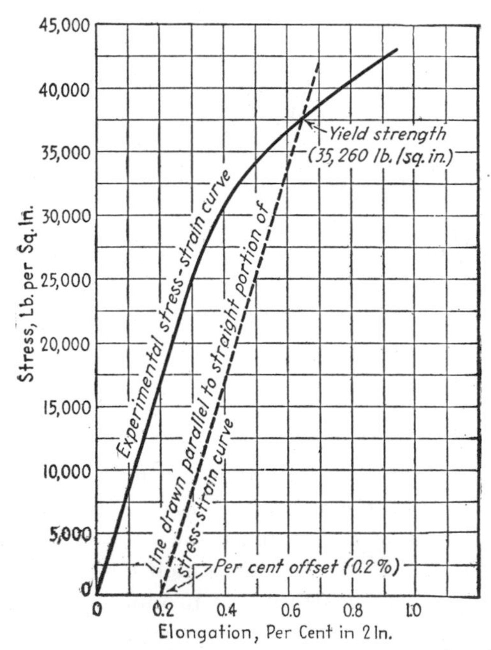

FIG. D-17.—Illustration of yield-strength determination by offset method.

far below the tensile strength if the stress is repeated many times. For this reason repeated stress tests are important. The results of such tests show that there is a limiting stress below which no failure will occur no matter how often applied. This *endurance* limit can be obtained graphically from a series of tests.

Obviously the above discussion has covered only a few tests in broad terms. These and many more tests have been carefully standardized by technical societies such as the American Society of Testing Materials (A.S.T.M.) and the American Society of Mechanical Engineers (A.S.M.E.). Most of the standards in use in the United States are catalogued and sold by the American Standards Association (A.S.A.).

In addition to data on properties of finished products, the design engineer requires information on the fabricating methods that may be employed—whether the metal is readily machined, welded, soldered, pressed, drawn, forged, etc.

Mechanical properties of metals depend on an extremely large number of variables. Some of these are listed below:

1. Chemical analysis (even small quantities of alloying element may be important).
2. Method of manufacture (cast, wrought, etc.).
3. Heat-treatment (annealing, quenching, drawing).
4. Mechanical processing (hot-rolling, cold-drawing, etc.).
5. Surface treatment (casehardening, finish, etc.).
6. Corrosion.

b. Wall Thickness of Vessels and Piping.—The fundamental purpose of mechanical tests is to afford a basis for mechanical design. The mechanical engineer wants to know how heavy the beams of a bridge must be to support the load. The chemical engineer wishes to know how heavy the shell of an autoclave must be to withstand the temperature, pressure, and corrosive nature of the contents. Here are two formulas for the minimum shell thickness of cylindrical walls:

$$t_m = \frac{PD}{2S} + C \text{ (Barlow's formula, modified)}$$

$$t_m = \frac{D}{2}\left(1 - \sqrt{\frac{S - P}{S + P}}\right) + C \text{ (Lamé's formula)}$$

where t_m = minimum wall thickness, in.
D = outside diameter of pipe, in.
S = design stress, psi
P = internal pressure, psig
C = allowances, in., for threading, mechanical strength, machining, corrosion

Barlow's formula is accurate for only thin wall vessels and may be used when the wall thickness is 10 per cent or less of the outside diameter. For thicker wall vessels Lamé's formula is applicable. (Even this formula fails, however, for high pressures where P is nearly as large as S.) In the case of piping the following very simple formula may be employed:

$$\text{Schedule number} = \frac{1{,}000P}{S}$$

where P = internal pressure, psig
S = design stress, psi

Some design stresses for use in the above formulas are given in the next section. Detailed discussion of the considerations involved in choice of design stresses is beyond the scope of this book. The A.P.I.-A.S.M.E. codes cover these for a variety of metals in different uses. In any event, the *code legal in the locality should be consulted for final engineering design.* For a great many purposes one-fourth of the ultimate tensile strength is used as the design stress. For brittle materials, higher factors of safety must be employed.

In general, elliptical ends, flanges, etc., must be heavier than the walls of a vessel. However, given formulas permit the process engineer to judge whether his process involves unduly heavy equipment.

c. Data Tables

TABLE D-1.—LISTING OF TABLES ON PROPERTIES OF METALS

Properties	Metals							
	Low-carbon steel	4/6 CrMo cast steel	Stainless steels	Monel	Copper alloys	Aluminum alloys	Lead	Other
Density, thermal conductivity, specific heat	D-2	....	D-2	D-2	D-2	D-2	D-2	D-2
Coefficient of expansion	D-2, D-3	....	D-2	D-2	D-2	D-2	D-2	D-2
Tensile strengths	D-3	D-4	....	D-10	....	D-11	....	D-7
Creep stress	D-3	D-4	....	D-10				
Temperature limits	D-8	....	....	....	....	....	....	D-8
Design stress	....	....	D-9	....	D-5	....	D-12	
Impact tests	D-6	....	....	....	....	....	....	D-7

TABLE D-2.—REPRESENTATIVE PROPERTIES OF METALS
(Compiled from several sources)

Metal	Specific gravity (water = 1 at 60°F)	Coefficient of expansion, ft/(1,000,000 ft)(°F)	Thermal conductivity, Btu/(hr)(sq ft)(°F/ft)	Specific heat, Btu/(lb)(°F)
Aluminum	2.7	12.7	118‡	0.226
Antimony	6.6	6.3	11	0.049
Bismuth	9.8	7.5	4.7	0.029
Chromium	7.2	4.7	157	0.12
Cobalt	8.9	6.7	40	0.098
Copper:				
Pure copper	8.9	9.3	220‡	0.092
Admiralty and 70% brass	8.5	10.3	58‡	
Muntz and yellow brass	8.4	11.6	70	0.092
Red brass	8.6–8.75	9.8–10.4	90–94	
Bronze	8.8	9.8	100	
Everdur	8.2	7.8	16	
Monel	8.8	7.8	15	0.127
Iron:				
Pure iron	7.9	6.6	38‡	0.1045
Wrought iron	7.7	....	33‡	
Low-carbon steel	7.8	6.1–7.3*	35‡	0.11
High-carbon steel	7.8	6.1–7.3*	26‡	0.12
Gray cast iron	7.0–7.7	10	24–30‡	0.13
18–8 Cr-Ni	7.9	9.5	9.6‡	0.117
25–12 Cr-Ni	7.9	8.3	8.8	0.12
12% Cr	7.7	5.8	14.3‡	0.11
Lead	11.3	16.4	19‡	0.030
Magnesium	1.74	14.3	92	0.249
Monel	8.8	7.8	15	0.127
Nickel	8.9	7.3	35‡	0.112
Silver	10.5	10.5	240	0.056
Sodium	0.97	39.5	26	0.295
Tantalum	16.6	3.6	35	0.036
Tin	7.3	11.2	34	0.054
Vanadium	19.3	2.2	115	0.034
Zinc	7.1	18.3	64	0.089

* Additional data given in Table D-3.
‡ Additional data given in Table G-4.

TABLE D-3.—PROPERTIES OF KILLED CARBON STEEL
(Reprinted from Hoyt, "Metals Data," 2d ed., Reinhold Publishing Corporation, New York, 1952, by permission of the author, publisher, and the Steel and Tube Division of The Timken Steel and Tube Company)
Composition: 0.10–0.20% C, 0.30–0.60% Mn, 0.25% Si max., 0.04% P max., 0.045% S max.

Temp., °F	Tensile strength	Yield strength 0.2% set	Propor-tional limit	Elonga-tion, % in 2 in.	Reduc-tion in area, %	Creep stress, 1%		Rupture stress*	Coefficient of expan-sion† in./(1,000,000 in.)(°F)
						100,000 hr	10,000 hr		
85	62.4	42.0	34.5	36	67				
750	58.0	24.6	13.1	34	67				
800	....	....	....	..	..	18.5	26.8	...	7.83
900	45.5	23.5	11.2	38	71	12.8	16.9	...	7.95
1000	36.5	20.1	8.7	42	77	2.7	5.7	2.8	8.02
1100	27.2	14.2	5.0	56	82	0.8	1.8	...	8.21
1200	20.0	10.2	1.9	54	89	0.3	0.6	0.6	8.36
1300	13.5	7.4	0	60	92	....	....	0.4	
1400	9.0	3.7	0	70	77	....	....	0.2	

NOTE: All stresses are in units of 1,000 psi.
* Stress to produce rupture in 100,000 hr.
† Mean coefficient from 70°F to stated temperature.

TABLE D-4.—HIGH-TEMPERATURE TESTS OF 4/6 Cr-Mo CAST STEELS (A.S.T.M.-A.S.M.E. "CREEP DATA")
(Hoyt, "Metals Data," 2d ed., Reinhold Publishing Corporation, New York, 1952)
Analysis, %: C, 0.31; Mn, 0.57; Si, 0.37; Cr, 4.86; Mo, 0.53. Normalized at 1650°F; drawn at 1300°F.

Temp., °F	Tensile strength, thousands of psi	Yield point, thousands of psi	Propor-tional limit, thousands of psi	Elonga-tion, % in 2 in.	Reduc-tion in area, %	Creep stress for 1% in 10,000 hr, thousands of psi
70	118	90	71	17	44	
300	111	89	77	14	43	
550	103	78	68	13	41	
800	101	70	52	12	37	
900	...	..	..	..	..	18
1000	73	56	38	21	58	10
1200	45	18	..	37	82	

NOTE: Yield point = yield strength for 0.2 per cent set.

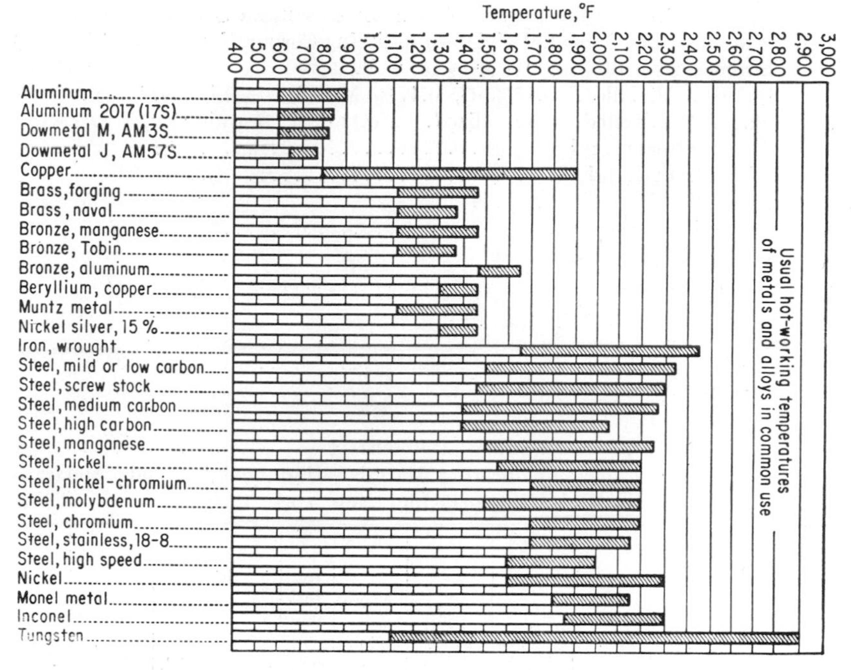

FIG. D-18.—Hot-working temperatures of metals. (*Reprinted from Mantell, "Engineering Materials Handbook," 1st ed., McGraw-Hill Book Company, Inc., New York, 1958.*)

TABLE D-5.—MAXIMUM ALLOWABLE STRESS VALUES FOR COPPER ALLOYS IN TENSION

(Abstracted from "Unfired Pressure Vessel Code," 1959, with permission of the publisher, The American Society of Mechanical Engineers, New York City)

Material: condition, annealed	Allowable maximum stresses, psi, for temperatures not exceeding:				
	150°F	250°F	350°F	400°F	450°F
Muntz metal, seamless condenser tubes (SB-111)	12,500	11,200	7,500	2,000	
Red brass pipe (SB-43) and tubes (SB-111)	8,000	8,000	6,000	3,000	2,000
Copper plates (SB-11) and rods (SB-12)	6,700	6,300	3,800	2,500	
Annealed copper seamless boiler tubes (SB-13), pipe (SB-42), and seamless tubes (SB-75)	6,000	5,800	3,800	2,500	
Copper-nickel, 70–30, seamless condenser tubes (SB-111)	11,600	11,000	10,600	10,300	10,100
Admiralty; A,B,C,D; seamless condenser tubes (SB-111) and tube plates (SB-171)	10,000	10,000	8,000	5,000	3,000

TABLE D-6.—CHARPY IMPACT TESTS ON BESSEMER STEEL, KILLED
(Hoyt, "Metals Data," 2d ed., Reinhold Publishing Corporation, New York, 1952)

Seamless pipe 10¾ in. OD × ½ in. wall.

Si, Mn, Al, killed. 0.15% C, 0.47% Mn, 0.22% Si, 0.074% P, 0.017% N, tested as normalized. McQuaid-Ehn grain size, 4–8.

Tensile strength, psi	67,500
Yield strength, psi	48,000
Elongation, %	47.5
Reduction in area, %	55.5
Charpy impact test, ft-lb at	
70°F	47
32°F	40
0°F	36
−25°F	34
—50°F	21

TABLE D-7.—MECHANICAL PROPERTIES OF WROUGHT IRON
(Reprinted from Mantell, "Engineering Materials Handbook," 1st ed., McGraw-Hill Book Company, Inc., New York, 1958)

Form or condition	Tensile strength, mpsi	Yield point, mpsi	Elongation 8 in., %	Reduction in area, %	Endurance limit: Rotating beam, mpsi	Endurance limit: Axial, mpsi	Charpy notched bar, ft-lb	Brinell hardness number
Longitudinal test	46.9	......	...	...	23	16	17.5	105
Transverse test	34.4	......	...	...	19	11	4.8	105
Minimum ASTM requirements								
Pipe, ASTM A72	40	24	12					
Plate, longitudinal, ASTM A42	48	27	14					
Shapes, bars ASTM A207 1⅝–2½ in.	47	0.55TS	22	35				
Bars, single refined, ASTM A189 1⅝–2½ in.	47	0.55TS	22	35				
Bars, double refined, ASTM A189 1⅝–2½ in.	47–54	0.55TS	25	40				
Rivets, ASTM A152 over 7⁄16–¾ in. incl	47	0.60TS	24					
Stay bolts, ASTM A84	47–52	0.60TS	30	48				
Forgings, ASTM A73	45	0.50TS	24	33				

TABLE D-8.—APPLICATION TEMPERATURE LIMITS OF METALS, °F
(Reprinted from Mantell, "Engineering Materials Handbook," 1st ed., McGraw-Hill Book Company, Inc., New York, 1958)

	Steam superheater	Refinery
Carbon steel	950	1050
0.50% Mo	975	1050
0.5 Cr, 0.5 Mo	1000	1075
1.25 Cr, 0.5 Mo	1050	1100
2 Cr, 0.5 Mo	1080	1150
2.25 Cr, 1 Mo	1100	1175
3 Cr, 1 Mo	1125	1175
5 Cr, 0.5 Mo	1150	1200
5 Cr, 0.5 Mo-Si	....	1300
5 Cr, 0.5 Mo-Ti	....	1250
7 Cr, 0.5 Mo	....	1250
9 Cr, 1 Mo	1200	1300

TABLE D-9.—DESIGN STRESSES FOR WROUGHT HEAT-RESISTING ALLOY STEELS
(Courtesy of the Midvale Company)
Analysis range: 25–30% Cr, 10–30% Ni

Temperature, °F	1,400	1,500	1,600	1,700	1,800	2,000
Design stress, psi	1,800	1,550	1,300	1,050	800	300

TABLE D-10.—HIGH-TEMPERATURE PROPERTIES OF MONEL
(Hoyt, "Metals Data," 2d ed., Reinhold Publishing Corporation, New York, 1952)

Temp., °F	Short-time tests			Creep tests		
	Yield strength, (0.2% set) psi	Tensile strength, psi	Elongation, % in 2 in.	0.01%/1,000 hr, psi	0.1%/1,000 hr, psi	1%/1,000 hr, psi
70	45,000	85,000	45			
600	34,000	80,000	44	26,000	36,000	46,000
800	30,000	70,000	40	19,000	23,500	29,000
1000	27,000	55,000	30	1,650	4,300	11,500
1200	23,000	36,000	18			1,700

TABLE D-11.—TYPICAL TENSILE PROPERTIES OF ALUMINUM ALLOYS
(Reprinted from Marks, "Mechanical Engineers' Handbook," 5th ed., McGraw-Hill Book Company, Inc., New York, 1951)

Temperature, °F	75	300	400	500	700
2SH:					
Tensile strength, psi	24,000	17,500	6,000	3,500	1,500
Yield strength, psi (0.2% set)	21,000	14,000	3,000	2,000	1,000
Elongation, % in 2 in	15	16	70	85	95
24S-T:					
Tensile strength, psi	68,000	46,000	27,000	14,000	5,000
Yield strength, psi (0.2% set)	44,000	38,000	21,000	9,500	3,500
Elongation, % in 2 in	22	22	25	45	100
53S-T:					
Tensile strength, psi	39,000	25,000	13,000	6,000	2,500
Yield strength, psi (0.2% set)	33,000	22,000	10,000	3,500	2,000
Elongation, % in 2 in	20	17	30	70	90

Designation 2SH is for commercially pure aluminum. 24S-T has 4.4 per cent Cu, 0.5 per cent Mn, and 1.5 per cent Mg. 53S-T has 0.7 per cent Si, 1.3 per cent Mg, and 0.25 per cent Cr.

TABLE D-12.—ALLOWABLE FIBER STRESS IN EXTRUDED LEAD PIPE
(Reprinted from Marks, "Mechanical Engineers' Handbook," 5th ed., McGraw-Hill Book Company, Inc., New York, 1951)

Temp., °F	Allowable stress, psi	
	Chemical lead	6% antimony lead
68	200	400
86	190	370
104	180	340
122	172	310
140	162	280
158	153	254
176	144	222
194	136	195
212	127	165
230	118	137
248	110	110
266	100	80
284	90	50
302	80	

3. Nonmetallics

a. Ceramics and Stone.—Properties of ceramics and stone are presented as follows:

TABLE D-13.—STRENGTH IN COMPRESSION OF BRICK AND OF TERRA-COTTA BLOCK PIERS AND WALLS

(Moore and Moore, "Textbook of the Materials of Engineering," 8th ed., McGraw-Hill Book Company, Inc., New York, 1953)

The values given are based on test data from Watertown (Mass.) Arsenal, Cornell University, U.S. Bureau of Standards, and the University of Illinois.

The weight of masonry may be taken as about 5 lb/cu ft less than the weight of the stone or brick used.

Brick or block used	Mortar	Ultimate in compression, psi
	Piers	
Vitrified brick	1 part portland cement, 3 parts sand	2,800
Pressed (face) brick	1 part portland cement, 3 parts sand	2,000
Pressed (face) brick	1 part lime, 3 parts sand	1,400
Common brick	1 part portland cement, 3 parts sand	1,000
Common brick	1 part lime, 3 parts sand	700
Terra-cotta block	1 part portland cement, 3 parts sand	3,000
	Walls	
Common brick	1 part portland cement, 1¼ parts lime, 6 parts sand	950
Common brick	1 part portland cement, 3 parts sand	1,150
Hollow-tile joints staggered	1 part portland cement, ¼ part lime, 3 parts sand	1,270

Test data for piers built of sand-lime brick are lacking, but judging from test data for individual brick, sand-lime brick piers might be expected to be about three-quarters as strong as piers built of common brick.

TABLE D-14.—PHYSICAL PROPERTIES OF REFRACTORY MATERIALS

(Perry, "Chemical Engineers' Handbook," 3d ed., McGraw-Hill Book Company, Inc., New York, 1950. Complete revision of earlier *Chem. & Met. Eng.* data, compiled by L. J. Trostel, General Refractories Co., Baltimore, with additional material on kaolin superduty and insulating refractories supplied by the Babcock & Wilcox Co., New York)

	Type of brick												
	Silica	High-heat duty (No. 1) fire clay	Super-duty fire clay	Super-duty kaolin	Alumina-diaspore, 70% Al_2O_3	Silli-manite (mullite)	Chrome	Unburned chrome[a]	Magne-site	Unburned magnesite[a]	Bonded silicon carbide, grade A	Bonded fused alumina	Kaolin insulat-ing re-fractory, 2600°F
Typical composition, %:													
SiO_2	96	50–57	52	52	22–26	35	6	5	3	5	7–9	8–10	57.7
Fe_2O_3	1	1.5–2.5	1	0.6	1–1.5	0.5	......	......	6	8.5	0.3–1	1–1.5	2.4
FeO	......	......	......	......	......	......	15	12					
Al_2O_3	1	36–42	43	45.4	68–72	62	23	18	2	7.5	2–4	85–90	36.8
TiO_2	......	1.5–2.5	2	1.7	3.5	1.5	......	......	......	......	1	1.5–2.2	1.5
CaO	2	......	......	0.1	......	......	......	......	3	2	......	......	0.6
MgO	......	......	......	0.2	......	......	17	32	86	64	......	......	0.5
Cr_2O_3	......	......	......	......	......	......	38	30	......	10			
SiC	......	......	......	......	......	......	......	......	......	......	85–90		
Flux[b]	......	1–3.5	2	......	1–1.5	0.5	......	......	......	......	1.5	0.8–1.3	
P.C.E.[c]	31–32	31–33	33–34	34	36	37–38	41+	41+	41+	41+	39	39+	29–30
Approx. equivalent temp., °F	(3056–3092°)	(3056–3173°)	(3173–3200°)	(3200°)	(3290°)	(3308–3335°)	(3578°+)	(3578°+)	(3578°+)	(3578°+)	(3389°)	(3389°+)	(2984–3002°)
Deformation under load,[d] % (at psi and temp., °F, shown)	Shears (25 psi, 2900°)	2.5–10* (25 psi 2460°)	2–4† (25 psi 2640°)	0.5† (25 psi 2640°)	1–4† (25 psi 2640°)	0.0–0.5† (25 psi 2640°)	Shears (28 psi 2740°)	Shears (28 psi 2955°)	Shears (28 psi 2765°)	Shears (28 psi 2940°)	0–1 (50 psi 2730°)	1 (50 psi 2730°)	0.3 (10 psi 2200°)
Resistance to spalling, % loss in appropriate A.S.T.M. panel test	Poor	5–20	0–4	No loss	No loss	No loss	Poor	Fair	Poor	Fair	Good	Good	Good
Permanent linear change on reheating[e] (after 5 hr at temp., °F, shown)	(+)0.5–0.8 (2640°)	(±)0–1.5 (2550°)	(±)0–1.5 (2910°)	(−)0.75–1 (2910°)	(−)2–4 (2910°)	(−)0–0.8 (2910°)	(−)0.5–1 (3000°)	(−)0.5–1.0 (3000°)	(−)1–2 (3000°)	(−)0.5–1.5 (3000°)	(+)2[f] (2910°)	(+)0.5 (2910°)	(−)0.2 (2600°)
Porosity (as open pores), %	20–30	15–25	12–15	18	34–38	20–25	20–26	10–12	20–26	10–12	13–28	20–26	75
Weight per brick (std. 9 in. straight), lb	6.5	7.5	8.5	7.7	7.5	8.5	11	11.3	10	10.7	8–9.3	9–10.6	2.25
Specific heat (60–1200°F)	0.23	0.23	0.23	0.22	0.23	0.23	0.20	0.21	0.27	0.26	0.20	0.20	0.22
Relative slag resistance:[g]													
Acid-steel slag	Good	Fair	Fair	Fair	Good	Good	Poor	Poor	Poor	Poor	Good	Good	Poor
Basic-steel slag	Poor	Poor	Poor	Poor	Fair	Fair	Good	Good	Good	Good	Good	Good	Poor
Mill scale	Fair	Poor	Fair	Good	Fair	Fair	Good	Good	Good	Good	Fair	Fair	Good
Coal-ash slag	Poor	Fair	Fair	Fair	Fair	Fair	Fair	Fair	Good	Good	Good	Good	Poor

[a] Made by hydraulic pressing.
[b] Includes CaO + MgO + alkalies.
[c] Pyrometric cone equivalent; terms "fusion," "softening," "deformation," and "melting points" heretofore loosely used.
[d] Data marked * are from A.S.T.M. test C 16-36 with high heat duty time-temperature schedule; those marked † are from same test with superduty time-temperature schedule; others determined by other commonly used tests.
[e] (+) means expansion; (−) means shrinkage.
[f] Oxidizing atmosphere.
[g] Ratings affected somewhat by varying temperatures and type of atmosphere prevailing. Resistance to coal-ash slag affected by furnace temperature as well as by analysis and fusion point of slag.

TABLE D-15.—VALUES FOR STRENGTH AND STIFFNESS OF AMERICAN BUILDING STONE

[Values based mainly on test data from the Watertown (Mass.) Arsenal]
(Moore and Moore, "Textbook of the Materials of Engineering," 8th ed.. McGraw-Hill Book Company, Inc., New York, 1953)

Stone	Ultimate in compression, psi		Modulus of rupture (computed ultimate in cross bending), psi		Shearing strength, psi	Modulus of elasticity (flexure), psi	Weight (av), lb/cu ft
	Range	Av	Range	Av	Av	Av	
Granite	15,000–26,000	20,200	1,200–2,200	1,600	2,300	7,500,000	165
Marble	10,300–16,100	12,600	850–2,300	1,500	1,300	8,200,000	170
Limestone	3,200–20,000	9,000	250–2,700	1,200	1,400	8,400,000	160
Sandstone	6,700–19,000	12,500	500–2,200	1,500	1,700	3,300,000	135
Slate	....	15,000	....	8,500	.....	14,000,000	175

TABLE D-16.—PROPERTIES OF COMMERCIAL GLASSES

(Reprinted from Mantell, "Engineering Materials Handbook," 1st ed., McGraw-Hill Book Company, Inc., New York, 1958)

Designation	Coeff. of expan/°C, 0–300°C	Density, g/cu cm	Young's modulus, psi
Silica glass (fused silica)	5.5×10^{-7}	2.20	10×10^{6}
96% silica glass—7900	8×10^{-7}	2.18	9.7×10^{6}
96% silica glass—7911	8×10^{-7}	2.18	9.7×10^{6}
Soda-lime—window sheet	85×10^{-7}	2.46–2.49	10×10^{6}
Soda-lime—plate glass	87×10^{-7}		
Soda-lime—containers	85×10^{-7}		
Soda-lime—electric lamp bulbs	92×10^{-7}	2.47	9.8×10^{6}
Lead silicate—electrical	91×10^{-7}	2.85	9.0×10^{6}
Lead silicate—high lead	91×10^{-7}	4.28	7.6×10^{6}
Aluminoborosilicate—apparatus	49×10^{-7}	2.36	
Borosilicate—low expansion	32×10^{-7}	2.23	9.8×10^{6}
Borosilicate—low electrical loss	32×10^{-7}	2.13	6.8×10^{6}
Borosilicate—tungsten seal	46×10^{-7}	2.25	
Aluminosilicate	42×10^{-7}	2.53	12.7×10^{6}

TABLE D-17.—PHYSICAL PROPERTIES OF CHEMICAL STONEWARE
(Perry, "Chemical Engineers' Handbook," 3d ed., McGraw-Hill Book Company, Inc., New York, 1950)

Specific gravity	2.2	Modulus of elasticity, psi	8×10^6
Hardness, scleroscope	100	Specific heat	0.2
Ultimate tensile strength, psi	2,000–3,000	Thermal conductivity, Btu/(hr)(sq ft)(°F/in.)	10–35
Ultimate compressive strength, psi	80,000	Linear thermal expansion per °F	2×10^{-6}
Modulus of rupture, psi	5,000–13,000	Water absorption, %	0–2

This table, which was prepared by the General Ceramics Co., gives the physical properties of an average grade of chemical stoneware. It should be emphasized here that "chemical stoneware" is not the name of a definite material, such as an alloy, but a generic term applied to a wide variety of ceramic compositions and hence that, in any particular composition designed to give optimum properties in one respect, it will ordinarily be impossible to secure optimum properties in all other respects.

b. Concrete.—The strength of even well-prepared concrete depends on age and on the proportion of water used in the mix. In general, low water contents yield higher strengths providing it is possible to compact the concrete so that the forms are completely filled. The proper choice of aggregate (sand, gravel, etc.) is important when high strength is required. Concrete should be kept moist during curing for development of full strength. Table D-18 gives data on strengths of concrete.

In the National Building Code recommended by the National Board of Fire Underwriters (1955), the following recommendations for concrete stresses are made: maximum working stress in direct compression, 500 psi; extreme fiber stress in compression, 900 psi; in shear, 60 psi; in shear when diagonal tension is resisted by properly designed steel, 240 psi; bond between concrete and steel, 60 psi; bond between concrete and steel with approved deformed bars, 140 psi. In special cases, for high-strength concrete, higher stresses are allowable as long as they conform to good practice.

Further information on stresses in shear, tension, and compression for columns, footings, and beams is available in the Building Code Requirements for Reinforced Concrete (ACI 318-56) published by the American Concrete Institute. The designer, however, should always check the above with local regulations, which may differ.

Most concrete structures of any consequence are reinforced with steel rods or netting. This reinforcing usually carries the bulk of the tension load because of the low tensile strength of concrete.

Concrete, though possessing a strong tendency toward porosity, may be made waterproof by proper attention to preparation, placing and curing. Nonetheless, concrete tanks are sometimes waterproofed as an added precaution. Because of porosity, concrete may suffer some deterioration from freezing and thawing of moisture. The deterioration is likely to be serious for poorly prepared concrete.

The density of gravel concrete is roughly 150 lb/cu ft or 2 tons/cu yd. The exact density, of course, depends on the densities of the aggregates and the proportions used in the mix.

TABLE D-18.—COMPRESSIVE AND TENSILE STRENGTH, AND MODULUS OF ELASTICITY OF CONCRETE

(Moore and Moore, "Textbook of the Materials of Engineering," 8th ed., McGraw-Hill Book Company, Inc., New York, 1953)

Tests made at the Research Laboratory of the Portland Cement Association, Chicago. Compression tests of 6- by 12-in. and tension tests of 6- by 18-in. concrete cylinders.

Aggregate: Elgin sand and gravel graded 0 to 1½ in.; aggregate grading and consistency constant. Specimens removed from molds after one day and cured in moist room at 70°F until test; tested damp. Deformation of concrete measured with a Martens mirror extensometer.

Mix by volume	Cement, sacks per cu yd	Net water-cement ratio	Strength, psi						Initial tangent modulus of elasticity, 1,000 psi					
			3 days	7 days	28 days	3 mo	1 yr	5 yr	3 days	7 days	28 days	3 mo	1 yr	5 yr
Compression														
1–2	10.5	0.57	3,220	4,650	5,960	7,810	9,170	10,160	2,870	3,340	3,770	4,460	5,530	5,750
1–3	7.8	0.64	2,160	3,250	5,140	6,500	7,570	8,820	2,490	2,930	3,480	4,200	5,400	5,650
1–4	6.3	0.76	1,560	2,620	4,460	5,710	6,720	7,750	2,310	2,660	3,340	3,740	5,340	5,580
1–5	5.2	0.85	1,340	2,120	3,510	5,130	6,010	6,860	2,230	2,590	3,130	3,610	5,340	5,570
1–8	3.4	1.13	480	945	1,740	2,550	3,140	3,930	1,710	1,990	2,770	3,420	4,970	5,530
Tension														
1–2	10.5	0.57	325	360	460	525	630	635	2,770	3,200	4,050	4,450	5,500	5,550
1–3	7.8	0.64	230	290	400	480	595	590	2,770	3,130	3,840	4,050	5,250	5,600
1–4	6.3	0.76	175	240	360	415	525	565	2,700	3,000	3,560	4,150	5,470	5,380
1–5	5.2	0.85	130	225	325	390	510	520	2,420	2,780	3,200	3,630	5,160	5,300
1–8	3.4	1.13	40	85	185	270	335	350	2,210	2,700	3,030	3,410	4,840	4,700

c. Wood.—Tables D-19 and D-20 give working stresses for woods. Table D-21 gives approximate densities of soft and hard woods. The disadvantages of wood as a construction material are nonuniformity and swelling with water absorption. Recently the latter has been eliminated by impregnating wood with plastics.

TABLE D-19.—BASIC STRESSES FOR CLEAR SOFTWOOD LUMBER UNDER LONG-TIME SERVICE AT FULL DESIGN LOAD* FOR USE IN DETERMINING WORKING STRESSES ACCORDING TO GRADE OF LUMBER AND OTHER APPLICABLE FACTORS

(From "Wood Handbook—Basic Information on Wood as a Material of Construction with Data for Its Use in Design and Specification," Handbook 72, U.S. Department of Agriculture)

Species†	Extreme fiber in bending or tension parallel to grain, psi	Maximum horizontal shear, psi	Compression perpendicular to grain,‡ psi	Compression parallel to grain L/d = 11 or less, psi	Modulus of elasticity in bending, 1,000 psi
Baldcypress (cypress)	1,900	150	220	1,450	1,200
Cedar:					
Alaska	1,600	130	185	1,050	1,200
Atlantic white- (southern whitecedar) and northern white-	1,100	100	130	750	800
Port-Orford	1,600	130	185	1,200	1,500
Western red cedar	1,300	120	145	950	1,000
Douglas-fir:					
Coast type	2,200	130	235	1,450	1,600
Coast type, close-grained	2,350	130	250	1,550	1,600
Rocky Mountain type	1,600	120	205	1,050	1,200
All types, dense	2,550	130	275	1,700	1,600
Fir:					
Balsam	1,300	100	110	950	1,000
California red, grand, noble, and white	1,600	100	220	950	1,100
Hemlock:					
Eastern	1,600	100	220	950	1,100
Western (west coast hemlock)	1,900	110	220	1,200	1,400
Larch, western	2,200	130	235	1,450	1,500
Pine:					
Eastern white (northern white), ponderosa, sugar, and western white (Idaho white)	1,300	120	185	1,000	1,000
Jack	1,600	120	160	1,050	1,100
Lodgepole	1,300	90	160	950	1,000
Red (Norway pine)	1,600	120	160	1,050	1,200
Southern yellow	2,200	160	235	1,450	1,600
Dense	2,550	160	275	1,700	1,600
Redwood	1,750	100	185	1,350	1,200
Close-grained	1,900	100	195	1,450	1,200
Spruce:					
Englemann	1,100	100	130	800	800
Red, white, and Sitka	1,600	120	185	1,050	1,200
Tamarack	1,750	140	220	1,350	1,300

* These stresses are based on the strength of green lumber and are applicable, with certain adjustments, to lumber of any degree of seasoning or lumber used under any conditions of duration of load.

† Species names approved by U.S. Forest Service. Commercial designations are shown in parentheses.

‡ Values given in previous editions of this handbook presumed some drying and were therefore at a higher level than these for green lumber.

TABLE D-20.—BASIC STRESSES FOR CLEAR HARDWOOD LUMBER UNDER LONG-TIME SERVICE AT FULL DESIGN LOAD* FOR USE IN DETERMINING WORKING STRESSES ACCORDING TO GRADE OF LUMBER AND OTHER APPLICABLE FACTORS

(From "Wood Handbook—Basic Information on Wood as a Material of Construction with Data for Its Use in Design and Specification," Handbook 72, U.S. Department of Agriculture)

Species†	Extreme fiber in bending or tension parallel to grain, psi	Maximum horizontal shear, psi	Compression perpendicular to grain,‡ psi	Compression parallel to grain L/d = 11 or less, psi	Modulus of elasticity in bending, 1,000 psi
Ash:					
Black	1,450	130	220	850	1,100
Commercial white	2,050	185	365	1,450	1,500
Aspen, bigtooth and quaking	1,300	100	110	800	800
Beech, American	2,200	185	365	1,600	1,600
Birch, sweet and yellow	2,200	185	365	1,600	1,600
Cottonwood, eastern	1,100	90	110	800	1,000
Elm:					
American and slippery (soft elm)	1,600	150	185	1,050	1,200
Rock	2,200	185	365	1,600	1,300
Hickory, true and pecan	2,800	205	440	2,000	1,800
Maple, black and sugar (hard maple)	2,200	185	365	1,600	1,600
Oak, commercial red and white	2,050	185	365	1,350	1,500
Sweetgum (gum, red gum, sap gum)	1,600	150	220	1,050	1,200
Tupelo, black (black gum) and water	1,600	150	220	1,050	1,200
Yellow poplar (poplar)	1,450	130	160	1,050	1,200

* These stresses are based on the strength of green lumber and are applicable, with certain adjustments, to lumber of any degree of seasoning or lumber used under any conditions of duration of load.

† Species names approved by U.S. Forest Service. Commercial designations are shown in parentheses.

‡ Values given in previous editions of this handbook presumed some drying and were therefore at a higher level than these for green lumber.

Table D-21.—Densities of Wood, Lb/Cu Ft
(Abstracted from Moore and Moore, "Textbook of the Materials of Engineering," 8th ed., McGraw-Hill Book Company, Inc., New York, 1953)

Wood	Density
Soft wood, full-size structural timber, ordinary defects, green:	
Southern pine	34
Douglas fir	28
Norway pine and white pine	24
Hemlock (eastern)	28
Tamarack	31
Redwood	22
Spruce	24
Cedar	23
Soft wood, small clear test specimens, air-dry (12% moisture content):	
Yellow pine	34
Douglas fir	28
Norway pine and white pine	24
Hemlock	28
Tamarack	31
Redwood	22
Spruce	39
Cedar	23
Cyprus	29
Hardwood, small clear test specimens, air-dry (12% moisture content):	
White oak	50
Red oak	45
Hickory	50
Ash	39
Elm	34
Maple	43

d. Plastics and Elastomers.—New developments in the field of plastics are so rapid that even experts on the subject would be unable to give a fully adequate discussion of their properties. Plastics are made from a wide variety of starting materials, and each of the different types exhibits a range in properties that may be controlled by the conditions of manufacture. When plastics are combined with fibrous materials to make laminated plastics, plastic treated woods, etc., the variety obtainable is increased enormously. Consequently, plastics are frequently "tailor-made" to secure properties most suited to the intended use.

Tables D-22 and D-25 give data on the properties of a few of the better known plastics; Tables D-23 and D-24 give data on rubber and related materials.

TABLE D-22.—PROPERTIES AND USES OF COMMON THERMOPLASTICS
(Abstracted from Moore and Moore, "Textbook of the Materials of Engineering," 8th ed., McGraw-Hill Book Company, Inc., New York, 1953)

	Shellac	Poly-ethylene	Polymono-chloro-trifluoro-ethylene	Vinylidene chloride molding	Poly-styrene	Methyl methac-rylate, cast	Polyamide (nylon) molding	Cellulose acetate molding	Cellulose nitrate (pyroxylin)
Specific gravity	1.1–2.7	0.92	2.10	1.65–1.72	1.05–1.07	1.18–1.20	1.14	1.27–1.37	1.35–1.40
Tensile strength, 1,000 psi	0.9–2	1.5–1.8	5.7	3–5	5–9	6–7	7–9	1.9–8.5	7–9
Elongation, % in 2 in.	....	50–400	28–36	20–250	1–3.6	2–7	40–100	6–50	40–45
Modulus of elasticity in tension, 100,000 psi	5–6	0.19	1.9	0.5–0.8	4–6	3.5–5	2.6–4	0.86–4.0	1.9–2.2
Compressive strength, 1,000 psi	10–17	....	32–80	7.5–8.5	11.5–16	11–19	7.2–13	13–36	22–35
Hardness, Rockwell*	....	R11	R110–115	M50–65	M65–90	M90–100	M111–118	R50–125	R95–115
Highest usable continuous temp. °F	150–190	212	390	160–200	150–205	140–200	270–300	140–220	140
Thermal conductivity, $10^{-4} \times$ cal/(sec)(cu cm)(°C)	....	8	1.4	3	2.4–3.3	4–6	5.2–5.5	4–8	3.1–5.5
Thermal expansion, $10^{-5} \times$ (in./in.°C)	....	16–18	4.5–7.0	19	6–8	9	10–15	8–16	8–12
Water absorption, % in 24 hr, ⅛ in. thick	0–0.1	<0.01	0.00	0–0.1	0.03–0.05	0.3–0.4	0.4–1.5	1.9–6.5	1.0–2.0
Effect of strong acids	Deteriorated	Attacked by oxid. acids	None	Highly resistant	Attacked by oxid. acids	Attacked by oxid. acids	Attacked	Decomposed	Decomposed
Color possibilities	Limited	Unlimited	Unlimited	Extensive	Unlimited	Unlimited	Unlimited	Unlimited	Unlimited
Common uses	Phonograph records, electrical insulation	Bottle stoppers, flexible bottles, wire insulation, textiles	Filter disks, insulators, gaskets	Screening, chemical tubing, automobile seat covers	Electrical insulators, battery boxes, lenses, toys, boxes	Windows, furniture, dentures, picture frames	Bearings, cups, fabrics, bristles	Fountain pens, tools, toys, spectacles, packaging	Packaging foils, glazing materials, photographic film

* Rockwell scales: M, ¼-in.-diameter ball, 100-kg major load; R, ½-in.-diameter ball, 60-kg major load.

TABLE D-23.—PHYSICAL PROPERTIES OF SYNTHETIC AND NATURAL RUBBERS
(Reprinted from McPherson and Klemin, "Engineering Uses of Rubber," Reinhold Publishing Corporation, New York, 1956)

Property	Type of rubber								
	Natural	GR-S	Butyl	Nitrile	Neoprene	Polyvinyl chloride	"Thiokol"	Silicone	Acrylate
Compounding range	Excellent	Good	Good	Good	Excellent	Fair	Fair	Fair	Fair
Mechanical properties:									
Resistance to flow	Excellent	Excellent	Fair	Excellent	Good	Fair	Poor	Fair	Fair
Resistance to abrasion	Excellent	Excellent	Fair	Excellent	Good	Fair	Poor	Poor	Fair
Resistance to tear	Excellent	Fair	Good	Good	Good	Fair	Poor	Poor	Good
Thermal behavior:									
Low temperature	Excellent	Good	Good	Poor	Poor	Poor	Poor	Excellent	Poor
High temperature	Good	Excellent	Excellent	Excellent	Excellent	Good	Poor	Excellent	Excellent
Resistance to flame	Poor	Poor	Poor	Poor	Good	Excellent	Poor	Fair	Poor
Resistance to gas permeation	Fair	Fair	Excellent	Good	Good	....	Excellent		
Resistance to solvents:									
Water	Good	Good	Good	Fair	Good	....	Fair		
Aliphatic hydrocarbons	Poor	Poor	Poor	Excellent	Good	Excellent	Excellent	Fair	Fair
Aromatic hydrocarbons	Poor	Poor	Poor	Fair	Poor	....	Good	Poor	Poor
Resistance to oxidizing agents	Poor	Poor	Good	Poor	Fair	....	Poor	....	Fair
Suitability for use with food	Excellent	Fair	Good	Fair	Fair	....	Poor	....	Excellent
Special characteristics	Excellent physical quality; ease of processing	General replacement for natural rubber	Resistance to gas diffusion and oxidation	Resistance to heat and oils	Resistance to heat, light, ozone, and oils	....	Excellent resistance to solvents and oils	Resistance to extremely high and extremely low temperatures	Resistance to ozone, light, and high temperatures

TABLE D-24.—SPECIFIC HEATS OF SOME POLYMERS AT 25°C
(McPherson and Klemin, "Engineering Uses of Rubber," Reinhold Publishing Corporation, New York, 1956)

Material	Specific heat (with reference to water at 25°C)	Specific heat, abs. joules/(°C)(g)	Rate of change of specific heat with temperature, abs. joules/(°C)²(g)
Natural rubber, unvulcanized	0.450	1.880	0.005
Acrylonitrile rubber	0.471	1.971	0.003
Polybutadiene (41°F)	0.471	1.969	0.003
Polybutadiene (122°F)	0.467	1.953	0.003
Butadiene-styrene copolymers:			
41°F, 8.58% bound styrene	0.462	1.931	0.003
122°F, 8.58% bound styrene	0.463	1.936	0.003
41°F, 22.61% bound styrene	0.452	1.892	0.003
122°F, 42.98% bound styrene	0.436	1.822	0.003
Polystyrene	0.293	1.223	0.004
Polyisobutene	0.464	1.948	0.004
Polyethylene (annealed, about 50% crystalline)	0.565	2.36	0.02
Polytetrafluoroethylene (measured at 37°C because of transition near 25°C)	0.231	0.996	0.0006
Polymonochlorotrifluoroethylene (annealed)	0.209	0.874	0.001

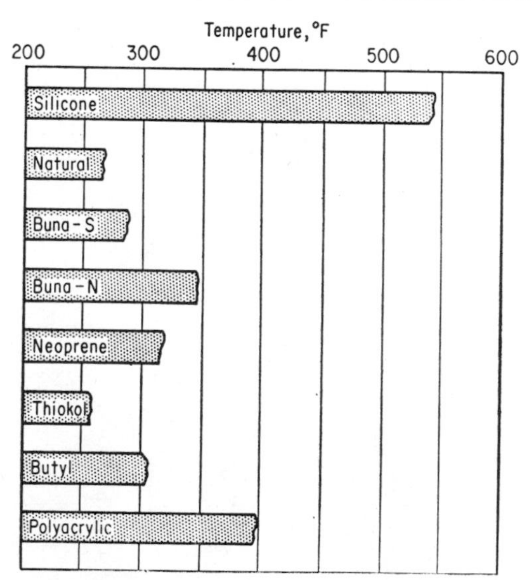

FIG. D-19.—High-temperature limits of various rubbers. [*From Charlotta and Hobein, Rubber Age*, **74,** 85 (1953) *New York. By permission.*]

TABLE D-25.—PROPERTIES AND USES OF THERMOSETTING PLASTICS
(Moore and Moore, "Textbook of the Materials of Engineering," 8th ed., McGraw-Hill Book Company, Inc., New York, 1953)

Property	Phenol-formaldehyde resin		Urea-formaldehyde, α-cellulose, molded	Melamine-formaldehyde, asbestos, paper or fabric, laminate	Polyester, glass fiber, mat, laminate	Silicone glass fabric, laminate	Cold-molded cement binder, asbestos-filled†	Hard rubber,‡ no filler
	Macerated cotton fabric or cord filler, molded	Mechanical grade, no filler, cast						
Compression molding pressure, 1,000 psi	2.00–8.00	0	2.00–8.00	1.00–1.80	0.01–0.15	1.00–2.00	1.00–10.00	1.20–1.80
Specific gravity	1.34–1.47	1.25–1.30	1.45–1.55	1.75–1.85	1.5–1.8	1.6–1.8	1.6–2.2	1.14
Tensile strength, 1,000 psi	2–9	4–7	6–13	6.5–12	10–20	10–25	1.6–2.2	8–10
Elongation, % in 2 in.	0.4–0.6	Very small	0.5–1.0	Very small	Very small	Very small	Very small	5–7.5
Modulus of elasticity in tension, 100,000 psi	9–13	5–7	12–15	16–39	10–19	20	....	3.0
Compressive strength, 1,000 psi	15–30	15–20	25–35	27–50	30–50	35–46	16	8–12
Impact strength, Izod test on ½ × ½ in. notched bar, ft.-lb. per in. width of notch	1–8	0.3–0.4	0.24–0.36	0.7–5.0	11–25	5–22	0.4	0.5
Hardness, Rockwell*	M 110–120	M 70–110	M 115–120	M 110–115	M 90–100	M 100	M 75–95	HR 95
Highest usable temperature, continuous °F	250	250	170	225–245	300–400	400–480	900–1300	....
Thermal conductivity, 10^{-4} cal/(sec)(cu cm)(°C)	4–7	3–5	7–10	10–17	8–12	3.5	....	2.9
Thermal expansion 10^{-5} in./(in.)(°C)	1–4	8–11	2.5–4.5	2.0–4.8	1.0–3.0	0.5	....	7.7
Dielectric strength, short time ⅛ in. thickness, volts/mil	200–400	....	300–400	40–150	250–400	200–480	45	470
Water absorption, 24 hr., ⅛ in. thick, %	0.04–1.8	0.2–0.4	0.4–0.8	1–5	0.3–1.0	0.2–0.7	0.5–15	0.02
Effect of strong acids	Decomposed by oxidizing acids	Decomposed by oxidizing acids	Decomposed	Decomposed	Some attack	Very slight	Decomposed	Attacked by oxidizing acids
Color possibilities	Limited	Limited	Unlimited	Limited	Unlimited	Limited	Gray and black	Limited
Common uses	Serving trays, radio cabinets, electrical parts	Punches and dies	Tablewear, electrical controls, housings	Aircraft structural parts, high-strength electrical insulation, electrical parts, stove switches	Aviation and automotive structures, decorative applications	High-temperature-resisting electrical insulation	Arc shields, terminal insulators, electric heater elements	Beakers, funnels, etc., for chemicals, combs

* Rockwell scales: M, ¼-in.-diameter ball, 100-kg. major load; HR (hard rubber), ¼-in.-diameter ball, 60-kg, major load.
† The cement binder is not strictly a thermosetting material but is set by chemical combination with water from steam (hydration).
‡ Hard rubber is not usually classified as thermosetting but as vulcanizing.

e. Carbon Products.

TABLE D-26.—PHYSICAL CHARACTERISTICS OF CARBON AND GRAPHITE PRODUCTS
(Abstracted from Perry, "Chemical Engineers' Handbook," 3d ed., McGraw-Hill Book Company, Inc., New York, 1950)

Material and form	Apparent density	Weight, lb/cu ft	Strength, psi			Elastic modulus, psi (multiply by 10^5)	Specific resistance, ohms/in.[3]	K (thermal expansion) (see Note)	Thermal conductivity, Btu/(hr)(sq ft)(°F/ft)
			Tensile	Compressive	Transverse				
Carbon tubes:									
½-4 in. ID	1.51	94.2	885	10,200	2,700	21.0	0.0014	15	3.0
5–10 in. ID	1.49	93.0	980	8,140	2,550	17.0	0.0016	21	3.0
Carbon brick:									
Dependent on application	1.56	96.7–97.8	970–1,530	5,340–8,320	1,950–3,070	8.9–10.3	0.0015–0.0016	13–14	3.0
Graphite tubes:									
½-4 in. ID	1.68	104.7	780	4,550	2,820	14.0	0.0003	12	94.0
5–10 in. ID	1.67	104.0	870	5,100	2,980	13.0	0.0003	12	84.0
Graphite brick, standard sizes	1.56	97.3	700	3,050	1,750	8.8	0.00036	5–12	84.0
Karbate No. 1 (impervious carbon):									
Tubes ½-2 in. ID	1.77	110.0	1,700	10,500	4,170	29.0	0.00164	27	3.0
Over 2 in. ID	1.76	110.0	2,000	10,500	4,640	26.0	0.0016	33	2.8
Karbate No. 2 (impervious graphite):									
Tubes ½-2 in. ID	1.86	116.0	2,600	8,900	4,650	23.0	0.00034	23	85.0
Over 2 in. ID	1.91	119.0	2,350	10,500	4,980	21.0	0.00033	24	75.0

NOTE: Coefficient of thermal expansion per degree: to temperature t°F = $[K + 0.0039t(°F)]\ 10^{-7}$; to temperature t°C = $[1.8K + 0.007t(°C)]10^{-7}$.

Carbon graphite products are resistant to most acids and alkalies.

4. Service of Materials Handling Process Fluids

a. General Notes on Corrosion.—Because corrosion and its twin, erosion, are extraordinarily complicated, no attempt will here be made to present detailed discussion. Rather a few of the complexities will be mentioned to forewarn the reader of the need of caution. *Inexperienced engineers are well advised to rely on advices of experts on corrosion.*

Corrosion may make itself known in several ways. The simplest is general surface attack, slowly reducing the thickness of the equipment. Instead of general surface attack, only isolated points may be affected, producing the familiar pitting. Microscopic examination of most materials of engineering shows them to be heterogeneous. The fluid may affect only certain of these component parts of the matrix. For example, in hydrogen and caustic embrittlements, only the intergranular material is attacked, but this may proceed until the cementing strength of this material is reduced and the metal becomes weak and brittle, even though unchanged in outward appearance. Failure of equipment may also result from other types of attack. For example, a plastic may absorb an organic solvent, swell, and deteriorate. Another example is the action of acetylene on copper. Though no appreciable attack is evident by inspection, enough copper carbide may form to cause an explosion and damage equipment.

Because process equipment generally handles moving rather than stagnant fluids, erosion is ordinarily associated with corrosion.

High stresses in metal usually increase corrosion rates. This is particularly true of "embrittlements," which rarely occur in unstressed metals. Conversely, corrosion affects the allowable stress and may reduce the endurance limit to zero.

Corrosion apparently proceeds by a wide variety of mechanisms. Of these one special case will be mentioned. When two metals are in contact with a solution containing even small amounts of electrolytes, an electric potential is set up between them which increases the rate of corrosion of the more basic (or electropositive) metal. The increase may cause a more rapid corrosion rate than would occur by the use of either metal alone. In special cases, such as galvanized iron, the net rate of corrosion may be lowered. However, untried metal combinations should be avoided, particularly metals widely separated in the electrochemical series; for example, iron and copper should not be used together.

If corrosion is complicated, the lack of corrosion is simple. The best method of assuring this is choice of materials that are inert toward the fluids handled. Other expedients to reduce or prevent corrosion are

1. Protective paints.
2. Protective metal coatings (galvanizing, electroplating, liquid spraying, calorizing, etc.).
3. Protective films produced on surface by chemical reaction (pickling of steel, Parkerizing, etc.).
4. Control of pH of water solution. (Usually corrosion is at a minimum when slightly alkaline.)
5. Application of electric potential to equipment.

If a metal that is corroded by the material being handled must be used, then the equipment should be designed for low mechanical stresses and low velocities of flow so that corrosion will not be unduly increased either by stress corrosion or by erosion.

b. Corrosion of Metals and Nonmetals.—The data in Fig. D-20 can materially assist in selecting a material of construction, either metal or plastic, for general areas of use. These charts do not show the effect of contaminants, aeration, galvanic coupling, or erosion, any one of which may change the selection indicated on the charts.

Most of the data in these charts were published in the Corrosion Forum department of *Chemical Engineering* prior to being consolidated and condensed into the following form. Actually, the charts should be used with caution,[1] tests under actual performance conditions being much more reliable.

Metallic corrosion rates exceeding 0.05 in. per year are usually prohibitive. Plastic materials usually fail by swelling and distortion, so that "ipy" ratings are not applied to them. When combinations of corrodents are met, if there are no other data, the corrosion values for the most severe corrodent should be used even if it is present in small amounts only.

[1] Mantell, "Engineering Materials Handbook," 1st ed., McGraw-Hill Book Company, Inc., New York, 1958.

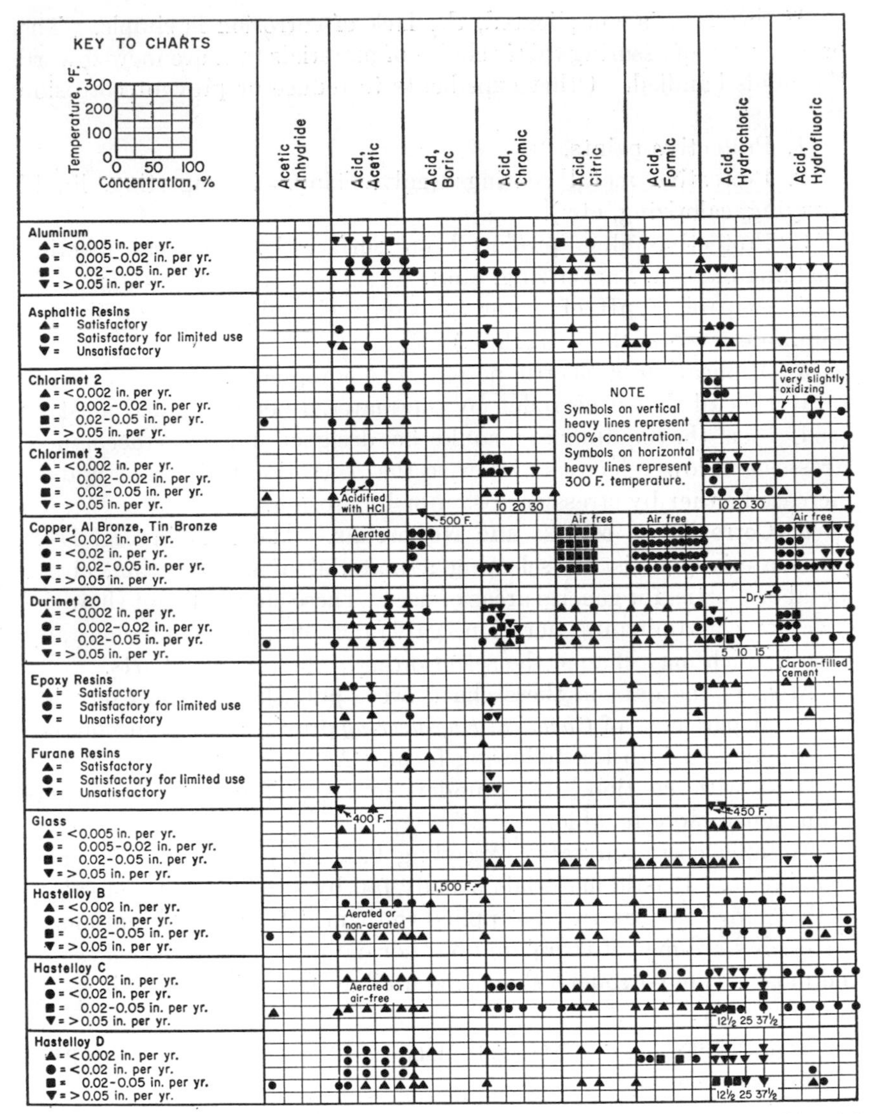

FIG. D-20.—Corrosion resistance of materials of construction. (*From 16th Biennial Materials of Construction Report, Chem. Eng., November,* 1954.)

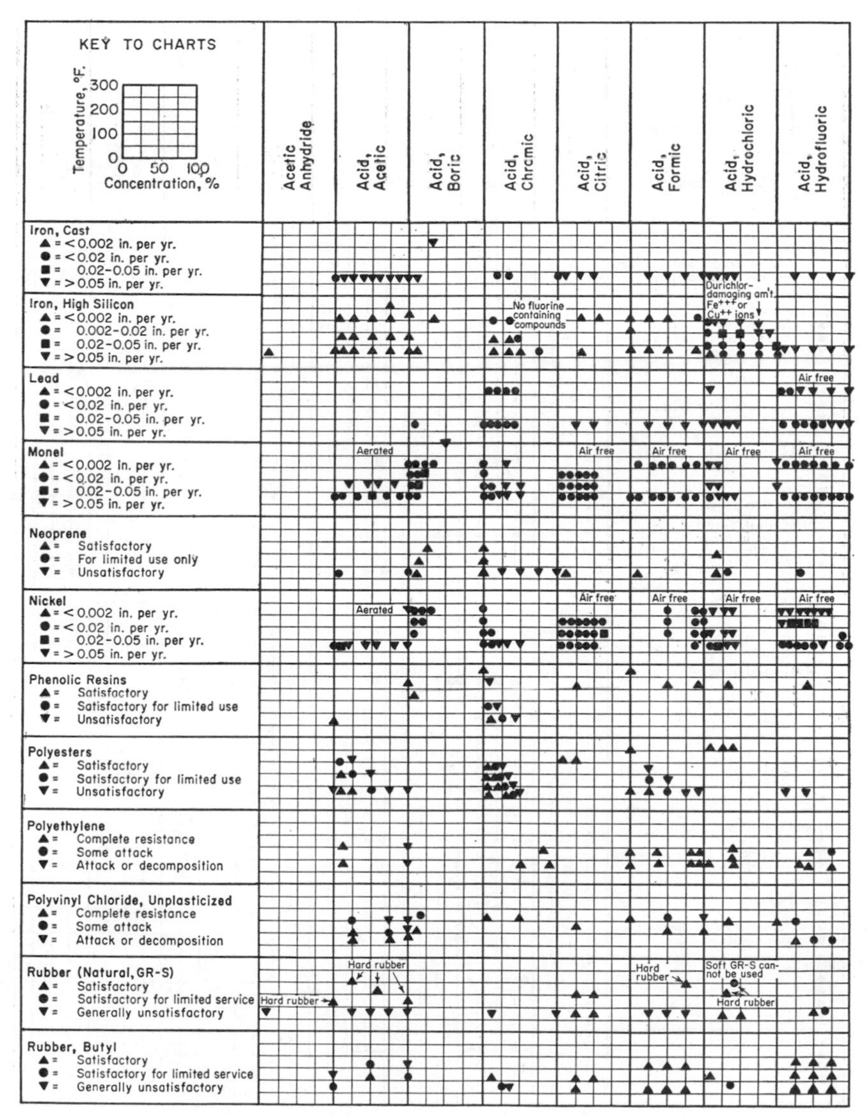

FIG. D-20 (*continued*).—Corrosion resistance of materials of construction. (*From 16th Biennial Materials of Construction Report, Chem. Eng., November*, 1954.)

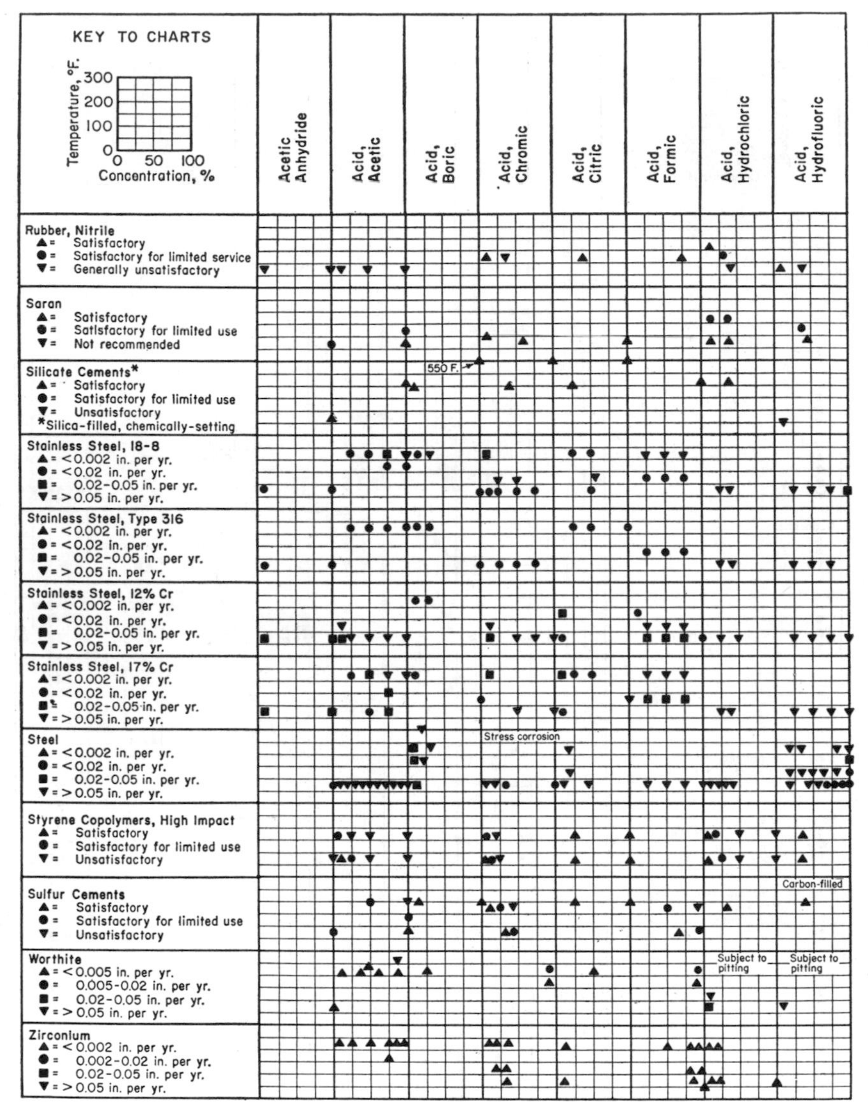

FIG. D-20 (*continued*).—Corrosion resistance of materials of construction. (*From 16th Biennial Materials of Construction Report, Chem. Eng., November,* 1954.)

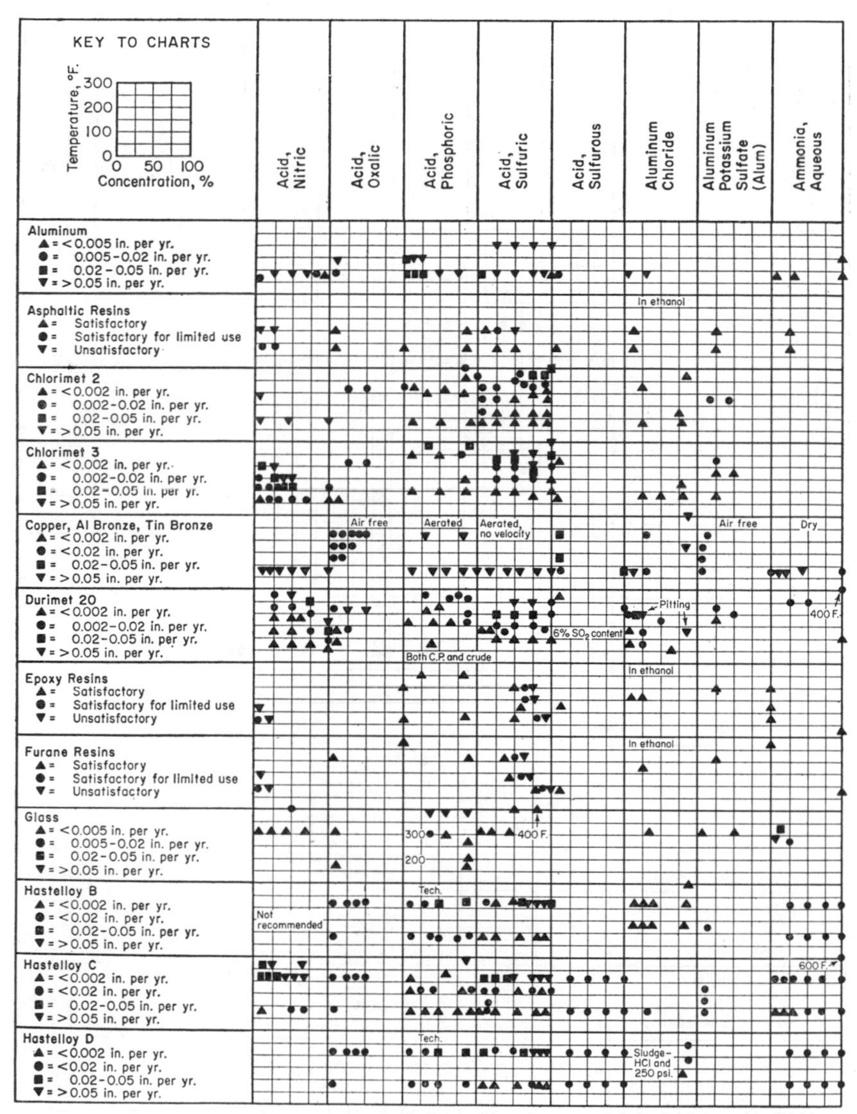

FIG. D-20 (*continued*).—Corrosion resistance of materials of construction. (*From 16th Biennial Materials of Construction Report, Chem. Eng., November,* 1954.)

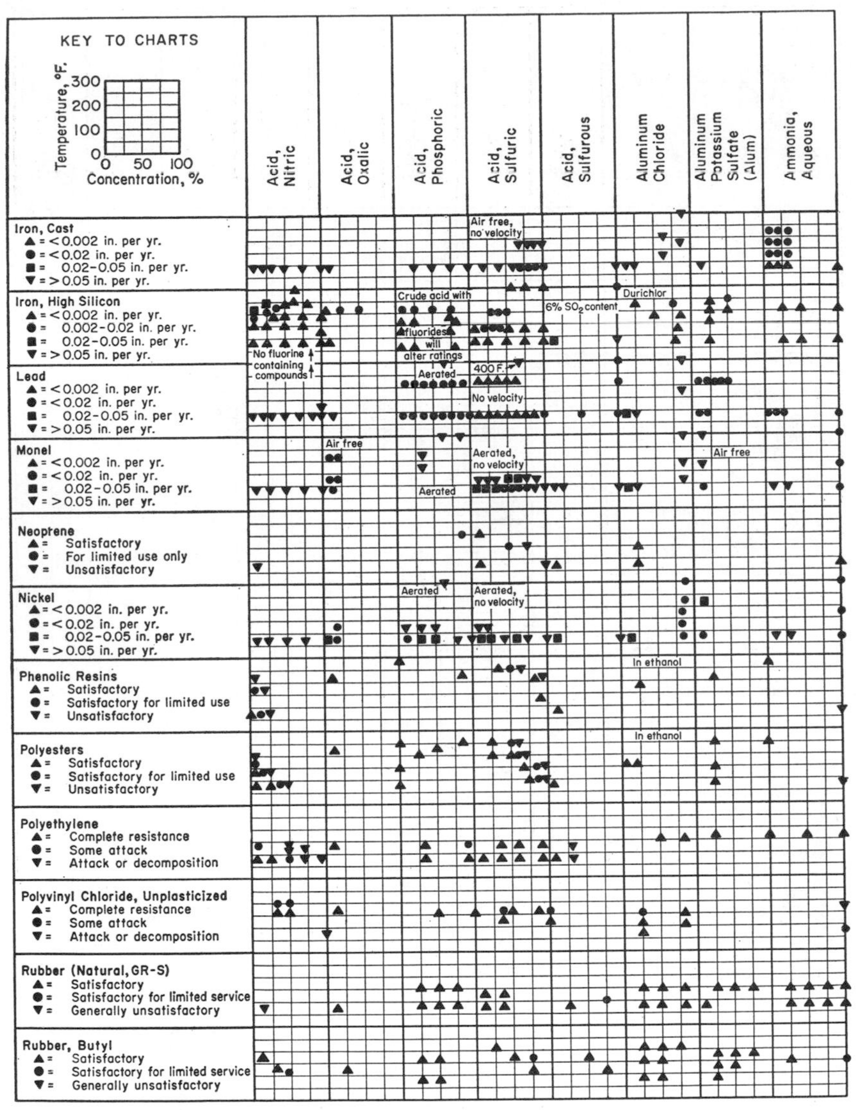

FIG. D-20 (*continued*).—Corrosion resistance of materials of construction. (*From 16th Biennial Materials of Construction Report, Chem. Eng., November*, 1954.)

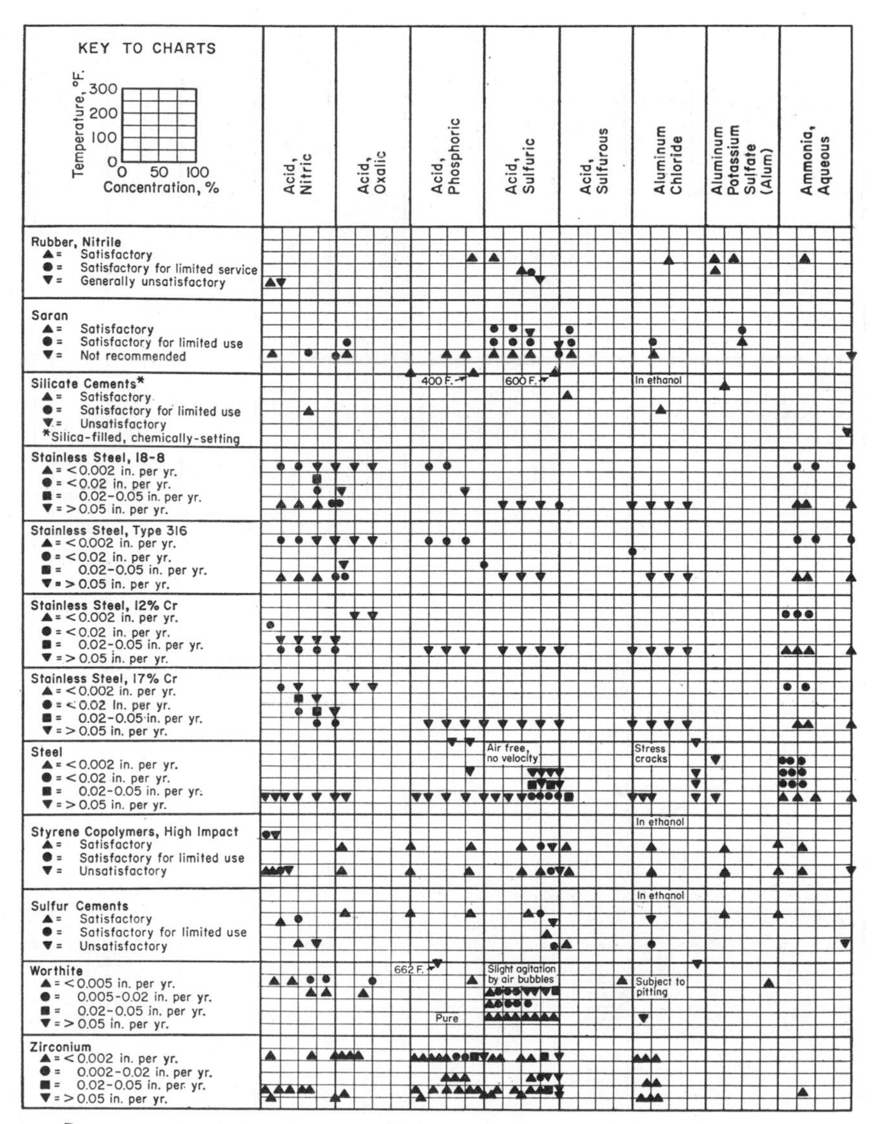

FIG. D-20 (*continued*).—Corrosion resistance of materials of construction. (*From 16th Biennial Materials of Construction Report, Chem. Eng., November,* 1954.)

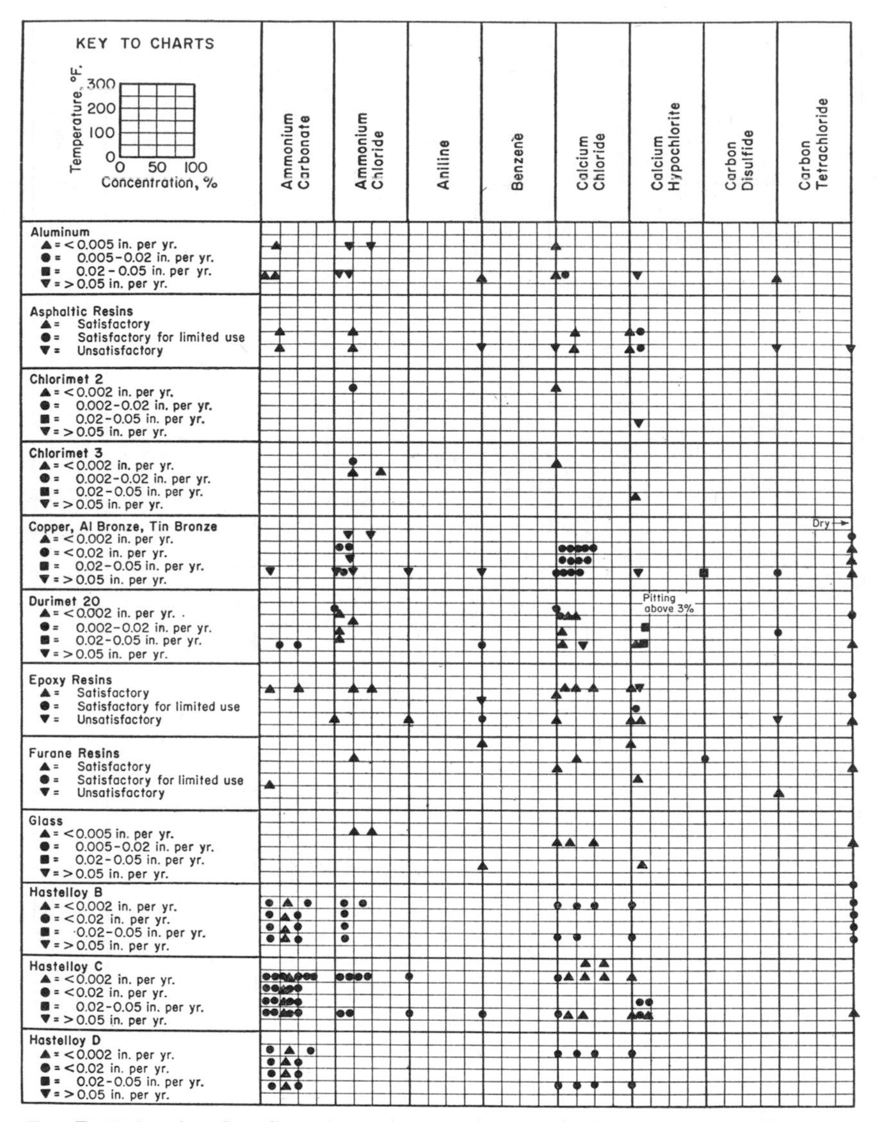

FIG. D-20 (*continued*).—Corrosion resistance of materials of construction. (*From 16th Biennial Materials of Construction Report, Chem. Eng., November,* 1954.)

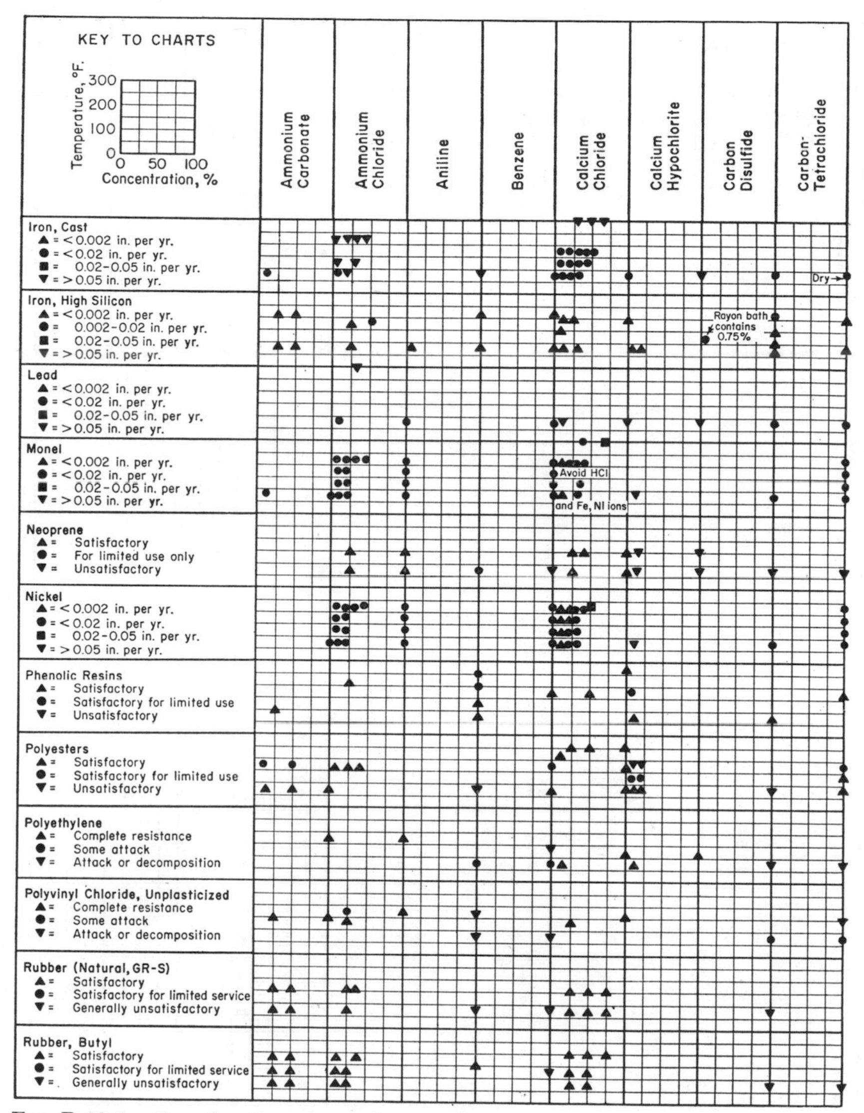

FIG. D-20 (*continued*).—Corrosion resistance of materials of construction. (*From 16th Biennial Materials of Construction Report, Chem. Eng., November,* 1954.)

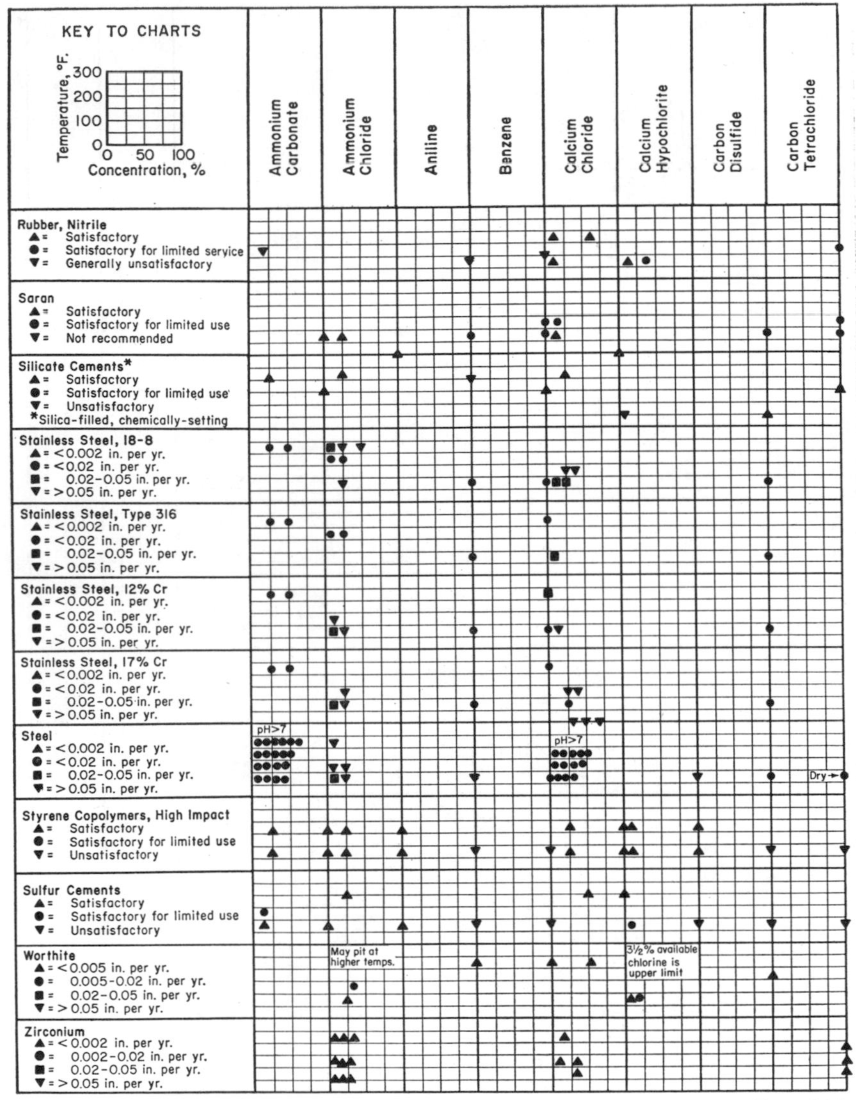

FIG. D-20 (*continued*).—Corrosion resistance of materials of construction. (*From 16th Biennial Materials of Construction Report, Chem. Eng., November,* 1954.)

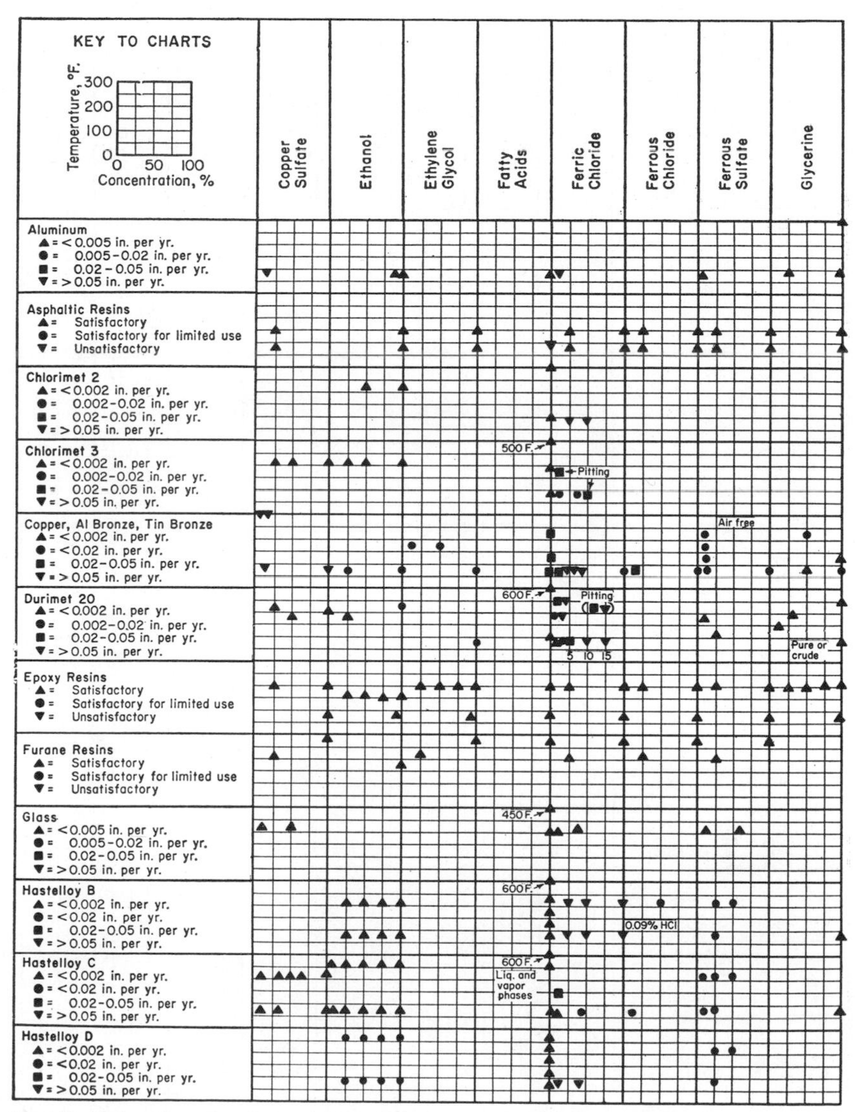

FIG. D-20 (*continued*).—Corrosion resistance of materials of construction. (*From 16th Biennial Materials of Construction Report, Chem. Eng., November, 1954.*)

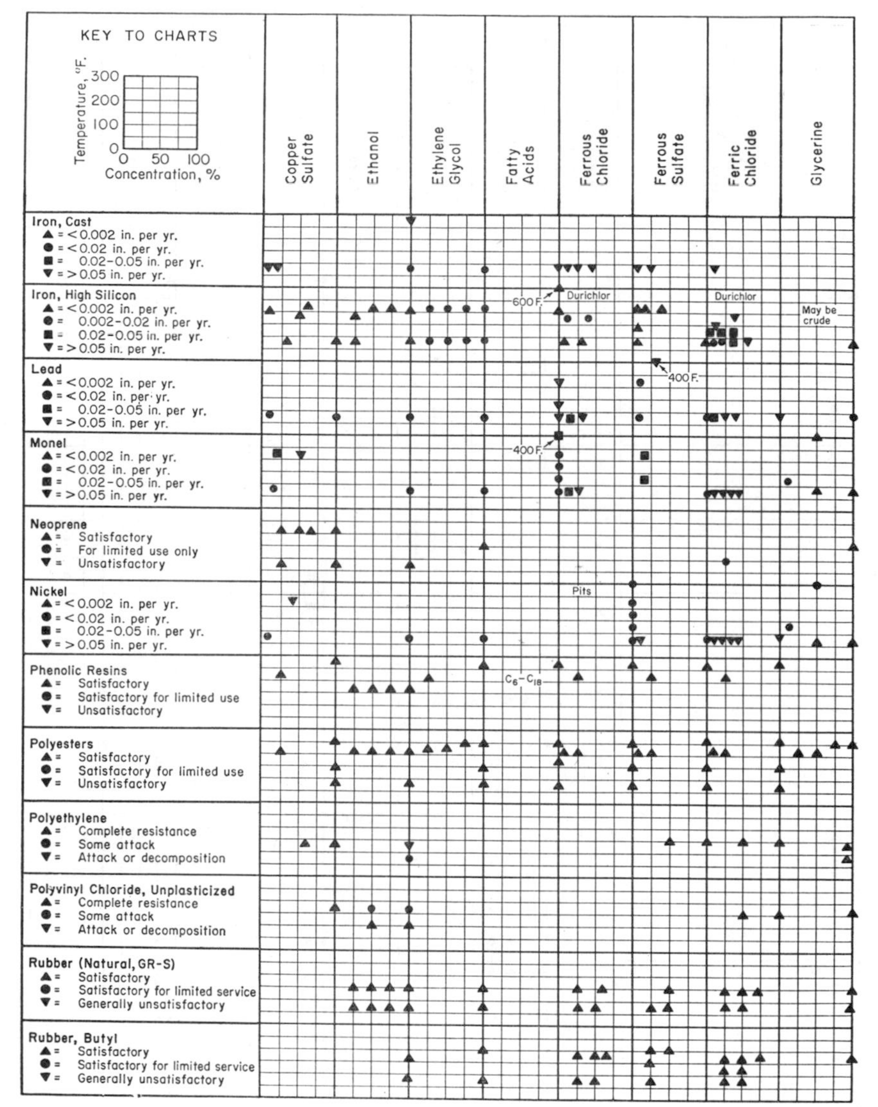

FIG. D-20 (*continued*).—Corrosion resistance of materials of construction. (*From 16th Biennial Materials of Construction Report, Chem. Eng., November,* 1954.)

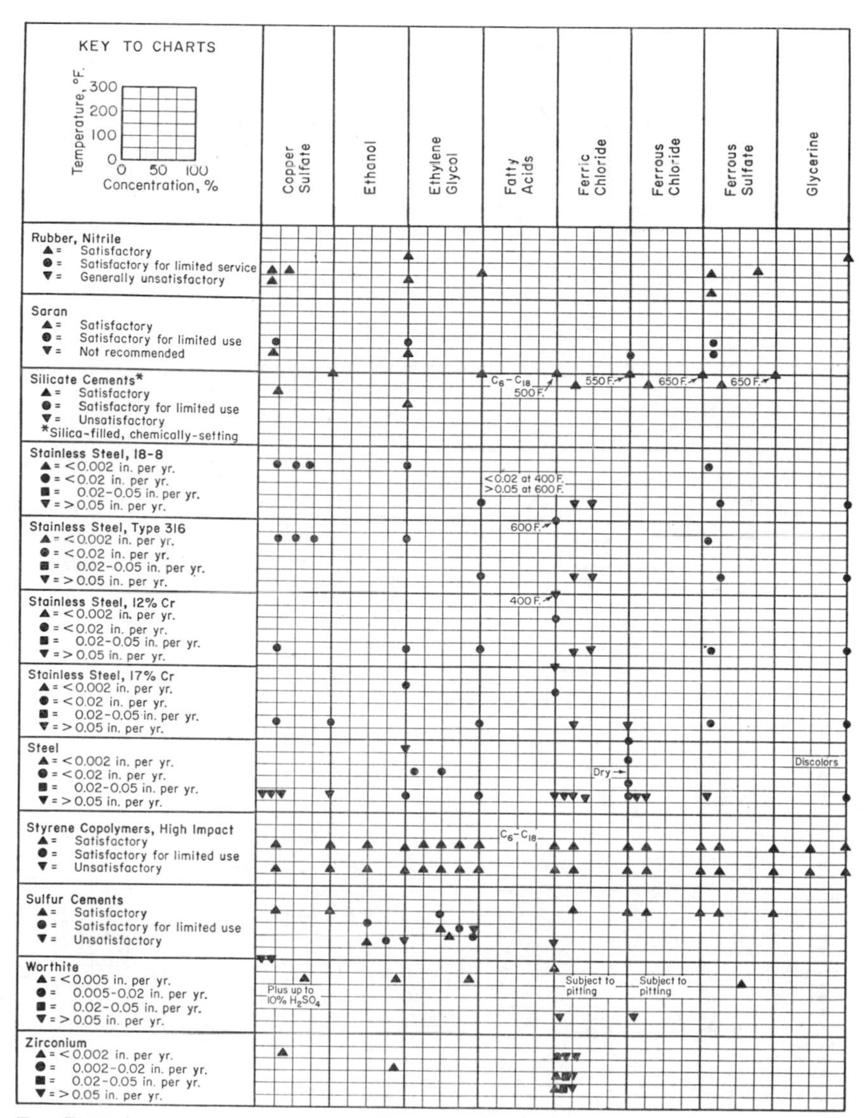

FIG. D-20 (*continued*).—Corrosion resistance of materials of construction. (*From 16th Biennial Materials of Construction Report, Chem. Eng., November*, 1954.)

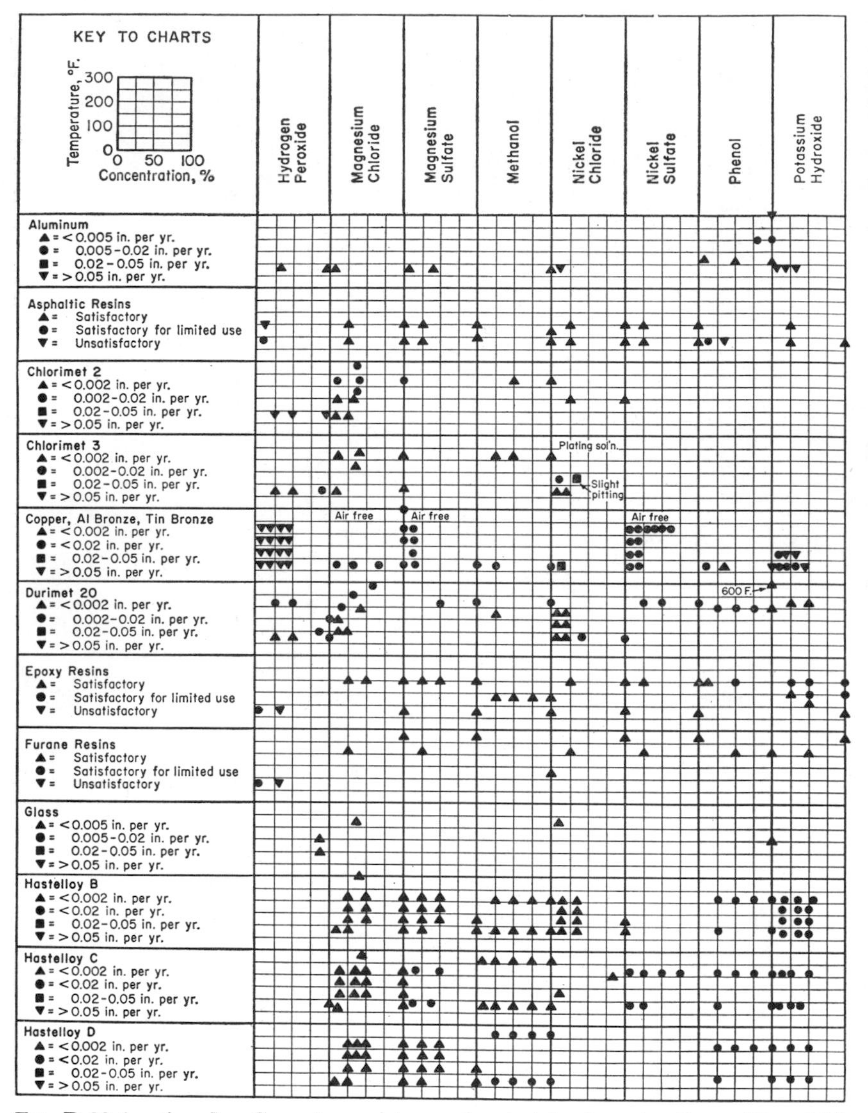

FIG. D-20 (*continued*).—Corrosion resistance of materials of construction. (*From 16th Biennial Materials of Construction Report, Chem. Eng., November,* 1954.)

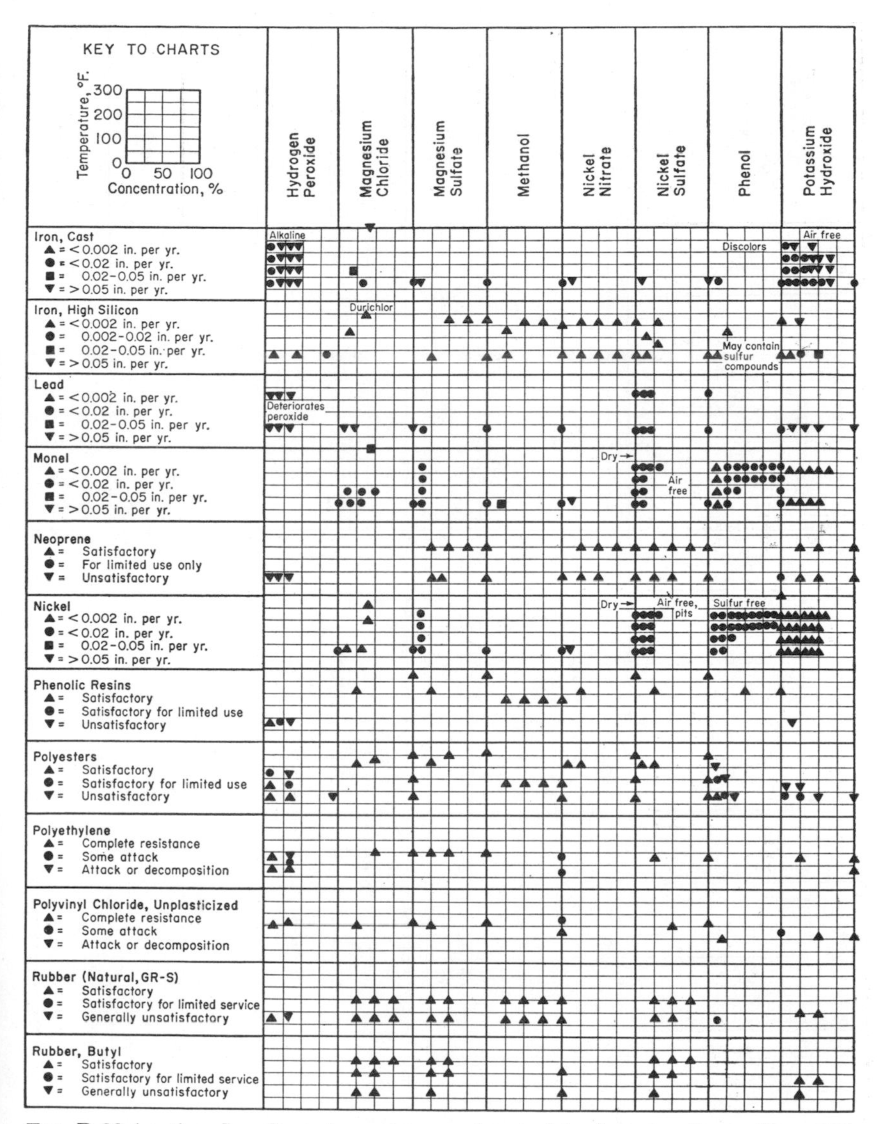

FIG. D-20 (*continued*).—Corrosion resistance of materials of construction. (*From 16th Biennial Materials of Construction Report, Chem. Eng., November*, 1954.)

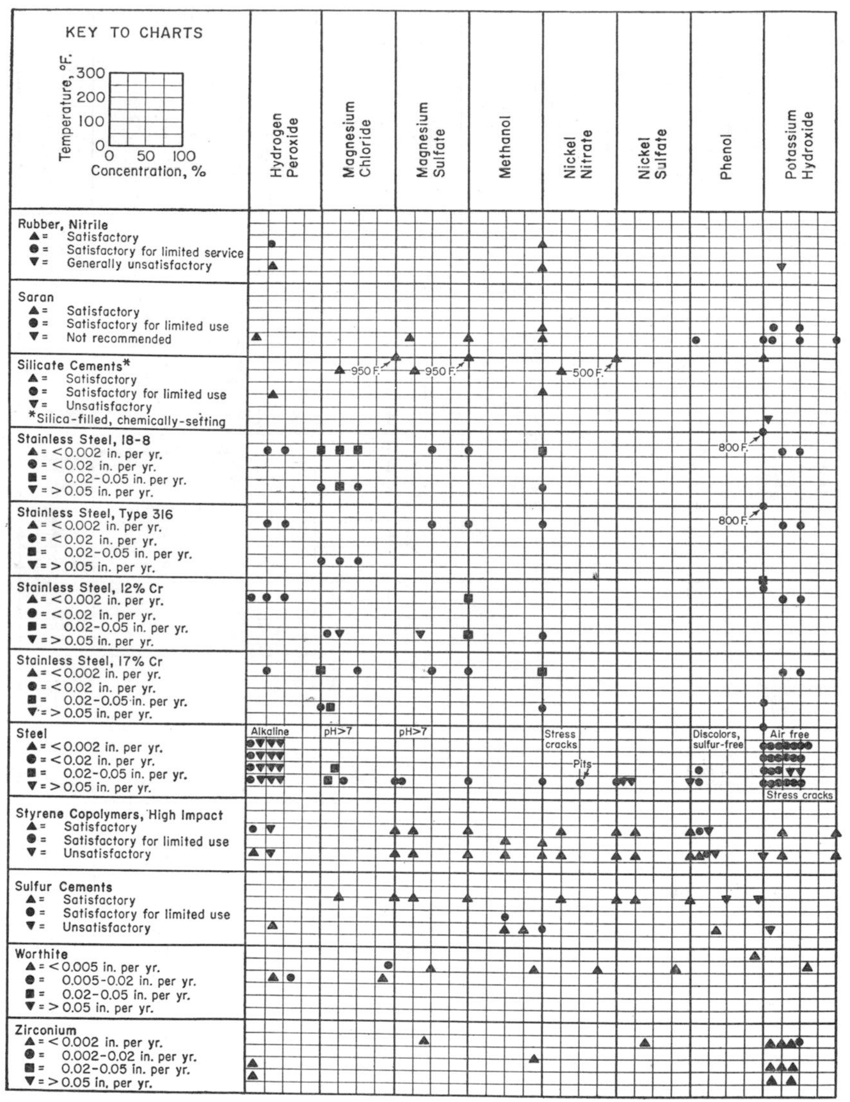

FIG. D-20 (*continued*).—Corrosion resistance of materials of construction. (*From 16th Biennial Materials of Construction Report, Chem. Eng., November,* 1954.)

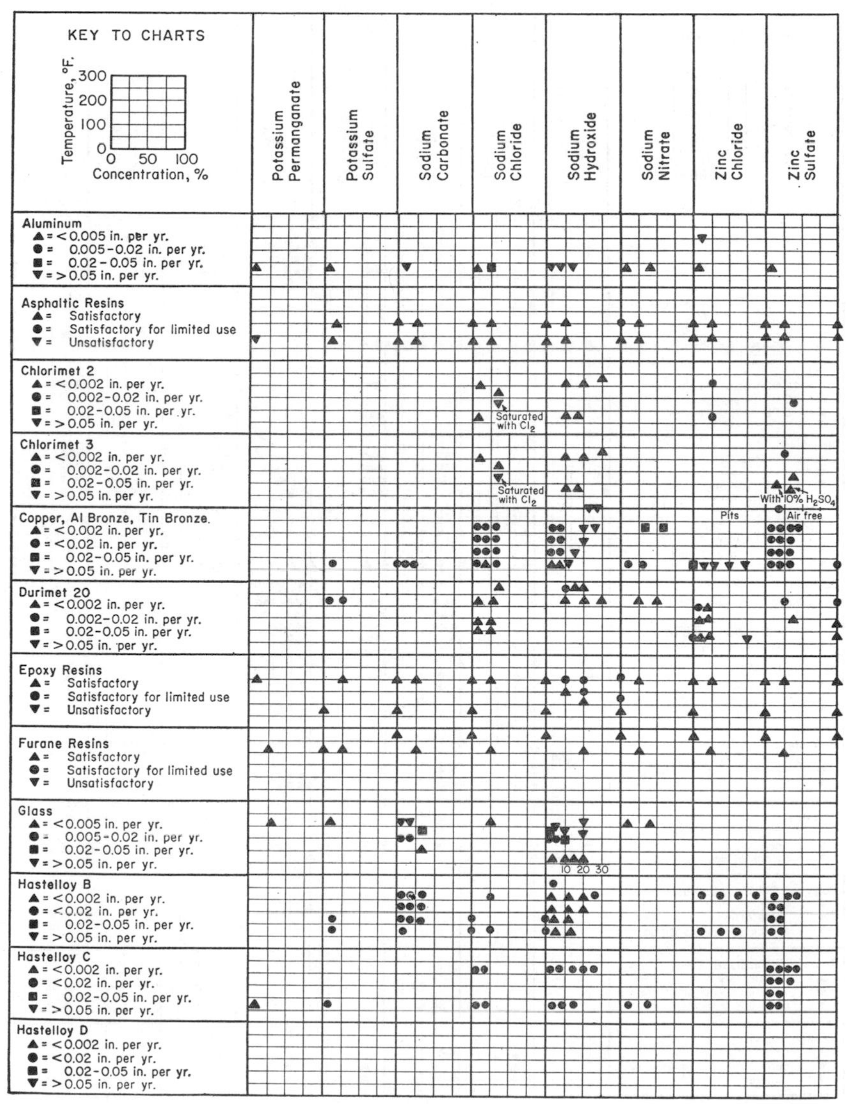

FIG. D-20 (*continued*).—Corrosion resistance of materials of construction. (*From 16th Biennial Materials of Construction Report, Chem. Eng., November,* 1954.)

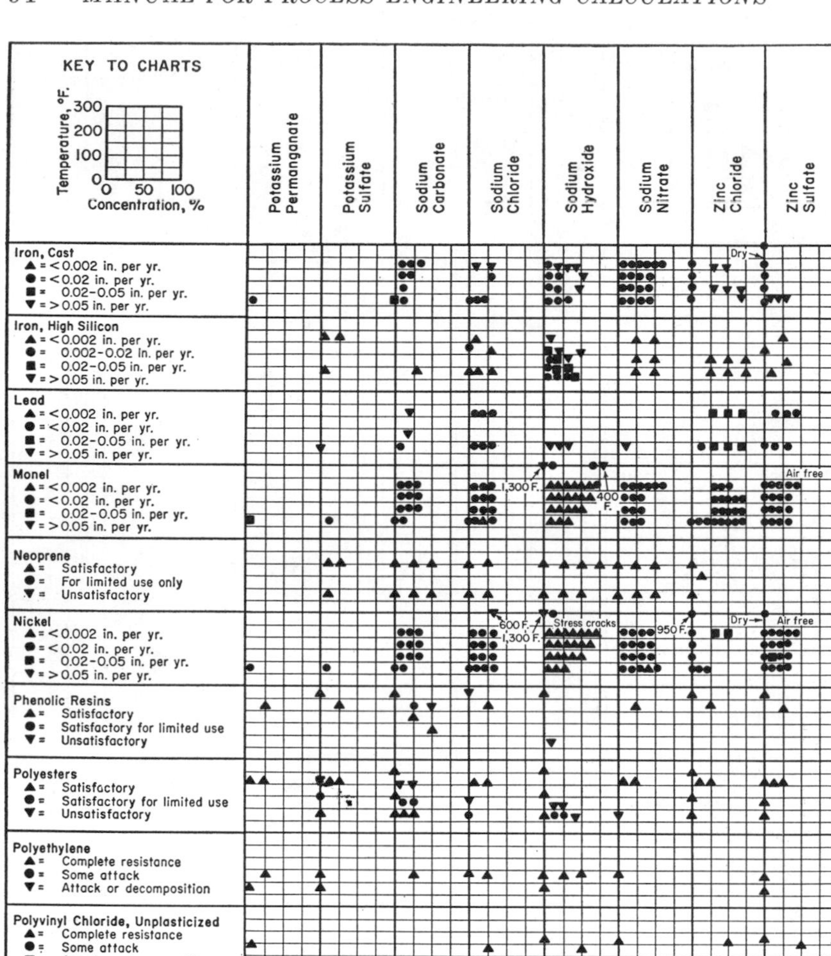

FIG. D-20 (*continued*).—Corrosion resistance of materials of construction. (*From 16th Biennial Materials of Construction Report, Chem. Eng., November*, 1954.)

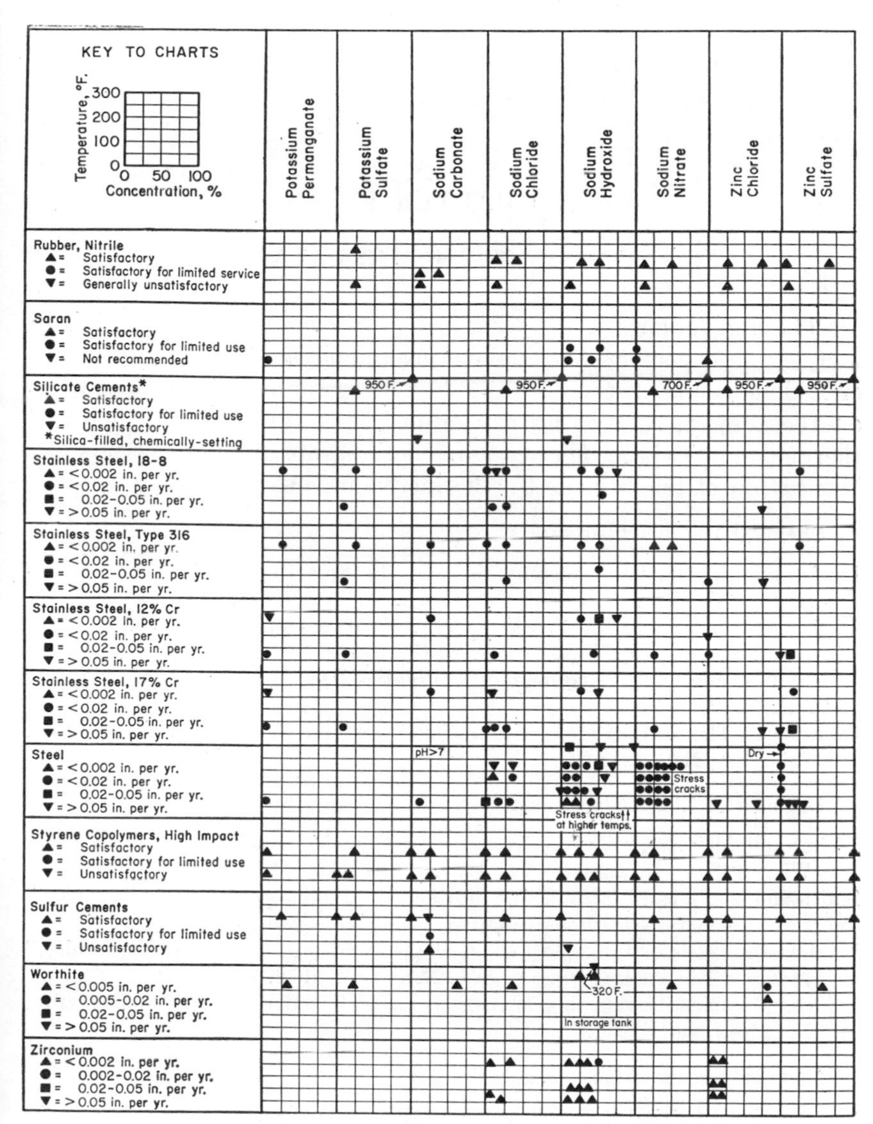

FIG. D-20 (*concluded*).—Corrosion resistance of materials of construction. (*From 16th Biennial Materials of Construction Report, Chem. Eng., November*, 1954.)

CHAPTER E

THERMODYNAMIC DATA

Thermodynamics is the backbone of process engineering and, in fact, of a large portion of all engineering and physical science. Because of its importance there are available a large number of excellent texts on the subject, so that an exposition of the abstract principles of thermodynamics is neither necessary nor appropriate here. Below are listed a few of the more useful thermodynamic source books.

Chemical Thermodynamics:

Lewis and Randall, "Thermodynamics and the Free Energy of Chemical Substances," McGraw-Hill Book Company, Inc., New York, 1923.

Taylor and Glasstone, "A Treatise on Physical Chemistry," 3d ed., D. Van Nostrand Company, Inc., Princeton, N.J., 1942.

Klotz, "Chemical Thermodynamics," Prentice-Hall, Inc., Englewood Cliffs, N.J., 1950

Glasstone, "Thermodynamics for Chemists," D. Van Nostrand Company, Inc., Princeton, N.J., 1947.

Engineering Thermodynamics:

Weber and Meissner, "Thermodynamics for Chemical Engineers," 2d ed., John Wiley & Sons, Inc., New York, 1957.

Smith and Van Ness, "Introduction to Chemical Engineering Thermodynamics," 2d ed., McGraw-Hill Book Company, Inc., New York, 1959.

Wilson and Ries, "Principles of Chemical Engineering Thermodynamics," McGraw-Hill Book Company, Inc., New York, 1956.

Dodge, "Chemical Engineering Thermodynamics," McGraw-Hill Book Company, Inc., New York, 1944.

Thermodynamic Data:

"International Critical Tables," McGraw-Hill Book Company, Inc., New York.

Landolt-Börnstein, "Physikalisch-Chemischen Tabellen," 5th ed., Springer-Verlag, Berlin, 1936. (Out of print.)

National Bureau of Standards, "Selected Values of Chemical Thermodynamic Properties," issued 1947–1953, Series I, II, and III; Series I and II reissued as Circular 500, 1952.

Hougen, Watson, and Ragatz, "Chemical Process Principles," 2d ed., Vol. II, Thermodynamics, John Wiley & Sons, Inc., New York, 1959.

Din, "Thermodynamic Functions of Gases," Butterworth & Co. (Publishers) Ltd., London, Vol. I, 1956; Vol. II, 1956; more volumes coming.

Latimer, "Oxidation States of the Elements and Their Potentials in Aqueous Solution," 2d ed., Prentice-Hall, Inc., Englewood Cliffs, N.J., 1952. (Free energy data.)

This chapter gives tabulations and charts of thermodynamic data which are useful for process calculations. At the end of the chapter some of the basic laws and equations of thermodynamics are given.

1. Equations of State

a. The Perfect Gas Law.—This law applies to all gases and vapors at very low pressures and may be applied further to most gases and vapors at moderate pressures without large error.

$$pV = RT \tag{E-1}$$

where p = pressure
V = volume
T = absolute temperature, °R or °K
R = gas-law constant (values in Table E-1)

TABLE E-1.—VALUES OF GAS-LAW CONSTANT, R
$pV = RT$

Units of				R
Pressure	Volume	Temperature	Quantity of gas	
Psi	Cu ft	°R = °F + 459.7	1 lb mole*	10.71
Psf	Cu ft	°R = °F + 459.7	1 lb mole	1,543
Psi	Cu ft	°R = °F + 459.7	1 SCF†	0.02838
Atm	Cu ft	°R = °F + 459.7	1 lb mole*	0.729
Atm	Cu cm	°K = °C + 273.16	1 g mole	82.06
Energy units				
Btu/(lb mole)(°R), or cal/(g mole)(°K)				1.9871
Btu/(SCF)(°F)				0.005247
Joule/(g mole)(°K)				8.3144

* The molecular weight expresses the number of pounds in a pound mole.

† One standard cubic foot (SCF) is the quantity of gas contained in 1 cu ft at 60°F and 30 in. of mercury (14.734 psi); 1 lb mole = 378.7 SCF. This is the definition used by the gas industry; it differs from that used by the *compressed* gas industry which defines the standard cubic foot at 68°F and atmospheric pressure (14.696 psi).

b. Reduced Equation of State.—A number of algebraic equations of state of varying degrees of accuracy and complexity have been developed. A simple one suitable for many thermodynamic equations[1] is

$$p = \frac{RT}{V - b} - \frac{a}{T^n(V + c)^2} \qquad \text{(E-2)}$$

Three of the four adjustable constants of this equation can be evaluated from the critical constants of the gas as follows:

$$a = \frac{27}{64} \frac{R^2 T_c^2 T_c^n}{p_c}$$

$$b = V_c - \frac{1}{4} \frac{RT_c}{p_c}$$

$$c = \frac{3}{8} \frac{RT_c}{p_c} - V_c$$

where p, p_c = pressure, critical pressure
V, V_c = volume, critical volume (per mole)
T, T_c = temperature, critical temperature (absolute units)

The fourth adjustable constant, the exponent n, is not a universal constant, but a value of 0.6 gives good results for a number of nonpolar low-molecular-weight compounds. Equation (E-2) is closely related to three familiar equations of state: Clausius ($n = 1$); Berthelot ($n = 1$, $c = 0$); and van der Waals ($n = 0$, $c = 0$). Equation (E-2) in reduced form is

$$p_R = \frac{T_R}{(V_R - 1)z_c + 0.25} - \frac{27}{64} \frac{1/T_R^n}{[(V_R - 1)z_c + 0.375]^2} \qquad \text{(E-3)}$$

where $p_R = p/p_c$, reduced pressure
$T_R = T/T_c$, reduced temperature
$V_R = V/V_c$, reduced volume
$z_c = p_c V_c / RT_c$, critical value of compressibility factor

For most engineering purposes, the reduced equation of state can be adequately expressed in the more convenient form

$$pV = zRT \qquad \text{(E-4)}$$

The values of z, the compressibility factor defined by this equation, have been correlated over a wide range as dependent on the reduced temperature and pressure. Figures E-1 to E-3 give numerical values of z. Figure E-1 is for vapors below the critical temperature, Fig. E-2

[1] Clarke, unpublished data.

is for gases at or above the critical temperature, and Fig. E-3 applies to the very high pressure range. Equation (E-4) may be employed with any units provided the appropriate value of R is chosen.

To aid in the use of Eq. (E-4), the critical pressures and temperatures of a number of substances are given in Table E-2.

The reduced equation of state also forms the basis of correlations of latent heats, viscosities, and other physical properties.

c. Gaseous Mixtures.—These may quite often be treated by either Dalton's law of additive pressures or by Amagat's law of additive volumes.

Dalton's law states that, in a mixture, each gas exerts the same (partial) pressure as if it alone occupied the volume and that the total pressure is the sum of the partial pressures of the individual components.

Amagat's law states that the volume occupied by a gaseous mixture is the sum of the volumes occupied by each component separately at the same temperature and pressure.

Both Dalton's law and Amagat's law are obeyed when the component gases all obey the perfect gas law. Although both fail for imperfect gases, Amagat's law is in the smaller error of the two.

Most imperfect gases may be treated with fair accuracy by the method of Kay.[1] He defines the pseudocritical pressure as the molal average of the individual critical pressures and the pseudocritical temperature as the molal average of the individual critical temperatures. Stating this in equation form,

$$p_c = N_1 p_{c1} + N_2 p_{c2} + \cdots \tag{E-5}$$

$$T_c = N_1 T_{c1} + N_2 T_{c2} + \cdots \tag{E-6}$$

where p_c = pseudocritical pressure
T_c = pseudocritical temperature
N_1 = mole fraction, component 1
N_2 = mole fraction, component 2, etc.
p_{c1} = critical pressure of component 1, etc.
T_{c1} = critical temperature of component 1, etc.

From these pseudocritical constants the reduced temperature and pressure may be computed for use in the reduced equation of state [Eq. (E-4)] as for a pure gas. This relation fails in the neighborhood of the critical point but is otherwise probably sufficiently accurate for engineering calculations.

[1] *Ind. Eng. Chem.*, **28**, 1014 (1936).

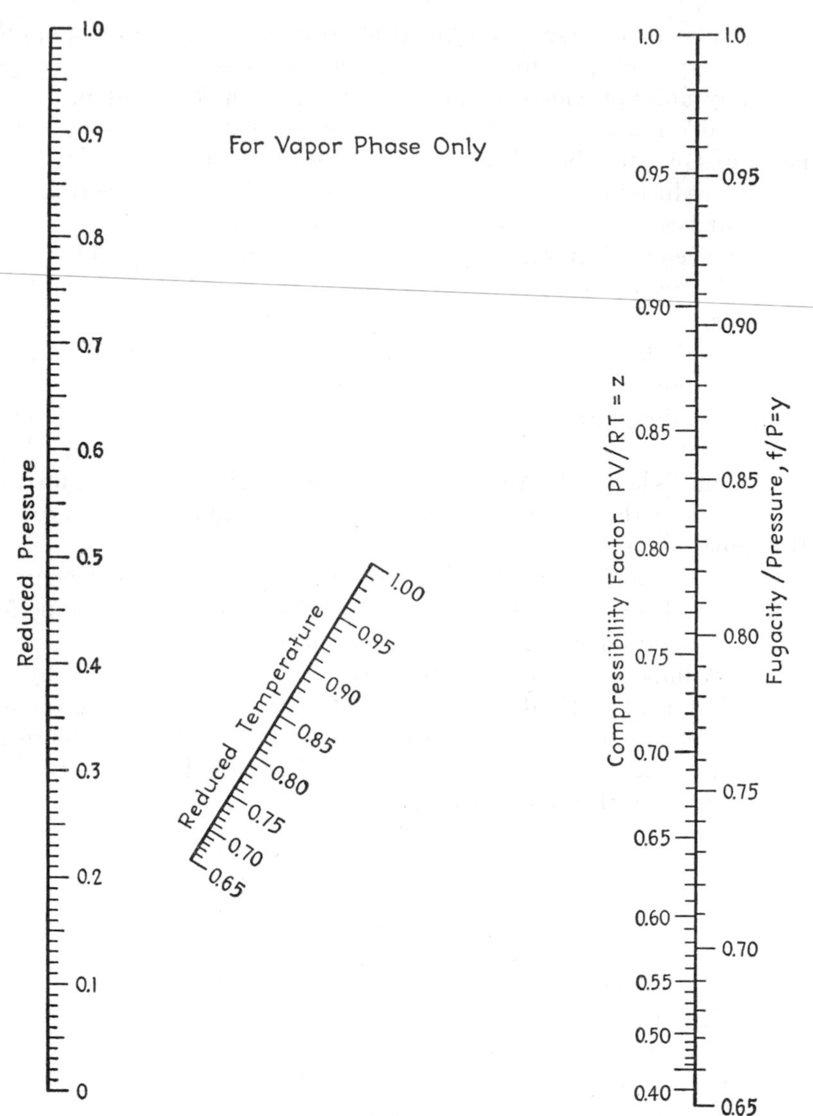

FIG. E-1.—Compressibility factors for vapors below the critical temperature. [*Thomson, Ind. Eng. Chem.*, **35,** 895, (1943). *By permission of the American Chemical Society.*]

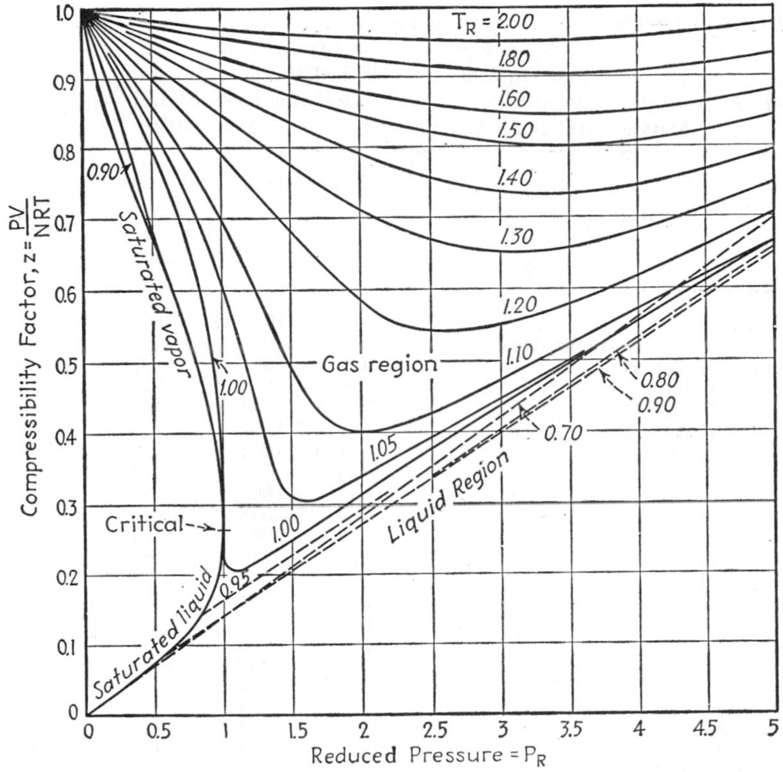

FIG. E-2.—Compressibility factors of gases. (*Reproduced by permission from "Thermodynamics for Chemical Engineers," by H. C. Weber, published by John Wiley & Sons, Inc., New York,* 1939.)

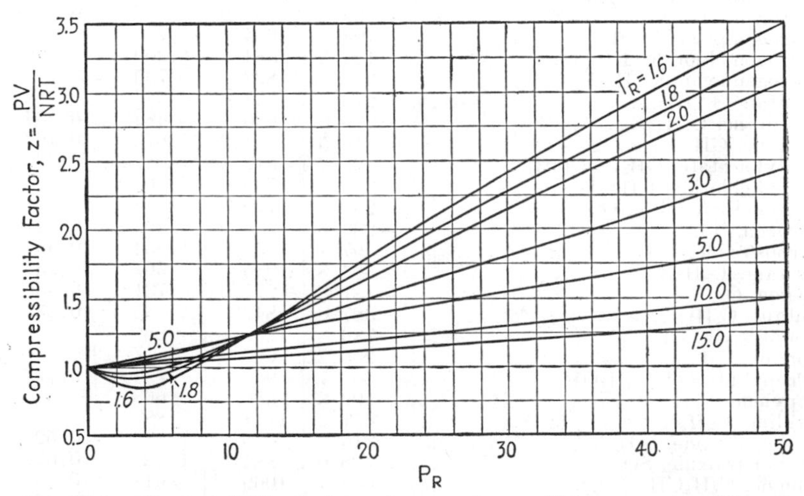

FIG. E-3.—Compressibility factors of gases at very high pressures. (*Reproduced by permission from "Thermodynamics for Chemical Engineers," by H. C. Weber, published by John Wiley & Sons, Inc., New York,* 1939.)

TABLE E-2.—CRITICAL TEMPERATURES, PRESSURES, AND DENSITIES OF SOME COMMON SUBSTANCES

(Based mainly on data from "International Critical Tables," Matteson and Hanna, *Oil Gas J.*, May 21, 1942; and Perry's "Chemical Engineers' Handbook," 3d ed., McGraw-Hill Book Company, Inc., New York, 1950)

Substance	Temperature		Pressure	Density
	°F	°R	Psi	g/cu cm
Acetic acid, CH_3COOH	611	1071	841	0.351
Acetic anhydride, $(CH_3CO)_2O$	565	1025	676	
Acetone, CH_3COCH_3	455	915	691	0.268
Acetylene, C_2H_2	96.8	556	911	0.231
Ammonia, NH_3	271.4	731	1,657	0.235
Aniline, $C_6H_5NH_2$	800	1260	770	
Benzene, C_6H_6	551.3	1011	700	0.304
n-Butane, C_4H_{10}	305.6	765	551	
n-Butyl alcohol, C_4H_9OH	549	1009	711	
Carbon dioxide, CO_2	88.0	548	1,073	0.460
Carbon disulfide, CS_2	523	983	1,117	0.441
Carbon monoxide, CO	−218.2	241.5	514	0.331
Carbon tetrachloride, CCl_4	539.8	1000	661.5	0.558
Chlorine, Cl_2	291.2	751	1,119	0.573
Chlorobenzene, C_6H_5Cl	678	1138	656	0.365
Cyclohexane, C_6H_{12}	537.8	998	594	0.270
Ethane, C_2H_6	90.1	550	708	0.21
Ethyl acetate, $CH_3COOC_2H_5$	482.2	942	556	0.308
Ethyl alcohol, C_2H_5OH	469.6	929	927	0.2755
Ethyl amine, $C_2H_5NH_2$	362	822	816	
Ethylene, C_2H_4	49.5	899	749	0.22
Ethyl ether, $(C_2H_5)_2O$	381	841	522	0.2625
n-Heptane, C_7H_{16}	512	972	397	0.234
n-Hexane, C_6H_{14}	454.6	914	434	0.234
Hydrazine, N_2H_4	716	1176	2,132	
Hydrogen, H_2	−399.8	59.9	188	0.0310
Hydrogen bromide, HBr	194.0	654	1,235	
Hydrogen chloride, HCl	124.5	584	1,200	
Hydrogen cyanide, HCN	362.3	822	735	
Isobutane, C_4H_{10}	273.2	733	544	
Isopentane, C_5H_{12}	370.0	830	482	0.234
Methane, CH_4	−116.5	343.2	673	0.162
Methyl acetate, CH_3COOCH_3	452.7	912	681	0.325
Methyl chloride, CH_3Cl	289.6	749	967	0.37
Methyl ether, $(CH_3)_2O$	260.4	720	764	0.271
Nitrogen, N_2	−232.8	226.9	492	0.3110
Nitrous oxide, N_2O	97.7	557	1,054	0.45
n-Octane, C_8H_{18}	564.8	1025	361	0.234
Oxygen, O_2	−181.8	277.9	731	0.430
Pentane, C_5H_{12}	387.0	847	485	0.232
Phenol, C_6H_5OH	786	1246	889	
Propane, C_3H_8	206.2	666	617	
n-Propyl alcohol, C_3H_7OH	506.7	966	734	0.273
Propylene, C_3H_6	196.8	656	668	
Pyridine, C_5H_5N	651.2	1111	882	
Sulphur dioxide, SO_2	315	775	1,142	0.52
Sulphur trioxide, SO_3	425.0	885	1,229	0.630
Toluene, $C_6H_5CH_3$	609.1	1069	611	0.292
Water, H_2O	705.4	1165	3,206	0.323

d. Estimation of Liquid Densities.—Although the simple form of the reduced equation of state is not accurate when applied to the liquid state, the properties of liquids may be correlated by the use of the reduced temperatures and reduced pressures. Watson[1] discusses several such correlations.

Figure E-4 is a chart for estimating liquid densities of saturated liquids (*i.e.*, under their own vapor pressures) from a single known density.

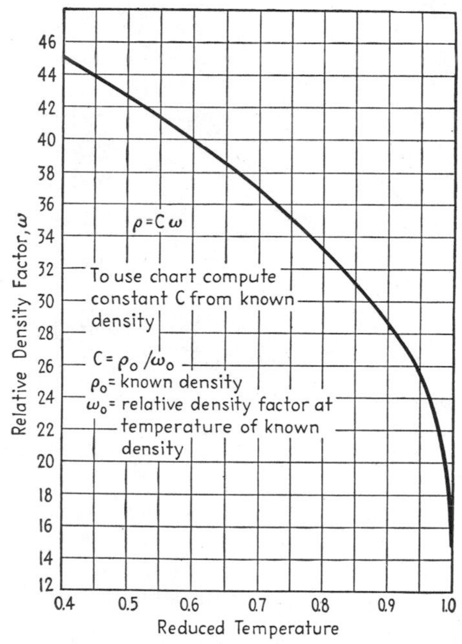

Fig. E-4.—Chart for estimating density of saturated liquids.

2. Specific Heats and Sensible Heats

a. Specific and Sensible Heats of Gases.—The heat capacities of a number of common gases are given in Fig. E-5. This figure was prepared from older data for the first edition of this book. With the exception of the acetylene line, which has been revised, this chart was in close agreement with the newer data of the National Bureau of Standards.[2]

[1] *Ind. Eng. Chem.*, **34**, 398 (1943).

[2] "Selected Values of Chemical Thermodynamic Properties," Series III, National Bureau of Standards, 1949–1953.

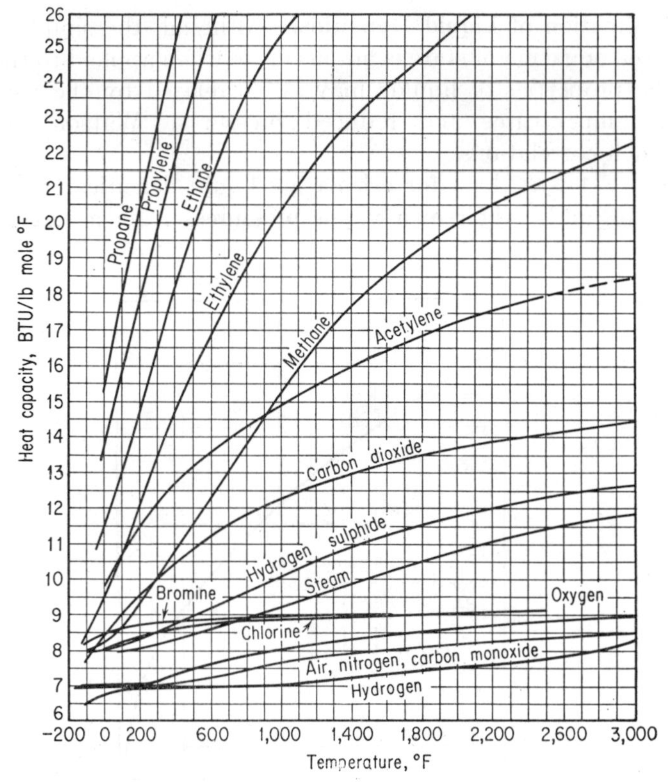

Fig. E-5.—Constant-pressure heat capacity of gases and vapors at low pressures.

Specific heat data may be summarized by empirical equations of the following type:

$$C_p = a + bT - cT^2 - \frac{d}{T^2} \qquad \text{(E-7)}$$

where C_p = specific heat or heat capacity
T = absolute temperature
a, b, c, d = empirical constants (usually one or more may be omitted)

Equations of this type are given for the constant-pressure heat capacities of a number of common gases in Table E-3.

The specific heat of a gas depends to some extent on the pressure. Figures E-6 to E-9 give the specific heats of air, hydrogen, carbon dioxide, and methane at various pressures. In the absence of specific data, the effect of pressure may be estimated by one or other of the following equations:

$$C_p = C_p^0 + 9\frac{p_R}{T_R^5} \qquad \text{(when } T_R \text{ is less than 1.2)} \qquad \text{(E-8)}$$

$$C_p = C_p^0 + 5\frac{p_R}{T_R^2} \qquad \text{(when } T_R \text{ is more than 1.2)} \qquad \text{(E-8}a\text{)}$$

where C_p^0 = heat capacity at low pressures, Btu/(lb mole)(°F)
C_p = heat capacity under pressure
p_R = the reduced pressure
T_R = the reduced temperature

Equation (E-8) is purely empirical but is sufficiently accurate for many purposes provided the correction for pressure is under 2.5 Btu/(lb mole)(°F). Equation (E-8*a*) is based on the Berthelot equation of state and may be used with a modest degree of accuracy provided the pressure correction is no more than 1 or 2 Btu/(lb mole)(°F).

The principal use of specific heats is for computing the heat change accompanying the heating or cooling of gases by the following relation:

$$Q = \Delta H = \int_{T_1}^{T_2} C_p\, dT$$
$$= a(T_2 - T_1) + \frac{b}{2}(T_2^2 - T_1^2) - \frac{c}{3}(T_2^3 - T_1^3) - d\left(\frac{1}{T_1} - \frac{1}{T_2}\right) \qquad \text{(E-9)}$$

where ΔH = increase in heat content of gas, Btu/lb or Btu/lb mole, etc.
Q = heat absorbed (at constant pressure)
T_1 = initial temperature, °*R*
T_2 = final temperature, °*R*
C_p = specific heat Btu/lb or Btu/lb mole, etc.
a, b, c, d = empirical constants, Eq. (E-7) (see Table E-3)

Often it is more convenient to compute the heat change from tabulations of heat contents usually referred to as sensible heats. Figures E-10 to E-12 (pp. 109 to 111) give sensible heats of six gases; Fig. E-13 (p. 112) gives the sensible heats of the hydrocarbon gases. These charts are based on atmospheric pressure.

Heat contents (like specific heats) are dependent on pressure as well as temperature. Figure E-14 (p. 113) permits an approximate computation of the heat content under pressure from the low pressure values.

TABLE E-3.—EMPIRICAL EQUATIONS FOR CONSTANT-PRESSURE HEAT CAPACITIES OF GASES AND VAPORS AT LOW PRESSURES

$$C_p = a + bT - cT^2 - \frac{d}{T^2}, \text{ Btu/(lb mole)(°F), where } T = °\text{R}$$

Substance	Molecular weight	*a*	*b*	*c*	*d*	Range, °R	References
Acetylene, C_2H_2	26.036	11.5	0.00255	....	700,000	500–2500	9
Ammonia, NH_3	17.032	7.12	0.00338	....	129,000	520–2500	4
Benzene, C_6H_6	78.108	6.16	0.0292	....	....	490–1000	7
n-Butane, C_4H_{10}	58.121	4.64	0.0310	....	....	490–1200	5
1-Butylene, C_4H_8	56.105	4.61	0.0285	....	....	490–1200	5
Bromine, Br_2	159.83	8.42	0.000541	0.11×10^{-6}	....	500–2500	1
Carbon dioxide, CO_2	44.010	10.34	0.00152	....	634,000	500–2500	2
Carbon monoxide, CO	28.010	6.60	0.00067	....	....	400–3000	2
Chlorine, Cl_2	70.914	7.58	0.00135	0.30×10^{-6}	....	500–2700	1
Ethane, C_2H_6	30.069	4.26	0.0158	....	....	490–1200	5
Ethylene, C_2H_4	28.052	3.25	0.0162	2.93×10^{-6}	....	500–2400	6, 9
Hydrogen, H_2	2.016	6.62	0.00045	....	....	490–4500	2
Hydrogen bromide, HBr	80.92	6.80	0.00047	....	....	360–2600	2
Hydrogen chloride, HCl	36.46	6.73	0.00024	0.11×10^{-6}	....	500–3000	1
Hydrogen cyanide, HCN	27.02	10.13	0.00115	....	807,000	760–2700	4
Hydrogen sulphide, H_2S	34.076	7.15	0.00184	....	....	500–3000	2
Methane, CH_4	16.042	3.36	0.0100	1.30×10^{-6}	....	520–3600	8, 9
Nitric oxide, NO	30.008	8.05	0.00013	....	507,000	490–9000	2
Nitrogen, N_2	28.016	6.45	0.00080	0.025×10^{-6}	....	500–2800	1
Oxygen, O_2	32.000	6.10	0.0018	0.314×10^{-6}	....	500–3000	1
Propane, C_3H_8	44.095	4.16	0.0240	....	....	490–1200	5
Propylene, C_3H_6	42.079	4.00	0.0207	....	....	490–1200	5
n-Pentane, C_5H_{12}	72.147	5.16	0.0382	....	....	490–1200	5
Steam, H_2O	18.016	6.95	0.00180	0.100×10^{-6}	....	500–3500	9, 8
Sulphur dioxide, SO_2	64.06	11.4	0.000786	....	640,000	520–3500	3

1. Spencer and Justice, *J. Am. Chem. Soc.*, **56,** 2311 (1934); cf. Spencer, *J. Am. Chem. Soc.*, **67,** 1859 (1945).
2. Kelley, *U.S. Bur. Mines Bull.* 371 (1934).
3. Kelley, *U.S. Bur. Mines Bull.* 406 (1937).
4. Kelley, *U.S. Bur. Mines Bull.* 407 (1937).
5. Thacker, Folkins, and Miller, *Ind. Eng. Chem.*, **33,** 584 (1941).
6. Wenner, "Thermochemical Calculations," McGraw-Hill Book Company, Inc., New York, 1941.
7. Edmister, *Ind. Eng. Chem.*, **30,** 352 (1940).
8. Bryant, *Ind. Eng. Chem.*, **25,** 820 (1933).
9. Privately derived.

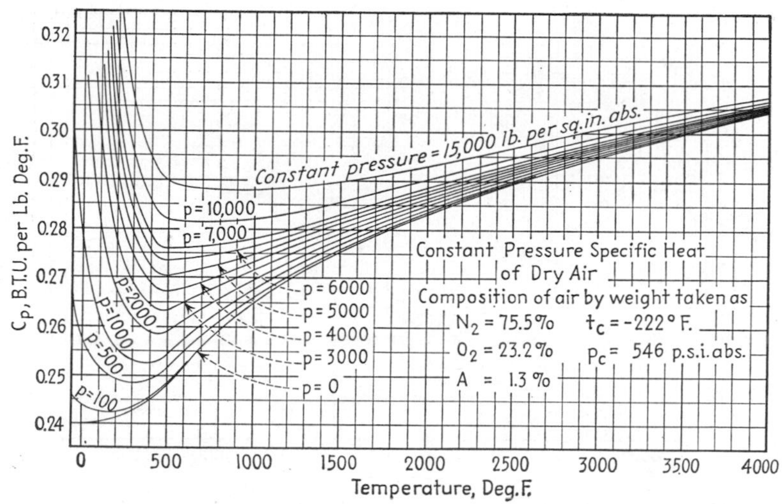

FIG. E-6.—The effect of temperature on C_p of air at various pressures. (*Ellenwood, Kulik, and Gay, The Specific Heats of Certain Gases over Wide Ranges of Pressures and Temperatures, Cornell Univ. Eng. Expt. Sta. Bull.* 30, 1942.)

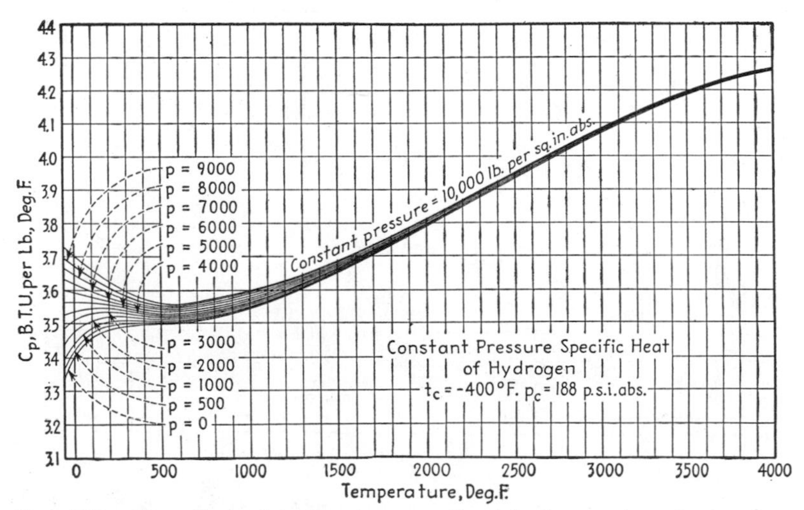

FIG. E-7.—The effect of temperature on C_p of hydrogen at various pressures. (*Ellenwood, Kulik, and Gay, The Specific Heats of Certain Gases over Wide Ranges of Pressures and Temperatures, Cornell Univ. Eng. Expt. Sta. Bull.* 30, 1942.)

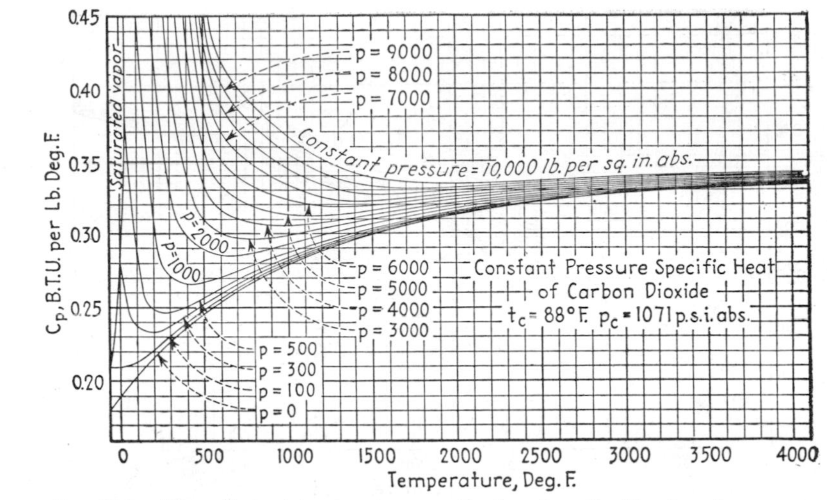

FIG. E-8.—The effect of temperature on C_p of carbon dioxide at various pressures. (*Ellenwood, Kulik, and Gay, The Specific Heats of Certain Gases over Wide Ranges of Pressures and Temperatures, Cornell Univ. Eng. Expt. Sta. Bull.* 30, 1942.)

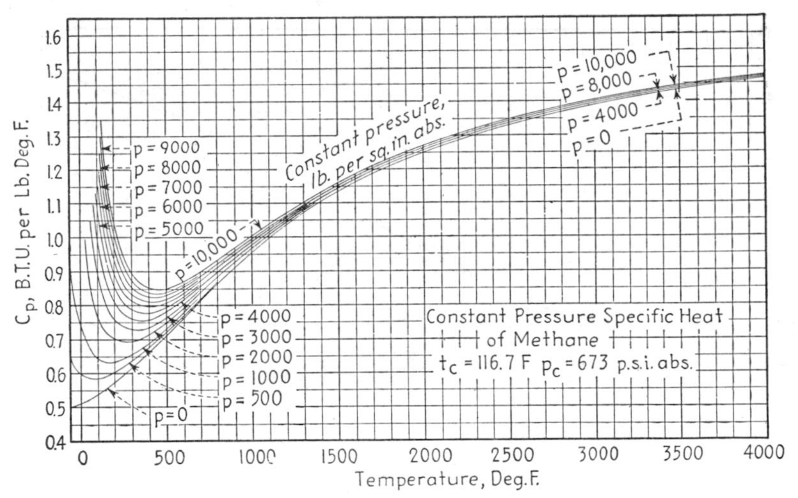

FIG. E-9.—The effect of temperature on C_p of methane at various pressures. (*Ellenwood, Kulik, and Gay, The Specific Heats of Certain Gases over Wide Ranges of Pressures and Temperatures, Cornell Univ. Eng. Expt. Sta. Bull.* 30, 1942.)

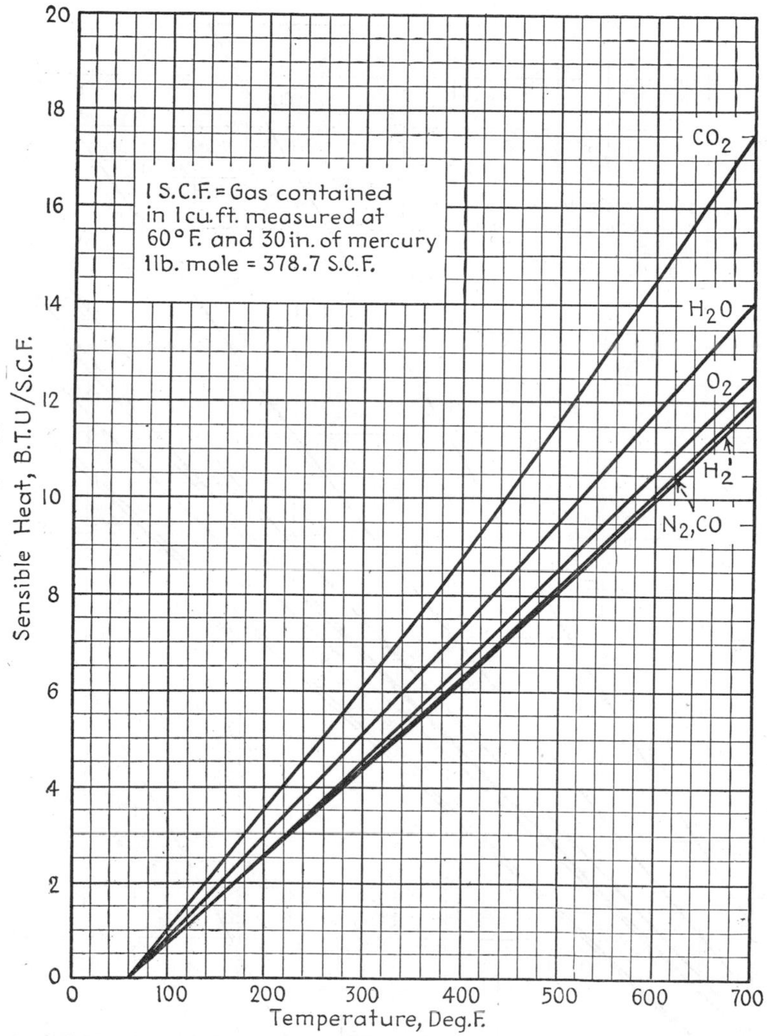

FIG. E-10.—Sensible heats above 60°F. Common gases at 1 atm, low range.

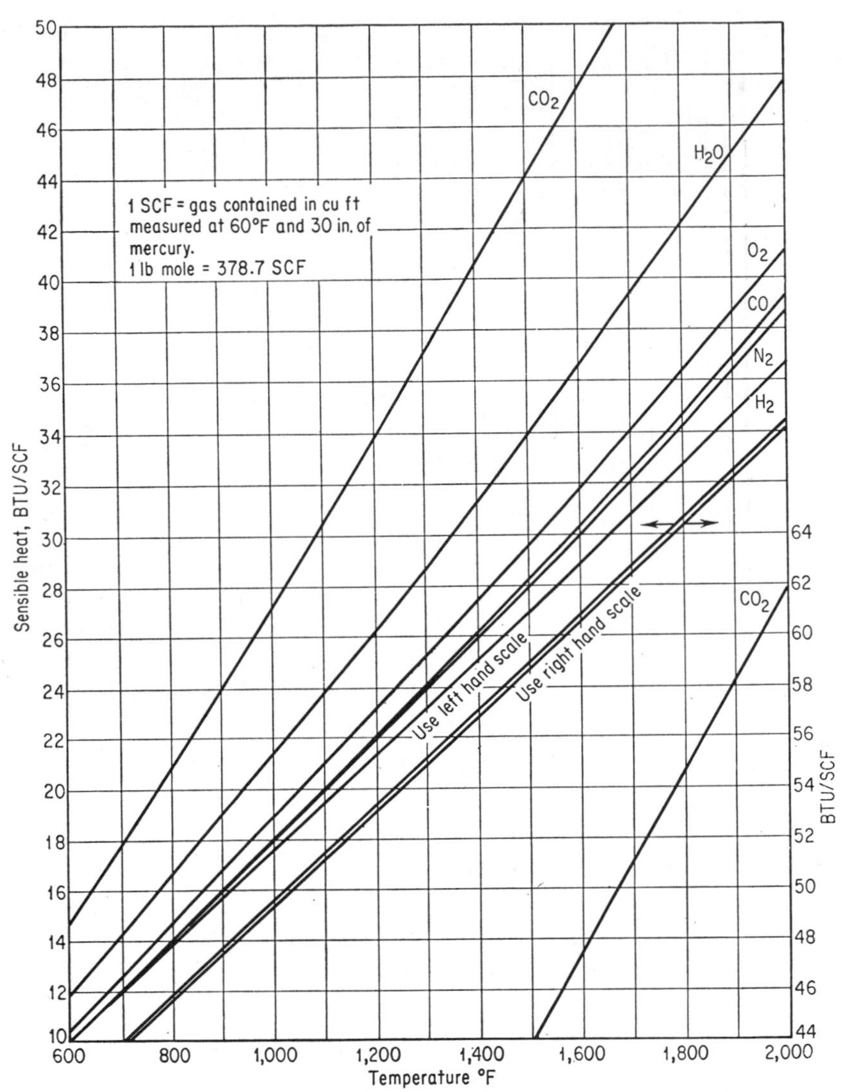

FIG. E-11.—Sensible heats above 60°F. Common gases at 1 atm, middle range.

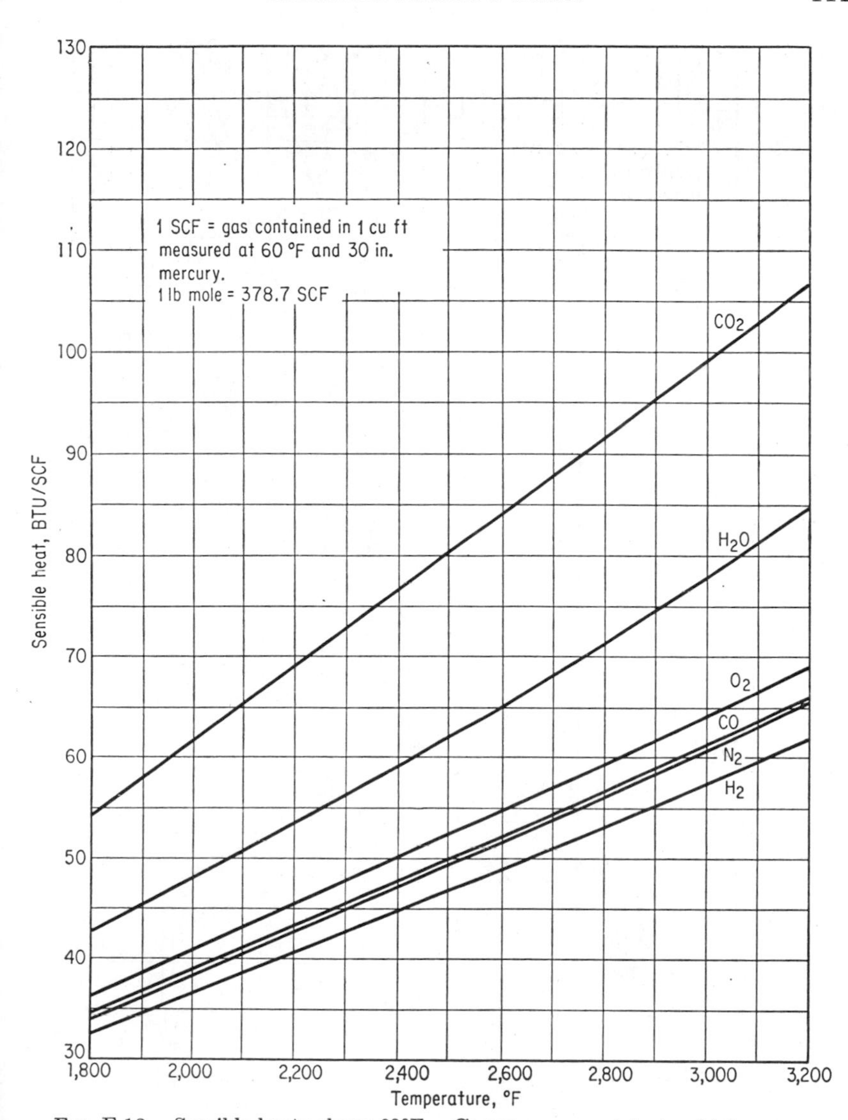

FIG. E-12.—Sensible heats above 60°F. Common gases at 1 atm, high range.

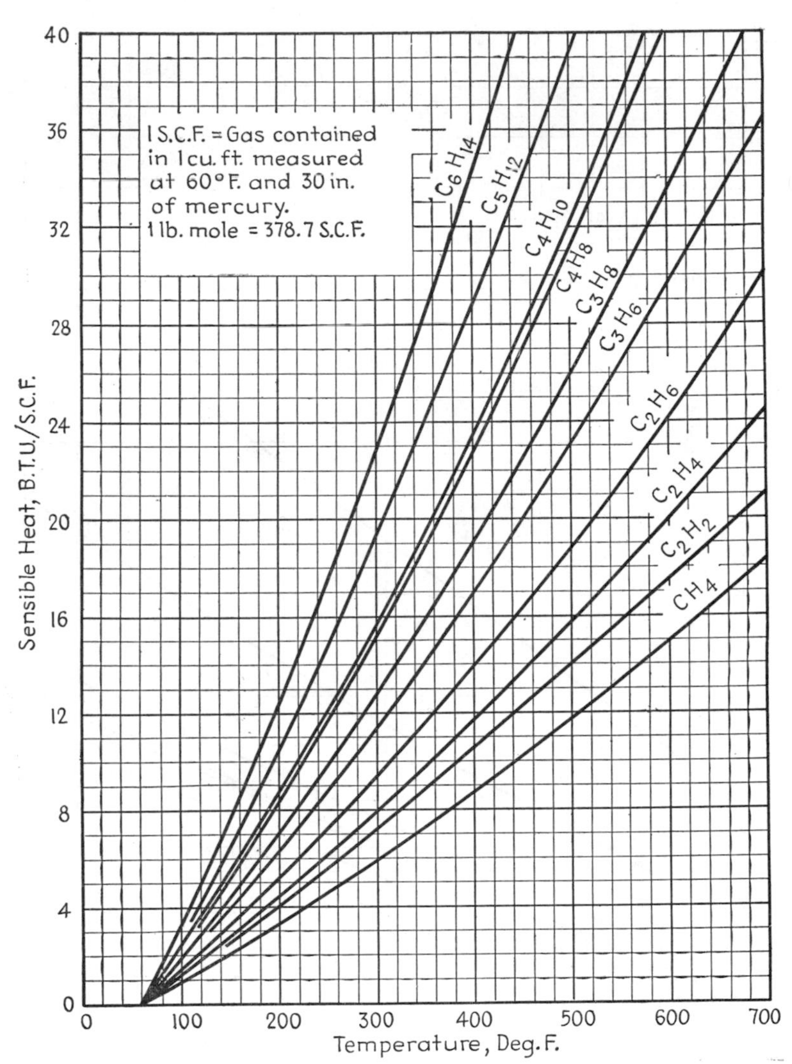

FIG. E-13.—Sensible heats above 60°F. Hydrocarbon gases at 1 atm.

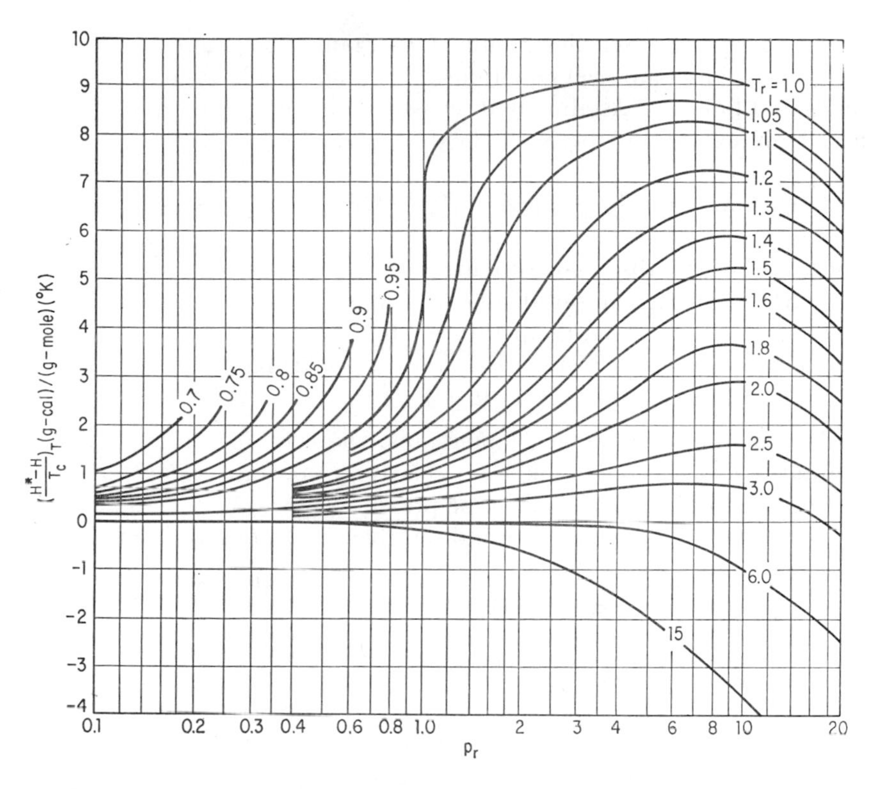

H^* = enthalpy of gas at low pressure
H = enthalpy of gas under pressure
T_c = reduced temperature of gas

FIG. E-14.—Enthalpy correction for gases (high range). (*Reprinted from Hougen and Watson, "Chemical Process Principles," Part 2, John Wiley & Sons, Inc., New York, 1947.*)

b. Specific and Sensible Heats of Inorganic Solids and Liquids.—Table E-5 gives empirical equations for the specific heats of inorganic solids and liquids. As already discussed under gases, these equations may be integrated to obtain the heat effects.

$$Q = \Delta H = a(T_2 - T_1) + \frac{b}{2}(T_2^2 - T_1^2) - d\left(\frac{1}{T_1} - \frac{1}{T_2}\right) \quad \text{(E-9}a\text{)}$$

where Q = heat absorbed, Btu/lb
ΔH = increase in sensible heat, Btu/lb
T_1, T_2 = initial, final temperature, °R
a, b, c, d = constants in Eq. (E-7) for C_p (see Table E-5)
C_p = specific heat, Btu/(lb)(°F)

Within the usual range of pressures, the specific heats and heat contents of solids and liquids are relatively constant, hence no correction need ordinarily be made for pressure effects.

Other data on specific heats and sensible heats are given as follows:

Sensible heats, inorganic solids............... Table E-4
Specific heats of organic liquids............... Fig. E-15
Specific heats of petroleum oils............... Table E-6
Sensible heats of petroleum liquids............ Table E-7
Specific heats of solutions..................... Tables E-8 to E-10
Miscellaneous materials....................... Table E-11
Metals.................................... Table D-2

TABLE E-4.—SENSIBLE HEATS ABOVE 60°F. INORGANIC SOLIDS
[Based on data of Kelley, *U.S. Bur. Mines Bull.* 371 (1934)]
Units, Btu/lb

	Temperature, °F												
	100	150	200	250	300	400	500	600	700	800	1000	1200	1500
Aluminum, Al........	8	18	30	42	52	75	98	122	148	178	236	286	
Cadmium, Cd........	2	5	8	10.5	13.5	19	25	31	61*	68*			
Calcium oxide, CaO..	7	15	25	35	46	66	87	110	123	158	208	268	
Copper, Cu..........	3.5	8	13	18	23	32	41	51	61	72	95	116	150
Carborundum, SiC...	7	16	26	36	46	67	89	114	139	167	220	278	365
Corundum, Al_2O_3.....	8	17	27	38	48	72	96	121	147	178	234	292	382
Graphite, C..........	8	18	29	40	52	79	107	138	171	215	283	370	506
Iron, Fe.............	4	10	16	21	27	39	51	65	80	97	130	168	224†
Lead, Pb............	1	2.5	4	5.5	7	10	13	16	30*	33*			
Magnesium, Mg......	10	22	34	47	60	86	112	141	169	200	264	322	
Salt, NaCl...........	8	18	29	40	51	72	94	116	138	164	216		
Fused silica, SiO_2.....	7	16	26	36	46	67	89	114	139	167	219	278	365
Zinc, Zn.............	4	8	13	18	23	33	43	53	64	75			

* Fused.
† β-iron.

TABLE E-5.—SPECIFIC HEATS OF INORGANIC SOLIDS AND LIQUIDS INCLUDING MOLECULAR WEIGHTS AND LATENT HEATS OF PHASE CHANGES
[Based on data of Kelley, *U.S. Bur. Mines Bull.* 371 (1934)]

$$c_p = a + bT - \frac{d}{T^2}, \text{ Btu/(lb)(°F)}$$

where a, b, d = constants given in this table
T = absolute temperature, °R

Substance*	Molecular weight	a	b	d	% error†	Absolute temp. range, °R	Enthalpy increase with phase change, Btu/lb
Aluminum:							
Al(*s*)	26.97	0.172	0.000066	......	1	490–1677	170 (fusion)
Al(*l*)	26.97	0.260	....	......	5	1677–3550	
Al_2O_3(*s*)	101.94	0.2166	0.0000489	16,600	3	490–2300	
$AlCl_3$(*s*)	133.25	0.0993	0.000117	......	3	490–838	
Mullite ($Al_6Si_2O_{13}$)	425.94	0.140	0.000874	......	5	490–1040	
Antimony:							
Sb(*s*)	121.76	0.0452	0.0000081	......	2	490–1626	70.5 (fusion)
Sb(*l*)	121.76	0.059	....	......	5	1626–2300	
Arsenic:							
As(*s*)	74.91	0.0690	0.0000174	......	5	490–2100	
As_2O_3(*s*)	197.82	0.042	0.000136	......	5	490–990	
Barium, $BaSO_4$(*s*)	233.42	0.09215	0.0000336	......	5	490–2380	
Beryllium, Be(*s*)	9.02	0.5208	0.000958	43,500	1	490–2100	
Bismuth, Bi(*s*)	209.00	0.0257	0.0000069	......	3	490–979	
Boron, B_2O_3(glass)	69.64	0.0738	0.000255	......	*ca.* 3	490–923	
Cadmium:							
Cd(*s*)	112.41	0.0486	0.000012	......	1	490–1069	23.4 (fusion)
Cd(*l*)	112.41	0.0634	....	......	5	1069–1750	
Calcium:							
CaO(*s*)	56.08	0.1783	0.000048	6,200	2	490–2100	
$CaCO_3$(*s*)	100.09	0.1966	0.000066	10,000	3	490–1860	
$CaSO_4$(*s*)	136.14	0.1360	0.0000897	3,730	5	490–2470	
$CaCl_2$(*s*)	110.99	0.152	0.000019	......	....	490–1900	
$Ca(OH)_2$(*s*)	74.10	0.289	....	......	....	500–670	
Carbon, C (graphite)	12.01	0.226	0.0001211	31,500	2	490–2470	
Chromium:							
Cr(*s*)	52.01	0.093	0.0000315	......	5	490–3281	136 (fusion)
Cr(*l*)	52.01	0.19	....	......	10	3281–3460	
Cobalt:							
Co(*s*)	58.94	0.0869	0.0000314	......	5	490–3173	112 (fusion)
Copper:							
Cu(*s*)	63.57	0.0856	0.0000128	......	1	490–2443	88 (fusion)
Cu(*l*)	63.57	0.118	....	......	3	2443–2830	
CuO(*s*)	79.57	0.1366	0.000025	6,130	2	490–1460	
$CuSO_4$(*s*)	159.63	0.151	....	......	....	508	
$CuSO_45H_2O$(*s*)	249.71	0.269	....	......	....	508	
Gold, Au(*s*)	197.2	0.00284	0.00000405	......	2	490–2460	27.6 (fusion)
Hydrogen, H_2O(*l*)	18.016	1.001	....	......	0.5	492–312	970 (vaporization)
Iron:							
Fe(α)	55.84	0.0740	0.0000634	......	3	490–1874	11.0 (α-β transition)
Fe(β)	55.84	0.1096	0.0000334	......	3	1874–2122	11.6 (β-γ transition)

* (*s*) denotes solid state, (*l*) liquid. (α), (β), etc., denote different crystalline forms.
† If no percentage of error given, values are estimated or are of uncertain accuracy.

TABLE E-5.—SPECIFIC HEATS OF INORGANIC SOLIDS AND LIQUIDS INCLUDING MOLECULAR WEIGHTS AND LATENT HEATS OF PHASE CHANGES (*Continued*)

Substance*	Molecular weight	*a*	*b*	*d*	% error†	Absolute temp. range, °R	Enthalpy increase with phase change, Btu/lb
Iron—(Cont.)							
$Fe(\gamma)$	55.84	0.150	....	......	5	2122–3013	6.4 (γ-δ transition)
$Fe(\delta)$	55.84	0.18	....	......	5	3013–3245	113 (fusion)
$Fe_2O_3(s)$	159.68	0.1550	0.0000558	8,600	2	490–1970	
$Fe_3O_4(s)$	231.52	0.1778	0.0000452	13,700	2	490–1920	
$FeS_2(s)$	119.96	0.089	0.000062	......	....	490–1390	
$FeSO_4(s)$	151.90	0.145	....	......	....	527–670	
$FeSO_47H_2O(s)$	278.01	0.35	....	......	....	524–574	
Lead:							
$Pb(s)$	207.21	0.0278	0.0000054	......	2	490–1081	10.6 (fusion)
$Pb(l)$	207.21	0.033	....	......	5	1081–2290	
$PbO(s)$	223.21	0.0463	0.0000079	......	2	490–980	
$PbCl_2(s)$	278.12	0.0571	0.0000161	......	2	490–1388	36.5 (fusion)
$PbO_2(s)$	239.21	0.053	0.000018	......	....	490	
Magnesium:							
$Mg(s)$	24.32	0.255	0.0000304	9,030	1	490–1661	86 (fusion)
$Mg(l)$	24.32	0.30	....	......	10	1661–1890	
$MgO(s)$	40.32	0.2693	0.0000165	16,800	2	490–3730	
$MgCl_2(s)$	95.23	0.182	0.000023	......	....	490–1780	
$MgCl_26H_2O(s)$	203.33	0.38	....	......	....	526–616	
$Mg(OH)_2(s)$	58.34	0.48	....	......	....	520–575	
$MgSO_4(s)$	120.38	0.222	....	......	....	530–670	
Manganese:							
$Mn(\alpha)$	54.93	0.0685	0.0000755	......	5	490–1994	2.2 (α-β transition)
$Mn(\beta)$	54.93	0.0921	0.0000400	......	5	1994–2371	3.9 (β-γ transition)
$Mn(\gamma)$	54.93	0.087	0.00427	......	5	2371–2687	113 (fusion)
$Mn(l)$	54.93	0.20	....	......	10	2687–3000	
$MnSO_4(s)$	150.99	0.182	....	......	....	527–670	
Mercury, $Hg(l)$	200.61	0.0329	....	......	1	490–1134	
Nickel:							
$Ni(\alpha)$	58.69	0.0726	0.0000605	......	2	490–1127	2.8 (α-β transition)
$Ni(\beta)$	58.69	0.119	0.000086	......	5	1127–3105	129 (fusion)
$Ni(l)$	58.69	0.146	....	......	10	3105–3430	
$NiSO_4(s)$	154.74	0.216	....	......	....	527–670	
Nitrogen:							
$NH_4Cl(\alpha)$	53.50	0.183	0.00038	......	5	490–824	32 (α-β transition)
$NH_4Cl(\beta)$	53.50	0.093	0.00035	......	5	824–941	
$NH_4Br(s)$	97.96	0.232	....	......	....	490–590	
$NH_4I(s)$	144.96	0.123	....	......	....	490–590	
$NH_4NO_3(s)$	85.05	0.37	....	......	....	490–527	
$(NH_4)_2SO_4(s)$	132.14	0.39	....	......	....	495–950	
Phosphorous:							
P(yellow)	30.98	0.177	....	......	5	490–571	8.5 (fusion)
$P(l)$	30.98	0.21	....	......	10	571–672	
$P_4O_{10}(s)$	284.08	0.0554	0.000212	......	2	490–1136	112 (sublimation)
Platinum, $Pt(s)$	195.23	0.0303	0.0000033	......	1	490–3370	

* (*s*) denotes solid state, (*l*) liquid. (α), (β), etc., denote different crystalline forms.
† If no percentage of error given, values are estimated or are of uncertain accuracy.

TABLE E-5.—SPECIFIC HEATS OF INORGANIC SOLIDS AND LIQUIDS INCLUDING MOLECULAR WEIGHTS AND LATENT HEATS OF PHASE CHANGES (*Concluded*)

Substance*	Molecular weight	a	b	d	% error†	Absolute temp. range, °R	Enthalpy increase with phase change, Btu/lb
Potassium:							
$KF(s)$	58.10	0.186	0.000027	......	....	490–2030	
$KCl(s)$	74.55	0.1466	0.000028	......	2	490–1877	154 (fusion)
$KBr(s)$	119.01	0.0965	0.0000168	......	2	490–977	
$KI(s)$	166.02	0.073	0.0000065	......	....	490–1720	
$KNO_3(\alpha)$	101.10	0.064	0.0003	......	10	490–722	25 (α-β transition)
$KNO_3(\beta)$	101.10	0.285	....	......	5	722–1010	51 (fusion)
$KNO_3(l)$	101.10	0.292	....	......	10	1010–1150	
$K_2CO_3(s)$	138.20	0.216	....	......	....	530–670	
$K_2SO_4(s)$	174 25	0.190	....	......	....	520–670	
Silicon:							
Glass (SiO_2)	60.06	0.213	0.0000413	16,300	3½	490–3550	
Quartz(α)(SiO_2)	60.06	0.1810	0.0000806	13,000	1	490–1526	
Silver:							
$Ag(s)$	107.88	0.0519	0.0000077	......	1	490–2221	45 (fusion)
$Ag(l)$	107.88	0.076	....	......	3	2221–2830	
$AgCl(s)$	143.34	0.067	0.000036	......	2	490–1310	40 (fusion)
$AgNO_3(\alpha)$	169.89	0.1108	0.0000524	......	2	490–779	
Sodium:							
$Na(s)$	23.00	0.218	0.000130	......	1½	490–668	49 (fusion)
$Na(l)$	23.00	0.326	....	......	2	668–810	
$NaF(s)$	42.00	0.248	0.000038	......	....	490–2270	
$NaCl(s)$	58 45	0.185	0.000040	......	2	490–1831	222 (fusion)
$NaCl(l)$	58 45	0.272	....	......	3	1831–2170	
$NaI(s)$	149.92	0.083	0.000006	......	....	490–1680	
$NaClO_3(s)$	106.45	0.089	0.000244	......	3	490–950	90 (fusion)
$NaNO_3(s)$	85.01	0.0536	0.00038	......	5	490–1049	80 (fusion)
$NaNO_3(l)$	85.01	0.438	....	......	10	1049–1265	
$Na_2SO_4(s)$	142.05	0.232	....	......	....	520–670	
Sulphur:							
S(rhombic)	32.06	0.113	0.000111	......	3	490–663	4.8 (transition: rhombic to monoclinic)
S(monoclinic)	32.06	0.137	0.000076	......	3	663–706	16 6 (fusion)
Tin:							
$Sn(s)$	118.70	0.0426	0.0000235	......	2	490–974	26 (fusion)
$Sn(l)$	118.70	0.056	....	......	10	974–2290	
Titanium:							
$TiO_2(s)$	79.90	0.148	0.0000525	1,700	3	490–1280	
Zinc:							
$Zn(s)$	65.38	0.0804	0.000023	......	1	490–1247	44 (fusion)
$Zn(l)$	65.38	0.116	0.0000047	......	3	1247–2020	
$ZnO(s)$	81.38	0.140	0.0000099	7,260	1	490–2830	
$ZnSO_4(s)$	161.44	0.173	....	......	....	525–670	

* (s) denotes solid state, (l) liquid. (α), (β), etc., denote different crystalline forms.

† If no percentage of error given, values are estimated or are of uncertain accuracy.

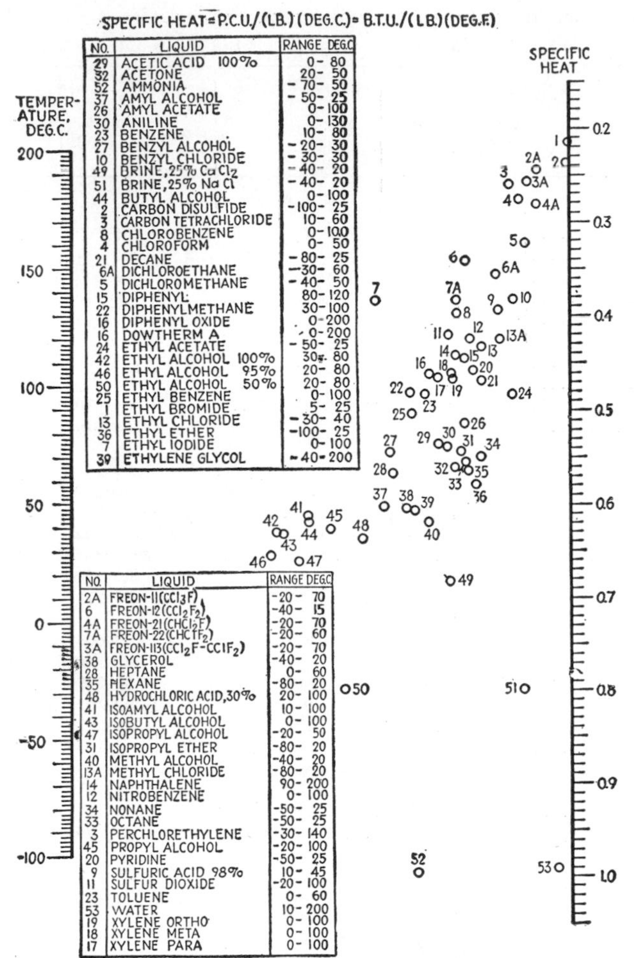

NO.	LIQUID	RANGE DEG.C
29	ACETIC ACID 100%	0- 80
32	ACETONE	20- 50
52	AMMONIA	-70- 50
37	AMYL ALCOHOL	-50- 25
26	AMYL ACETATE	0-100
30	ANILINE	0-130
23	BENZENE	10- 80
27	BENZYL ALCOHOL	-20- 30
10	BENZYL CHLORIDE	-30- 30
49	BRINE, 25% $CaCl_2$	-40- 20
51	BRINE, 25% NaCl	-40- 20
44	BUTYL ALCOHOL	0-100
2	CARBON DISULFIDE	-100- 25
3	CARBON TETRACHLORIDE	10- 60
8	CHLOROBENZENE	0-100
4	CHLOROFORM	0- 50
21	DECANE	-80- 25
6A	DICHLOROETHANE	-30- 60
5	DICHLOROMETHANE	-40- 50
15	DIPHENYL	80-120
22	DIPHENYLMETHANE	30-100
16	DIPHENYL OXIDE	0-200
16	DOWTHERM A	0-200
24	ETHYL ACETATE	-50- 25
42	ETHYL ALCOHOL 100%	30- 80
46	ETHYL ALCOHOL 95%	20- 80
50	ETHYL ALCOHOL 50%	20- 80
25	ETHYL BENZENE	0-100
1	ETHYL BROMIDE	5- 25
13	ETHYL CHLORIDE	-30- 40
36	ETHYL ETHER	-100- 25
7	ETHYL IODIDE	0-100
39	ETHYLENE GLYCOL	-40-200

NO.	LIQUID	RANGE DEG.C
2A	FREON-11 (CCl_3F)	-20- 70
6	FREON-12 (CCl_2F_2)	-40- 15
4A	FREON-21 ($CHCl_2F$)	-20- 70
7A	FREON-22 ($CHClF_2$)	-20- 60
3A	FREON-113 (CCl_2F-$CClF_2$)	-20- 70
38	GLYCEROL	-40- 20
28	HEPTANE	0- 60
35	HEXANE	-80- 20
48	HYDROCHLORIC ACID, 30%	20- 100
41	ISOAMYL ALCOHOL	10- 100
43	ISOBUTYL ALCOHOL	0- 100
47	ISOPROPYL ALCOHOL	-20- 50
31	ISOPROPYL ETHER	-80- 20
40	METHYL ALCOHOL	-40- 20
13A	METHYL CHLORIDE	-80- 20
14	NAPHTHALENE	90- 200
12	NITROBENZENE	0- 100
34	NONANE	-50- 25
33	OCTANE	-50- 25
3	PERCHLORETHYLENE	-30- 140
45	PROPYL ALCOHOL	-20- 100
20	PYRIDINE	-50- 25
9	SULFURIC ACID 98%	10- 45
11	SULFUR DIOXIDE	-20- 100
23	TOLUENE	0- 60
53	WATER	10- 200
19	XYLENE ORTHO	0- 100
18	XYLENE META	0- 100
17	XYLENE PARA	0- 100

FIG. E-15.—Specific heats of liquids. (*Chilton, Colburn, and Vernon, in Perry, "Chemical Engineers' Handbook," 3d ed., McGraw-Hill Book Company, Inc., New York,* 1950.)

TABLE E-6.—SPECIFIC HEAT OF PETROLEUM OILS OF VARIOUS GRAVITIES IN BTU/(LB)(°F) OR CAL/(G)(°C)

[Reprinted from *Nat. Bur. Standards Misc. Pub.* 97 (1933)]

Temp., °F	Degrees A.P.I. at 60°F							
	10	20	30	40	50	60	70	80
	Specific gravity at 60°/60°F							
	1.000	0.9340	0.8762	0.8251	0.7796	0.7389	0.7022	0.6690
0	0.388	0.401	0.415	0.427	0.439	0.451	0.463	0.474
20	0.397	0.411	0.424	0.437	0.450	0.462	0.474	0.485
40	0.406	0.420	0.434	0.447	0.460	0.472	0.485	0.496
60	0.415	0.429	0.443	0.457	0.470	0.483	0.495	0.507
80	0.424	0.439	0.453	0.467	0.480	0.493	0.506	0.518
100	0.433	0.448	0.463	0.477	0.490	0.504	0.517	0.529
120	0.442	0.457	0.472	0.487	0.501	0.514	0.527	0.540
140	0.451	0.467	0.482	0.497	0.511	0.525	0.538	0.551
160	0.460	0.476	0.491	0.506	0.521	0.535	0.549	0.562
180	0.469	0.485	0.501	0.516	0.531	0.546	0.560	0.573
200	0.478	0.495	0.511	0.526	0.541	0.556	0.570	0.584
220	0.487	0.504	0.520	0.536	0.552	0.567	0.581	
240	0.496	0.513	0.530	0.546	0.562	0.577	0.592	
260	0.505	0.523	0.540	0.556	0.572	0.588	0.603	
280	0.514	0.532	0.549	0.566	0.582	0.598	0.613	
300	0.523	0.541	0.559	0.576	0.592	0.609	0.624	
320	0.532	0.550	0.568	0.586	0.603	0.619		
340	0.541	0.560	0.578	0.596	0.613	0.629		
360	0.550	0.569	0.588	0.606	0.623	0.640		
380	0.559	0.578	0.597	0.615	0.633	0.650		
400	0.568	0.588	0.607	0.625	0.643	0.661		
420	0.577	0.597	0.616	0.635	0.653			
440	0.586	0.606	0.626	0.645	0.664			
460	0.595	0.616	0.636	0.655	0.674			
480	0.604	0.625	0.645	0.665	0.684			
500	0.613	0.634	0.655	0.675	0.694			
520	0.622	0.644	0.665	0.685				
540	0.631	0.653	0.674	0.695				
560	0.640	0.662	0.684	0.705				
580	0.649	0.672	0.693	0.715				
600	0.658	0.681	0.703	0.724				
620	0.667	0.690	0.713					
640	0.676	0.699	0.722					
660	0.685	0.709	0.732					
680	0.694	0.718	0.741					
700	0.703	0.727	0.751					
720	0.712	0.737	0.761					
740	0.721	0.746	0.770					
760	0.730	0.755	0.780					
780	0.739	0.765	0.790					
800	0.748	0.774	0.799					

TABLE E-7.—HEAT CONTENT OF PETROLEUM LIQUIDS OF VARIOUS GRAVITIES IN BTU/GAL* ABOVE 32°F

[*Nat. Bur. Standards Misc. Pub.* 97 (1933)]

Temp., °F	Degrees A.P.I. at 60°F							
	10	20	30	40	50	60	70	80
	Specific gravity at 60°/60°F							
	1.000	0.9340	0.8762	0.8251	0.7796	0.7389	0.7022	0.6690
0	−105	−102	−99	−96	−93	−91	−88	−86
10	−73	−70	−68	−66	−64	−63	−61	−60
20	−40	−39	−37	−36	−35	−34	−34	−33
40	+27	+26	+25	+24	+24	+23	+23	+22
50	61	59	57	55	54	52	51	50
60	95	92	89	86	84	82	80	78
70	130	126	122	118	115	112	109	106
80	165	160	155	150	146	142	138	135
90	201	194	188	182	177	173	168	164
100	237	229	222	215	209	204	198	194
110	273	264	256	248	241	235	229	223
120	310	300	290	281	273	267	260	253
130	347	335	325	315	306	299	291	284
140	384	371	360	349	339	331	322	314
150	422	408	395	383	372	363	354	345
160	460	445	431	418	406	396	386	376
170	499	482	467	453	440	429	418	408
180	538	520	503	488	475	462	451	440
190	577	558	540	524	509	496	484	472
200	617	596	577	560	544	530	517	504
210	657	635	615	596	580	564	550	537
220	697	674	652	633	615	599	584	570
230	738	713	691	670	651	634	618	603
240	779	753	729	707	688	669	653	637
250	820	793	768	745	724	705	688	671
260	862	833	807	783	761	741	723	705
270	904	874	847	822	799	778	758	740
280	947	915	887	861	836	814	794	775
290	990	957	927	900	874	851	830	810
300	1034	999	968	939	913	889	866	846
310	1078	1041	1009	979	952	926	903	881
320	1122	1084	1050	1019	991	964	940	917
330	1166	1127	1092	1059	1030	1002	977	954
340	1211	1170	1134	1100	1070	1041	1015	981
350	1256	1214	1176	1141	1110	1080	1053	1028
360	1302	1258	1219	1183	1150	1119	1091	1065
370	1348	1303	1262	1225	1190	1159	1130	1103
380	1395	1348	1306	1267	1231	1199	1169	1141
390	1441	1393	1349	1309	1273	1239	1208	1179

* The unit used here is 1 gal of oil at 60°F.

TABLE E-7.—HEAT CONTENT OF PETROLEUM LIQUIDS OF VARIOUS GRAVITIES IN BTU/GAL* ABOVE 32°F (*Concluded*)

Temp. °F	Degrees A.P.I. at 60°F							
	10	20	30	40	50	60	70	80
	Specific gravity at 60°/60°F							
	1.000	0.9340	0.8762	0.8251	0.7796	0.7389	0.7022	0.6690
400	1489	1439	1393	1352	1314	1280	1247	1217
410	1536	1485	1438	1395				
420	1584	1531	1483	1439				
430	1632	1578	1528	1483				
440	1681	1625	1573	1527				
450	1730	1672	1619	1571				
460	1779	1720	1666	1616				
470	1829	1768	1712	1661				
480	1879	1816	1759	1707				
490	1930	1865	1806	1753				
500	1981	1914	1854	1799				
510	2032	1964	1902	1846				
520	2084	2014	1950	1893				
530	2136	2064	1999	1940				
540	2188	2115	2048	1988				
550	2241	2166	2097	2036				
560	2294	2217	2147	2084				
570	2348	2269	2197	2133				
580	2402	2321	2248	2182				
590	2456	2373	2299	2231				
600	2511	2426	2350	2281				
610	2566	2479	2402	2331				
620	2621	2533	2454	2381				
630	2677	2587	2506	2432				
640	2733	2641	2558	2483				
650	2789	2696	2611	2534				
660	2846	2751	2665	2586				
670	2903	2806	2718	2638				
680	2961	2862	2772	2690				
690	3019	2918	2826	2743				
700	3078	2974	2881	2796				
710	3137	3031	2936	2849				
720	3196	3088	2991	2903				
730	3255	3146	3047	2957				
740	3315	3204	3103	3011				
750	3376	3262	3159	3066				
760	3436	3321	3216	3121				
770	3497	3380	3273	3177				
780	3559	3440	3331	3232				
790	3621	3499	3389	3289				
800	3683	3559	3447	3345				

* The unit used here is 1 gal of oil at 60°F.

c. Specific Heats of Solutions.—Table E-10 gives data on specific heats of water solutions of a few inorganic compounds. Tables E-8 and E-9 give specific heats of aqueous methyl and ethyl alcohol, respectively.

TABLE E-8.—SPECIFIC HEAT OF METHYL ALCOHOL–WATER MIXTURES
BTU/(LB)(°F)
(Based on data from "International Critical Tables")

Weight % alcohol	40°F	60°F	80°F	100°F
0	1.005	1.000	0.999	0.999
10	1.025	1.005	1.00	0.995
20	0.97	0.98	0.98	0.98
30	0.92	0.945	0.95	0.95
40	0.87	0.91	0.915	0.92
50	0.82	0.855	0.87	0.975
60	0.77	0.805	0.82	0.83
70	0.72	0.75	0.77	0.78
80	0.67	0.70	0.715	0.725
90	0.62	0.645	0.665	0.675
100	0.57	0.595	0.615	0.625

TABLE E-9.—SPECIFIC HEAT OF ETHYL ALCOHOL–WATER MIXTURES,
BTU/(LB)(°F)
(Based on data from "International Critical Tables")

Weight % alcohol	40°F	60°F	80°F	100°F
0	1.005	1.000	0.999	0.999
10	1.04	1.02	1.02	1.02
20	1.04	1.04	1.04	1.04
30	1.00	1.01	1.01	1.01
40	0.94	0.96	0.97	0.97
50	0.87	0.90	0.92	0.93
60	0.81	0.84	0.86	0.87
70	0.74	0.77	0.80	0.81
80	0.62	0.71	0.73	0.74
90	0.61	0.64	0.66	0.68
100	0.54	0.56	0.58	0.62

TABLE E-10.—SPECIFIC HEATS OF SOME WATER SOLUTIONS
(Data given in equation form based on data from "International Critical Tables")
P denotes weight %; T, absolute temperature, °R

Solution of	Specific heat, Btu/(lb)(°F)	Range	Accuracy, %
Hydrochloric acid, HCl	$C_p = 1 - 35\sqrt{P}/T$	40–140°F, 16–42%	4
Salt, NaCl	$C_p = 1 - 0.008P$	40–140°F, 0–25%	2
Ammonia, NH_3	$C_p = 1.0$	40–140°F, 0–40%	6
Sodium hydroxide, NaOH	$C_p = 1 - 0.032\sqrt{P}$	68°F, 0–57%	2
Potassium hydroxide, KOH	$C_p = 1 - 0.0115P$	66°F, 0–24%	2
Potassium chloride, KCl	$C_p = 1 - 0.0113P$	40–110°F, 0–31%	2

d. Specific Heats of Miscellaneous Materials

TABLE E-11.—SPECIFIC HEATS OF MISCELLANEOUS MATERIALS
(Abstracted from Perry, "Chemical Engineers' Handbook," 3d ed., McGraw-Hill Book Company, Inc., New York, 1950)

Material	Specific Heat, Btu/(lb)(°F)
Alumina	0.2(212°F); 0.274(2732°F)
Asbestos	0.25
Asphalt	0.22
Bakelite	0.3–0.4
Cellulose	0.32
Cement, portland clinker	0.186
Charcoal (wood)	0.242
Chrome brick	0.17
Clay	0.224
Coal	0.26–0.37
Coal tars	0.35(104°F); 0.45(392°F)
Coke	0.265(70–752°F)
	0.359(70–1472°F)
	0.403(70–2372°F)
Concrete	0.156(70–312°F); 0.219(72–1472°F)
Fire-clay brick	0.198(212°F); 0.298(2732°F)
Glass wool	0.157
Granite	0.20(68–212°F)
Limestone	0.217
Magnesite brick	0.222(212°F); 0.195(2732°F)
Sand	0.191
Steel	0.12
Wood (oak)	0.570
Most woods	0.45–0.65

3. Vapor Pressure

a. Vapor-pressure Charts.—Vapor pressures for a number of substances are summarized in Figs. E-16 to E-18. These charts, though in nomograph form, are based on the familiar Cox[1] chart, using water as the reference liquid. The accuracy is about the same as the Cox relation in the conventional line plot. The basic data are taken from numerous sources.[2]

It will be noted at once that the vapor pressures may be extrapolated by these charts to above critical temperatures. Such values are, of course, fictitious and entirely devoid of physical reality except to permit a doubtful application of Raoult's law to solution of gases above the critical temperature.

The reader may readily add points for vapor pressures of other compounds to the charts, provided vapor pressures at two or more temperatures are available. To do this, draw lines each connecting an observed vapor pressure at the corresponding temperature. The intersection of these lines is the point representing the vapor pressure of the substance. Although valid for most pure liquids, the chart is not necessarily applicable to all liquid mixtures or to solids; further, it is not applicable to a very few substances like hydrogen fluoride and nitrogen dioxide which exhibit varying degrees of polymerization at different temperatures and pressures.

b. Other Correlations for Vapor Pressures.—The Clausius-Clapeyron equation in its integrated form is the most widely used

[1] *Ind. Eng. Chem.*, **15**, 592 (1923).

[2] Lange, "Handbook of Chemistry," 9th ed., McGraw-Hill Book Company, Inc., New York, 1959.

Hodgman, "Handbook of Chemistry and Physics," Chemical Rubber Publishing Co., Cleveland, Ohio, 1940.

Perry, "Chemical Engineers' Handbook," 3d ed., McGraw-Hill Book Company, Inc., New York, 1950.

Laverty and Edmister, *Oil Gas J.*, 1938.

Matteson and Hanna, *Oil Gas J.*, May 21, 1942.

Wilson, Walker, Rinelli, and Mars, *Chem. Eng. News*, **21**, 1254 (1943).

Stearns and George, *Ind. Eng. Chem.*, **35**, 602 (1943).

Garner, Adams, and Stuchell, *Petroleum Refiner*, October, 1942.

Gillespie and Fraser, *J. Am. Chem. Soc.*, **58**, 2260 (1936).

Sage and Lacey, *Ind. Eng. Chem.*, **34**, 730 (1942).

Kay, *Ind. Eng. Chem.*, **33**, 590 (1941).

Meyers and Van Dusen, *Nat. Bur. Standards J. Research*, **10**, 381 (1933).

Nat. Bur. Standards Circ. 142 (1923).

Keenan and Keyes, "Thermodynamic Properties of Steam," John Wiley & Sons, Inc., New York, 1936.

method of correlating vapor pressures. This Eq. (E-10) assumes (1) that the latent heat of vaporization is constant, (2) that the vapor obeys the perfect gas law, and (3) that the volume of the liquid is much smaller than that of its vapor. These assumptions are all approximations.

$$\log p = \frac{A}{T} + B \qquad \text{(E-10)}$$

where p = vapor pressure, any convenient units
T = absolute temperature, °R or °K
A, B = empirical constants whose values depend on units of T and p

Values of A and B for a variety of substances are given in most handbooks and permit computation of vapor pressures with moderate accuracy, provided the pressure is not too high. An alternate form of this equation is

$$\log p_1 = \frac{\Delta H}{2.303R}\left(\frac{T_1 - T_2}{T_1 T_2}\right) + \log p_2 \qquad \text{(E-11)}$$

where p_1 = vapor pressure at T_1, any units such as psi
p_2 = vapor pressure at T_2, same units as p_1
T_1, T_2 = absolute temperatures, °R or °K
ΔH = heat of vaporization, any units desired such as Btu/lb mole (value of ΔH is approximately that at the mean temperature)
R = gas-law constant, units same as $\Delta H/T$, such as Btu/(lb mole)(°R) (see Table E-1)

This may be used for computing vapor pressure at one temperature from a known vapor pressure (such as the boiling point) and the heat of vaporization.

c. Vapor Pressure of Solutions.—Figures of the type of E-16 to E-18 are probably very good ways of presenting vapor-pressure data for solutions, but their use has not been employed to any extent. Data on the vapor pressures for aqua ammonia are given in Tables L-14 and L-15. The same chapter gives additional equilibrium data related to vapor pressures.

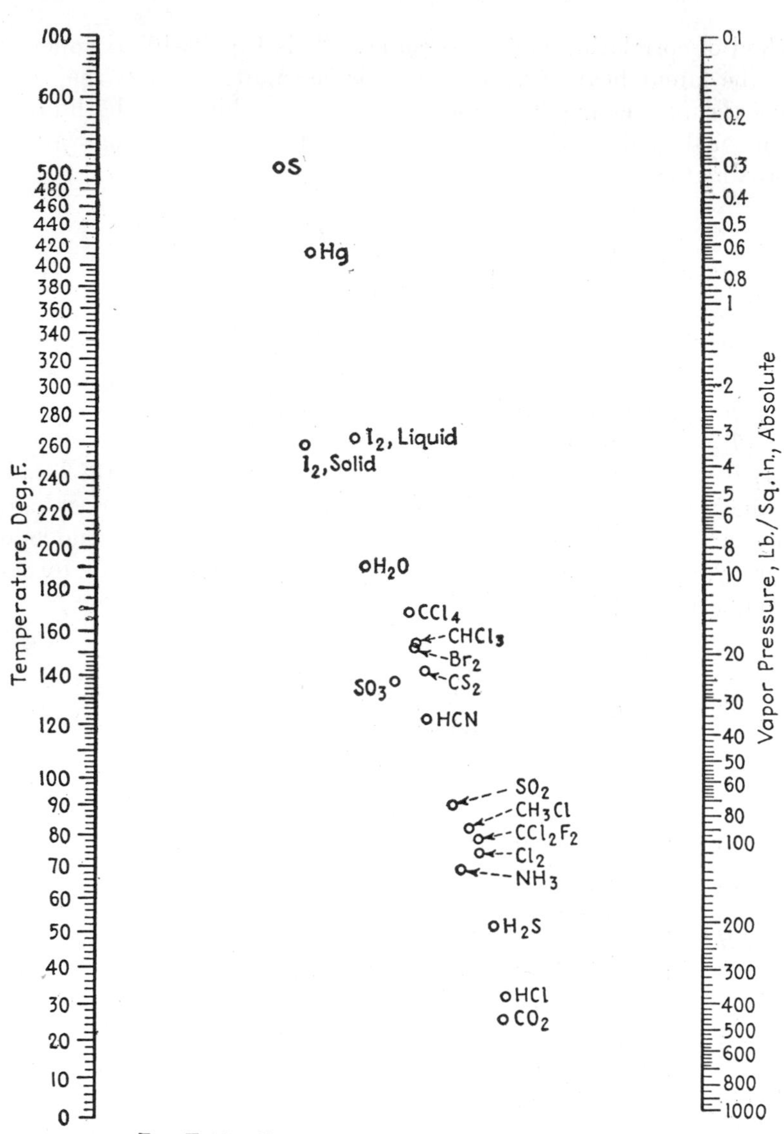

FIG. E-16.—Vapor pressures of inorganic compounds.

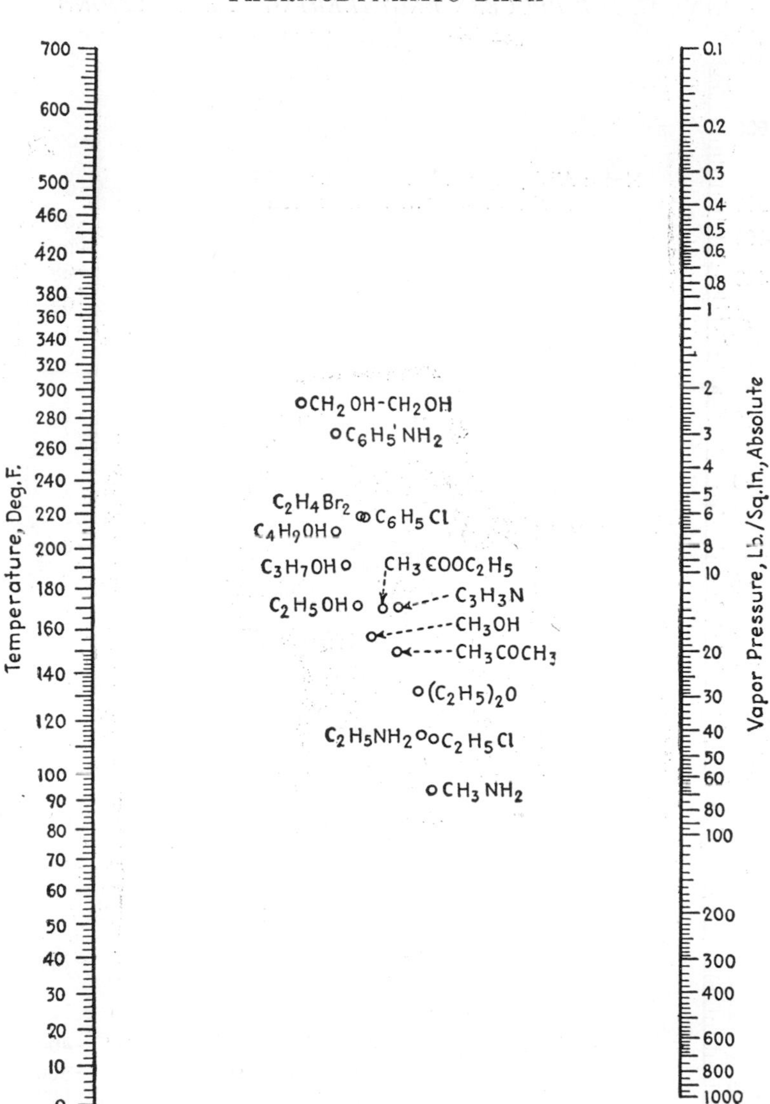

FIG. E-17.—Vapor pressures of organic compounds.

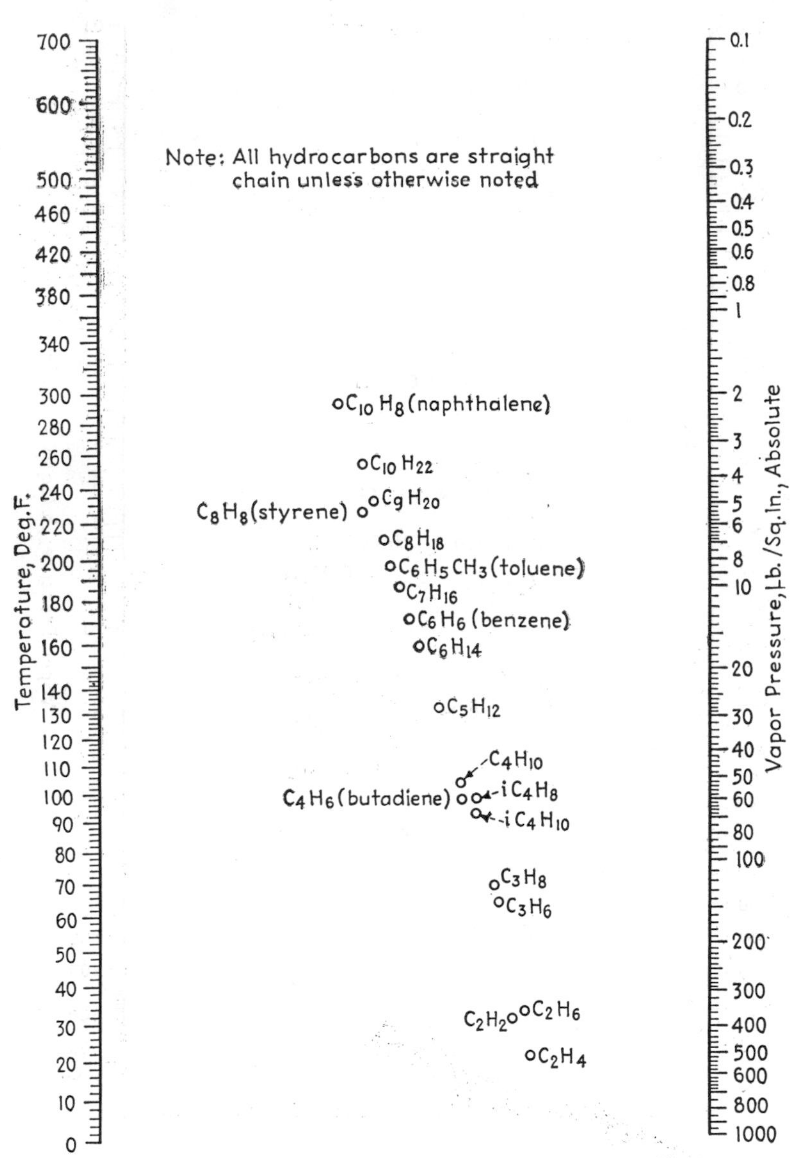

FIG. E-18.—Vapor pressures of hydrocarbons.

4. Latent Heats of Vaporization

a. Heat of Vaporization Chart.—Heats of vaporization for a number of substances are summarized in Fig. E-19 and the accompanying key Table E-12. The chart actually reads the molal entropy of vaporizations, but it is only necessary to multiply by the absolute temperature and divide by the molecular weight to secure the heat of vaporization, that is,

$$L = \frac{T\Delta S}{m} = \frac{\Delta H}{m} \tag{E-12}$$

where L = latent heat of vaporization, Btu/lb
T = absolute temperature, °R
ΔS = molal entropy of vaporization, Btu/(°R)(lb mole)
ΔH = molal heat of vaporization, Btu/lb mole
m = molecular weight

b. Estimation of Latent Heats of Vaporization.—The material following gives several methods for estimating latent heats. The choice of method depends on the type of data available and the accuracy required. Figure E-19 may be used for estimating latent heats provided the latent heat is known at two temperatures by constructing the point on the diagram in the same method as given for the vapor-pressure charts. The latent heat of any substance is of course zero at the critical temperature, and this will establish one of the two required values. If the critical pressures and temperatures are known, the latent heat may be estimated conveniently by use of Eq. (E-13) due to Watson.[1]

$$L = L_K \left(\frac{T_c - T}{T_c - T_K} \right)^{0.38} \tag{E-13}$$

where L = latent heat desired at temperature T (same units as L_K)
L_K = known latent heat at temperature, T_K
T, T_K = absolute temperatures, °R
T_c = critical temperature, °R

(T, T_K, T_c may also be in °K provided they are all in the same units.)

Watson also gives an equation for estimating the latent heat at the boiling point; however, Eq. (E-14), proposed by Pollara,[2] is more convenient and only slightly less accurate.

$$\Delta H_B = 4.577 \left(\frac{T_c T_B}{T_c - T_B} \right) \log p_c \tag{E-14}$$

where ΔH_B = latent heat at atmospheric boiling point, Btu/lb mole
T_c = critical temperature, °R
T_B = atmospheric boiling point, °R
p_c = critical pressure atmospheres

[1] *Ind. Eng. Chem.*, **35**, 398 (1943).
[2] *J. Phys. Chem.*, **46**, 1163 (1942).

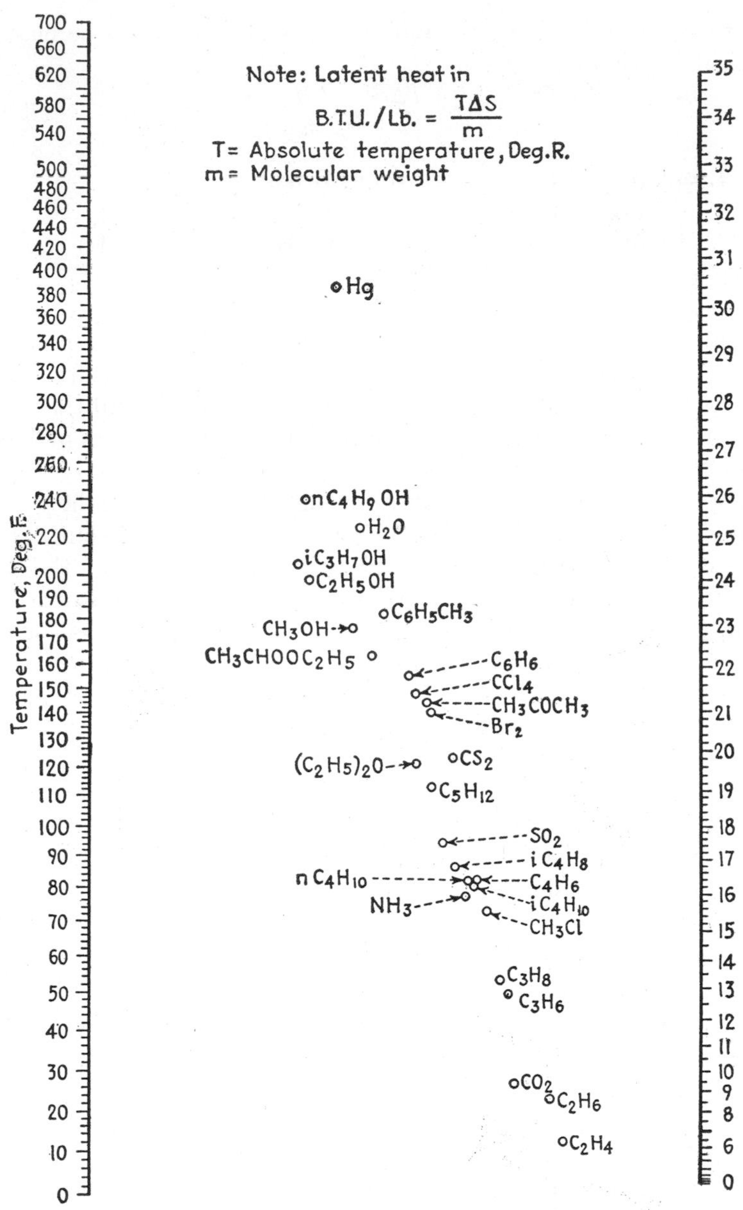

FIG. E-19.—Heats and entropies of vaporization.

TABLE E-12.—KEY FOR FIG. E-19. HEATS OF VAPORIZATION

Substance	Formula	Molecular weight	Principal references
Sulphur dioxide	SO_2	64.06	Perry, "Chemical Engineers' Handbook," 2d ed., 1941.
Isobutylene	$i\text{-}C_4H_8$	56.10	Garner, Adams, and Stuchell, *Petroleum Refiner*, October, 1942.
Normal butane	$n\text{-}C_4H_{10}$	58.10	Dana, Jenkins, Burdick, and Timm, *Refrig. Eng.*, **12,** No. 12, 403 (1926).
1-3, Butadiene	C_4H_6	54.09	Garner, Adams, and Stuchell, *Petroleum Refiner*, October, 1942.
Ammonia	NH_3	17.03	*Bur. Standards Circ.* 142 (1923).
Isobutane	$i\text{-}C_4H_{10}$	58.10	Dana, Jenkins, Burdick, and Timm, *Refrig. Eng.*, **12,** No. 12, 403 (1926).
Methyl chloride	CH_3Cl	50.48	Tanner, Banning, and Mathewson, *Ind. Eng. Chem.*, **31,** 878 (1939).
Propane	C_3H_8	44.09	Dana, Jenkins, Burdick, and Timm, *Refrig. Eng.*, **12,** No. 12, 403 (1926).
Propylene	C_3H_6	42.08	Private source.
Carbon dioxide	CO_2	44.01	*Am. Soc. Refrig. Eng., Circ.* 9 (1926).
Ethane	C_2H_6	30.07	Beall, *Refiner Natural Gasoline Mfr.*, **14,** No. 12, 588 (1935).
Ethylene	C_2H_4	28.05	Private source.
Mercury	Hg	200.61	Kelley, *U.S. Bur. Mines. Bull.* 383 (1935).
Normal butyl alcohol	C_4H_9OH	74.08	Hodgman, "Handbook of Chemistry and Physics," 1940.
Water	H_2O	18.016	Keenan and Keyes, "Thermodynamic Properties of Steam," 1936.
Isopropyl alcohol	$i\text{-}C_3H_7OH$	60.76	Hodgman, "Handbook of Chemistry and Physics," 1940.
Ethyl alcohol	C_2H_5OH	46.05	Fisk, Ginnings, and Holton, *Bur. Standards, J. Research*, **6,** No. 5, 895 (1931).
Toluene	$C_6H_5CH_3$	92.21	Doss, "Physical Properties of the Principal Hydrocarbons," 1940.
Methyl alcohol	CH_3OH	32.03	Hodgman, "Handbook of Chemistry and Physics," 1940.
Ethyl acetate	$CH_3CHOOC_2H_5$	88.06	Hodgman, "Handbook of Chemistry and Physics," 1940.
Benzene	C_6H_6	78.05	Doss, "Physical Properties of the Principal Hydrocarbons," 1940.
Carbon tetrachloride	CCl_4	153.83	Hodgman, "Handbook of Chemistry and Physics," 1940.
Acetone	CH_3COCH_3	58.05	Hodgman, "Handbook of Chemistry and Physics," 1940.
Bromine	Br_2	159.83	Hodgman, "Handbook of Chemistry and Physics," 1940.
Ethyl ether	$(C_2H_5)_2O$	74.08	*Am. Soc. Refrig. Eng., Circ.* 9 (1926).
Carbon disulphide	CS_2	76.12	Hodgman, "Handbook of Chemistry and Physics," 1940.
Normal pentane	C_5H_{12}	72.15	Sage and Lacey, *Ind. Eng. Chem.*, **34,** 730 (1942).

The latent heat of vaporization, at any temperature, may be estimated from the boiling point; the critical pressure and temperature by means of these Eqs. (E-13) and (E-14), or Eq. (E-15) which results from their combination. This equation is believed to be sufficiently accurate for many engineering calculations.

$$\Delta H = 4.577 \left(\frac{T_c T_B}{T_c - T_B} \right) \left(\frac{T_c - T}{T_c - T_B} \right)^{0.38} \log p_c \qquad \text{(E-15)}$$

where ΔH = latent heat (at temperature T), Btu/lb mole
T_c = critical temperature, °R
T_B = boiling point, °R
T = temperature at which latent heat is required, °R
p_c = critical pressure atmospheres

The latent heat may be estimated from vapor-pressure data alone by the Clausius-Clapeyron Eq. (E-11) given under Vapor Pressure. Repeating this in altered form,

$$\Delta H = \frac{2.303 R T_1 T_2}{T_1 - T_2} \log \frac{p_1}{p_2} \qquad \text{(E-11}a\text{)}$$

where ΔH = molal heat of vaporization, Btu/lb mole
p_1 = vapor pressure at T_1, any units
p_2 = vapor pressure at T_2, units same as p_1
T_1, T_2 = absolute temperature, °R
R = 1.987, the gas-law constant in Btu/(lb mole)(°R)

If only the boiling point of the substance is known, the latent heat may be estimated by the Kistiakowski[1] equation:

$$\Delta H = 8.75 T_B + 4.571 T_B \log T_B \qquad \text{(E-16)}$$

where ΔH = molal heat of vaporization at boiling point, Btu/lb mole
T_B = boiling point, °R

Despite the simplicity of this relation, it is surprisingly accurate for compounds classified by the chemist as nonpolar, but it does give low results for polar compounds such as water, ammonia, most acids, bases, and alcohols.

c. Exact Determination of Latent Heats of Vaporization.—The Clapeyron equation expresses the exact relation between vapor pressures and latent heats.

$$\frac{dp}{dT} = \frac{JL}{T(V_g - V_e)} \qquad \text{(E-17)}$$

[1] *Z. physik. Chem.*, **107**, 65 (1923).

where p = vapor pressure, lb/sq ft
T = absolute temperature, °R
J = 778.1, the mechanical equivalent of heat
L = latent heat, Btu/lb
V_g = volume of vapor, cu ft/lb
V_e = volume of liquid, cu ft/lb

To compute L, the specific volumes of the gas and liquid must be known and dp/dT obtained by either graphical or analytical differentiation of the vapor pressure–temperature curve. This equation is the basis of the Clausius-Clapeyron and numerous other relations.

The method of Gordon[1] appears quite accurate and permits computation of the latent heats without knowledge of vapor density, though the critical pressure and temperature must be known.

5. Tables and Charts of Thermodynamic Properties of Steam, Propane, Ammonia, and Carbon Dioxide

List of Tables and Charts:

The uses of thermodynamic data presented in this section are manifold. For simple heating and cooling problems the heat supplied or liberated is given by the difference between the final and initial heat content.

$$Q = \Delta H = H_2 - H_1$$

where Q = heat absorbed, Btu/lb
H_2 = heat content, final state, Btu/lb
H_1 = heat content, initial state, Btu/lb

Example E-1 gives the calculation of a refrigeration cycle to illustrate the use of tables and charts. See also the discussion of compression in Chapter K and engines in Chapter I.

Example E-1. Find the power and cooling water requirements for a propane refrigerator supplying 80,000 Btu/hr refrigeration at −20°F (temperature of refrigerant). Cooling water may be assumed to be at 80°F condensing the propane at 100°F.

[1] *Ind. Eng. Chem.*, **35**, 851 (1943).

SOLUTION:

1. Conditions at exit of condenser, 100°F liquid propane is saturated at 188.7 psia

$$h_1 = 264.6 \text{ Btu/lb from Table E-16}$$

2. After expansion, −20°F propane is saturated at 25.05 psia

$$h_2 = h_1 = 264.6 \text{ Btu/lb (since no heat removed or work performed)}$$

3. After evaporation, −20°F propane saturated vapor at 25.05 psia

$$h_3 = 371.5 \text{ Btu/lb (Table E-16)}$$

4. Refrigeration supplied,

$$h_3 - h_2 = 371.5 - 264.6 = 106.9 \text{ Btu/lb}$$

and

$$\frac{80{,}000}{106.9} = 748 \text{ lb/hr propane circulated}$$

5. Ideal work of compression under adiabatic (no-heat-loss) conditions, $\Delta S = 0$; follow constant entropy line from −20°F saturated propane to 188.7 psi on Fig. E-20

$$h_5 = 413 \text{ Btu/lb}$$

$$\text{Work} = h_5 - h_3 = 413 - 371.5 = 41.5 \text{ Btu/lb}$$

6. Actual work of compression, assume compressor efficiency = 75 per cent:

$$\text{Work} = \frac{41.5}{0.75} = 55.3 \text{ Btu/lb}$$

Bhp required to compress 748 lb/hr,

$$\frac{55.3 \times 748}{2544} = 16.25$$

$$(1 \text{ hp} = 2544 \text{ Btu/hr})$$

7. Conditions at exit of compressor,

$$\text{Heat content} = h_6 = h_3 + \text{work of compression,}$$
$$371.5 + 55.3 = 426.8$$

(This corresponds to 151°F if no heat is removed by compressor jackets.)

8. Heat to be removed by cooling and condensing,

$$h_6 - h_1 = 426.8 - 264.6 = 162.2 \text{ Btu/lb}$$

or, for 748 lb/hr of propane,

$$\text{Heat to be removed} = 748 \times 162.2 = 121{,}000 \text{ Btu/hr}$$

9. Cooling water required, assume 10°F rise,

$$\text{Gpm} = \frac{121{,}000}{500 \times 10} = 24$$

$$(1 \text{ gpm} = 500 \text{ lb/hr})$$

Table E-13.—Properties of Saturated Steam: Temperature Table (Abstracted by permission from J. H. Keenan and F. G. Keyes, "Thermodynamic Properties of Steam," published by John Wiley & Sons, Inc.)

Temp., °F	Abs. press., psi	Specific volume		Heat content			Entropy	
		Sat. solid, V_i	Sat. vapor, V_g	Sat. solid, h_i	Subl. L_s	Sat. vapor, h_g	Sat. solid, S_i	Sat. vapor, S_g
−40	0.0019	0.01737	133,900	−177.00	1221.2	1044.2	−0.3654	2.5433
−30	0.0035	0.01738	74,100	−172.63	1221.2	1048.6	−0.3551	2.4860
−20	0.0062	0.01739	42,200	−168.16	1221.2	1053.0	−0.3448	2.4316
−10	0.0108	0.01741	24,670	−163.59	1221.0	1057.4	−0.3346	2.3797
0	0.0185	0.01742	14,770	−158.93	1220.7	1061.8	−0.3241	2.3305
10	0.0309	0.01744	9,050	−154.17	1220.4	1066.2	−0.3141	2.2836
20	0.0505	0.01745	5,658	−149.31	1219.9	1070.6	−0.3038	2.2387
32	0.0885	0.01747	3,306	−143.35	1219.1	1075.8	−0.2916	2.1877
Temp., °F	Abs. press., psi	Specific volume		Heat content			Entropy	
		Sat. liquid, V_l	Sat. vapor, V_g	Sat. liquid, h_l	Evap., L	Sat. vapor, h_g	Sat. liquid, S_l	Sat. vapor, S_g
32	0.08854	0.01602	3306	0.00	1075.8	1075.8	0.000	2.1877
40	0.12170	0.01602	2444	8.05	1071.3	1079.3	0.0162	2.1597
50	0.17811	0.01603	1703.2	18.07	1065.6	1083.7	0.0361	2.1264
60	0.2563	0.01604	1206.7	28.06	1059.9	1088.0	0.0555	2.0948
65	0.3056	0.01605	1021.4	33.05	1057.1	1090.2	0.0651	2.0796
70	0.3631	0.01606	867.9	38.04	1054.3	1092.3	0.0745	2.0647
75	0.4208	0.01607	740.0	43.03	1051.5	1094.5	0.0839	2.0502
80	0.5069	0.01608	633.1	48.02	1048.6	1096.6	0.0932	2.0360
85	0.5959	0.01609	543.5	53.00	1045.8	1098.8	0.1024	2.0222
90	0.6982	0.01610	468.0	57.99	1042.9	1100.9	0.1115	2.0087
95	0.8153	0.01612	404.3	62.98	1040.1	1103.1	0.1205	1.9955
100	0.9492	0.01613	350.4	67.97	1037.2	1105.2	0.1295	1.9826
105	1.1016	0.01615	304.5	72.95	1034.3	1107.3	0.1383	1.9700
110	1.2748	0.01617	265.4	77.94	1031.6	1109.5	0.1471	1.9577
115	1.4709	0.01618	231.9	82.93	1028.7	1111.6	0.1559	1.9457
120	1.6924	0.01620	203.27	87.92	1025.8	1113.7	0.1645	1.9339
130	2.2225	0.01625	157.34	97.90	1020.0	1117.9	0.1816	1.9112
140	2.8886	0.01629	123.01	107.89	1014.1	1122.0	0.1984	1.8894
150	3.718	0.01634	97.07	117.89	1008.2	1126.1	0.2149	1.8685
160	4.741	0.01639	77.29	127.89	1002.3	1130.2	0.2311	1.8485
170	5.992	0.01645	62.06	137.90	996.3	1134.2	0.2472	1.8293
180	7.510	0.01651	50.23	147.92	990.2	1138.1	0.2630	1.8109
190	9.339	0.01657	40.96	157.95	984.1	1142.0	0.2785	1.7932
200	11.526	0.01663	33.64	167.99	977.9	1145.9	0.2938	1.7762
210	14.123	0.01670	27.82	178.05	971.6	1149.7	0.3090	1.7598
220	17.186	0.01677	23.15	188.13	965.2	1153.4	0.3239	1.7440
230	20.780	0.01684	19.382	198.23	958.8	1157.0	0.3387	1.7288
240	24.969	0.01692	16.323	208.34	952.2	1160.5	0.3531	1.7140
250	29.825	0.01700	13.821	218.48	945.5	1164.0	0.3675	1.6998
260	35.429	0.01709	11.763	228.64	938.7	1167.3	0.3817	1.6860
270	41.858	0.01717	10.061	238.84	931.8	1170.6	0.3958	1.6727
280	49.203	0.01726	8.645	249.06	924.7	1173.8	0.4096	1.6597
290	57.556	0.01735	7.461	259.31	917.5	1176.8	0.4234	1.6472
300	67.013	0.01745	6.466	269.59	910.1	1179.7	0.4369	1.6350
310	77.68	0.01755	5.626	279.92	902.6	1182.5	0.4504	1.6231
320	89.66	0.01765	4.914	290.28	894.9	1185.2	0.4637	1.6115
330	103.06	0.01776	4.307	300.68	887.0	1187.7	0.4769	1.6002
340	118.01	0.01787	3.788	311.13	879.0	1190.1	0.4900	1.5891
350	134.63	0.01799	3.342	321.63	870.7	1192.3	0.5029	1.5783
360	153.04	0.01811	2.957	332.18	862.2	1194.4	0.5158	1.5677
370	173.37	0.01823	2.625	342.79	853.5	1196.3	0.5286	1.5573
380	195.77	0.01836	2.335	353.45	844.6	1198.1	0.5413	1.5471
390	220.37	0.01850	2.0836	364.17	835.4	1199.6	0.5539	1.5371
400	247.31	0.01864	1.8633	374.97	826.0	1201.0	0.5664	1.5272
450	422.6	0.0194	1.0993	430.1	774.5	1204.6	0.6280	1.4793
500	680.8	0.0204	0.6749	487.8	713.9	1201.7	0.6887	1.4325
550	1045.2	0.0218	0.4240	549.3	640.8	1190.0	0.7497	1.3843
600	1542.9	0.0236	0.2668	617.0	548.5	1165.5	0.8131	1.3307
650	2208.2	0.0268	0.1616	695.7	422.8	1118.5	0.8828	1.2637
700	3093.7	0.0369	0.0761	823.3	172.1	995.4	0.9905	1.1389
705.4	3206.2	0.0503	0.0503	902.7	0	902.7	1.0580	1.0580

TABLE E-14.—PROPERTIES OF SATURATED STEAM: PRESSURE TABLE
(Abstracted by permission from J. H. Keenan and F. G. Keyes, "Thermodynamic Properties of Steam," published by John Wiley & Sons, Inc.)

Abs. press., psi	Temp., °F	Specific volume		Heat content			Entropy	
		Sat. liquid, V_l	Sat. vapor, V_g	Sat. liquid, h_l	Evap., L	Sat. vapor, h_g	Sat. liquid, S_l	Sat. vapor, S_g
1.0	101.74	0.01614	333.6	69.70	1036.3	1106.0	0.1326	1.9782
5.0	162.24	0.01640	73.52	130.13	1001.0	1131.1	0.2347	1.8441
10	193.21	0.01659	38.42	161.17	982.1	1143.3	0.2835	1.7876
14.7	212.00	0.01672	26.80	180.07	970.3	1150.4	0.3120	1.7566
20	227.96	0.01683	20.089	196.16	960.1	1156.3	0.3356	1.7319
25	240.07	0.01692	16.303	208.42	952.1	1160.6	0.3533	1.7139
30	250.33	0.01701	13.746	218.82	945.3	1164.1	0.3680	1.6993
35	259.28	0.01708	11.898	227.91	939.2	1167.1	0.3807	1.6870
40	267.25	0.01715	10.498	236.03	933.7	1169.7	0.3919	1.6763
45	274.44	0.01721	9.401	243.36	928.6	1172.0	0.4019	1.6669
50	281.01	0.01727	8.515	250.09	924.0	1174.1	0.4110	1.6585
55	287.07	0.01732	7.787	256.30	919.6	1175.9	0.4193	1.6509
60	292.71	0.01738	7.175	262.09	915.5	1177.6	0.4270	1.6438
70	302.92	0.01748	6.206	272.61	907.9	1180.6	0.4409	1.6315
80	312.03	0.01757	5.472	282.02	901.1	1183.1	0.4531	1.6207
90	320.27	0.01766	4.896	290.56	894.7	1185.3	0.4641	1.6112
100	327.81	0.01774	4.432	298.40	888.8	1187.2	0.4740	1.6026
110	334.77	0.01782	4.049	305.66	883.2	1188.9	0.4832	1.5948
120	341.25	0.01789	3.728	312.44	877.9	1190.4	0.4916	1.5878
130	347.32	0.01796	3.455	318.81	872.9	1191.7	0.4995	1.5812
140	353.02	0.01802	3.220	324.82	868.2	1193.0	0.5069	1.5751
160	363.53	0.01815	2.834	335.93	859.2	1195.1	0.5204	1.5640
180	373.06	0.01827	2.532	346.03	850.8	1196.9	0.5325	1.5542
200	381.79	0.01839	2.288	355.36	843.0	1198.4	0.5435	1.5453
250	400.95	0.01865	1.8438	376.00	825.1	1201.1	0.5675	1.5263
300	417.33	0.01890	1.5433	393.84	809.0	1202.8	0.5879	1.5104
350	431.72	0.01913	1.3260	409.69	794.2	1203.9	0.6056	1.4966
400	444.59	0.0193	1.1613	424.0	780.5	1204.5	0.6214	1.4844
500	467.01	0.0197	0.9278	449.4	755.0	1204.4	0.6487	1.4634
600	486.21	0.0201	0.7698	471.6	731.6	1203.2	0.6720	1.4454
800	518.23	0.0209	0.5687	509.7	688.9	1198.6	0.7108	1.4153
1,000	544.61	0.0216	0.4456	542.4	649.4	1191.8	0.7430	1.3897
1,500	596.23	0.0235	0.2765	611.6	556.3	1167.9	0.8082	1.3351
2,000	635.82	0.0257	0.1878	671.7	463.4	1135.1	0.8619	1.2849
2,500	668.13	0.0287	0.1307	730.6	360.5	1091.1	0.9126	1.2322
3,000	695.36	0.0346	0.0858	802.5	217.8	1020.3	0.9731	1.1615
3,206.2	705.40	0.0503	0.0503	902.7	0	902.7	1.0580	1.0580

TABLE E-15.—PROPERTIES OF SUPERHEATED STEAM

(Abstracted by permission from J. H. Keenan and F. G. Keyes, "Thermodynamic Properties of Steam," published by John Wiley & Sons, Inc.)

Abs. press., psi (sat. temp.)		Temperature, degrees Fahrenheit																			
		220°	240°	260°	280°	300°	320°	340°	360°	380°	400°	420°	440°	460°	500°	600°	800°	1000°	1200°	1400°	1600°
1	*v*	404.5	416.5	428.4	440.4	452.3	464.2	476.2	488.1	500.0	512.0	523.9	535.8	547.7	571.6	631.2	750.4	869.5	988.7	1107.8	1227.0
(101.74)	*h*	1159.5	1168.5	1177.6	1186.7	1195.8	1204.9	1214.1	1223.3	1232.5	1241.7	1251.0	1260.3	1269.6	1288.3	1335.7	1432.8	1533.5	1637.7	1745.7	1857.5
	s	2.0647	2.0779	2.0907	2.1031	2.1153	2.1271	2.1387	2.1501	2.1612	2.1720	2.1827	2.1931	2.2034	2.2233	2.2702	2.3542	2.4283	2.4952	2.5566	2.6137
5	*v*	80.59	83.01	85.43	87.85	90.25	92.66	95.06	97.46	99.86	102.26	104.65	107.04	109.44	114.22	126.16	150.03	173.87	197.71	221.6	245.4
(162.24)	*h*	1158.1	1167.3	1176.5	1185.7	1195.0	1204.2	1213.4	1222.7	1231.9	1241.2	1250.5	1259.8	1269.2	1288.0	1335.4	1432.7	1533.4	1637.7	1745.7	1857.4
	s	1.8857	1.8991	1.9121	1.9247	1.9370	1.9490	1.9607	1.9721	1.9833	1.9942	2.0049	2.0154	2.0256	2.0456	2.0927	2.1767	2.2509	2.3178	2.3792	2.4363
10	*v*	40.09	41.33	42.56	43.78	45.00	46.21	47.42	48.63	49.84	51.04	52.24	53.45	54.65	57.05	63.03	74.98	86.92	98.84	110.77	122.69
(193.21)	*h*	1156.2	1165.7	1175.1	1184.5	1193.9	1203.2	1212.5	1221.9	1231.2	1240.6	1249.9	1259.3	1268.7	1287.5	1335.1	1432.5	1533.2	1637.6	1745.6	1857.3
	s	1.8071	1.8208	1.8341	1.8470	1.8595	1.8716	1.8834	1.8950	1.9062	1.9172	1.9280	1.9385	1.9488	1.9689	2.0160	2.1002	2.1744	2.2413	2.3028	2.3598
14.696	*v*	27.15	28.00	28.85	29.70	30.53	31.37	32.20	33.03	33.85	34.68	35.50	36.32	37.14	38.78	42.86	51.00	59.13	67.25	75.37	83.48
(212.00)	*h*	1154.4	1164.2	1173.8	1183.3	1192.8	1202.3	1211.7	1221.1	1230.5	1239.9	1249.3	1258.8	1268.2	1287.1	1334.8	1432.3	1533.1	1637.5	1745.5	1857.3
	s	1.7624	1.7766	1.7902	1.8033	1.8160	1.8283	1.8402	1.8518	1.8631	1.8743	1.8850	1.8956	1.9060	1.9261	1.9734	2.0576	2.1319	2.1989	2.2603	2.3174
20	*v*	......	20.48	21.11	21.74	22.36	22.98	23.60	24.21	24.82	25.43	26.04	26.65	27.25	28.46	31.47	37.46	43.44	49.41	55.37	61.34
(227.96)	*h*	......	1162.3	1172.2	1182.0	1191.6	1201.2	1210.8	1220.3	1229.7	1239.2	1248.7	1258.2	1267.6	1286.6	1334.4	1432.1	1533.0	1637.4	1745.4	1857.2
	s	......	1.7405	1.7545	1.7679	1.7808	1.7932	1.8053	1.8170	1.8285	1.8396	1.8505	1.8612	1.8716	1.8918	1.9392	2.0235	2.0978	2.1648	2.2263	2.2834
25	*v*	......	......	16.819	17.330	17.836	18.337	18.834	19.329	19.821	20.31	20.80	21.29	21.77	22.74	25.16	29.96	34.74	35.92	44.30	49.07
(240.07)	*h*	......	......	1170.7	1180.6	1190.5	1200.2	1209.8	1219.4	1229.0	1238.5	1248.1	1257.6	1267.1	1286.2	1334.1	1431.9	1532.8	1637.3	1745.3	1857.2
	s	......	......	1.7282	1.7418	1.7550	1.7676	1.7798	1.7917	1.8032	1.8144	1.8254	1.8361	1.8465	1.8668	1.9144	1.9988	2.0732	2.1402	2.2017	2.2587
30	*v*	......	......	13.957	14.390	14.816	15.238	15.657	16.072	16.485	16.897	17.306	17.714	18.121	18.933	20.95	24.96	28.95	32.93	36.91	40.89
(250.33)	*h*	......	......	1169.1	1179.3	1189.3	1199.1	1208.9	1218.6	1228.3	1237.9	1247.5	1257.0	1266.6	1285.7	1333.8	1431.7	1532.7	1637.2	1745.3	1857.1
	s	......	......	1.7063	1.7203	1.7336	1.7464	1.7588	1.7708	1.7824	1.7937	1.8047	1.8155	1.8260	1.8464	1.8940	1.9786	2.0530	2.1201	2.1815	2.2386
35	*v*	......	......	11.911	12.288	12.659	13.025	13.387	13.746	14.103	14.457	14.810	15.162	15.512	16.210	17.943	21.38	24.81	28.22	31.63	35.04
(259.28)	*h*	......	......	1167.5	1177.9	1188.1	1198.1	1208.0	1217.8	1227.5	1237.2	1246.8	1256.4	1266.1	1285.3	1333.5	1431.5	1532.6	1637.1	1745.2	1857.0
	s	......	......	1.6875	1.7018	1.7154	1.7284	1.7409	1.7530	1.7647	1.7761	1.7872	1.7980	1.8086	1.8290	1.8768	1.9615	2.0360	2.1030	2.1645	2.2216
40	*v*	......	......	......	10.711	11.040	11.364	11.684	12.001	12.315	12.628	12.938	13.247	13.555	14.168	15.688	18.702	21.70	24.69	27.68	30.66
(267.25)	*h*	......	......	......	1176.5	1186.8	1197.0	1207.0	1216.9	1226.7	1236.5	1246.2	1255.9	1265.5	1284.8	1333.1	1431.3	1532.4	1637.0	1745.1	1857.0
	s	......	......	......	1.6855	1.6994	1.7126	1.7252	1.7375	1.7493	1.7608	1.7719	1.7828	1.7934	1.8140	1.8619	1.9467	2.0212	2.0883	2.1498	2.2069
50	*v*	......	......	......	......	8.773	9.038	9.299	9.557	9.812	10.065	10.317	10.567	10.815	11.309	12.532	14.950	17.352	19.747	22.14	24.53
(281.01)	*h*	......	......	......	......	1184.3	1194.8	1205.0	1215.2	1225.2	1235.1	1244.9	1254.7	1264.5	1283.9	1332.5	1430.9	1532.1	1636.8	1745.0	1856.8
	s	......	......	......	......	1.6721	1.6857	1.6987	1.7112	1.7233	1.7349	1.7462	1.7572	1.7680	1.7887	1.8368	1.9219	1.9964	2.0636	2.1251	2.1822

TABLE E-15.—PROPERTIES OF SUPERHEATED STEAM (*Continued*)

Abs. press., psi (sat. temp.)		Temperature, degrees Fahrenheit																			
		220°	240°	260°	280°	300°	320°	340°	360°	380°	400°	420°	440°	460°	500°	600°	800°	1000°	1200°	1400°	1600°
55	v	......	......	......	......	7.947	8.192	8.432	8.668	8.902	9.134	9.364	9.592	9.819	10.269	11.384	13.586	15.771	17.950	20.12	22.30
	h	......	......	......	......	1183.0	1193.6	1204.0	1214.3	1224.4	1234.3	1244.3	1254.1	1263.9	1283.4	1332.1	1430.7	1532.0	1636.7	1744.9	1856.8
(287.07)	s	......	......	......	......	1.6602	1.6740	1.6872	1.6999	1.7120	1.7238	1.7351	1.7462	1.7570	1.7778	1.8261	1.9112	1.9859	2.0530	2.1146	2.1717
60	v	......	......	......	......	7.259	7.486	7.708	7.927	8.143	8.357	8.569	8.779	8.988	9.403	10.427	12.449	14.454	16.451	18.446	20.44
	h	......	......	......	......	1181.6	1192.5	1203.0	1213.4	1223.6	1233.6	1243.6	1253.5	1263.4	1283.0	1331.8	1430.5	1531.9	1636.6	1744.8	1856.7
(292.71)	s	......	......	......	......	1.6492	1.6633	1.6766	1.6894	1.7017	1.7135	1.7250	1.7361	1.7470	1.7678	1.8162	1.9015	1.9762	2.0434	2.1049	2.1621
70	v	......	......	......	......	......	6.376	6.571	6.762	6.950	7.136	7.320	7.502	7.683	8.041	8.924	10.662	12.383	14.097	15.808	17.516
	h	......	......	......	......	......	1190.1	1201.0	1211.5	1221.9	1232.1	1242.3	1252.3	1262.2	1282.0	1331.1	1430.1	1531.6	1636.3	1744.6	1856.6
(302.92)	s	......	......	......	......	......	1.6438	1.6576	1.6707	1.6832	1.6952	1.7068	1.7181	1.7291	1.7501	1.7988	1.8843	1.9591	2.0263	2.0879	2.1451
80	v	......	......	......	......	......	5.543	5.718	5.888	6.055	6.220	6.383	6.544	6.704	7.020	7.797	9.322	10.830	12.332	13.830	15.325
	h	......	......	......	......	......	1187.6	1198.8	1209.7	1220.3	1230.7	1240.9	1251.1	1261.1	1281.1	1330.5	1429.7	1531.3	1636.2	1744.5	1856.5
(312.03)	s	......	......	......	......	......	1.6266	1.6407	1.6541	1.6669	1.6791	1.6909	1.7023	1.7134	1.7346	1.7836	1.8694	1.9442	2.0115	2.0731	2.1303
90	v	......	......	......	......	......	......	5.053	5.208	5.359	5.508	5.654	5.799	5.942	6.225	6.920	8.279	9.623	10.959	12.291	13.621
	h	......	......	......	......	......	......	1196.6	1207.7	1218.6	1229.1	1239.5	1249.8	1260.0	1280.1	1329.8	1429.3	1531.0	1635.9	1744.3	1856.4
(320.27)	s	......	......	......	......	......	......	1.6256	1.6393	1.6523	1.6648	1.6767	1.6883	1.6995	1.7209	1.7702	1.8562	1.9311	1.9984	2.0601	2.1173
100	v	......	......	......	......	......	......	4.521	4.663	4.801	4.937	5.071	5.202	5.333	5.589	6.218	7.446	8.656	9.860	11.060	12.258
	h	......	......	......	......	......	......	1194.3	1205.7	1216.8	1227.6	1238.1	1248.6	1258.8	1279.1	1329.1	1428.9	1530.8	1635.7	1744.2	1856.2
(327.81)	s	......	......	......	......	......	......	1.6117	1.6258	1.6391	1.6518	1.6639	1.6756	1.6869	1.7085	1.7581	1.8443	1.9193	1.9867	2.0484	2.1056
110	v	......	......	......	......	......	......	4.085	4.216	4.345	4.470	4.593	4.714	4.834	5.069	5.644	6.763	7.866	8.961	10.053	11.142
	h	......	......	......	......	......	......	1192.0	1203.7	1215.0	1226.0	1236.7	1247.3	1257.7	1278.2	1328.4	1428.5	1530.5	1635.5	1744.0	1856.1
(334.77)	s	......	......	......	......	......	......	1.5988	1.6133	1.6269	1.6398	1.6521	1.6640	1.6754	1.6972	1.7471	1.8336	1.9087	1.9761	2.0378	2.0951
120	v	......	......	......	......	......	......	......	3.844	3.964	4.081	4.195	4.307	4.418	4.636	5.165	6.195	7.207	8.212	9.214	10.213
	h	......	......	......	......	......	......	......	1201.6	1213.2	1224.4	1235.3	1246.0	1256.5	1277.2	1327.7	1428.1	1530.2	1635.3	1743.9	1856.0
(341.25)	s	......	......	......	......	......	......	......	1.6017	1.6156	1.6287	1.6413	1.6533	1.6649	1.6869	1.7370	1.8237	1.8990	1.9664	2.0281	2.0854
130	v	......	......	......	......	......	......	......	3.529	3.641	3.751	3.858	3.963	4.066	4.269	4.761	5.714	6.650	7.579	8.503	9.426
	h	......	......	......	......	......	......	......	1199.5	1211.3	1222.7	1233.8	1244.6	1255.3	1276.2	1327.0	1427.7	1529.9	1635.1	1743.7	1855.9
(347.32)	s	......	......	......	......	......	......	......	1.5908	1.6050	1.6184	1.6312	1.6433	1.6550	1.6773	1.7277	1.8147	1.8900	1.9575	2.0193	2.0765
140	v	......	......	......	......	......	......	......	3.258	3.365	3.468	3.569	3.667	3.764	3.954	4.413	5.301	6.172	7.035	7.895	8.752
	h	......	......	......	......	......	......	......	1197.3	1209.4	1221.1	1232.3	1243.3	1254.1	1275.2	1326.4	1427.3	1529.7	1634.9	1743.5	1855.7
(353.02)	s	......	......	......	......	......	......	......	1.5804	1.5950	1.6087	1.6217	1.6340	1.6458	1.6683	1.7190	1.8063	1.8817	1.9493	2.0110	2.0683
160	v	......	......	......	......	......	......	......	......	2.914	3.008	3.098	3.187	3.273	3.443	3.849	4.631	5.396	6.152	6.906	7.656
	h	......	......	......	......	......	......	......	......	1205.5	1217.6	1229.3	1240.6	1251.6	1273.1	1325.0	1426.4	1529.1	1634.5	1743.2	1855.5
(363.53)	s	......	......	......	......	......	......	......	......	1.5766	1.5908	1.6042	1.6169	1.6291	1.6519	1.7033	1.7911	1.8667	1.9344	1.9962	2.0535
180	v	......	......	......	......	......	......	......	......	2.563	2.649	2.732	2.813	2.891	3.044	3.411	4.110	4.792	5.466	6.136	6.804
	h	......	......	......	......	......	......	......	......	1201.4	1214.0	1226.1	1237.8	1249.1	1271.0	1323.5	1425.6	1528.6	1634.1	1742.9	1855.2
(373.06)	s	......	......	......	......	......	......	......	......	1.5596	1.5745	1.5884	1.6015	1.6139	1.6373	1.6894	1.7776	1.8534	1.9212	1.9831	2.0404

200 (381.79)	v	......	......	......	......	......	......	......	......	......	2.361	2.438	2.513	2.585	2.726	3.060	3.693	4.309	4.917	5.521	6.123
	h	......	......	......	......	......	......	......	......	......	1210.3	1222.9	1234.9	1246.5	1268.9	1322.1	1424.8	1528.0	1633.7	1742.6	1855.0
	s	......	......	......	......	......	......	......	......	......	1.5594	1.5738	1.5873	1.6001	1.6240	1.6767	1.7655	1.8415	1.9094	1.9713	2.0287
250 (400.95)	v	......	......	......	......	......	......	......	......	......	......	1.9077	1.9717	2.033	2.151	2.427	2.942	3.439	3.928	4.413	4.896
	h	......	......	......	......	......	......	......	......	......	......	1214.2	1227.3	1239.7	1263.4	1318.5	1422.7	1526.6	1632.7	1741.8	1854.4
	s	......	......	......	......	......	......	......	......	......	......	1.5414	1.5560	1.5697	1.5949	1.6495	1.7397	1.8162	1.8843	1.9464	2.0039
300 (417.33)	v	......	......	......	......	......	......	......	......	......	......	1.5513	1.6090	1.6638	1.7675	2.005	2.442	2.859	3.269	3.674	4.078
	h	......	......	......	......	......	......	......	......	......	......	1204.8	1219.1	1232.5	1257.6	1314.7	1420.6	1525.2	1631.7	1741.0	1853.7
	s	......	......	......	......	......	......	......	......	......	......	1.5126	1.5286	1.5434	1.5701	1.6268	1.7184	1.7954	1.8638	1.9260	1.9835
350 (431.72)	v	......	......	......	......	......	......	......	......	......	......	......	1.3478	1.3984	1.4923	1.7036	2.084	2.445	2.798	3.147	3.493
	h	......	......	......	......	......	......	......	......	......	......	......	1210.3	1224.8	1251.5	1310.9	1418.5	1523.8	1630.7	1740.3	1853.1
	s	......	......	......	......	......	......	......	......	......	......	......	1.5037	1.5197	1.5481	1.6070	1.7002	1.7777	1.8463	1.9086	1.9663
400 (444.59)	v	......	......	......	......	......	......	......	......	......	......	......	......	1.1978	1.2851	1.4770	1.8161	2.134	2.445	2.751	3.055
	h	......	......	......	......	......	......	......	......	......	......	......	......	1216.5	1245.1	1306.9	1416.4	1522.4	1629.6	1739.5	1852.5
	s	......	......	......	......	......	......	......	......	......	......	......	......	1.4977	1.5281	1.5894	1.6842	1.7623	1.8311	1.8936	1.9513
500 (467.01)	v	......	......	......	......	......	......	......	......	......	......	......	......	......	0.9927	1.1591	1.4405	1.6996	1.9504	2.197	2.442
	h	......	......	......	......	......	......	......	......	......	......	......	......	......	1231.3	1298.6	1412.1	1519.6	1627.6	1737.9	1851.3
	s	......	......	......	......	......	......	......	......	......	......	......	......	......	1.4919	1.5588	1.6571	1.7363	1.8056	1.8683	1.9262
600 (486.21)	v	......	......	......	......	......	......	......	......	......	......	......	......	......	0.7947	0.9463	1.1899	1.4096	1.6208	1.8279	2.033
	h	......	......	......	......	......	......	......	......	......	......	......	......	......	1215.7	1289.9	1407.7	1516.7	1625.5	1736.3	1850.0
	s	......	......	......	......	......	......	......	......	......	......	......	......	......	1.4586	1.5323	1.6343	1.7147	1.7846	1.8476	1.9056
800 (518.23)	v	......	......	......	......	......	......	......	......	......	......	......	......	......	......	0.6779	0.8763	1.0470	1.2088	1.3662	1.5214
	h	......	......	......	......	......	......	......	......	......	......	......	......	......	......	1270.7	1398.6	1511.0	1621.4	1733.2	1847.5
	s	......	......	......	......	......	......	......	......	......	......	......	......	......	......	1.4863	1.5972	1.6801	1.7510	1.8146	1.8729
1000 (544.61)	v	......	......	......	......	......	......	......	......	......	......	......	......	......	......	0.5140	0.6878	0.8294	0.9615	1.0893	1.2146
	h	......	......	......	......	......	......	......	......	......	......	......	......	......	......	1248.8	1389.2	1505.1	1617.3	1730.0	1845.0
	s	......	......	......	......	......	......	......	......	......	......	......	......	......	......	1.4450	1.5670	1.6525	1.7245	1.7886	1.8474
2000 (635.82)	v	......	......	......	......	......	......	......	......	......	......	......	......	......	......	......	0.3074	0.3935	0.4668	0.5352	0.6011
	h	......	......	......	......	......	......	......	......	......	......	......	......	......	......	......	1335.5	1474.5	1596.1	1714.1	1832.5
	s	......	......	......	......	......	......	......	......	......	......	......	......	......	......	......	1.4576	1.5603	1.6384	1.7055	1.7660
3206.2 (705.40)	v	......	......	......	......	......	......	......	......	......	......	......	......	......	......	......	0.1583	0.2288	0.2806	0.3267	0.3703
	h	......	......	......	......	......	......	......	......	......	......	......	......	......	......	......	1250.5	1434.7	1569.8	1694.6	1817.2
	s	......	......	......	......	......	......	......	......	......	......	......	......	......	......	......	1.3508	1.4874	1.5742	1.6452	1.7080
4000	v	......	......	......	......	......	......	......	......	......	......	......	......	......	......	......	0.1052	0.1743	0.2192	0.2581	0.2943
	h	......	......	......	......	......	......	......	......	......	......	......	......	......	......	......	1174.8	1406.8	1552.1	1681.7	1807.2
	s	......	......	......	......	......	......	......	......	......	......	......	......	......	......	......	1.2757	1.4482	1.5417	1.6154	1.6795

TABLE E-16.—THERMODYNAMIC PROPERTIES OF SATURATED PROPANE
[Stearns and George, *Ind. Eng. Chem.*, **35**, 602 (1943). By permission of the American Chemical Society]

Temp. t, °F	Pressure		Volume, cu ft/lb		Heat content, Btu/lb		Latent heat L, Btu/lb	Entropy, Btu/(lb)(°F)	
	Psia p	Psig gp	Liquid v	Vapor V	Liquid h	Vapor H		Liquid s	Vapor S
−80	5.65	......	0.0265	16.2	162.6	354.0	191.4	0.8794	1.3832
−78	6.00	......	0.0266	15.3	163.6	354.6	191.0	0.8821	1.3822
−76	6.32	......	0.0266	14.6	164.6	355.2	190.6	0.8847	1.3812
−74	6.71	......	0.0267	13.8	165.6	355.8	190.2	0.8874	1.3801
−72	7.18	......	0.0268	13.0	166.6	356.4	189.8	0.8900	1.3791
−70	7.48	......	0.0268	12.5	167.6	357.0	189.4	0.8927	1.3781
−68	7.91	......	0.0268	11.9	168.6	357.6	189.0	0.8954	1.3773
−66	8.33	......	0.0269	11.3	169.7	358.2	188.5	0.8980	1.3765
−64	8.80	......	0.0269	10.8	170.7	358.8	188.1	0.9007	1.3756
−62	9.28	......	0.0270	10.3	171.7	359.4	187.7	0.9033	1.3748
−60	9.78	......	0.02703	9.77	172.7	360.0	187.3	0.9060	1.3740
−58	10.29	......	0.02709	9.33	173.7	360.6	186.9	0.9086	1.3732
−56	10.80	......	0.02715	8.92	174.7	361.1	186.4	0.9111	1.3725
−54	11.36	......	0.02721	8.51	175.8	361.7	185.9	0.9137	1.3717
−52	11.95	......	0.02727	8.12	176.8	362.3	185.5	0.9162	1.3710
−50	12.60	......	0.02733	7.73	177.8	362.8	185.0	0.9188	1.3702
−48	13.20	......	0.02739	7.50	178.8	363.4	184.6	0.9213	1.3696
−46	13.85	......	0.02745	7.06	179.9	364.0	184.1	0.9238	1.3689
−44	14.52	......	0.02751	6.74	180.9	364.6	183.8	0.9264	1.3683
−42	15.28	......	0.02757	6.42	181.9	365.2	183.3	0.9289	1.3676
−40	16.00	1.30	0.02763	6.16	183.0	365.7	182.7	0.9315	1.3670
−38	16.79	2.09	0.02769	5.92	184.0	366.3	182.3	0.9340	1.3664
−36	17.56	2.86	0.02775	5.66	185.1	366.9	181.8	0.9365	1.3658
−34	18.40	3.70	0.02782	5.44	186.2	367.5	181.3	0.9391	1.3652
−32	19.30	4.60	0.02788	5.22	187.3	368.0	180.7	0.9416	1.3646
−30	20.18	5.48	0.02794	5.02	188.4	368.6	180.2	0.9441	1.3640
−28	21.05	6.35	0.02800	4.82	189.4	369.2	179.7	0.9467	1.3634
−26	22.01	7.31	0.02807	4.63	190.5	369.8	179.3	0.9492	1.3628
−24	22.98	8.28	0.02813	4.44	191.6	370.3	178.7	0.9517	1.3622
−22	23.98	9.28	0.02820	4.25	192.7	370.9	178.2	0.9543	1.3616
−20	25.05	10.35	0.02826	4.06	193.8	371.5	177.7	0.9568	1.3610
−18	26.15	11.45	0.02833	3.90	194.9	372.1	177.2	0.9592	1.3604
−16	27.30	12.60	0.02839	3.76	196.0	372.7	176.6	0.9617	1.3599
−14	28.50	13.80	0.02846	3.61	197.1	373.2	176.1	0.9641	1.3593
−12	29.70	15.00	0.02852	3.47	198.2	373.8	175.6	0.9666	1.3588

TABLE E-16.—THERMODYNAMIC PROPERTIES OF SATURATED PROPANE (*Continued*)

Temp. t, °F	Pressure		Volume, cu ft/lb		Heat content, Btu/lb		Latent heat L, Btu/lb	Entropy, Btu/(lb)(°F)	
	Psia p	Psig gp	Liquid v	Vapor V	Liquid h	Vapor H		Liquid s	Vapor S
−10	30.95	16.25	0.02859	3.33	199.4	374.4	175.0	0.9690	1.3582
− 8	32.23	17.53	0.02866	3.20	200.5	375.0	174.5	0.9714	1.3577
− 6	33.55	18.85	0.02873	3.08	201.6	375.5	173.9	0.9739	1.3571
− 4	35.00	20.30	0.02879	2.98	202.7	376.1	173.4	0.9763	1.3566
− 2	36.40	21.70	0.02886	2.86	203.8	376.7	172.8	0.9788	1.3560
0	37.81	23.11	0.02893	2.74	205.0	377.2	172.2	0.9812	1.3555
+ 2	39.30	24.60	0.02900	2.66	206.1	377.8	171.6	0.9836	1.3550
4	40.85	26.15	0.02908	2.56	207.2	378.3	171.1	0.9860	1.3545
6	42.50	27.80	0.02915	2.47	208.4	378.9	170.5	0.9884	1.3541
8	44.13	29.43	0.02923	2.38	209.6	379.5	169.9	0.9908	1.3536
10	45.85	31.15	0.02930	2.30	210.7	380.0	169.3	0.9932	1.3531
12	47.55	32.85	0.02938	2.22	211.9	380.6	168.7	0.9956	1.3527
14	49.35	34.65	0.02946	2.14	213.1	381.1	168.0	0.9979	1.3523
16	51.20	36.50	0.02954	2.08	214.2	381.6	167.4	1.0003	1.3518
18	53.10	38.40	0.02962	2.00	215.4	382.1	166.7	1.0026	1.3514
20	55.00	40.30	0.02970	1.93	216.6	382.6	166.0	1.0050	1.3510
22	57.05	42.35	0.02978	1.86	217.7	383.1	165.4	1.0073	1.3506
24	59.10	44.40	0.02986	1.79	218.8	383.6	164.8	1.0097	1.3502
26	61.25	46.55	0.02995	1.73	220.0	384.1	164.2	1.0120	1.3499
28	63.45	48.75	0.03003	1.67	221.2	384.6	163.5	1.0144	1.3495
30	65.70	51.00	0.03011	1.60	222.3	385.1	162.8	1.0167	1.3491
32	67.95	53.25	0.03020	1.54	223.4	385.6	162.2	1.0190	1.3487
34	70.33	55.63	0.03029	1.48	224.5	386.1	161.5	1.0213	1.3484
36	72.75	58.05	0.03037	1.43	225.6	386.6	160.8	1.0237	1.3480
38	75.20	60.50	0.03046	1.39	226.8	387.1	160.2	1.0260	1.3477
40	77.80	63.10	0.03055	1.33	227.9	387.5	159.6	1.0283	1.3473
42	80.40	65.70	0.03064	1.28	229.1	388.0	158.9	1.0306	1.3470
44	83.05	68.35	0.03073	1.25	230.2	388.5	158.2	1.0329	1.3466
46	85.83	71.13	0.03083	1.21	231.4	389.0	157.5	1.0352	1.3463
48	88.65	73.95	0.03092	1.17	232.6	389.5	156.8	1.0375	1.3459
50	91.50	76.80	0.03101	1.14	233.8	389.9	156.1	1.0398	1.3456
52	94.50	79.80	0.03111	1.10	234.9	390.4	155.4	1.0421	1.3453
54	97.5	82.80	0.03121	1.07	236.1	390.8	154.7	1.0443	1.3450
56	100.6	85.9	0.03130	1.04	237.3	391.3	154.0	1.0466	1.3447
58	103.7	89.0	0.03140	1.01	238.5	391.7	153.3	1.0488	1.3444

TABLE E-16.—THERMODYNAMIC PROPERTIES OF SATURATED PROPANE (*Continued*)

Temp. t, °F	Pressure		Volume, cu ft/lb		Heat content, Btu/lb		Latent heat L, Btu/lb	Entropy, Btu/(lb)(°F)	
	Psia p	Psig gp	Liquid v	Vapor V	Liquid h	Vapor H		Liquid s	Vapor S
60	106.9	92.2	0.03150	0.984	239.6	392.2	152.6	1.0511	1.3441
62	110.2	95.5	0.03162	0.958	240.8	392.7	151.8	1.0534	1.3438
64	113.6	98.9	0.03174	0.932	242.0	393.1	151.1	1.0556	1.3435
66	117.1	102.4	0.03185	0.906	243.2	393.5	150.3	1.0579	1.3433
68	120.6	105.9	0.03197	0.880	244.4	394.0	149.5	1.0601	1.3430
70	124.3	109.6	0.03209	0.854	245.7	394.4	148.7	1.0624	1.3427
72	127.9	113.2	0.03221	0.832	246.9	394.8	147.8	1.0647	1.3424
74	131 7	117.0	0.03233	0.810	248 1	395.2	147.0	1.0669	1.3421
76	135.6	120 9	0 03245	0.788	249.4	395.6	146.2	1.0692	1.3419
78	139.6	124.9	0.03257	0.766	250.2	396.0	145.4	1.0714	1.3416
80	143.6	128.9	0.03269	0.745	251.9	396.4	144.5	1.0737	1.3413
82	147.7	133.0	0.03281	0.725	253.1	396.8	143.7	1.0760	1.3410
84	151.8	137.1	0.03293	0.704	254.4	397.2	142.8	1.0782	1.3408
86	156.2	141.5	0.03305	0.684	255.6	397.6	141.9	1.0805	1.3405
88	160.6	145.9	0.03317	0.663	256.9	398.0	141.0	1.0827	1.3403
90	165.0	150.3	0.03329	0.643	258.2	398.3	140.1	1.0850	1.3400
92	169.6	154.9	0.03341	0.626	259.5	398.7	139.2	1.0873	1.3398
94	174.2	159.5	0.03353	0.609	260.8	399.1	138.3	1.0895	1.3395
96	178.9	164 2	0.03366	0.592	262.1	399.5	137.4	1.0918	1.3393
98	183.7	169.0	0.03378	0.575	263.3	399.9	136.5	1.0940	1.3390
100	188.7	174.0	0.03390	0.558	264.6	400.2	135.6	1.0963	1.3388
102	193.8	179.1	0.03402	0.544	265.9	400.5	134.6	1.0986	1.3386
104	198.9	184.2	0.03415	0.530	267.2	400.9	133.7	1.1010	1.3384
106	204.1	189.4	0.03427	0.516	268.5	401.2	132.7	1.1033	1.3382
108	209.3	194.6	0.03439	0.502	269.8	401.6	131.8	1.1057	1.3380
110	214.8	200.1	0.03452	0.487	271.1	401.9	130.8	1.1080	1.3378
112	220.4	205.7	0.03468	0.475	272.5	402.3	129.8	1.1103	1.3376
114	226.0	211.3	0.03484	0.463	273.9	402.7	128.8	1.1126	1.3374
116	231.6	216.9	0.03500	0.451	275.2	403.0	127.8	1.1149	1.3372
118	237.3	222.6	0.03516	0.439	276.6	403.4	126.8	1.1172	1.3370
120	243.4	228.7	0.03532	0.426	278.0	403.8	125.8	1.1195	1.3368
122	249.7	235.0	0.03548	0.415	279.4	404.1	124.7	1.1218	1.3366
124	255.7	241.0	0 03564	0.404	280.9	404.5	123.6	1.1241	1.3363
126	261.7	247.0	0.03580	0.393	282.3	404.8	122.5	1.1264	1.3361
128	267.9	253.2	0.03596	0.382	283.8	405.2	121.4	1.1287	1.3358

TABLE E-16.—THERMODYNAMIC PROPERTIES OF SATURATED PROPANE (*Concluded*)

Temp. t, °F	Pressure		Volume, cu ft/lb		Heat content, Btu/lb		Latent heat L, Btu/lb	Entropy, Btu/(lb)(°F)	
	Psia p	Psig gp	Liquid v	Vapor V	Liquid h	Vapor H		Liquid s	Vapor S
130	274.5	259.8	0.03612	0.370	285.2	405.4	120.2	1.1310	1.3356
132	281.1	266.4	0.03630	0.360	286.7	405.7	119.0	1.1334	1.3353
134	287.9	273.2	0.03648	0.350	288.2	406.1	117.9	1.1358	1.3350
136	294.7	280.0	0.03666	0.340	289.7	406.4	116.7	1.1382	1.3348
138	301.4	286.7	0.03684	0.330	291.2	406.7	115.5	1.1406	1.3345
140	308.4	293.7	0.03702	0.320	292.7	407.0	114.3	1.1430	1.3347
142	315.5	300.8	0.03725	0.312	294.2	407.3	113.1	1.1454	1.3339
144	322.8	308.1	0.03748	0.303	295.7	407.6	111.9	1.1479	1.3336
146	330.2	315.5	0.03771	0.295	297.2	407.8	110.6	1.1503	1.3332
148	337.6	322.9	0.03794	0.286	298.6	408.0	109.4	1.1528	1.3329
150	345.4	330.7	0.03817	0.278	300.2	408.2	108.0	1.1552	1.3326
152	352.9	338.2	0.03846	0.270	301.8	408.4	106.6	1.1578	1.3321
154	360.8	346.1	0.03875	0.263	303.4	408.6	105.2	1.1603	1.3317
156	368.6	353.9	0.03904	0.255	305.1	408.7	103.6	1.1629	1.3312
158	376.6	361.9	0.03933	0.248	306.8	408.8	102.0	1.1654	1.3308
160	385.0	370.3	0.03962	0.240	308.4	408.8	100.4	1.1680	1.3303
162	392.9	378.2	0.03996	0.234	310.2	408.8	98.6	1.1707	1.3297
164	401.0	386.3	0.04030	0.227	312.0	408.9	96.9	1.1734	1.3291
166	409.3	394.6	0.04064	0.221	313.9	408.9	95.0	1.1762	1.3284
168	417.8	403.1	0.04098	0.214	315.7	408.8	93.1	1.1789	1.3278
170	426.0	411.3	0.04132	0.208	317.5	408.6	91.1	1.1816	1.3272
172	463.4	421.7	0.04179	0.202	319.5	408.6	89.1	1.1847	1.3262
174	445.9	431.2	0.04226	0.197	321.5	408.6	87.1	1.1878	1.3252
176	455.2	440.5	0.04273	0.191	323.5	408.4	84.9	1.1908	1.3243
178	464.1	449.4	0.04320	0.186	325.5	408.2	82.7	1.1939	1.3233
180	473.2	458.5	0.04367	0.180	327.5	407.6	80.1	1.1970	1.3223
182	483.0	468.3	0.04436	0.174	329.8	407.4	77.6	1.2004	1.3210
184	492.9	478.2	0.04505	0.168	332.2	407.1	74.9	1.2038	1.3196
186	503.1	488.4	0.04574	0.161	334.5	406.6	72.1	1.2072	1.3183
188	512.8	498.1	0.04643	0.155	336.9	405.9	69.0	1.2106	1.3169
190	523.4	508.7	0.04712	0.149	339.2	404.6	65.7	1.2140	1.3156
200	575.0	560.3	0.0521	0.113	353.5	398.3	44.8	1.2360	1.3040

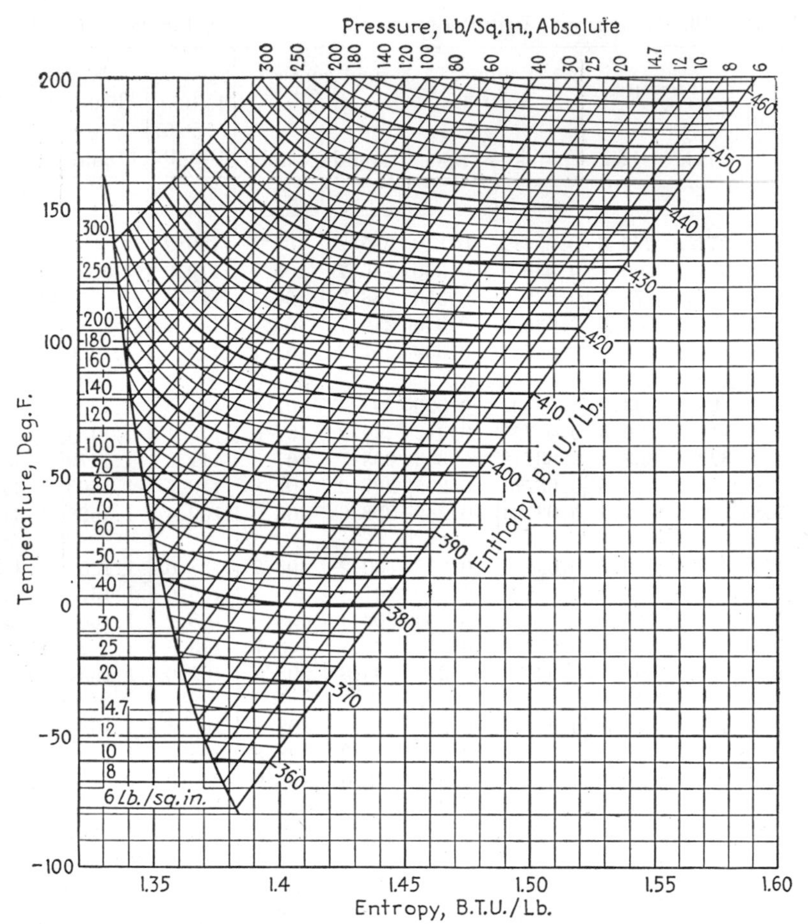

FIG. E-20.—Temperature-entropy chart for propane vapor. [*Stearns and George, Ind. Eng. Chem.*, **35,** 602 (1943). *By permission of the American Chemical Society.*]

TABLE E-17.—SATURATED AMMONIA: TEMPERATURE TABLE
[*Bur. Standards Circ.* 142 (1923)]

Temp. t, °F	Pressure		Volume vapor, cu ft/lb	Density vapor, lb/cu ft	Heat content		Latent heat, Btu/lb	Entropy	
	Psia	Psig			Liquid, Btu/lb	Vapor, Btu/lb		Liquid, Btu/(lb)(°F)	Vapor, Btu/(lb)(°F)
	p	gp	V	$1/V$	h	H	L	s	S
−60	5.55	18.6*	44.73	0.02235	−21.2	589.6	610.8	−0.0517	1.4769
−58	5.93	17.8*	42.05	0.02378	−19.1	590.4	609.5	−0.0464	1.4713
−56	6.33	17.0*	39.56	0.02528	−17.0	591.2	608.2	−0.0412	1.4658
−54	6.75	16.2*	37.24	0.02685	−14.8	592.1	606.9	0.0360	1.4604
−52	7.20	15.3*	35.09	0.02850	−12.7	592.9	605.6	−0.0307	1.4551
−50	7.67	14.3*	33.08	0.03023	−10.6	593.7	604.3	−0.0256	1.4497
−48	8.16	13.3*	31.20	0.03205	− 8.5	594.4	602.9	−0.0204	1.4445
−46	8.68	12.2*	29.45	0.03395	− 6.4	595.2	601.6	−0.0153	1.4393
−44	9.23	11.1*	27.82	0.03595	− 4.3	596.0	600.3	−0.0102	1.4342
−42	9.81	10.0*	26.29	0.03804	− 2.1	596.8	598.9	−0.0051	1.4292
−40	10.41	8.7*	24.86	0.04022	0.0	597.6	597.6	0.0000	1.4242
−38	11.04	7.4*	23.53	0.04251	2.1	598.3	596.2	0.0051	1.4193
−36	11.71	6.1*	22.27	0.04489	4.3	599.1	594.8	0.0101	1.4144
−34	12.41	4.7*	21.10	0.04739	6.4	599.9	593.5	0.0151	1.4096
−32	13.14	3.2*	20.00	0.04999	8.5	600.6	592.1	0.0201	1.4048
−30	13.90	1.6*	18.97	0.05271	10.7	601.4	590.7	0.0250	1.4001
−28	14.71	0.0	18.00	0.05555	12.8	602.1	589.3	0.0300	1.3955
−26	15.55	0.8	17.09	0.05850	14.9	602.8	587.9	0.0350	1.3909
−24	16.42	1.7	16.24	0.06158	17.1	603.6	586.5	0.0399	1.3863
−22	17.34	2.6	15.43	0.06479	19.2	604.3	585.1	0.0448	1.3818
−20	18.30	3.6	14.68	0.06813	21.4	605.0	583.6	0.0497	1.3774
−18	19.30	4.6	13.97	0.07161	23.5	605.7	582.2	0.0545	1.3729
−16	20.34	5.6	13.29	0.07522	25.6	606.4	580.8	0.0594	1.3686
−14	21.43	6.7	12.66	0.07898	27.8	607.1	579.3	0.0642	1.3643
−12	22.56	7.9	12.06	0.08289	30.0	607.8	577.8	0.0690	1.3600
−10	23.74	9.0	11.50	0.08695	32.1	608.5	576.4	0.0738	1.3558
− 8	24.97	10.3	10.97	0.09117	34.3	609.2	574.9	0.0786	1.3516
− 6	26.26	11.6	10.47	0.09555	36.4	609.8	573.4	0.0833	1.3474
− 4	27.59	12.9	9.991	0.1001	38.6	610.5	571.9	0.0880	1.3433
− 2	28.98	14.3	9.541	0.1048	40.7	611.1	570.4	0.0928	1.3393

* Inches of mercury below 1 standard atmosphere (29.92 in.).

TABLE E-17.—SATURATED AMMONIA: TEMPERATURE TABLE (*Continued*)

Temp. t, °F	Pressure, Psia p	Pressure, Psig gp	Volume vapor, cu ft/lb V	Density vapor, lb/cu ft $1/V$	Heat content, Liquid, Btu/lb h	Heat content, Vapor, Btu/lb H	Latent heat, Btu/lb L	Entropy, Liquid, Btu/(lb)(°F) s	Entropy, Vapor, Btu/(lb)(°F) S
0	30.42	15.7	9.116	0.1097	42.9	611.8	568.9	0.0975	1.3352
2	31.92	17.2	8.714	0.1148	45.1	612.4	567.3	0.1022	1.3312
4	33.47	18.8	8.333	0.1200	47.2	613.0	565.8	0.1069	1.3273
6	35.09	20.4	7.971	0.1254	49.4	613.6	564.2	0.1115	1.3234
8	36.77	22.1	7.629	0.1311	51.6	614.3	562.7	0.1162	1.3195
10	38.51	23.8	7.304	0.1369	53.8	614.9	561.1	0.1208	1.3157
12	40.31	25.6	6.996	0.1429	56.0	615.5	559.5	0.1254	1.3118
14	42.18	27.5	6.703	0.1492	58.2	616.1	557.9	0.1300	1.3081
16	44.12	29.4	6.425	0.1556	60.3	616.6	556.3	0.1346	1.3043
18	46.13	31.4	6.161	0.1623	62.5	617.2	554.7	0.1392	1.3006
20	48.21	33.5	5.910	0.1692	64.7	617.8	553.1	0.1437	1.2969
22	50.36	35.7	5.671	0.1763	66.9	618.3	551.4	0.1483	1.2933
24	52.59	37.9	5.443	0 1837	69.1	618.9	549.8	0.1528	1.2897
26	54.90	40.2	5.227	0.1913	71.3	619.4	548.1	0 1573	1.2861
28	57.28	42.6	5.021	0.1992	73.5	619.9	546.4	0.1618	1.2825
30	59.74	45.0	4.825	0.2073	75.7	620.5	544.8	0.1663	1.2790
32	62.29	47.6	4.637	0.2156	77.9	621.0	543.1	0.1708	1.2755
34	64.91	50.2	4.459	0.2243	80.1	621.5	541.4	0.1753	1.2721
36	67.63	52.9	4.289	0.2332	82.3	622.0	539.7	0.1797	1.2686
38	70.43	55.7	4.126	0.2423	84.6	622.5	537.9	0.1841	1.2652
40	73.32	58.6	3.971	0.2518	86.8	623.0	536.2	0.1885	1.2618
42	76.31	61.6	3.823	0.2616	89.0	623.4	534.4	0.1930	1.2585
44	79.38	64.7	3.682	0.2716	91.2	623.9	532.7	0.1974	1.2552
46	82.55	67.9	3.547	0.2819	93.5	624.4	530.9	0.2018	1.2519
48	85.82	71.1	3.418	0.2926	95.7	624.8	529.1	0.2062	1.2486
50	89.19	74.5	3.294	0.3036	97.9	625.2	527.3	0.2105	1.2453
52	92.66	78.0	3.176	0.3149	100.2	625.7	525.5	0.2149	1.2421
54	96.23	81.5	3.063	0.3265	102.4	626.1	523.7	0.2192	1.2389
56	99.91	85.2	2.954	0.3385	104.7	626.5	521.8	0.2236	1.2357
58	103.7	89.0	2.851	0.3508	106.9	626.9	520.0	0.2279	1.2325
60	107.6	92.9	2.751	0.3635	109.2	627.3	518.1	0.2322	1.2294
62	111.6	96.9	2.656	0.3765	111.5	627.7	516.2	0.2365	1.2262
64	115.7	101.0	2.565	0.3899	113.7	628.0	514.3	0.2408	1.2231
66	120.0	105.3	2.477	0.4037	116.0	628.4	512.4	0.2451	1.2201
68	124.3	109.6	2.393	0.4179	118.3	628.8	510.5	0.2494	1.2170

TABLE E-17.—SATURATED AMMONIA: TEMPERATURE TABLE (*Concluded*)

Temp. t, °F	Pressure		Volume vapor, cu ft/lb	Density vapor, lb/cu ft	Heat content		Latent heat, Btu/lb	Entropy	
	Psia	Psig			Liquid, Btu/lb	Vapor, Btu/lb		Liquid, Btu/(lb)(°F)	Vapor, Btu/(lb)(°F)
	p	gp	V	$1/V$	h	H	L	s	S
70	128.8	114.1	2.312	0.4325	120.5	629.1	508.6	0.2537	1.2140
72	133.4	118.7	2.235	0.4474	122.8	629.4	506.6	0.2579	1.2110
74	138.1	123.4	2.161	0.4628	125.1	629.8	504.7	0.2622	1.2080
76	143.0	128.3	2.089	0.4786	127.4	630.1	502.7	0.2664	1.2050
78	147.9	133.2	2.021	0.4949	129.7	630.4	500.7	0.2706	1.2020
80	153.0	138.3	1.955	0.5115	132.0	630.7	498.7	0.2749	1.1991
82	158.3	143.6	1.892	0.5287	134.3	631.0	496.7	0.2791	1.1962
84	163.7	149.0	1.831	0.5462	136.6	631.3	494.7	0.2833	1.1933
86	169.2	154.5	1.772	0.5643	138.9	631.5	492.6	0.2875	1.1904
88	174.8	160.1	1.716	0.5828	141.2	631.8	490.6	0.2917	1.1875
90	180.6	165.9	1.661	0.6019	143.5	632.0	488.5	0.2958	1.1846
92	186.6	171.9	1.609	0.6214	145.8	632.2	486.4	0.3000	1.1818
94	192.7	178.0	1.559	0.6415	148.2	632.5	484.3	0.3041	1.1789
96	198.9	184.2	1.510	0.6620	150.5	632.6	482.1	0.3083	1.1761
98	205.3	190.6	1.464	0.6832	152.9	632.9	480.0	0.3125	1.1733
100	211.9	197.2	1.419	0.7048	155.2	633.0	477.8	0.3166	1.1705
102	218.6	203.9	1.375	0.7270	157.6	633.2	475.6	0.3207	1.1677
104	225.4	210.7	1.334	0.7498	159.9	633.4	473.5	0.3248	1.1649
106	232.5	217.8	1.293	0.7732	162.3	633.5	471.2	0.3289	1.1621
108	239.7	225.0	1.254	0.7972	164.6	633.6	469.0	0.3330	1.1593
110	247.0	232.3	1.217	0.8219	167.0	633.7	466.7	0.3372	1.1566
112	254.5	239.8	1.180	0.8471	169.4	633.8	464.4	0.3413	1.1538
114	262.2	247.5	1.145	0.8730	171.8	633.9	462.1	0.3453	1.1510
116	270.1	255.4	1.112	0.8996	174.2	634.0	459.8	0.3495	1.1483
118	278.2	263.5	1.079	0.9269	176.6	634.0	457.4	0.3535	1.1455
120	286.4	271.7	1.047	0.9549	179.0	634.0	455.0	0.3576	1.1427
122	294.8	280.1	1.017	0.9837	181.4	634.0	452.6	0.3618	1.1400
124	303.4	288.7	0.987	1.0132	183.9	634.0	450.1	0.3659	1.1372
125	307.8	293.1	0.973	1.028	185.1	634.0	448.9	0.3679	1.1358

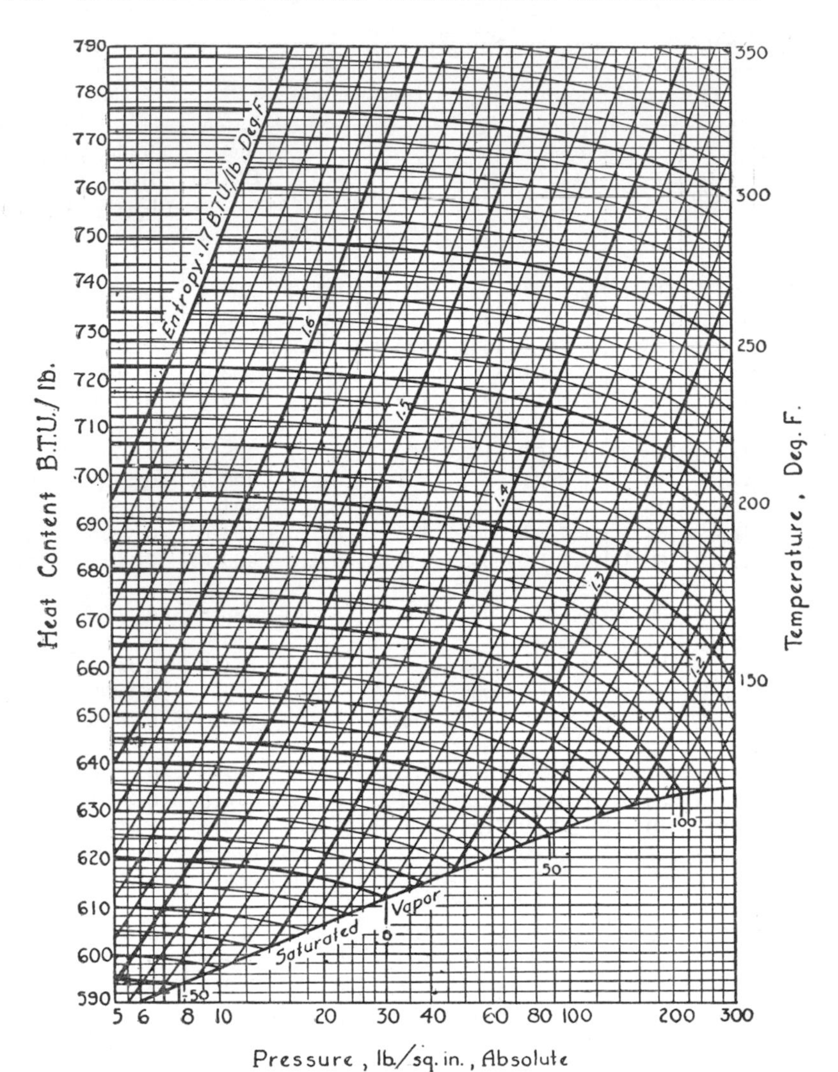

FIG. E-21.—Thermodynamic chart for ammonia vapor. (*Natl. Bur. Stand. Circ.* 142, 1923.)

TABLE E-18.—CARBON DIOXIDE PROPERTIES, SOLID, LIQUID, AND SATURATED VAPOR

("Refrigerant Tables, Charts, and Characteristics, 1959," published by the American Society of Heating, Refrigerating, and Air Conditioning Engineers, by permission)

Temperature, °F	Pressure		Specific volume, cu ft/lb		Enthalpy, datum −40°F, Btu/lb		Entropy, datum −40°F, Btu/lb	
	Psia	Psig	Solid or liquid	Vapor	Solid or liquid	Vapor	Solid or liquid	Vapor
Solid and vapor:								
−147	2.14	25.56*	0.01004	35.80	−123.3	128.2	−0.3214	0.4832
−145	2.43	24.97*	0.01005	32.40	−122.8	128.5	−0.3196	0.4792
−140	3.19	23.41*	0.01007	24.50	−121.4	129.3	−0.3153	0.4691
−135	4.16	21.43*	0.01009	18.70	−120.1	130.0	−0.3110	0.4593
−130	5.39	18.92*	0.01012	14.74	−118.7	130.7	−0.3068	0.4500
−125	6.90	15.08*	0.01015	11.50	−117.4	131.3	−0.3027	0.4409
−120	8.85	11.87*	0.01018	9.13	−116.0	132.1	−0.2986	0.4318
−115	11.20	7.08*	0.01022	7.27	−114.5	132.7	−0.2944	0.4230
−110	14.22	0.92*	0.01024	5.85	−113.1	133.3	−0.2904	0.4145
−109.4	14.70	0.00*	0.01025	5.69	−112.9	133.4	−0.2898	0.4134
−105	17.80	3.13	0.01028	4.72	−111.5	133.9	−0.2860	0.4062
−100	22.34	7.67	0.01032	3.80	−110.0	134.4	−0.2815	0.3981
− 95	27.63	12.96	0.01036	3.09	−108.3	134.9	−0.2768	0.3902
− 90	34.05	19.38	0.01040	2.52	−106.5	135.3	−0.2720	0.3822
− 85	41.67	27.00	0.01044	2.07	−104.5	135.6	−0.2667	0.3742
− 80	50.70	36.03	0.01049	1.70	−102.3	135.8	−0.2610	0.3665
− 79	52.80	38.13	0.01050	1.63	−101.9	135.8	−0.2599	0.3649
− 78	55.00	40.33	0.01051	1.56	−101.4	135.9	−0.2587	0.3633
− 77	57.20	42.53	0.01052	1.51	−101.0	135.9	−0.2575	0.3617
− 76	59.44	44.77	0.01053	1.46	−100.5	135.9	−0.2563	0.3601
− 75	61.75	47.08	0.01054	1.40	−100.1	135.9	−0.2551	0.3585
− 74	64.25	49.58	0.01055	1.35	− 99.7	135.9	−0.2539	0.3570
− 73	66.75	52.08	0.01056	1.30	− 99.2	135.9	−0.2528	0.3554
− 72	69.40	54.73	0.01057	1.25	− 98.8	136.0	−0.2516	0.3539
− 71	72.10	57.43	0.01058	1.21	− 98.4	136.0	−0.2505	0.3523
− 70	74.90	60.23	0.01059	1.17	− 98.0	136.0	−0.2494	0.3508
− 69.9	75.1	60.4	0.01059	1.16	− 97.8	136.0	−0.2493	0.3506
			Triple Point					
Liquid and vapor:								
− 69.9	75.1	60.4	0.01360	1.1570	− 13.7	136.0	−0.0333	0.3506
− 68	78.59	63.92	0.01363	1.1095	− 12.8	136.2	−0.0312	0.3491
− 66	82.42	67.75	0.01369	1.0590	− 11.9	136.3	−0.0290	0.3475
− 64	86.39	71.72	0.01373	1.0100	− 10.9	136.4	−0.0266	0.3460
− 62	90.49	75.82	0.01378	0.9650	− 10.1	136.6	−0.0243	0.3444
− 60	94.75	80.08	0.01384	0.9250	− 9.1	136.7	−0.0221	0.3429
− 58	99.15	84.48	0.01389	0.8875	− 8.2	136.8	−0.0198	0.3413
− 56	103.69	89.02	0.01393	0.8520	− 7.3	137.0	−0.0175	0.3398
− 54	108.40	93.73	0.01398	0.8180	− 6.4	137.1	−0.0153	0.3383
− 52	113.25	98.58	0.01403	0.7840	− 5.5	137.2	−0.0131	0.3368

* Inches of mercury below 1 atm.

TABLE E-18.—CARBON DIOXIDE PROPERTIES, SOLID, LIQUID, AND SATURATED VAPOR (*Concluded*)

Temperature, °F	Pressure		Specific volume, cu ft/lb		Enthalpy, datum −40°F, Btu/lb		Entropy, datum −40°F, Btu/lb	
	Psia	Psig	Solid or liquid	Vapor	Solid or liquid	Vapor	Solid or liquid	Vapor
−50	118.27	103.60	0.01409	0.7500	−4.6	137.3	−0.0109	0.3354
−48	123.45	108.78	0.01414	0.7200	−3.6	137.5	−0.0087	0.3339
−46	128.80	114.13	0.01419	0.6930	−2.7	137.6	−0.0065	0.3325
−44	134.31	119.64	0.01425	0.6660	−1.8	137.7	−0.0043	0.3311
−42	140.00	125.33	0.01430	0.6380	−0.9	137.8	−0.0021	0.3297
−40	145.87	131.20	0.01437	0.6113	0.0	137.9	0.0000	0.3285
−38	151.92	137.25	0.01442	0.5880	1.0	138.0	0.0022	0.3271
−36	158.15	143.48	0.01447	0.5650	1.9	138.1	0.0043	0.3258
−34	164.56	149.89	0.01454	0.5420	2.8	138.2	0.0065	0.3245
−32	171.17	156.50	0.01458	0.5210	3.8	138.3	0.0085	0.3232
−30	177.97	163.30	0.01465	0.5025	4.7	138.3	0.0106	0.3219
−28	184.97	170.30	0.01472	0.4845	5.6	138.4	0.0126	0.3205
−26	192.17	177.50	0.01478	0.4670	6.5	138.5	0.0147	0.3193
−24	199.57	184.90	0.01485	0.4500	7.4	138.6	0.0168	0.3180
−22	207.19	192.58	0.01491	0.4325	8.3	138.6	0.0190	0.3167
−20	215.02	200.35	0.01498	0.4165	9.2	138.7	0.0210	0.3155
−18	223.06	208.39	0.01504	0.4015	10.3	138.7	0.0231	0.3142
−16	231.32	216.65	0.01511	0.3865	11.2	138.8	0.0252	0.3130
−14	239.81	225.14	0.01518	0.3725	12.1	138.8	0.0272	0.3117
−12	248.52	233.85	0.01525	0.3590	12.9	138.8	0.0293	0.3104
−10	257.46	242.79	0.01533	0.3465	13.9	138.9	0.0314	0.3091
−8	266.63	251.96	0.01540	0.3345	15.0	138.9	0.0335	0.3079
−6	276.05	261.38	0.01547	0.3228	15.9	138.9	0.0356	0.3067
−4	285.70	271.03	0.01555	0.3118	16.9	138.9	0.0376	0.3054
−2	295.61	280.94	0.01563	0.3012	17.9	138.9	0.0397	0.3042
0	305.76	291.09	0.01571	0.2905	18.8	138.9	0.0419	0.3030

6. Heats of Reaction

a. Heats of Formation.—Heats of reaction are conveniently summarized in tabular form by the heats of formation from the elements. The heat of formation of a compound is usually defined as the heat that would be absorbed by the spontaneous formation of 1 mole of that substance from the elements when the reaction is conducted at 25°C (77°F) and atmospheric pressure. Actually, of course, very few substances may be so formed, and most heats of formation are, therefore, determined indirectly.

Heats of formation for a few substances are given in Table E-19. For additional heats of formation, the reader is referred to reference texts.[1]

The procedure for using these heats of formation is illustrated by Example E-2.

Heats of reaction computed as in this example are precise, provided the reactants and products are at the standard conditions specified. If high pressures or concentrated solutions are involved, corrections may be necessary.

Very often a heat of reaction may be desired at temperatures other than 77°F. In this case, correction is necessary but can be readily made from specific-heat data by use of Hess' law of constant heat summation. This law states that the heat of reaction is independent of the manner in which it is conducted and depends only on the initial and final states.

If the heat of reaction is desired at some temperature T, it may be computed in three steps: (1) Compute the heat change ΔH_1 required to heat or cool the reactants to 77°F; (2) compute the heat change accompanying reaction ΔH_2 at 77°F from the heats of formation; (3) compute the heat change ΔH_3 required to restore the reaction products to T, the initial temperature. The heat of reaction ΔH will then be obtained by securing the algebraic sum

$$\Delta H = \Delta H_1 + \Delta H_2 + \Delta H_3$$

and reversing the sign. The procedure is illustrated by the second part of Example E-2.

Example E-2. Compute the heat of the reaction $C(s) + CO_2(g) = 2CO(g)$ at 77°F and at 1500°F.

SOLUTION:

1. Heat of reaction at 77°F,

Heat of formation of product,	$H^\circ = 2(-47{,}548) =$		$-95{,}096$ Btu
Heat of formation of reactants,			
1 mole graphite.	$\Delta H^\circ =$	0.0 Btu	
1 mole CO_2.	$\Delta H^\circ =$	$-169{,}143$	
	Total =		$-169{,}143$

[1] Selected Values of Chemical Thermodynamic Properties, *Natl. Bur. Standards Circ.* 500; Stull and Sinke, "Thermodynamic Properties of the Elements"; for values at temperatures other than 77°F see Series III of "Selected Chemical Thermodynamic Properties," issued in sheet form by the National Bureau of Standards, Washington, D.C., 1947–1952.

Enthalpy change accompanying reaction, by difference: products less reactants, ΔH = +74,047 Btu

or

Heat of reaction = $-\Delta H$ = $-74{,}047$ Btu ($-37{,}523$ Btu/lb mole CO formed)

In this case the heat of reaction is negative, *i.e.*, heat is absorbed rather than released by the reaction.

2. Cooling reactants from 1500°F to 77°F,

$$CO_2,\ \Delta H = 0.2 - 44.0 = -43.8 \text{ Btu/SCF (Figs. E-10 and E-11)}$$
$$C,\ \Delta H = (17/60) \times 8 - 506 = -504 \text{ Btu/lb (Table E-4)}$$

Conversion to molar units,

$$CO_2,\ \Delta H = -43.8 \times 378.7 = -16{,}600 \text{ Btu/lb mole}$$
$$C,\ \Delta H = -504 \times 12 = -6{,}050$$
$$\text{Total, } \Delta H = -22{,}650$$

3. Heating product CO from 77 to 1500°F,

$$\Delta H = 29.3 - 0.2 = 29.1 \text{ Btu/SCF}$$

or

$$2 \times 378.7(29.1) = 22{,}040 \text{ Btu}$$

4. Reaction at 1500°F,

Cooling (from 2), ΔH_1 = $-22{,}650$ Btu
Reaction at 77°F (from 1), ΔH_2 = +74,050
Reheating (from 3), ΔH_3 = +22,040
Reaction at 1500°F, ΔH = +73,440
Heat of reaction = $-\Delta H$ = $-73{,}440$ Btu (or, $-36{,}720$ Btu/mole of CO formed)

b. Heat of Combustion.—For combustible materials and combustion products (water, carbon dioxide, sulphur dioxide), the heats of combustion may be used for computation of heats of reaction as if they were the heats of formation. That is, the heat of reaction $-\Delta H$ is the heat of combustion of the reactants less the heat of combustion of the products. Either the gross or net heating values may be so employed. However, if water is one of the reactants or products, the gross value must be used; conversely, if water vapor is involved, the net heat must be employed. Numerical values and definitions of heating values are discussed under Fuels in Chapter H.

TABLE E-19.—HEATS AND FREE ENERGIES OF FORMATION
(Abstracted from "Selected Values of Chemical Thermodynamic Properties," National Bureau of Standards, Washington, D.C., 1947–1952)

Substance, formula, and state*	Heat of formation at 25°C (77°F) from elements		Free energy of formation at 25°C, Kg-cal/g-mole
	Kg-cal/g-mole	MBtu/lb-mole	
Aluminum:			
Metal, Al (*s*)	0.00	0.00	0.00
Chloride, $AlCl_3$	−166.2	−299.2	−152.2
$AlCl_3$ (*aq*, 600)	−245.5	−441.90	
Oxide (α)	−399.09	−718.36	−376.77
Calcium:			
Metal, Ca (*s*)	0.00	0.00	0.00
Carbide, CaC_2 (*s*)	−15.0	−27.0	−16.2
Chloride, $CaCl_2$ (*s*)	−190.0	−342.0	−179.3
$CaCl_2$ (*aq*)	−209.82	−377.67	−194.88
Carbonate, $CaCO_3$ (*s*)	−288.45	−519.21	−269.78
Oxide, CaO (*s*)	−151.9	−273.42	−144.4
Hydroxide, $Ca(OH)_2$ (*s*)	−235.80	−424.44	−214.33
Sulphate, $CaSO_4 \cdot 2H_2O$ (*s*)	−483.06	−869.51	−429.19
Carbon compounds:			
Graphite, C (*s*)	0.00	0.00	0.00
Methane, CH_4 (*g*)	−17.889	−32.200	−12.140
Acetylene, C_2H_2 (*g*)	54.194	97.549	50.000
Ethylene, C_2H_4 (*g*)	12.496	22.493	16.482
Ethane, C_2H_6 (*g*)	−20.236	−36.425	−7.860
Propylene, C_3H_6 (*g*)	4.879	8.782	−14.990
Propane, C_3H_8 (*g*)	−24.820	−44.676	−5.612
Isobutylene, C_4H_8 (*g*)	−3.343	−6.017	14.582
1-Butylene, C_4H_8 (*g*)	0.280	0.504	17.217
Isobutane, C_4H_{10} (*g*)	−31.452	−56.614	−4.296
n-Butane, C_4H_{10} (*g*)	−29.812	−53.662	−3.754
n-Pentane, C_5H_{12} (*g*)	−35.0	−63.00	−1.96
Methyl alcohol, CH_3OH (*l*)	−57.02	−102.64	−39.73
Ethyl alcohol, C_2H_5OH (*l*)	−66.356	−119.441	−41.77
Carbon monoxide, CO (*g*)	−26.4157	−47.548	−32.8079
Carbon dioxide, CO_2 (*g*)	−93.9686	−169.143	−94.0518
Chlorine, Cl_2 (*g*)	0.00	0.00	0.00
Hydrogen compounds (including acids):			
Element, H_2 (*g*)	0.00	0.00	0.00
Acetic acid, CH_3COOH (*l*)	−116.0	−208.8	−93.8
Hydrogen chloride, HCl (*g*)	−22.063	−39.713	−22.769
HCl (*aq*)	−40.023	−72.041	−31.350
Hydrogen cyanide, HCN (*l*)	25.2	45.4	29.0

* See end of Table E-19, page 155.

TABLE E-19.—HEATS AND FREE ENERGIES OF FORMATION (*Continued*)

Substance, formula, and state*	Heat of formation at 25°C (77°F) from elements		Free energy of formation at 25°C, Kg-cal/g-mole
	Kg-cal/g-mole	MBtu/lb-mole	
Hydrogen sulphide, H_2S (*g*)	−4.815	−8.667	−7.892
Nitric acid, HNO_3 (*l*)	−41.404	−74.527	−19.100
Phosphoric acid, H_3PO_4 (*l*)	−229.1	−412.4	
H_3PO_4 (*aq*, 400)	−309.32	−556.78	
Sulphuric acid, H_2SO_4 (*l*)	−193.91	−349.04	
H_2SO_4 (*aq*)	−216.90	−390.42	−177.34
Water, H_2O (*l*)	−68.3174	−122.971	−56.6902
H_2O (*g*)	−57.7979	−104.036	−54.6357
Iron compounds:			
Metal, Fe (*α*)	0.00	0.00	0.00
Ferric chloride, $FeCl_3$ (*s*)	−96.8	−174.2	
$FeCl_3$ (*aq*)	−127.9	−230.2	
Ferrous chloride, $FeCl_2$ (*s*)	−81.5	−146.7	−72.2
$FeCl_2$ (*aq*)	−101.0	−181.8	
Ferric oxide, Fe_2O_3 (*s*)	−196.5	−353.7	−177.1
Magnetic oxide, Fe_3O_4 (*s*)	−267.0	−480.6	−242.4
Magnesium compounds:			
Metal, Mg (*s*)	0.00	0.00	0.00
Chloride, $MgCl_2$ (*s*)	−153.40	−276.12	−141.57
$MgCl_2$ (*aq*)	−190.46	−342.83	−171.69
Oxide, MgO (*s*)	−143.84	−258.91	−136.13
Sulphate, $MgSO_4$ (*s*)	−305.5	−549.9	−280.5
$MgSO_4$ (*aq*)	−327.31	−589.16	−286.33
Nitrogen compounds:			
Element, N_2 (*g*)	0.00	0.00	0.00
Ammonia, NH_3 (*g*)	−11.04	−19.87	−3.976
NH_3 (*aq*)	−19.32	−34.78	−6.37
Ammonium chloride, NH_4Cl (*s*)	−75.38	−135.68	−48.73
NH_4Cl (*aq*, 400)	−71.664	−128.995	
Nitric oxide, NO (*g*)	21.600	38.880	20.719
Nitrogen dioxide, NO_2 (*g*)	8.091	14.564	12.390
Oxygen, O_2 (*g*)	0.00	0.00	0.00
Phosphorus, white, P_4 (*s*)	0.00	0.00	0.00
Potassium compounds:			
Metal, K (*s*)	0.00	0.00	0.00
Chloride, KCl (*s*)	−104.175	−187.515	−97.596
KCl (*aq*)	−100.06	−180.11	−98.816
Hydroxide, KOH, (*s*)	−101.78	−183.20	
KOH (*aq*)	−115.09	−207.16	−105.061
Sodium compounds:			
Metal, Na (*s*)	0.00	0.00	0.00

* See end of Table E-19, page 155.

TABLE E-19.—HEATS AND FREE ENERGIES OF FORMATION (*Concluded*)

Substance, formula, and state*	Heat of formation at 25°C (77°F) from elements		Free energy of formation at 25°C, Kg-cal/g-mole
	Kg-cal/g-mole	MBtu/lb-mole	
Chloride, NaCl (*s*)	−98.232	−176.818	−91.785
NaCl (*aq*)	−97.302	−175.144	−93.939
Carbonate, Na_2CO_3 (*s*)	−270.3	−486.5	−250.4
Na_2CO_3 (*aq*, 400)	−275.9	−496.6	
Bicarbonate, $NaHCO_3$ (*s*)	−226.5	−407.7	−203.6
$NaHCO_3$ (*aq*)	−222.5	−400.5	
Hydroxide, NaOH (*s*)	−101.99	−183.58	
NaOH (*aq*)	−112.236	−202.025	−100.184
Nitrate, $NaNO_3$ (*s*)	−111.54	−200.77	−87.45
$NaNO_3$ (*aq*)	−106.651	−191.972	−89.00
Sulphate, Na_2SO_4 (*s*)	−330.95	−595.710	−302.78
Na_2SO_4 (*aq*)	−331.46	−596.63	−302.52
Sulphide, Na_2S (*s*)	−89.2	−160.6	
Na_2S (*aq*, 400)	−104.3	−187.7	
Sulphur compounds:			
Rhombic, S (*s*)	0.00	0.00	0.00
Dioxide, SO_2 (*g*)	−70.96	−127.73	−71.79
Trioxide, SO_3 (*g*)	−94.45	−170.01	−98.52

* Symbol in parenthesis indicates state: (*s*) denotes solid; (*g*) gas; (α) crystal form; (*aq*) dilute water solution; and (*aq*, 400) solution of 1 mole in 400 moles of water. Values of free energies for solution based on one molal activity.

7. Heat Balances

The principle of heat balance is the conservation of energy; roughly stated, "energy in = energy out." This principle is used in nearly all process calculations of steam, power, and cooling-water requirements. The term "heat balance" is usually reserved for calculations made in a formal manner. (The term heat balance is used by common consent because heat is usually the predominant item; however, it is really the energy that is balanced.) There are many systems of computing the Btu's in and out. One such system of bookkeeping for a flowing system maintained at steady temperatures and pressures is roughly indicated below.

Heat input includes

1. Heat content (relative to same standard temperature and pressure) of all materials entering system
2. Heat of reactions taking place in system (heat computed at standard temperature and pressure used for materials entering)

3. Heat supplied by steam, fuel, etc.
4. Work done on system—compression pumping, etc.

(1 hp-hr = 2544 Btu; 1 watt-hr = 3.345 Btu)

5. Kinetic energy of materials entering system,

$$E_K = \frac{Wv^2}{50{,}000}$$

where E_K = kinetic energy, Btu/hr
W = mass of flowing material, lb/hr
v = velocity, fps

6. Potential energy of all materials entering system relative to some fixed height (datum plane) such as the lowest point of system

Heat output includes

1. Heat content of all materials leaving the system
2. Work done by system—expansion engines, etc.
3. Heat leaks to atmosphere
4. Heat removed by cooling water or refrigeration
5. Kinetic energy of all materials leaving
6. Potential energy of all materials leaving

Other items may have to be included in special cases. Frequently, different systems of accounting will be found more suitable. For example, in cracking units a large number of reactions proceed simultaneously and make direct computation difficult. However, the net effect of heats of all reactions may be taken care of by including the heat of formation of materials charged in "heat in" and the heat of formation of materials leaving in "heat out." If only combustible materials and their combustion products are handled, the heat of combustion may be used in place of the heat of formation.

8. Equilibrium Data and Free Energies

Equilibrium involved in a chemical reaction is conveniently expressed as the equilibrium constant, and equilibrium constants are related to the free-energy change as follows:

$$-\frac{\Delta F}{T} = R \ln K = 4.577 \log K \qquad \text{(E-18)}$$

where ΔF = change in free energy accompanying reaction (for the number of moles indicated by the reaction equation)
K = equilibrium constant for reaction

T = absolute temperature
R = gas-law constant, same units as $\Delta F/T$
$R \ln K = 4.577 \log K$ when $\Delta F/T$ is in cal/(g moles)(°K) or in Btu/(lb moles)(°R)

Figure E-22 may be used instead of Eq. (E-18). The accuracy is low, but possibility of large error is reduced, and the reader who makes many such computations may prepare a chart of larger scale on semilogarithmic paper.

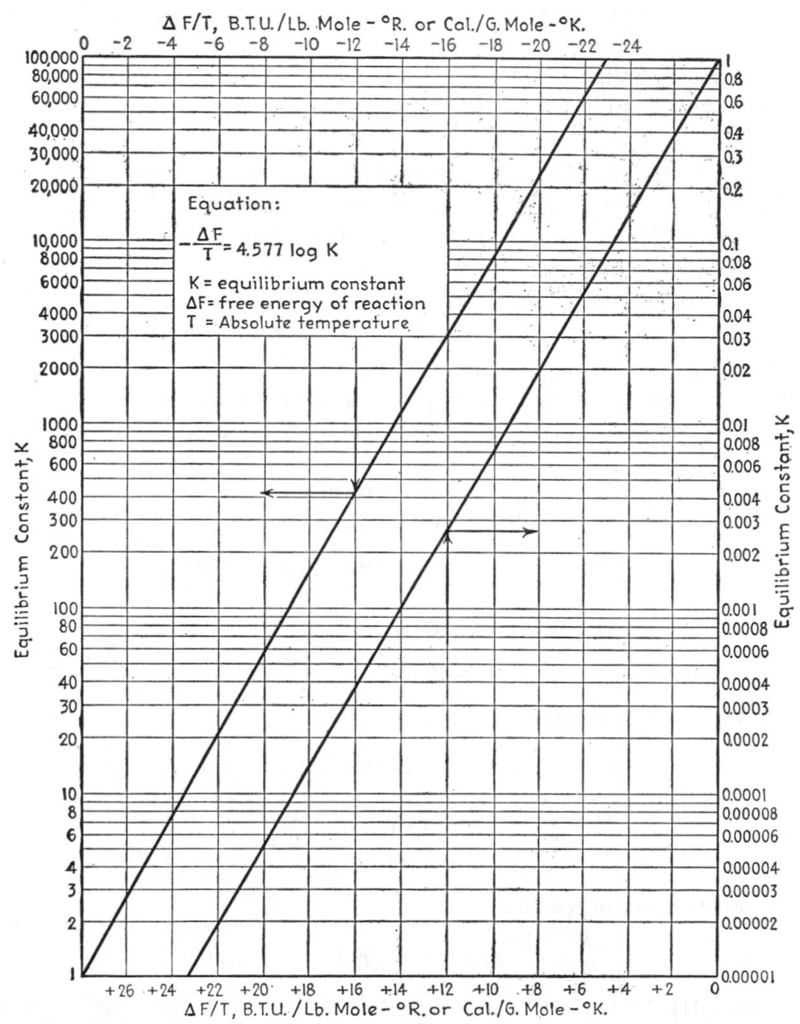

FIG. E-22.—Chart for computing equilibrium constants from free energies.

Thus, free-energy data can be used as a convenient method of summarizing chemical equilibrium constants. Both equilibrium constants and free energies are defined in the next section covering basic equations. For engineering purposes, interest in free energies is largely confined to changes accompanying reactions, and these can be summarized as free energies of formation of individual substances, defined in the same way as heats of formation. Table E-19 tabulates the free energies of formation of a number of substances along with heats of formation. The standard states used in this tabulation are 25°C (77°F) and 1 atm for pure substances, 25°C and hypothetical ideal one molal for aqueous solutions. Thus equilibrium constants calculated therefrom will be in units as follows: gases and vapors, atmospheres partial pressure; liquids, mole fraction; aqueous solutions, molalities.

Equilibrium constants are defined in terms of ideal phase behavior, and correction is frequently required for actual deviation through use of activity coefficients (see p. 164, also Chapter L).

Free-energy data for other temperatures are available[1] or can be obtained from available data by use of Eqs. (E-56) and (E-57). The use of free-energy data is illustrated by Example E-3.

Example E-3. Compute the equilibrium constant for the reaction $CH_4(g) + H_2O(g) = CO(g) + 3H_2(g)$ at 1700°F (1200°K), and find the amount of conversion achieved when 2 moles of steam per mole of methane are reacted at 1 atm.

SOLUTION: Free-energy data from literature

1. Free energy of formation of products,

$$\text{CO}: \frac{\Delta F^\circ}{T} = -43.63 \text{ Btu/(lb mole)(°R)}$$

$$3\text{H}_2: \frac{\Delta F^\circ}{T} = 0.00$$

$$\text{Total} = -43.63$$

2. Free energy of formation of reactants,

$$\text{CH}_4: \frac{\Delta F^\circ}{T} = 8.14 \text{ Btu/(lb mole)(°R)}$$

$$\text{H}_2\text{O}: \frac{\Delta F^\circ}{T} = -36.15$$

$$\text{Total} = -28.01$$

3. Free energy of reaction,

$$\frac{\Delta F^\circ}{T} = -43.63 - (-28.01) = -15.62$$

[1] Series III, "Selected Values of Chemical Thermodynamic Properties," issued in sheet form by the National Bureau of Standards, Washington, D.C., 1947–1952.

4. Equilibrium constant,

$$\log K = \frac{1}{4.577} \frac{-\Delta F}{T} = \frac{15.62}{4.577} = 3.410$$

$$K = 2{,}570 = \frac{(CO)(H_2)^3}{(CH_4)(H_2O)}$$

5. Methane conversion:

Trial assumption, conversion of methane is nearly complete after reaction of 100 moles CH_4 + 200 moles H_2O. Moles of unconverted methane, $CH_4 = y$, are to be computed.

$H_2O = 200 - (100 - y) = 100 + y$	100
$CO = 100 - y$	100
$H_2 = 300 - 3y$	300
Total $= 500 - 3y$	500

Partial pressure in atmospheres = mole fractions.

$$CH_4 = y/500 = 0.002y$$
$$H_2O = 100/500 = 0.2$$
$$CO = 100/500 = 0.2$$
$$H_2 = 300/500 = 0.6$$

$$2{,}570 = \frac{(CO)(H_2)^3}{(CH_4)(H_2O)} = \frac{0.2 \times (0.6)^3}{0.002y \times 0.2} = \frac{108}{y}$$

or

$$y = \frac{108}{2{,}570} = 0.04 \text{ mole/100 moles } CH_4 \text{ charged}$$

This is nearly complete conversion justifying the assumption made and the neglect of y in computing the number of moles of H_2O, CO, and H_2.

DISCUSSION: This computation shows that *if* this reaction is rapid and *if* no side reactions are involved, complete conversion of methane may be effected at 1700°F. This reaction is the basis of one commercial process for production of hydrogen. In practice a catalyst (such as nickel) is employed to speed the reaction. Some CO_2 is also formed in accordance with the reaction

$$CO + H_2O = CO_2 + H_2$$

9. Basic Laws and Equations of Physical Thermodynamics

Although laws of thermodynamics are universally applicable, they may be expressed in relatively simple forms when no *changes* in composition are involved. Hence, the distinction between physical and chemical thermodynamics.

a. Law of Conservation of Mass.—This law states that mass is indestructible and may not be destroyed or created by any process.

b. First Law of Thermodynamics. The Law of Conservation of Energy.—This law states that energy is also indestructible, though one form of energy may disappear and reappear as a different form. For example, if a weight is dragged along the ground, the work expended is not lost but is converted to heat.

c. Second Law of Thermodynamics.—There are many ways of stating this law, all of which appear unrelated except that they are uniformly confusing to the novice, and nothing can be done about it. One such statement is that heat cannot be transferred from one temperature to a higher temperature without the expenditure of work. Equation (E-32) is a mathematical statement of this law. The various statements of the law are related to each other as the hen to the egg; the others all follow from the one chosen (arbitrarily) as first.

d. Thermodynamic Functions

Quantity	Symbol	Definition
Energy content (internal energy)	U	By properties of system
Heat content (enthalpy)	H	$H = U + pV$
Entropy	S	By properties of system
Helmholz free energy	A	$A = U - TS$
Free energy (chemical potential)	F	$F = U + pV - TS$
Absolute temperature	T	Thermodynamic scale
Pressure (absolute)	p	External force/area
Volume	V	By dimensions of system
Specific heat at constant volume	C_v	$C_v = \left(\frac{\partial U}{\partial T}\right)_v = \left(\frac{\partial Q}{\partial T}\right)_v$
Specific heat at constant pressure	C_p	$C_p = \left(\frac{\partial H}{\partial T}\right)_p = \left(\frac{\partial Q}{\partial T}\right)_p$
Heat absorbed by system	Q	Experiment
Work done by system	w	Experiment

The absolute values of the quantities U, H, A, and F are not specified but must be arbitrarily assigned a given value at some standard state of the system. For example, the values of heat contents and entropies given in steam tables are based on $H = S = 0$ for liquid water at 32°F and 1 atm pressure. Entropy does have an absolute value, but this is of no consequence for physical thermodynamics, though it is of value in chemical thermodynamics and statistical mechanics. It may be noted that pure thermodynamics defines the quantities only *mathematically*. If some of the quantities are found experimentally, the thermodynamic equations establish others.

e. Basic Equations, Static Systems.—The following equations are rigorously applicable to any stationary system whether homogeneous or heterogeneous.

Change in energy content at constant volume:

$$\left(\frac{\partial U}{\partial T}\right)_v = C_v = \left(\frac{\partial Q}{\partial T}\right)_v \tag{E-19}$$

or

$$U = U_o + \int C_v \, dT \tag{E-20}$$

Change in heat content at constant pressure:

$$\left(\frac{\partial H}{\partial T}\right)_p = C_p = \left(\frac{\partial Q}{\partial T}\right)_p \tag{E-21}$$

or

$$H = H_0 + \int C_p \, dT \tag{E-22}$$

if

$$C_p = a + bT - cT^2 - \frac{d}{T^2} \tag{E-23}$$

$$H = H_0 + aT + \tfrac{1}{2}bT^2 - \tfrac{1}{3}cT^3 + \frac{d}{T} \tag{E-24}$$

Change in heat content at constant temperature:

$$\left(\frac{\partial H}{\partial p}\right)_T = -\mu C_p \tag{E-25}$$

where $\mu = \left(\frac{\partial T}{\partial p}\right)_H$, the Joule-Thomson coefficient determined by measuring the temperature change accompanying a free expansion under adiabatic (perfectly insulated) conditions

Change in entropy at constant pressure:

$$\left(\frac{\partial S}{\partial T}\right)_p = \frac{c_p}{T} \tag{E-26}$$

$$S = S_o + \int \frac{C_p}{T} \, dT \tag{E-27}$$

Change of entropy at constant temperature:

$$\left(\frac{\partial S}{\partial V}\right)_T = \left(\frac{\partial p}{\partial T}\right)_v \tag{E-28}$$

$$\left(\frac{\partial S}{\partial p}\right)_T = -\left(\frac{\partial V}{\partial T}\right)_p \tag{E-29}$$

Thermodynamic equation of state:

$$p = T\left(\frac{\partial p}{\partial T}\right)_v - \left(\frac{\partial U}{\partial V}\right)_T \tag{E-30}$$

$$V = T\left(\frac{\partial V}{\partial T}\right)_p + \left(\frac{\partial H}{\partial p}\right)_T \tag{E-31}$$

Condition for maximum mechanical efficiency of any cyclical process:

$$\oint \frac{dQ}{T} = 0 \tag{E-32}$$

This is a statement of the second law. (The symbol $\oint$ denotes integration for one complete cycle which returns system to its original state.)

Condition for maximum efficiency of any process:

$$\frac{dQ}{T} = dS \tag{E-33}$$

(Such a process is said to be a "reversible" one.)

Work produced by any reversible process:

$$w = \int p\, dV = -\Delta U + Q \tag{E-34}$$
$$Q = \int T\, dS \tag{E-35}$$

Work produced by a reversible process at constant temperature:

$$w = \int p\, dv = -\Delta A = -\Delta U + Q \tag{E-36}$$
$$Q = T\, \Delta S \tag{E-37}$$

Work produced by a reversible process at adiabatic conditions (no heat loss or gain):

$$w = \int p\, dv = \Delta H \tag{E-38}$$
$$Q = 0, \qquad \Delta S = 0 \tag{E-39}$$

Relation of work produced to heat absorbed for any process:

$$Q - w = \Delta U \tag{E-40}$$

Equilibrium pressure for phase change:

$$\frac{dp}{dT} = \frac{\Delta H}{T\, \Delta V} \qquad \text{(Clapeyron equation)} \tag{E-17a}$$

f. Basic Equations, Flowing Systems.

Kinetic energy of flow:

$$\text{K.E.} = \frac{mu^2}{2g} \tag{E-41}$$

where u = linear velocity (or root of mean square velocity if system is flowing at a nonuniform velocity)

m = weight of system

g = gravitational acceleration

Potential energy:

$$\text{P.E.} = mZ$$

where Z = height above datum plane

Bernoulli's equation:

$$H_1 + mZ_1 + \frac{mu_1^2}{2g} + Q - w = H_2 + mZ_2 + \frac{mu_2^2}{2g} \qquad \text{(E-42)}$$

This equation applies to any flowing system when steady flow is established. The subscript (1) refers to material entering system; the subscript (2) refers to material leaving. Q is the heat added to the flowing material between the entrance and exit points; w is the external work done by flowing material between the entrance and exit points.

10. Basic Laws and Equations of Chemical Thermodynamics

Because of the need of speed even at the expense of *last-place* accuracy, many process calculations are made without the benefit of the precise but time-consuming methods of chemical thermodynamics, except by employing thermochemical tabulations supplied by the research worker. However, such refined calculations are desirable when they can be made.

a. Laws of Chemical Thermodynamics.—The basic laws are identical with physical thermodynamics. An added principle is the third law of thermodynamics which states that the absolute entropy of a perfect crystal is zero when reduced to a temperature of absolute zero. Most substances do form substantially perfect crystals at absolute zero, but many such as carbon monoxide, ice, and spinel minerals form imperfect crystals and do not obey the "law."

Thermodynamic equations may readily be modified to include composition as a variable and thereby treat problems involving composition change (*i.e.*, chemical thermodynamics).

b. Partial Molal Quantities.—These are defined by the following

$$\bar{G}_1 = \frac{\partial G}{\partial n_1}, \qquad \bar{G}_2 = \frac{\partial G}{\partial n_2}, \text{ etc.} \qquad \text{(E-43)}$$

$$G = n_1\bar{G}_1 + n_2\bar{G}_2 + n_3\bar{G}_3 + \cdots \qquad \text{(E-44)}$$

where subscripts designate components 1, 2, 3, etc.

G = any extensive property[1] such as V, H, U, A, F, C_p, C_v

n_1 = number of moles, component 1, etc.

Partial molal heat contents, free-energy volumes, heat capacities, etc., may be used directly in the equations of physical thermodynamics, provided that all components are treated in computing over-all effects.

c. Fugacity and Activity.—The equations just given are precise but not always convenient because of the difficulty of experimentally

[1] An extensive property is proportional to quantity of material. T, p, $\bar{V}_1$, $\bar{H}_1$, etc., are independent of quantity and are called 'intensive" properties.

determining the partial molal volumes, heats, etc. The use of partial pressures is much more convenient than the use of partial volumes. Consequently the fugacity, an idealized partial pressure, has been devised. The partial pressure of a gas or vapor is defined by:

$$p_1 = N_1 p, \qquad p_2 = N_2 p, \text{ etc.} \tag{E-45}$$

where p_1 = partial pressure, component 1

N_1 = mole fraction of component 1 in the gaseous phase

When the perfect gas law is obeyed, as it is at low pressures,

$$f_1 = p_1 \tag{E-46}$$

where f_1 = fugacity of component 1.

For other conditions (including the solid and liquid phases) the fugacity is defined by the following relation:

$$\bar{F}_1 = RT \ln f_1 + B_1 \tag{E-47}$$

where R = gas-law constant

B_1 = arbitrary constant whose value is so chosen that Eq. (E-46) holds when the conditions for validity are fulfilled

For liquid and solid solutions, it may be more convenient to use the activity, which is an idealized mole fraction and is defined in the same way as the fugacity, except that the additive constant is different:

$$\bar{F}_1 = RT \ln a_1 + F_1^0 \tag{E-48}$$

where a_1 = activity of component 1

F_1^0 = free energy per mole of component in its pure liquid state at same temperature (or solid state, if a solid solution)

For aqueous solutions of inorganic solids the activity is defined as the idealized molality.

For convenience, activity and fugacity data are usually given as activity coefficients, which are defined below:

$$\gamma_1 = \frac{f_1}{p_1} \qquad \text{(gases)} \tag{E-49}$$

$$\gamma_1 = \frac{a_1}{N_1} \qquad \text{(solutions)} \tag{E-50}$$

$$\gamma_1 = \frac{a_1}{m_1} \qquad \text{(aqueous solutions)} \tag{E-51}$$

where γ_1 = activity coefficient of component 1

m_1 = molality (g moles/kg of water)

The basic law of ideal solution is Raoult's law:

$$p_1 = p_1^0 N_1 \tag{E-52}$$

where p_1^0 = vapor pressure, pure liquid component 1

N_1 = mole fraction, component 1

This law is more precisely stated,

$$f_1 = f_1^0 N_1$$

where f_1^0 = fugacity of pure liquid, component 1
f_1 = fugacity of component 1 in mixture

d. Thermodynamics of Chemical Reactions.—A thermodynamic equation of a reaction states the thermodynamic state of the substance in parenthesis after the formula as follows:

(g) gaseous state; standard state at fugacity of 1 atm, unless otherwise stated
(l) liquid or liquid solution; standard state is pure liquid at 1 atm, unless otherwise stated
(aq) aqueous solution; standard state at activity = 1 molal
(s) solid; standard state is pure solid unless otherwise stated
(α), (β), etc., denote different crystalline forms of solids

Below are given three such equations along with their equilibrium constants.

Equation	Equilibrium constant	
	Exact	Approximate
A. $H_2O(l) + C_2H_4(g) = C_2H_5OH(l)$	$K_A = \frac{a_{C_2H_5OH}}{a_{H_2O} f_{C_2H_4}}$	$K_A = \frac{N_{C_2H_5OH}}{N_{H_2O} p_{C_2H_4}}$
*B.** $2Cu^{++}(aq) + 5I^-(aq) = 2CuI(s) + I_3^-(aq)$	$K_B = \frac{a_{I_3^-}}{a_{Cu^{++}}^2 a_{I^-}^5}$	$K_B = \frac{m_{I_3^-}}{m_{Cu^{++}}^2 m_{I^-}^5}$
C. $\frac{1}{2}N_2(g) + \frac{3}{2}H_2(g) = NH_3(g)$	$K_C = \frac{f_{NH_3}}{f_{N_2}^{1/2} f_{H_2}^{3/2}}$	$K_C = \frac{p_{NH_3}}{p_{N_2}^{1/2} p_{H_2}^{3/2}}$

* CuI does not form solid solutions with any component present; therefore $a_{CuI} = 1$ and does not appear.

The thermodynamic reaction equations are regarded as expressing the number of moles of the stated substances, that is, $2H_2O(l)$ signifies 2 moles of water. Since the thermodynamic reaction equations specify both the amount and state of the reactants and products, one may speak of the thermodynamic properties of a reaction much the same as one speaks of the thermodynamic properties of a substance.

For example, the entropy of reaction B, ΔS_B is the sum of the entropies of the products less the entropies of the reactants

$$\Delta S_B = 2S_{CuI} + S_{I_3^-} - 5S_{I^-} - 2S_{Cu^{++}}$$

Similarly,

ΔC_p = heat capacity of reaction
ΔH = heat content of reaction ($-\Delta H$ = heat of reaction)
ΔF = free energy of reaction
ΔV = volume change of reaction

These Δ quantities obey the same equations for their mutual relations as do H, F, S, V, and C_p for single substances.

Because H and F have no absolute, but only relative, values ΔH and ΔF cannot be computed from the properties of reactants and products until a standard state is established which includes chemical as well as physical factors. Fortunately such a standard may be obtained by arbitrarily taking $H = F = 0$ for the elements at 1 atm in their most stable state. The standard heat of formation, $\Delta H°$, is the heat of formation of the substance in question from the elements. The standard free energy of formation is similarly defined. For example, reaction C is the reaction for formation of ammonia. Hence,

$$\Delta H^{\circ}_{NH_3} = \Delta H_c$$
$$\Delta F^{\circ}_{NH_3} = \Delta F_c$$

From these standard quantities the change in heat content and free energy accompanying any reaction may be computed as follows,

$$\Delta H_A = \Delta H^{\circ}_{C_2H_5OH} - \Delta H^{\circ}_{C_2H_4} - \Delta H^{\circ}_{H_2O}$$
$$\Delta F_A = \Delta F^{\circ}_{C_2H_5OH} - \Delta F^{\circ}_{C_2H_4} - \Delta F^{\circ}_{H_2O}$$

Heat of reaction:

The heat liberated by spontaneous reaction at constant temperature and pressure is the *decrease* in heat content:

$$Q_R = \Delta H = \text{heat absorbed}$$
$$-Q_R = \text{heat of reaction}$$

Variation of heat of reaction with temperature:

$$\frac{d\Delta H}{dT} = \Delta C_p = \Delta a + \Delta bT - \Delta cT^2 - \frac{\Delta d}{T^2} \quad \text{(E-53)}$$

$$\Delta H = \Delta H_0 + \int \Delta C_p \, dT \quad \text{(E-54)}$$

$$= \Delta H_0 + \Delta aT + \frac{\Delta b}{2} T^2 - \frac{\Delta c}{3} T^2 + \frac{\Delta d}{T}$$

Variation of free energy of reaction with temperature:

$$\frac{d\Delta F}{dT} = -\Delta S \quad \text{(E-55)}$$

$$\frac{d(\Delta F/T)}{dT} = \frac{-\Delta H}{T^2} \quad \text{(E-56)}$$

$$\frac{\Delta F}{T} = \frac{\Delta H_0}{T} - \Delta a \ln T - \frac{\Delta b}{2} T + \frac{\Delta c}{6} T^2 - \frac{\Delta d}{2T^2} + I \quad \text{(E-57)}$$

where I is the constant of integration.

Relation between the free energy and equilibrium constant:

$$-\frac{\Delta F}{T} = R \ln K \quad \text{(E-18)}$$

CHAPTER F

PIPING, HYDRAULICS, AND FLOW MEASUREMENTS

Here are presented data for sizing pipes, determining pressure drops, and for flow measurements. A number of excellent texts are available on these subjects. A few are listed below:

Goldstein (editor), "Modern Developments in Fluid Dynamics," Oxford University Press, London, 1938.

Vennard, "Elementary Fluid Mechanics," 3d ed., John Wiley & Sons, Inc., New York, 1954.

Russell, "Textbook on Hydraulics," 5th ed., Henry Holt and Company, Inc., New York, 1942.

Crocker, "Piping Handbook," 4th ed., McGraw-Hill Book Company, Inc., New York, 1945.

M. W. Kellogg Co., "Design of Piping Systems," 2d ed., John Wiley & Sons, Inc., New York, 1956.

Littlejohn, "Industrial Piping," McGraw-Hill Book Company, Inc., New York, 1951.

"Pipe Friction Manual," Hydraulic Institute, New York, 1954.

1. Viscosity Data

Viscosity data are presented as follows:

Liquids	Table F-1 and Fig. F-1
Gases at atmospheric pressure	Table F-2 and Fig. F-2
Steam	Table F-3
Compressed gases	Fig. F-3
Hydrocarbon liquids and vapors	Fig. F-4

Figure F-3 is based on the reduced equation of state. It is in fair agreement with most available data but should not be used when direct measurements are available. The critical pressure and temperature must be known for use of this figure. Values for a few substances are given in Table E-2.

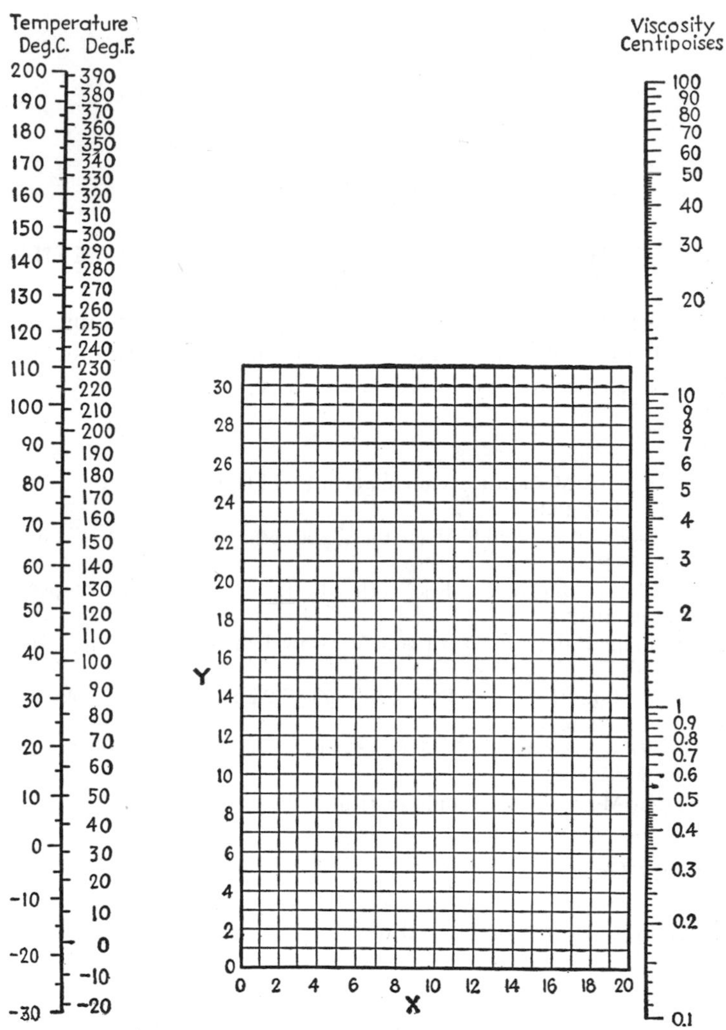

FIG. F-1.—Viscosities of liquids at 1 atm. For coordinates, see Table F-1. (*Perry, "Chemical Engineers' Handbook," 3d ed., McGraw-Hill Book Company, Inc., New York,* 1950.)

TABLE F-1.—VISCOSITIES OF LIQUIDS. COORDINATES FOR FIG. F-1
(Perry, "Chemical Engineers' Handbook," 3d ed., McGraw-Hill Book Company, Inc., New York, 1950)

No.	Liquid	*X*	*Y*	No.	Liquid	*X*	*Y*
1	Acetaldehyde	15.2	4.8	56	Freon-22	17.2	4.7
2	Acetic acid, 100%	12.1	14.2	57	Freon-113	12.5	11.4
3	Acetic acid, 70%	9.5	17.0	58	Glycerol, 100%	2.0	30.0
4	Acetic anhydride	12.7	12.8	59	Glycerol, 50%	6.9	19.6
5	Acetone, 100%	14.5	7.2	60	Heptene	14.1	8.4
6	Acetone, 35%	7.9	15.0	61	Hexane	14.7	7.0
7	Allyl alcohol	10.2	14.3	62	Hydrochloric acid, 31.5%	13.0	16.6
8	Ammonia, 100%	12.6	2.0	63	Isobutyl alcohol	7.1	18.0
9	Ammonia, 26%	10.1	13.9	64	Isobutyric acid	12.2	14.4
10	Amyl acetate	11.8	12.5	65	Isopropyl alcohol	8.2	16.0
11	Amyl alcohol	7.5	18.4	66	Kerosene	10.2	16.9
12	Aniline	8.1	18.7	67	Linseed oil, raw	7.5	27.2
13	Anisole	12.3	13.5	68	Mercury	18.4	16.4
14	Arsenic trichloride	13.9	14.5	69	Methanol, 100%	12.4	10.5
15	Benzene	12.5	10.9	70	Methanol, 90%	12.3	11.8
16	Brine, $CaCl_2$, 25%	6.6	15.9	71	Methanol, 40%	7.8	15.5
17	Brine, NaCl, 25%	10.2	16.6	72	Methyl acetate	14.2	8.2
18	Bromine	14.2	13.2	73	Methyl chloride	15.0	3.8
19	Bromotoluene	20.0	15.9	74	Methyl ethyl ketone	13.9	8.6
20	Butyl acetate	12.3	11.0	75	Naphthalene	7.9	18.1
21	Butyl alcohol	8.6	17.2	76	Nitric acid, 95%	12.8	13.8
22	Butyric acid	12.1	15.3	77	Nitric acid, 60%	10.8	17.0
23	Carbon dioxide	11.6	0.3	78	Nitrobenzene	10.6	16.2
24	Carbon disulfide	16.1	7.5	79	Nitrotoluene	11.0	17.0
25	Carbon tetrachloride	12.7	13.1	80	Octane	13.7	10.0
26	Chlorobenzene	12.3	12.4	81	Octyl alcohol	6.6	21.1
27	Chloroform	14.4	10.2	82	Pentachloroethane	10.9	17.3
28	Chlorosulphonic acid	11.2	18.1	83	Pentane	14.9	5.2
29	Chlorotoluene, ortho	13.0	13.3	84	Phenol	6.9	20.8
30	Chlorotoluene, meta	13.3	12.5	85	Phosphorus tribromide	13.8	16.7
31	Chlorotoluene, para	13.3	12.5	86	Phosphorus trichloride	16.2	10.9
32	Cresol, meta	2.5	20.8	87	Propionic acid	12.8	13.8
33	Cyclohexanol	2.9	24.3	88	Propyl alcohol	9.1	16.5
34	Dibromoethane	12.7	15.8	89	Propyl bromide	14.5	9.6
35	Dichloroethane	13.2	12.2	90	Propyl chloride	14.4	7.5
36	Dichloromethane	14.6	8.9	91	Propyl iodide	14.1	11.6
37	Diethyl oxalate	11.0	16.4	92	Sodium	16.4	13.9
38	Dimethyl oxalate	12.3	15.8	93	Sodium hydroxide, 50%	3.2	25.8
39	Diphenyl	12.0	18.3	94	Stannic chloride	13.5	12.8
40	Dipropyl oxalate	10.3	17.7	95	Sulphur dioxide	15.2	7.1
41	Ethyl acetate	13.7	9.1	96	Sulphuric acid, 110%	7.2	27.4
42	Ethyl alcohol, 100%	10.5	13.8	97	Sulphuric acid, 98%	7.0	24.8
43	Ethyl alcohol, 95%	9.8	14.3	98	Sulphuric acid, 60%	10.2	21.3
44	Ethyl alcohol, 40%	6.5	16.6	99	Sulphuryl chloride	15.2	12.4
45	Ethyl benzene	13.2	11.5	100	Tetrachloroethane	11.9	15.7
46	Ethyl bromide	14.5	8.1	101	Tetrachloroethylene	14.2	12.7
47	Ethyl chloride	14.8	6.0	102	Titanium tetrachloride	14.4	12.3
48	Ethyl ether	14.5	5.3	103	Toluene	13.7	10.4
49	Ethyl formate	14.2	8.4	104	Trichloroethylene	14.8	10.5
50	Ethyl iodide	14.7	10.3	105	Turpentine	11.5	14.9
51	Ethylene glycol	6.0	23.6	106	Vinyl acetate	14.0	8.8
52	Formic acid	10.7	15.8	107	Water	10.2	13.0
53	Freon-11	14.4	9.0	108	Xylene, ortho	13.5	12.1
54	Freon-12	16.8	5.6	109	Xylene, meta	13.9	10.6
55	Freon-21	15.7	7.5	110	Xylene, para	13.9	10.9

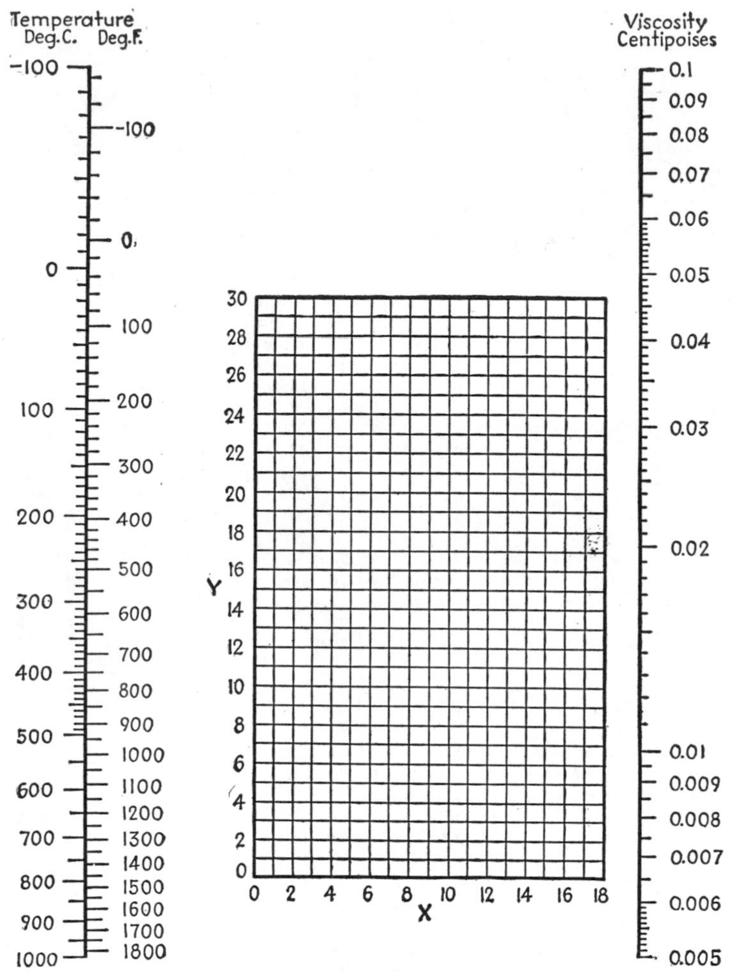

FIG. F-2.—Viscosities of gases at 1 atm. For coordinates, see Table F-2. (*Perry, "Chemical Engineers' Handbook,"* 3d *ed., McGraw-Hill Book Company, Inc., New York,* 1950.)

TABLE F-2.—VISCOSITIES OF GASES. COORDINATES FOR FIG. F-2
(Perry, "Chemical Engineers' Handbook," 3d ed., McGraw-Hill Book Company, Inc., New York, 1950)

No.	Gas	X	Y	No.	Gas	X	Y
1	Acetic acid	7.7	14.3	29	Freon-113	11.3	14.0
2	Acetone	8.9	13.0	30	Helium	10.9	20.5
3	Acetylene	9.8	14.9	31	Hexane	8.6	11.8
4	Air	11.0	20.0	32	Hydrogen	11.2	12.4
5	Ammonia	8.4	16.0	33	$3H_2 + 1N_2$	11.2	17.2
6	Argon	10.5	22.4	34	Hydrogen bromide	8.8	20.9
7	Benzene	8.5	13.2	35	Hydrogen chloride	8.8	18.7
8	Bromine	8.9	19.2	36	Hydrogen cyanide	9.8	14.9
9	Butane	9.2	13.7	37	Hydrogen iodide	9.0	21.3
10	Butylene	8.9	13.0	38	Hydrogen sulphide	8.6	18.0
11	Carbon dioxide	9.5	18.7	39	Iodine	9.0	18.4
12	Carbon disulfide	8.0	16.0	40	Mercury	5.3	22.9
13	Carbon monoxide	11.0	20.0	41	Methane	9.9	15.5
14	Chlorine	9.0	18.4	42	Methyl alcohol	8.5	15.6
15	Chloroform	8.9	15.7	43	Nitric oxide	10.9	20.5
16	Cyanogen	9.2	15.2	44	Nitrogen	10.6	20.0
17	Cyclohexane	9.2	12.0	45	Nitrosyl chloride	8.0	17.6
18	Ethane	9.1	14.5	46	Nitrous oxide	8.8	19.0
19	Ethyl acetate	8.5	13.2	47	Oxygen	11.0	21.3
20	Ethyl alcohol	9.2	14.2	48	Pentane	7.0	12.8
21	Ethyl chloride	8.5	15.6	49	Propane	9.7	12.9
22	Ethyl ether	8.9	13.0	50	Propyl alcohol	8.4	13.4
23	Ethylene	9.5	15.1	51	Propylene	9.0	13.8
24	Fluorine	7.3	23.8	52	Sulphur dioxide	9.6	17.0
25	Freon-11	10.6	15.1	53	Toluene	8.6	12.4
26	Freon-12	11.1	16.0	54	2,3,3-Trimethylbutane	9.5	10.5
27	Freon-21	10.8	15.3	55	Water	8.0	16.0
28	Freon-22	10.1	17.0	56	Xenon	9.3	23.0

TABLE F-3.—VISCOSITY OF STEAM
[Data of Hawkins, Solberg, and Potter, *Trans. Am. Soc. Mech. Engrs.*, **62**, 677 (1940)]
Values in centipoise

Temp.		Pressure, lb force/sq in. abs					
°C	°F	100	200	400	500	600	800
204	400	0.0198	0.0230				
260	500	0.0213	0.0236	0.0272	0.0289	0.0311	
316	600	0.0228	0.0246	0.0279	0.0294	0.0314	0.0350
371	700	0.0243	0.0259	0.0290	0.0304	0.0321	0.0357
427	800	0.0260	0.0275	0.0304	0.0318	0.0334	0.0370
482	900	0.0278	0.0292	0.0320	0.0335	0.0352	0.0390
538	1000	0.0296	0.0310	0.0338	0.0354	0.0372	0.0414

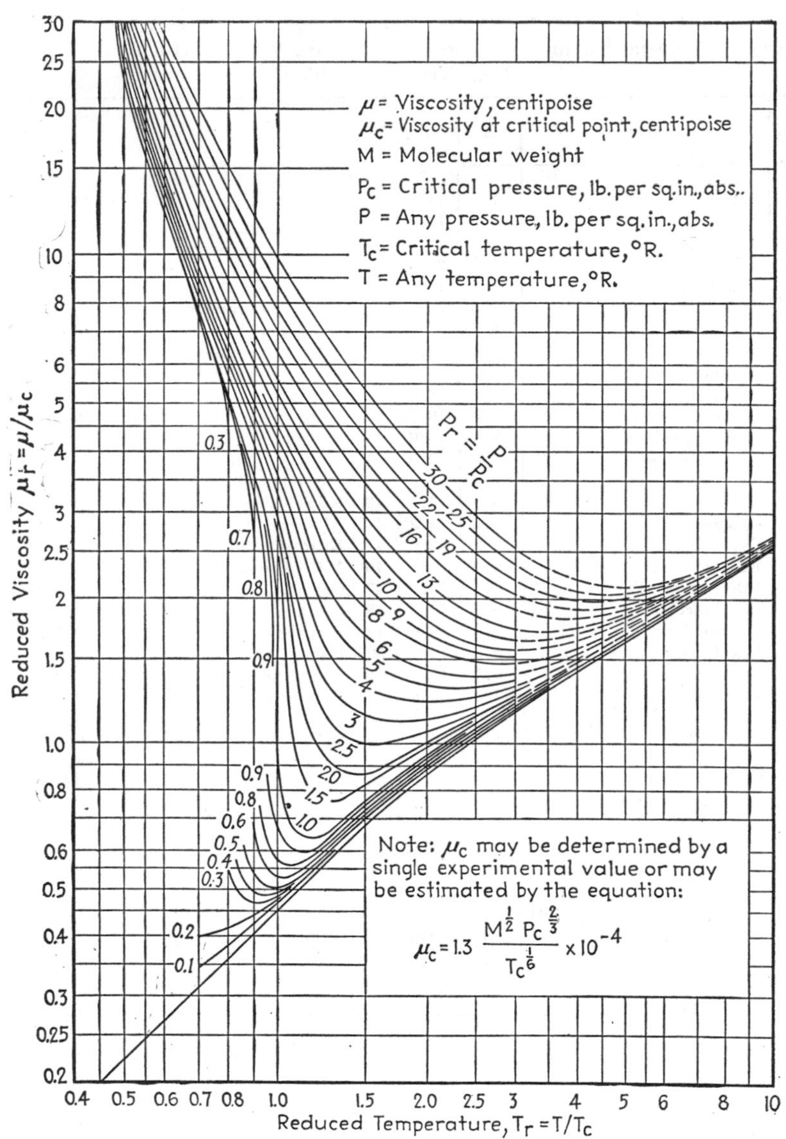

FIG. F-3.—A universal viscosity correlation. (*Uyehara and Watson, Natl. Petroleum News, Tech. Sec., Oct. 4, 1944.*)

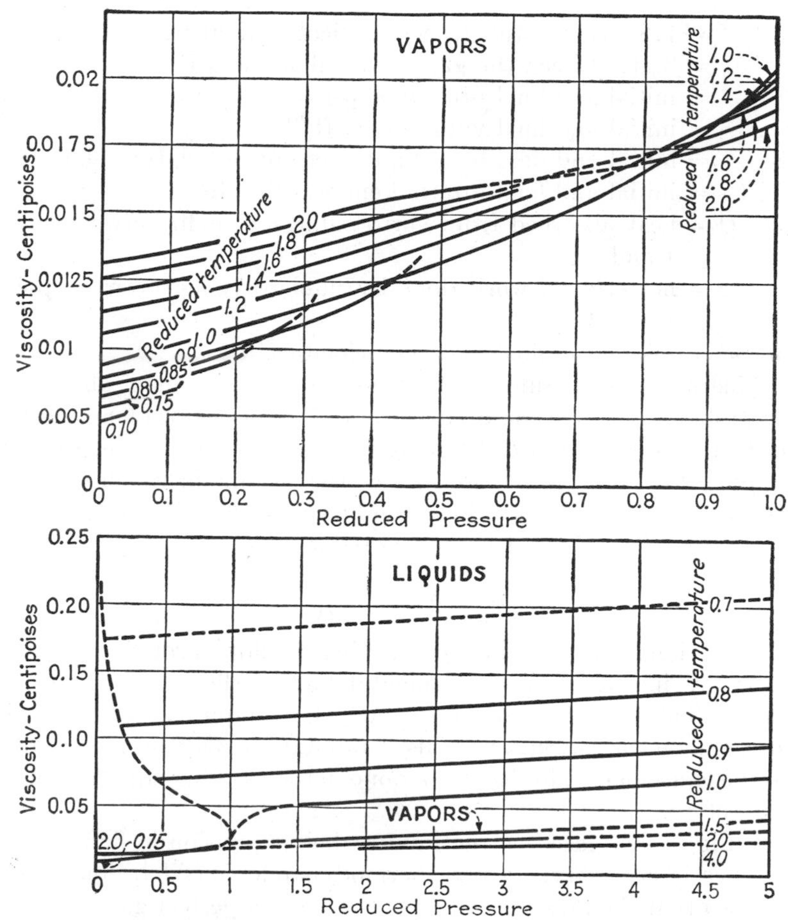

FIG. F-4.—Viscosity of hydrocarbon liquids and vapors. (*Nelson, "Petroleum Refinery Engineering," 2d ed., McGraw-Hill Book Company, Inc., New York,* 1941.)

2. Flow Measurements

a. Basic Equations.—Bernoulli's theorem is a convenient starting point in the study of hydraulics. This theorem results from the application of the first law of thermodynamics to moving fluids, that is, from an energy balance for the liquid entering "the system at point 1" and leaving at "point 2." The theorem is stated below for 1 lb of flowing fluid and units of foot-pounds.

$$JU_2 + p_2V_2 + \frac{u_2^2}{2g} + Z_2 + JQ - w = JU_1 + p_1V_1 + \frac{u_1^2}{2g} + Z_1 \quad \text{(F-1)}$$

where J = 778.1 ft-lb/Btu, the mechanical equivalent of heat
g = 32.17 ft/sec^2, the gravitational acceleration
p_1, p_2 = initial and final pressures, psf
V_1, V_2 = initial and final volumes, cu ft/lb
Z_1, Z_2 = initial and final heights, in feet above a datum plane, ft
U_1, U_2 = initial and final internal energy, Btu/lb
Q = heat absorbed from surroundings, Btu/lb between points 1 and 2
w = net *external* work done by fluid traveling between points 1 and 2
u_1, u_2 = initial and final average velocities, fps

Equation (F-1) is sufficiently general for treatment of almost any flow problem and may be greatly simplified for particular problems. We shall consider the case of horizontal flow of an incompressible fluid through a "perfect" orifice or venturi, that is, one in which friction is entirely negligible. Under these conditions, $U_2 = U_1$, $V_2 = V_1 = V$, $Q = w = 0$, $Z_2 = Z_1$, and Eq. (F-1) reduces to

$$p_1 - p_2 = \frac{u_2^2 - u_1^2}{2gV} \tag{F-2}$$

This equation may be used for computing the flow through orifices and nozzles by including an experimental factor, the coefficient of discharge. The velocities, of course, may be computed from the flows and diameters involved, and this computation may be incorporated into the equation. This has been done in the working equations in the next section.

In the case of gases, correction for their expansion in flow through an orifice is necessary. From thermodynamic considerations it may be shown that the pressure-volume relation of a perfect gas in frictionless flow, when no work is done and no heat is lost or gained, is

$$p_2V_2^\gamma = p_1V_1^\gamma$$

where p_1, p_2 = upstream and downstream pressures
V_1, V_2 = upstream and downstream specific volumes
$\gamma = \dfrac{C_p - R}{C_p} = \dfrac{C_v}{C_p}$ (for perfect gases)
C_p = specific heat at constant pressure
C_v = specific heat at constant volume
R = perfect gas constant, 1.968 Btu/(lb mole)(°F)

This may be introduced in Eq. (F-1), and the result leads to the theoretical value of the expansion factor given by Eq. (F-3). The addition of this factor to flow equations for incompressible fluids makes the result applicable to gases.

$$Y_2 = \sqrt{\frac{r}{n}\left(\frac{1 - r^n}{1 - r}\right)\left(\frac{1 - \beta^4}{r^{2n} - \beta^4 r^2}\right)} \tag{F-3}$$

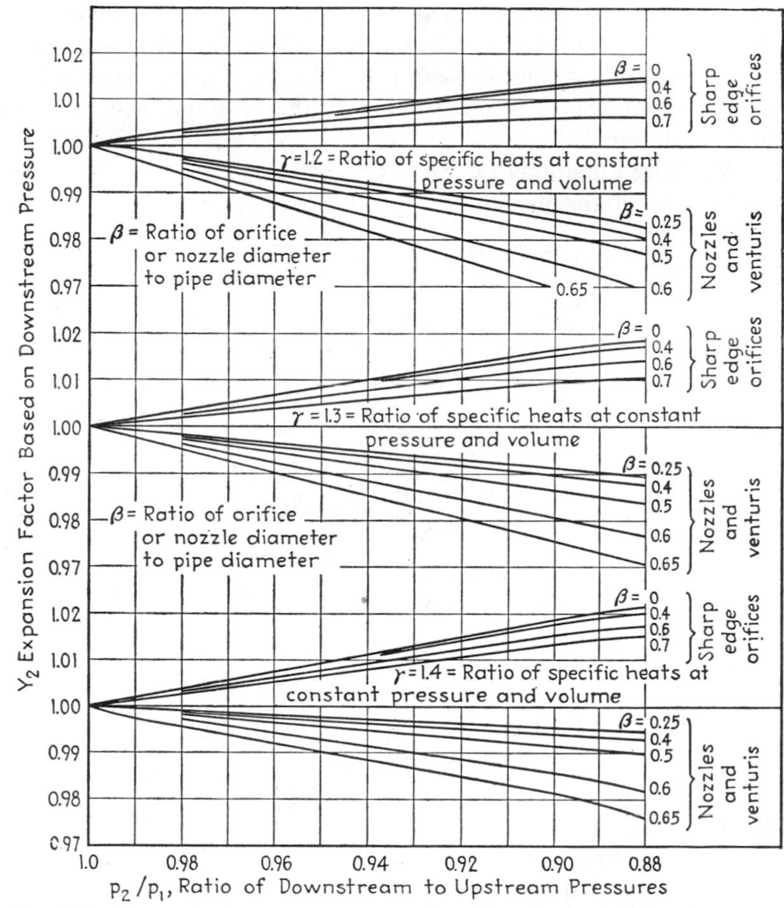

FIG. F-5.—Expansion factors for gas flow through orifices, nozzles, and venturis.

where Y_2 = expansion factor based on *downstream* pressure, Fig. F-5

$r = \dfrac{p_2}{p_1}$

p_1, p_2 = upstream and downstream pressures

$n = \dfrac{C_p - C_v}{C_p} = \dfrac{R}{C_p} = \dfrac{\gamma - 1}{\gamma}$ (for perfect gases)

C_p, C_v = specific heats at constant pressure, volume

$\gamma = \dfrac{C_p}{C_v}$

R = gas constant

β = ratio of orifice to pipe diameter

It will be noted that the expansion factor Y_2, as given, is based on the downstream pressure. As this results in values near unity, the factor may be neglected entirely for most cases. Numerical values of

Y_2 are given in Fig. F-5. The values given for nozzles and venturis are computed from Eq. (F-3). For sharp-edge orifices experimental values are given.[1] Some workers prefer to use the *upstream* pressure as the basis for orifice calculations. If this is done, the expansion factors are different.

b. Working Equations for Flow Calculations

Weight discharge for any fluid:

$$W = \frac{358.9CY_2d^2}{\sqrt{1-\beta^4}}\sqrt{\rho h_w} = 358.9KY_2d^2\sqrt{\rho h_w}$$

$$= \frac{358.9CY_2d^2}{\sqrt{1-\beta^4}}\sqrt{\frac{h_w}{V}} = 358.9KY_2d^2\sqrt{\frac{h_w}{V}} \qquad \text{(F-4)}$$

Volume discharge of a liquid:

$$m = \frac{44.75Cd^2}{\sqrt{1-\beta^4}}\sqrt{\frac{h_w}{\rho}} = 44.75Kd^2\sqrt{\frac{h_w}{\rho}}$$

$$= \frac{5.667Cd^2}{\sqrt{1-\beta^4}}\sqrt{\frac{h_w}{s}} = 5.667Kd^2\sqrt{\frac{h_w}{s}} \qquad \text{(F-5)}$$

Volume flow of gases:

$$M = \frac{128.5CY_2d^2}{\sqrt{1-\beta^4}}\sqrt{\frac{p_2h_w}{zTS}} = 128.5KY_2d^2\sqrt{\frac{p_2h_w}{zTS}} \qquad \text{(F-6)}$$

where W = flow rate, lb/hr
m = flow rate, gpm
M = flow rate, SCF/min (1 SCF = cu ft of gas at 60°F and 30 in. Hg)
C = coefficient of discharge (see Figs. F-6, F-8, and F-11)
= ratio of actual discharge to discharge of perfect orifice
$K = \frac{C}{\sqrt{1-\beta^4}}$ = coefficient of discharge including velocity of approach (see Figs. F-7, F-9, and F-10)
h_w = differential head, in. water (60°F)
$\beta = \frac{d}{D}$
d = internal diameter of orifice, nozzle, or venturi, in.
D = internal diameter of upstream pipe, in.

[1] "Fluid Meters," Part I, 4th ed., American Society of Mechanical Engineers, New York, 1937.

ρ = fluid density, lb/cu ft (for gases, measured at T and p_2)
T = absolute temperature, °R (upstream conditions)
p_2 = absolute pressure, psi, at downstream tap
s = specific gravity, referred to water at 60°F
S = specific gravity of gas relative to air, $\dfrac{\text{molecular weight}}{28.9}$
V = specific volume, cu ft/lb, (for gases, measured at T and p_2)
z = compressibility factor for gas, the ratio of actual to perfect gas law volumes (see Figs. E-1 to E-3)
Y_2 = expansion factor for gases based on downstream pressure (see Fig. F-5)

These equations all give the differential head in terms of inches of water. For other manometer fluids, multiply the head by the specific gravity of the manometer liquid. Many meters use mercury as the manometer fluid but are calibrated in inches of water, or in square root of inches of water. Such readings may be used directly when metering gases without the use of a seal fluid. When a liquid is measured, a correction must be made for the density of fluid. This may be done by multiplying the recorded head by the factor

$$\frac{13.6 - s}{13.6}$$

where s is the specific gravity of the liquid and 13.6 is the specific gravity of mercury. In the case of gases that attack mercury a seal liquid is usually employed. The same factor may be used if the seal pots are large in diameter compared to the diameter of the mercury pots; otherwise a more complicated correction taking care of the change in level of the seal liquid in the seal pots must be made.

Coefficients of discharge for use in these equations are given as follows:

Venturi meters	Fig. F-6
I. S. A. nozzles	Fig. F-7
A. S. M. E. nozzles	Fig. F-8
Sharp-edge orifices	Figs. F-9 to F-11

From these figures it will be noted that coefficients of discharge depend on the ratio of orifice to pipe diameter and on the Reynolds number. This number [defined by Eq. (F-12)] is basic to the friction accompanying flow and will be discussed in the next section. In the case of sharp-edge orifices two sets of data are given. Figures F-9 and

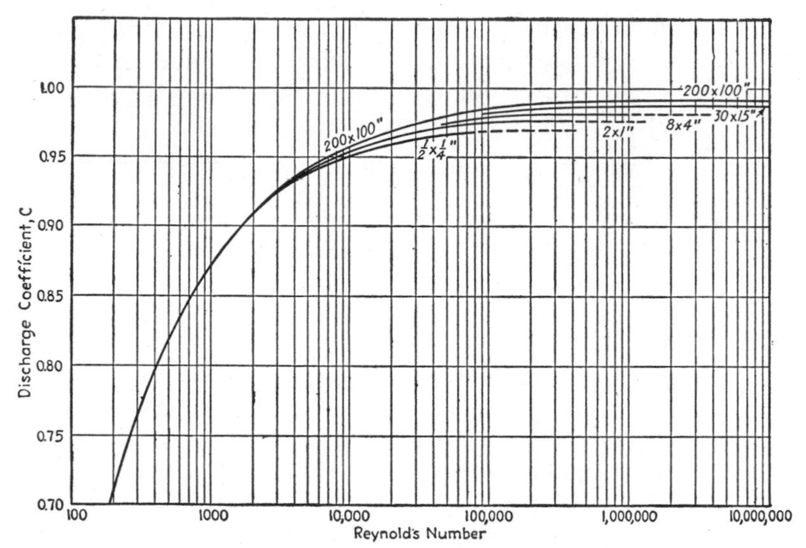

FIG. F-6.—Discharge coefficients for Herschel's design venturi tubes. New tubes with cast-iron body and bronze throat have a diameter ratio of 0.50. (*Data from "Fluid Meters," Part I, 4th ed., American Society of Mechanical Engineers, New York,* 1937.)

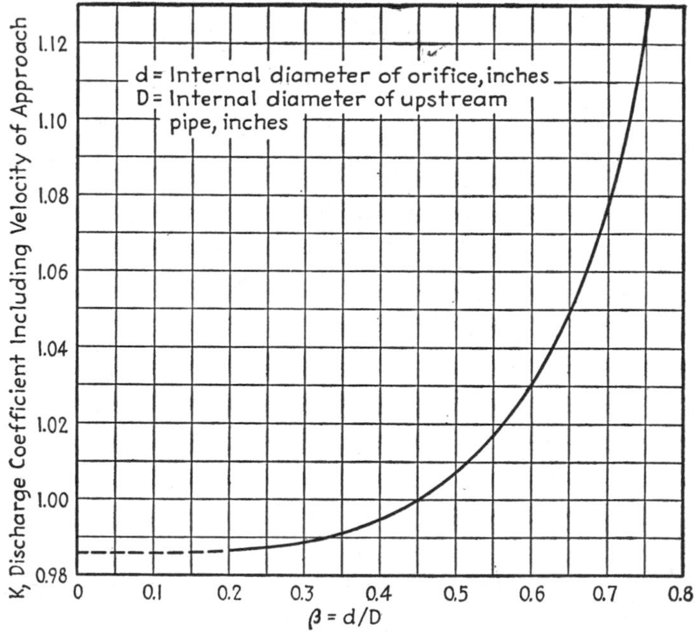

FIG. F-7.—Discharge coefficients for I.S.A. (German Standard) rounded entrance nozzles. (*Adapted by permission from "Elementary Fluid Mechanics," by J. K. Vennard, published by John Wiley & Sons, Inc., New York,* 1940.)

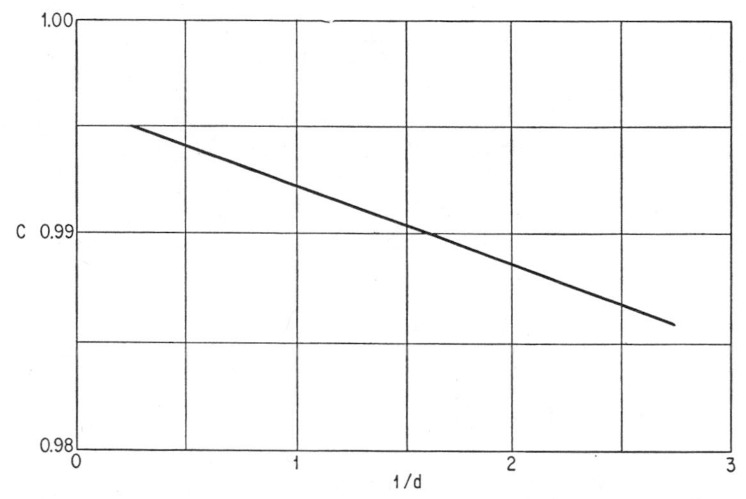

FIG. F-8.—Tentative values of discharge coefficients of A.S.M.E. long-radius nozzles, where d = orifice diameter, in.; C = coefficient of discharge; D = internal diameter of upstream pipe. Valid when (1) Reynolds number (at orifice) is between 400,000 and $100{,}000/d$, (2) d/D is below 0.51, and (3) d is between 0.37 and 5.25. (*"Fluid Meters," Part* I, *4th ed., American Society of Mechanical Engineers, New York,* 1937.)

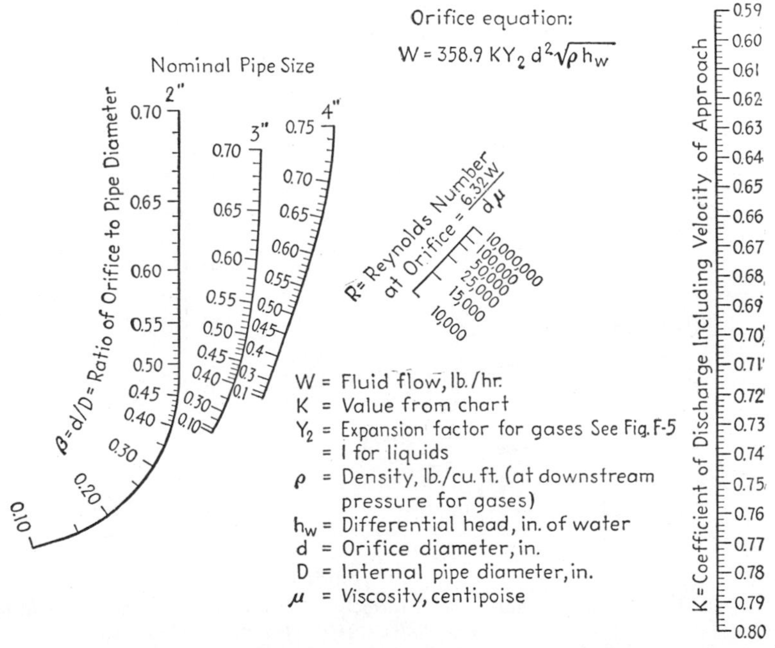

FIG. F-9.—Discharge coefficients including velocity of approach for sharp-edge orifices with flange taps. (*Based on data from "Fluid Meters," Part* I, *4th ed., American Society of Mechanical Engineers, New York,* 1937.)

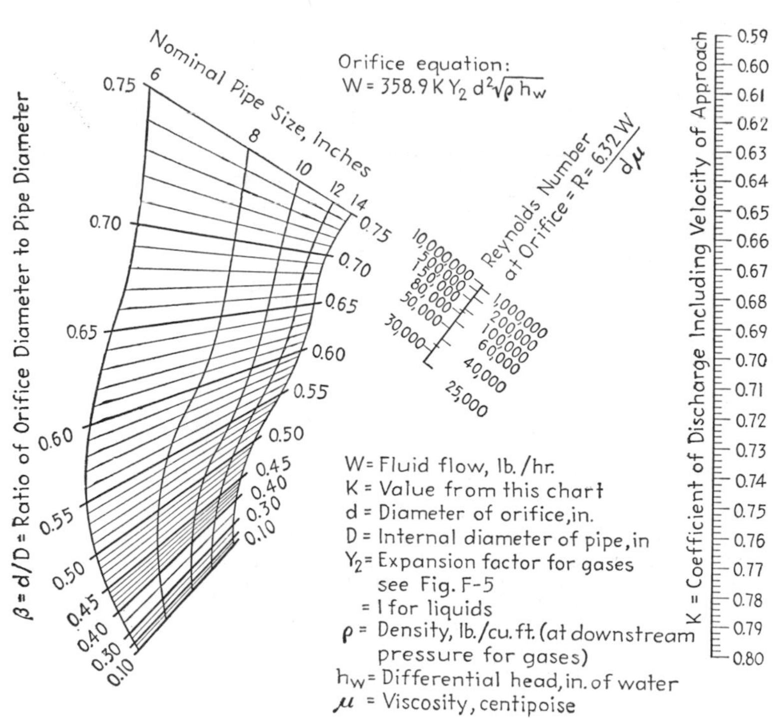

FIG. F-10.—Discharge coefficients including velocity of approach for sharp-edge orifices using flange taps. (*Based on data from "Fluid Meters," Part* I, 4*th ed., American Society of Mechanical Engineers, New York,* 1937.)

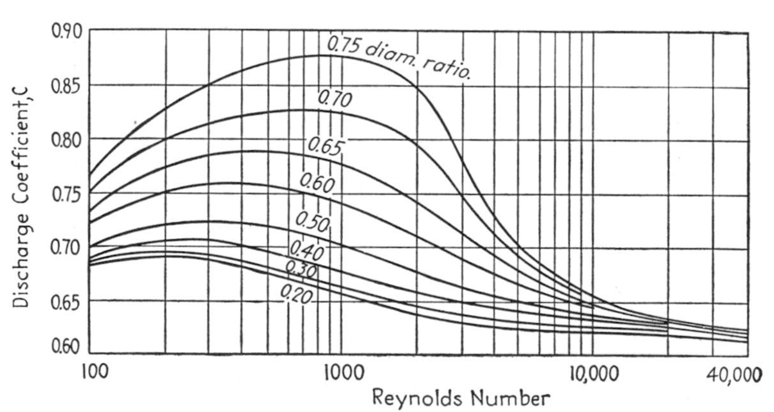

FIG. F-11.—Coefficients of sharp-edge orifices. Applicable to flange or vena contracta taps. [*Tuve and Sprenkle, Instruments,* **6,** 201 (1933).]

F-10 cover the usual range; Fig. F-11 covers the range of low Reynolds numbers which are encountered in metering viscous liquids.

In theory, it would be desirable to correct every single measurement for the Reynolds number at that flow. Practically, such refinement is not desirable, as a constant factor corresponding to average values of Reynolds number may be employed for the relatively narrow range of a flow meter without real loss of accuracy. To illustrate the use of these working equations Example F-1 is presented.

Example F-1. Size an orifice and compute the meter coefficient for flow of 1,000 SCF/min of carbon dioxide in an 8-in. pipe at a temperature of 90°F and a pressure of 10 psig. Assume that a 50-in. meter is to be used and that the meter gives satisfactory accuracy down to a head of 10 in. of water.

SOLUTION:

1. Size of orifice,

$$M = 128.5KY_2d^2\sqrt{\frac{p_2h_w}{zTS}} \qquad \text{[Eq. (F-6)]}$$

where $M = 1{,}000$ SCF/min

$KY_2 = 0.62$ (trial assumption)

$T = 90 + 460 = 550°R$

$S = \dfrac{44}{28.9} = 1.519$

(44 = molecular weight of CO_2)

$h_w = 25$ in. of water (½ of head at full scale)

$p_2 = 10 + 14.7 - {}^{25}\!/_{12} \times 0.434 = 24.7 - 0.9 = 23.8$ psi

$z = 0.97$ (from Fig. E-1)

$$1{,}000 = 128.5 \times 0.62d^2\sqrt{\frac{23.8 \times 25}{0.97 \times 550 \times 1.519}} = 79.7d^2\sqrt{0.734}$$

$$d^2 = \frac{1{,}000}{68.2} = 14.66$$

$d = 3.83$ in., say 3⅞ as a tentative value
(subject to change if assumed value of KY_2 is in error)

2. Meter coefficient,

Approximate mass flow,

$$W = \frac{1{,}000 \times 44}{379} = 116 \text{ lb/min or } 7{,}000 \text{ lb/hr (round value)}$$

(1 mole = 378.7 SCF)

Viscosity = 0.015 centipoise, from Fig. F-2

Reynolds number,

$$R = \frac{6.32W}{d\mu} = \frac{6.32 \times 7{,}000}{0.015 \times 3.875} = 760{,}000 \qquad \text{[Eq. (F-12)]}$$

Orifice coefficient,

$$\beta = \frac{d}{D} = \frac{3.875}{8.07} = 0.48$$

$$K = 0.620, \text{ from Fig. F-10}$$

Expansion ratio,

$$\frac{p_2}{p_1} = \frac{23.8}{24.7} = 0.964$$

Specific heat,

$$C_p = 9.0 \text{ Btu/(lb mole)(°F)} \qquad \text{(Fig. E-5)}$$

$$\gamma = \frac{C_p}{C_p - R} = \frac{9.0}{9.0 - 1.986} = \frac{9.0}{7.01} = 1.29$$

Expansion factor,

$$\gamma = 1.3, \qquad Y_2 = 1.006 \qquad \text{(Fig. F-5)}$$

Flow equation,

$$M = 128.5KY_2d^2\sqrt{\frac{p_2h_w}{zTS}}$$

$$= 128.5 \times 0.620 \times 1.006 \times (3.875)^2\sqrt{\frac{p_2h_w}{0.97 \times 1.519T}}$$

$$= 989\sqrt{\frac{p_2h_w}{T}}$$

(This equation may be used for all further calculations for this installation as the factors KY_2 and z do not vary much within the range of the meter.)

3. Check on orifice size,

Maximum meter reading, $h_w = 50$

$$M = 989\sqrt{\frac{23.6 \times 50}{550}}$$

$$= 1{,}450 \text{ SCF/min}$$

Minimum meter reading for good accuracy, $h_w = 10$

$$M = 648 \text{ SCF/min}$$

This justifies the choice of orifice size as the normal reading is well within the useful range.

DISCUSSION: This example is considerably more detailed than is frequently necessary. Some instrument companies recommend coefficients dependent only on the diameter ratio, β, and whether the fluid metered is a liquid or a gas. Since fluids are bought and sold on the basis of these factors, it may be appreciated that only a moderate error is introduced.

c. Accuracy of Measurements.—The coefficients given for venturis and nozzles are based on data accurate to 1 per cent. The accuracy of the coefficients for sharp-edge orifices, as read in Figs. F-9 and F-10, is better than 1 per cent for all but a small portion of the range. The maximum error is probably no more than 1.3 per cent for extreme values of Reynolds numbers. The coefficients presented in Fig. F-11 are stated to be accurate to 1.5 per cent. For more detailed information on accuracy, the reader is referred to the references accompanying the figures. It should be noted that these accuracies make no allowance for added error introduced by the measuring instrument. The accuracies quoted by instrument companies usually refer to full-scale reading where error is at a minimum. Taking account of inevitable minor imperfections of commercial installations and inaccuracy of the metering instrument, over-all accuracies may be expected to be in the range of 2 to 3 per cent. In any event, careful installations are essential, and the directions of instrument companies or standard texts should be followed closely to avoid large errors.

d. Critical Flow Nozzles.—Flow nozzles may, of course, be used at large pressure drops beyond the range just discussed. The more general equation given below applies:

$$w = 575Cd^2 \sqrt{\frac{M}{zT} \frac{\gamma}{\gamma - 1} \left[\left(\frac{p_2}{p_1}\right)^{\frac{2}{\gamma}} - \left(\frac{p_2}{p_1}\right)^{\frac{\gamma+1}{\gamma}} \right]} \qquad \text{(F-7)}$$

where w = rate of flow, lb/hr
C = coefficient of discharge
d = orifice diameter, in.
p_1, p_2 = upstream and downstream pressure, psia
M = molecular weight of gas
z = compressibility factor (upstream)
T = upstream temperature, °R
γ = ratio of specific heats at constant pressure, volume

This equation is valid for relatively perfect gases and pressure ratios (p_1/p_2) greater than the critical value. At this critical ratio the calculated discharge velocity reaches a maximum value which is the velocity of sound prevailing at the nozzle throat. This is critical flow, and it is sustained for downstream pressures below those in critical ratio to the upstream. Thus, for this condition, the upstream pressure alone serves as an accurate flow measurement and Eq. (F-7) reduces to a simpler form. The result can be expressed as follows:

$$w = BCd^2p_1 \sqrt{\frac{M}{zT}} \qquad \text{(F-8)}$$

where w = flow rate, lb/hr
B = constant from Fig. F-12
C = coefficient of discharge
d = orifice diameter, in.
M = molecular weight
p_1 = upstream pressure, psia
z = compressibility factor
T = temperature, °R

The numerical values of the term B are given in Fig. F-12 together with the critical pressure ratios. For well-rounded nozzles the coefficient of discharge is between 0.97 and 1.00. Sharp-edge orifices can be used with considerable sacrifice in accuracy, since observed coefficients of discharge range from 0.8 to 0.9 under conditions of critical flow for nozzles.

e. Other Methods of Flow Measurement.—Formerly a scientific curiosity, the rotameter has become a popular method of flow measure-

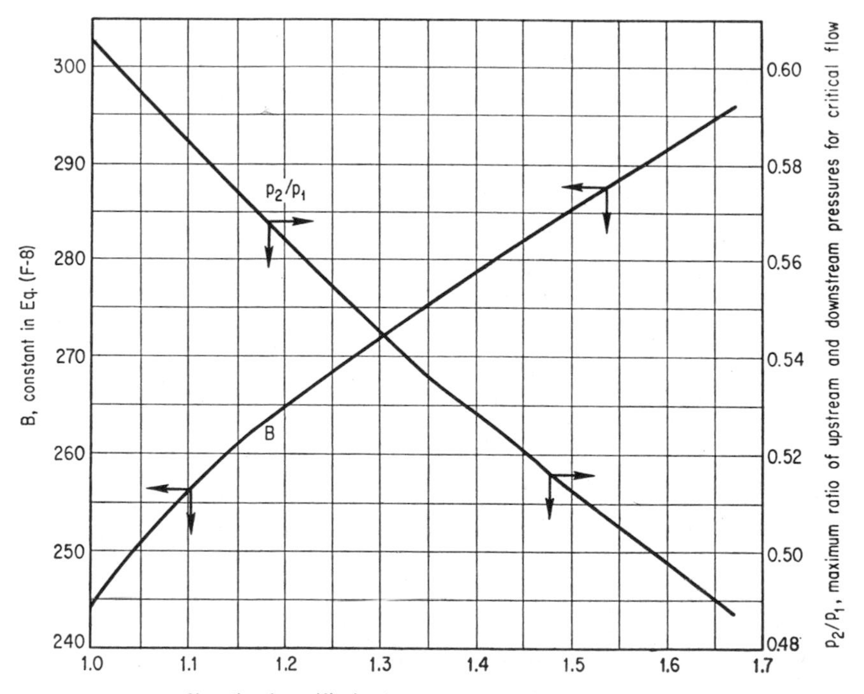

FIG. F-12.—Numerical values of B [for Eq. (F-8)] and upstream-downstream pressure ratios (p_2/p_1) for critical flow nozzles.

ment. It presents the advantage of visibility so that fouling is at once detected and corrected before causing large error. Further, rotameters permit accurate measurement over a wide range; and scales of infinite variety may be obtained for special uses. They are particularly advantageous for liquids, since a proper choice of float makes meter reading directly proportional to the weight flow without correction for density or viscosity. The large variety of rotameters prevents giving coefficients here, but such information is readily available from the manufacturer. The theory of rotameter is presented by Schoenborn and Colburn[1] and, to some extent, in standard texts.

The flow of liquids from an open channel over, or through, a wide variety of dams, notches, or other openings has been studied. Here will be given equations for sharp-edge rectangular and triangular weirs where the velocity of liquid approaching the weir is low. For rectangular weirs with vertical sides and one edge horizontal the Francis formula is usually employed:

$$m = 3.0(L + 0.2h)h^{3/2} \tag{F-9}$$

where m = flow, gpm
L = length of crest, in.
h = head of liquid, in.
= difference in level of body of liquid and bottom of weir

According to Perry[2] this equation is accurate 3 per cent or less when

1. L is greater than $2h$
2. Height of crest above the bottom of upstream channel is at least $3h$
3. Velocity of approach is 1 fps or less
4. h is not less than 3.6 in.

For triangular V-notch weirs cut with the vertex downward Greve[3] gives

$$w = 610\left(\frac{L}{H}\right)^{0.996} h^{2.47} \tag{F-10}$$

where w = flow, lb/hr
L = breadth of notch at top, in.
H = height of notch, in.
h = head of liquid, in.

[1] *Trans. Am. Inst. Chem. Engrs.*, **35**, 359 (1939).

[2] Perry, "Chemical Engineers' Handbook," 3d ed., McGraw-Hill Book Company, Inc., New York, 1950.

[3] *Eng. News-Record*, **105**, 166 (1930).

This equation is accurate for water. For other liquids Perry[1] suggests multiplying the flow [computed by Eq. (F-10)] by the factor

$$\left(\frac{\mu}{s}\right)^{0.02}$$

where μ = viscosity, centipoise

s = specific gravity referred to water at 60°F

Positive-displacement meters are in wide use for metering fluids but require no particular discussion since they are customarily calibrated to read the volume flow directly. The volume measured is, of course, that under the conditions prevailing at the meter; if the volume of the fluid is desired at other temperatures and pressures, a correction must be made.

3. Pressure Drops in Pipes

a. General Notes.—Piping usually represents a very important fraction of the cost of constructing a chemical plant. Consequently the proper selection of pipe sizes is important. Usually it is necessary to make a preliminary guess as to the proper size and estimate the resulting pressure drop, before adopting the final size of the line in question. Figure F-17, given under special methods, and Fig. F-16 will be found useful for making preliminary guesses. This section presents, first, the general method of estimating pressure drops, followed by special methods which are less time-consuming.

The problem of estimating pressure drops is complicated by the existence of two types of fluid flow: laminar (also called "streamline" or "viscous") and turbulent. Laminar flow is characteristic of low velocities, and its nature may be likened to the orderly march of soldiers. Turbulent flow resembles herding of sheep and occurs at higher velocities. Most commercial piping is designed for high velocities in the turbulent range unless viscous materials are being handled.

b. Laminar Flow and the Reynolds Number.—The pressure drop through straight pipe arises almost entirely from friction since velocity is constant. Since the pipe walls are dense compared to fluids, it would be expected that friction between the walls and the liquid in immediate contact would be somewhat larger than between two layers of liquid. This is usually the case and, consequently, even though the fluid may be moving at high velocity in the center of the tube, the liquid very near the walls will be nearly stationary. If one carries this to the extreme, the molecules in direct contact with the walls may best be regarded as at rest. The measure of friction between liquid

[1] Perry, *op. cit.*

layers is the viscosity. One unit of viscosity, the poise, is defined as the force in dynes required to maintain a centimeter per second velocity difference between adjacent liquid "films" of 1 sq cm area. This definition is highly theoretical and does not at once present a clear picture of the property. Nevertheless, using this "definition" and the idea that the molecules in contact with the pipe are stationary, one may set up a differential equation for laminar flow. This may be integrated to obtain Eq. (F-11). This is called the "Poiseuille equation" and is the basis for the Ostwald type of viscometer. Equation (F-11) may, therefore, be regarded as a *practical* definition of viscosity. Conversely, given the viscosity of a fluid, the equation may be used to predict the pressure drops for laminar flow.

$$\Delta p = \frac{0.000034 L \mu W}{\rho d^4} = \frac{0.000273 L \mu q}{d^4} \tag{F-11}$$

where Δp = pressure drop, psi
L = length of pipe, ft
q = flow, gpm
W = flow, lb/hr
ρ = fluid density, lb/cu ft
d = inside diameter of pipe, in.
μ = fluid viscosity, centipoise

The transition from laminar to turbulent flow is complex. Nevertheless, the more important factors are adequately covered by the Reynolds number. This number is a dimensionless quantity; that is, its value is the same when calculated from any *consistent* set of units whether based on pounds, seconds, and feet, or on grams, seconds, and centimeters. Equation (F-12) defines Reynolds number in terms of a variety of convenient though *inconsistent* units.

$$R = \frac{D_e u \rho}{\mu_e} = \frac{124 d_e u \rho}{\mu} = \frac{50.7 m \rho}{d \mu} = \frac{6.32 W}{d \mu} \tag{F-12}$$

where R = Reynolds number
D_e, d_e = equivalent diameter, ft, in.
d = diameter of circular pipe, in.
u = average velocity, fps
μ_e = viscosity, lb/(ft)(sec)
μ = viscosity, centipoise
m = flow, gpm
W = flow, lb/hr
ρ = density, lb/cu ft

For relatively long, straight, circular conduits, the flow is streamline when the Reynolds number is below 2,000 and turbulent when the Reynolds number exceeds 2,300, though the friction factors are not well correlated for Reynolds numbers between 2,300 and 4,000. For curved pipes, streamline flow may persist even above 2,700, and for noncircular conduits, transition begins at lower Reynolds numbers.

As noted under Eq. (F-12), the equivalent diameter is employed for computing the Reynolds number when the cross section is not circular. This is defined as four times the ratio of cross-sectional area of flowing fluid to the wetted perimeter of the channel. The equivalent diameter is four times the hydraulic radius. Table F-4 gives the equivalent diameters for several shapes.

TABLE F-4.—VALUES OF EQUIVALENT DIAMETERS FOR SEVERAL CROSS SECTIONS

$$\text{Equivalent diameter} = \frac{4 \times \text{cross section}}{\text{wetted perimeter}}$$

Shape of Cross Section	Equivalent Diameter
Pipes and ducts running full:	
Circle, diameter $= D$	D
Annulus, inner diameter $= d$, outer $= D$	$D - d$
Square, side $= D$	D
Rectangle, sides $= a, b$	$\dfrac{2ab}{a + b}$
Open channels or partly filled ducts:	
Rectangle, depth $= D$, width $= W$	$\dfrac{4WD}{W + 2D}$
Semicircle, free surface on diameter D	D

c. General Method—Turbulent Flow.—The results of experimental data and its theoretical analysis are expressed by the dimensionless equations

$$n = \frac{2gh}{u^2} = \frac{fL}{D} \tag{F-13}$$

$$n = \frac{f'A_t}{A_f} \tag{F-14}$$

where n = number of velocity heads
g = acceleration due to gravity
u = average linear velocity of fluid
L, D = length, diameter of pipe
h = pressure loss due to friction as feet of liquid flowing
A_t, A_f = frictional area traversed, flow area
f = friction factor used in this chapter (Figs. F-13 to F-15)
f' = $\frac{1}{4}f$ = Fanning friction factor

The friction factor f given by Eq. (F-13) differs from that used by many chemical engineering books including the first edition of this book. The competing factor, frequently called the Fanning friction factor, is one-fourth as large, and a choice between the two is largely a matter of taste. The friction factor is primarily a function of the Reynolds number, but with secondary dependence on pipe surface.

Moody[1] has summarized friction factors on the basis of "roughness-to-diameter" ratios. A slightly modified version of his correlation is given in Figs. F-14 and F-15. The second of these gives roughness-to-diameter ratios for different types of pipe. Figure F-14 is based on this parameter and the Reynolds number. Figure F-13 reads factors directly for steel pipes of various sizes.

For many purposes, the dimensional forms given below will be more convenient than Eq. (F-13):

$$\Delta p = \frac{fLG^2}{772\rho d} = \frac{fLu^2\rho}{772d} = \frac{fLW^2}{296{,}000\rho d^5} = \frac{fLm^2\rho'}{620d^5} \qquad \text{(F-15)}$$

where Δp = pressure drop, psi
f = friction factors (see Figs. F-13 to F-15)
L = length of pipe, ft
G = mass velocity, lb/(sq ft)(sec)
ρ = fluid density, lb/cu ft
ρ' = fluid density, lb/gal
d = inside diameter of pipe, in.
u = average linear velocity, fps
W = flow, lb/hr
m = flow, gpm

This equation gives reliable results for normal liquids and for gases but is not, in general, valid for two-phase flow or for semiplastic materials. It will be noted that it is based on single values (*i.e.*, constant values) of the physical properties—a condition normally met within acceptable engineering accuracy but one that is not fulfilled for gas flow involving large pressure drops. The error is small if the total pressure drop is no more than 10 per cent of entrant absolute pressure. Use of average, rather than inlet, properties is helpful in extending the useful range. The validity of the equation can be extended to very large drops by replacing the term Δp in Eq. (F-15) by the following:

$$\frac{p_1^2 - p_2^2}{p}$$

[1] *Trans. Am. Soc. Mech. Engrs.*, 1944, pp. 671–684.

where p_1, p_2 = upstream and downstream pressure
p = pressure used to evaluate the density and Reynolds number, preferably the average pressure

Thin slurries, even though two-phase, can usually be handled by Eq. (F-15) through use of slurry density and liquid viscosity providing, of course, that line velocity is rapid enough to prevent solids from settling. Thick slurries and highly viscous or semiplastic materials exhibit non-Newtonian viscosity dependent on rate of shear, and special methods are required.

Pressure drops through piping systems include change in static head and losses through valves and fittings. This subject is treated later in the chapter.

Pressure drops for noncircular ducts and across tubes are frequently required, especially for heat-exchanger design. Accordingly, some data on friction factors for these cases are presented in Chapter G. It may be noted that the friction factor for noncircular tubes is based on use of the equivalent diameter in Eq. (F-13), and that Eq. (F-14) is unchanged.

Figure F-16 gives calculated values of pressure drops of liquids flowing in clean-steel piping. These values were calculated for liquids of several kinematic viscosities in schedule 40 pipe of various sizes. The values were calculated using friction factors as presented in Fig. F-13. The results are expressed in feet of liquid for each 1,000 ft of pipe for a flow rate expressed in cubic feet per hour. This type of plot is consistent with dimensionless analysis. Useful conversion factors are given below:

$$1 \text{ gpm} = 8.02 \text{ cfh}$$

$$1 \text{ cfh} = 0.1247 \text{ gpm}$$

$$\Delta p = 0.433sh = \frac{\rho h}{144} = \frac{\rho' h}{19.25} \qquad \text{(F-16)}$$

where Δp = pressure drop, psi/1,000 ft
s = specific gravity, 60/60
ρ = liquid density, lb/cu ft
ρ' = liquid density, lb/gal
h = feet of liquid (values given in Fig. F-16)

The next section Special Methods gives additional convenient charts and tables based on approximations, but fully adequate for selection of line sizes and preliminary estimate of pressure drops.

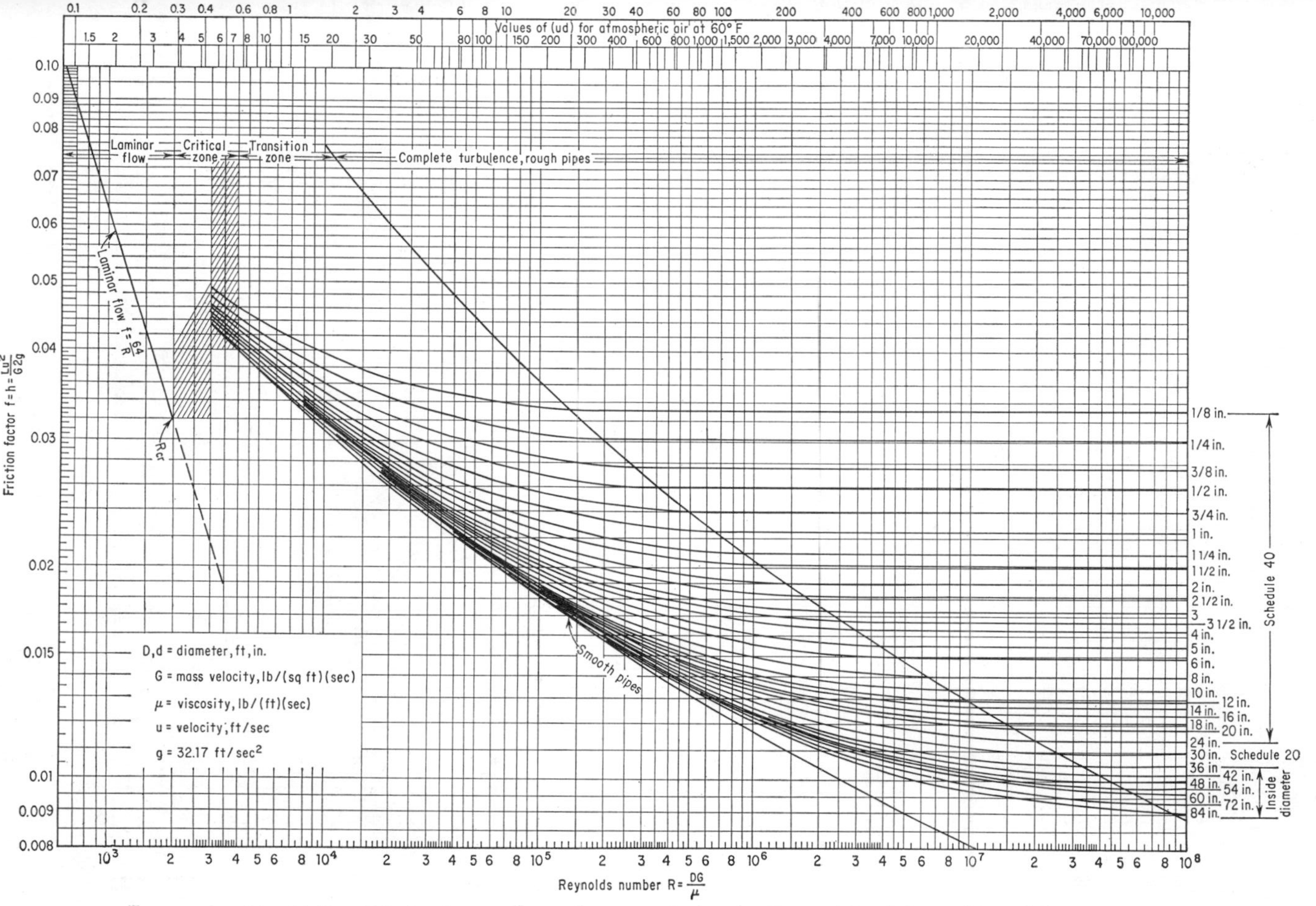

FIG. F-13.—Friction factors for steel or wrought iron. (*From "Pipe Friction Manual," Hydraulics Institute, New York,* 1954.)

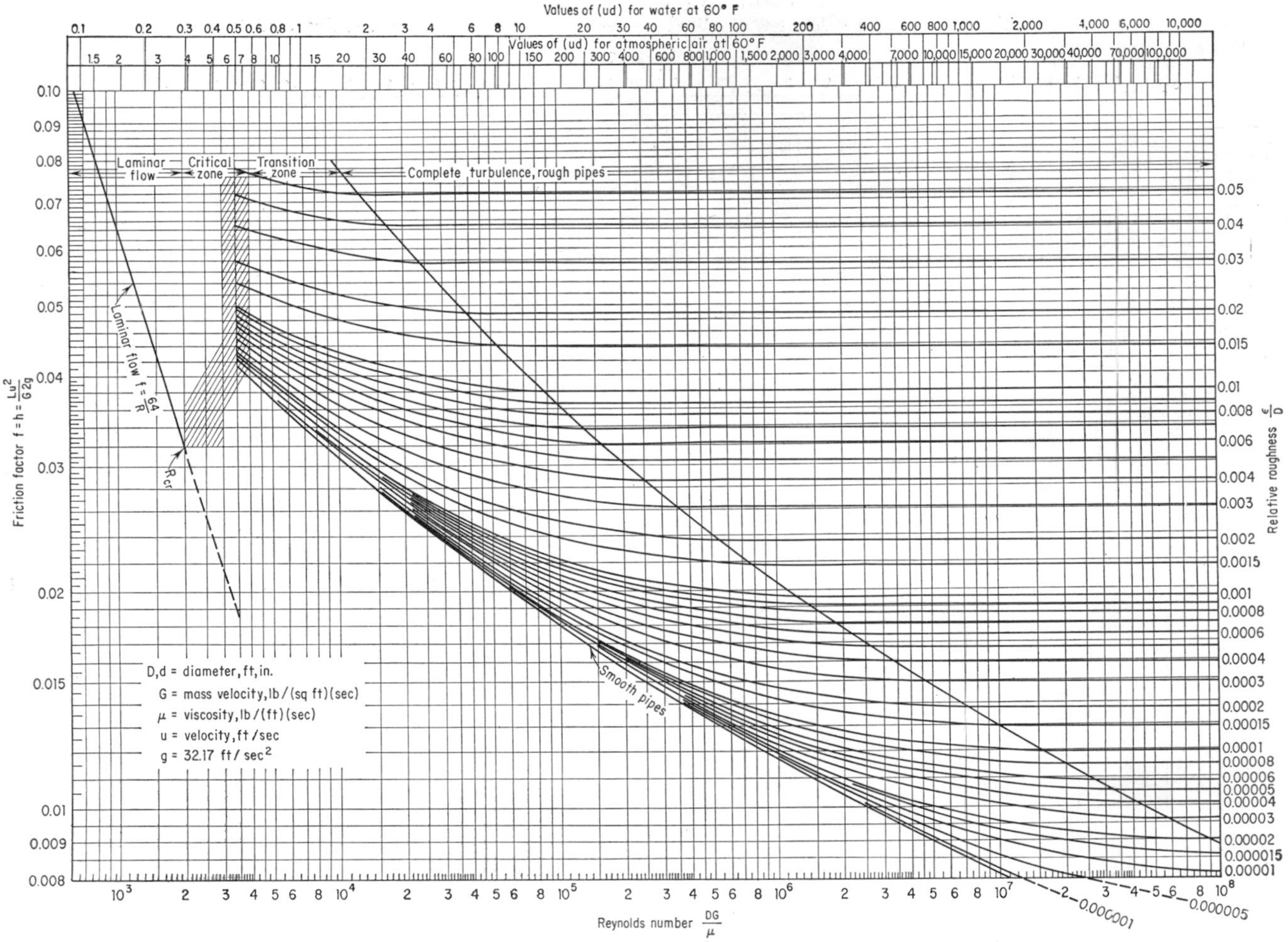

FIG. F-14.—Friction factors for any kind and size of pipe. (*From "Pipe Friction*

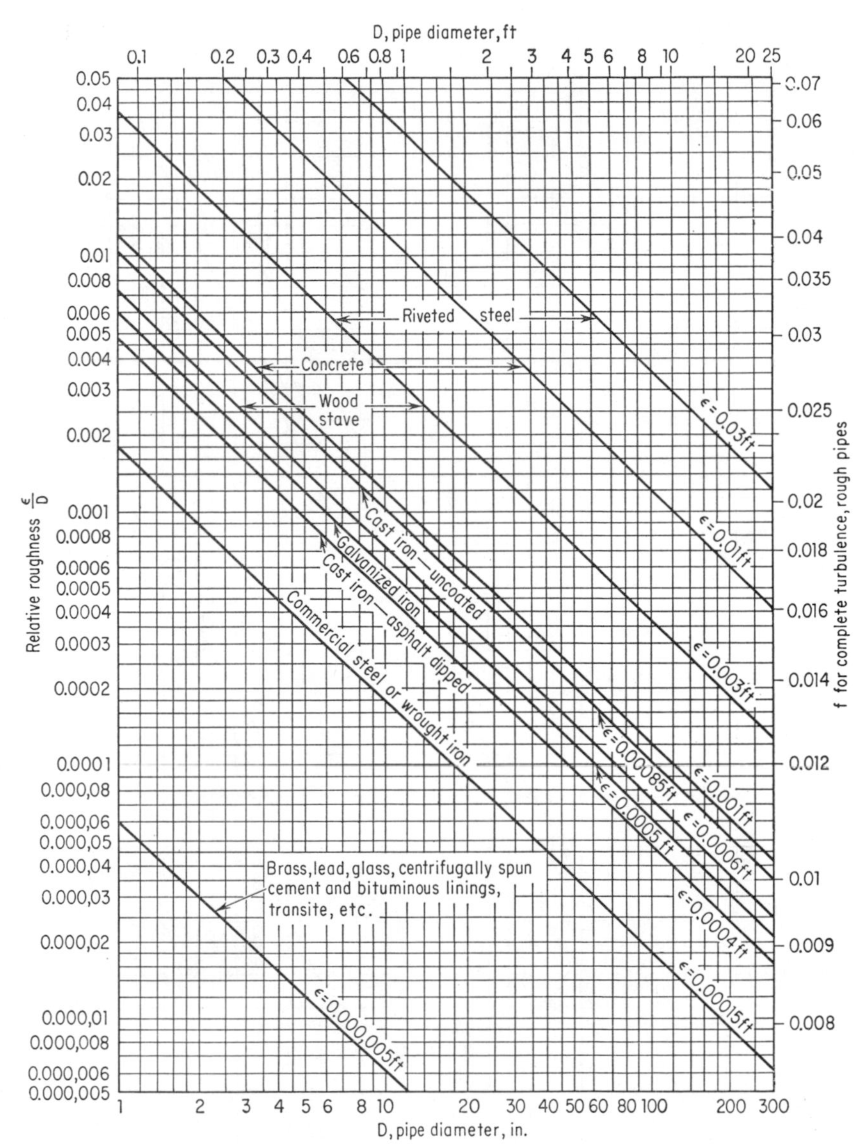

FIG. F-15.—Relative roughness factors for new clean pipe. (*From "Pipe Friction Manual," Hydraulics Institute, New York,* 1954.)

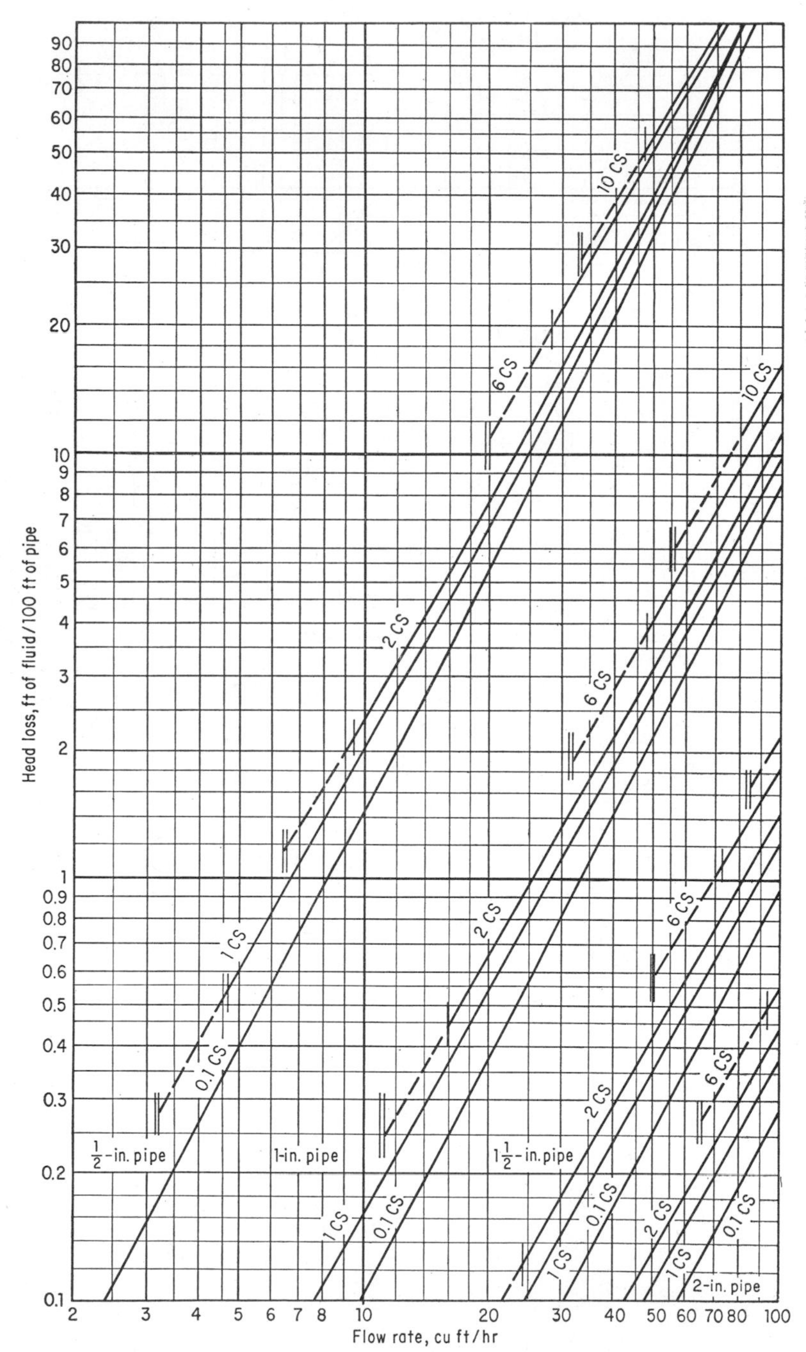

FIG. F-16.—Pressure drop of fluids in schedule 40 piping. Calculated from Moody friction factors as published in the "Tentative Standards of the Hydraulics Institute." Parameters are kinematic viscosity (centistokes) $v = \mu/\rho$, where μ = viscosity, centipoises, and ρ = density, g/cu cm. (*Continued on pp.* 195, 196.)

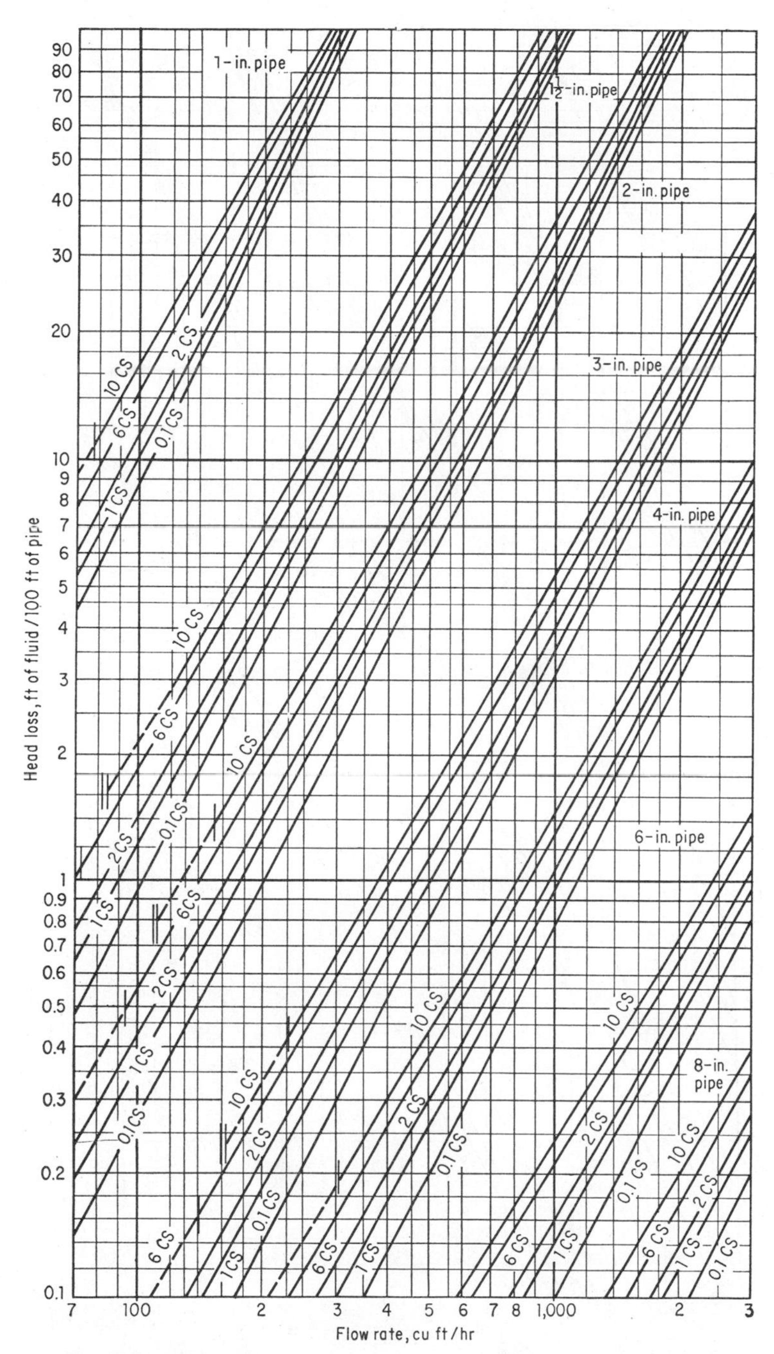

FIG. F-16.—Pressure drop of fluids in schedule 40 piping (*continued*).

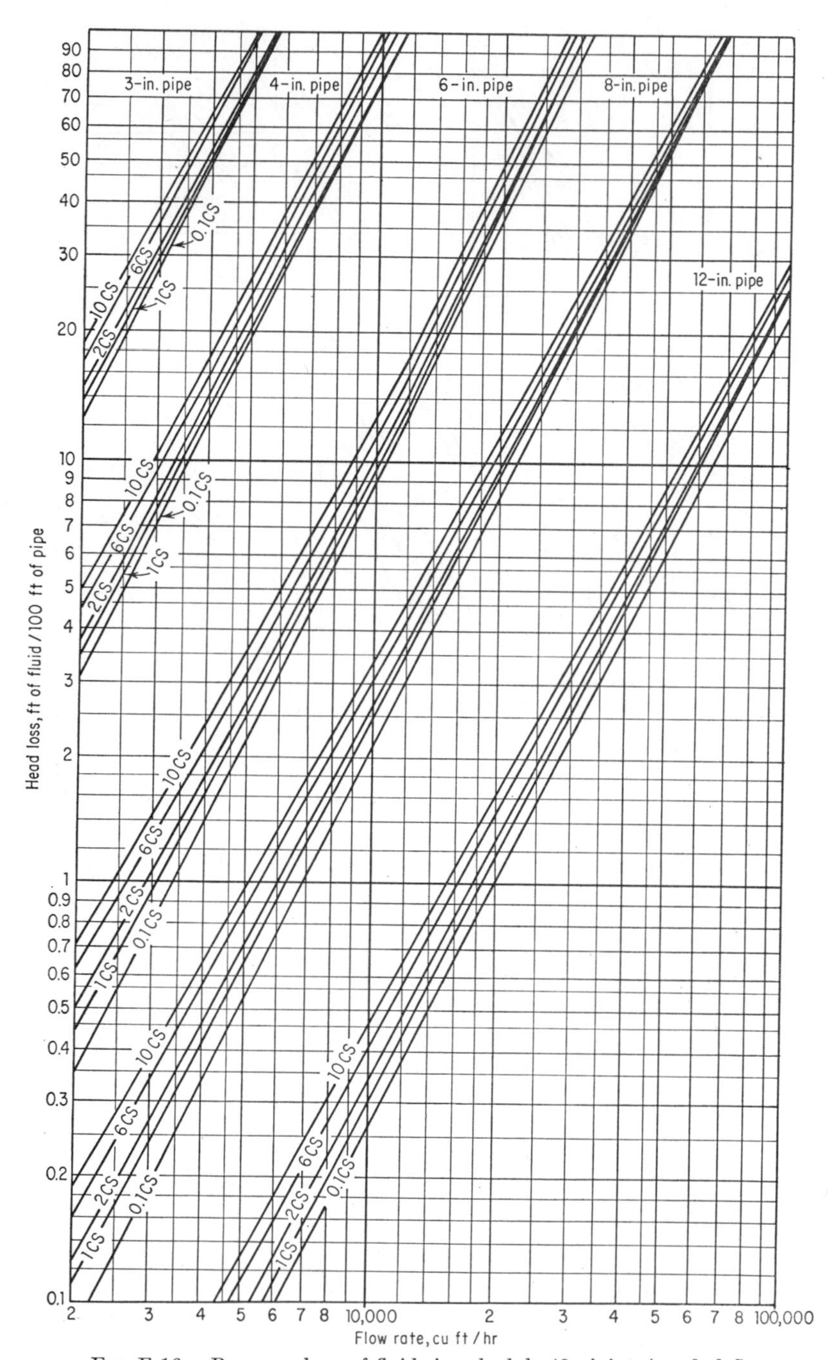

FIG. F-16.—Pressure drop of fluids in schedule 40 piping (*concluded*).

d. Special Methods.—The main use of pressure-drop data is selection of the most economic pipe size. Such a choice is dependent on a number of factors, particularly on the relative costs of piping and power. Figure F-17 permits a preliminary choice of pipe size. This chart is based on relative costs typical of the Atlantic seaboard before the war. The basic equation for this chart is given by Drew and Genereaux[1] and the reader may prepare similar charts based on costs and depreciation rates typical of his own cost studies. Usually it will be found that pipe sizes, as read by this chart, may be multiplied by a constant correction factor to secure a result corrected for revised factors. Table F-5 gives typical fluid velocities used in service.

TABLE F-5.—REASONABLE VELOCITIES OF STEAM AND WATER IN POWER-PLANT PIPING WHERE ACTUAL PRESSURE DROPS ARE NOT COMPUTED
(Sabin Crocker, "Piping Handbook," 4th ed., McGraw-Hill Book Company, Inc., New York, 1945)

Fluid	Pressure, psig	Use	Reasonable velocity, fpm
Water	25–40	City water	120–300
Water	50–150	General service	300–600
Water	150 up	Boiler feed	600
Saturated steam	0–15	Heating	4,000–6,000
Saturated steam	50 up	Miscellaneous	6,000–10,000
Superheated steam	200 up	Large turbine and boiler leads	10,000–15,000

Pressure drops for liquids conveniently can be sized by use of Fig. F-16 immediately preceding. This figure is based directly on the precise general method and was discussed under that heading. See Eq. (F-16) for conversion factors.

Pressure drops for flow of gases through steel pipes may be conveniently made by either the Spitzglass or the Weymouth formulas. Spitzglass formula for low-pressure gas, at 1 psig or less:

$$M = 3{,}550 \sqrt{\frac{h_w d^5}{SL(1 + 3.6/d + 0.03d)}} \qquad \text{(F-17)}$$

where M = SCF/hr

h_w = pressure drop, measured as waterhead, in./ft

S = specific gravity relative to air

$= \dfrac{\text{molecular weight}}{28.9}$

L = length of pipe, ft

d = inside diameter of pipe, in.

[1] Perry, *op. cit.*

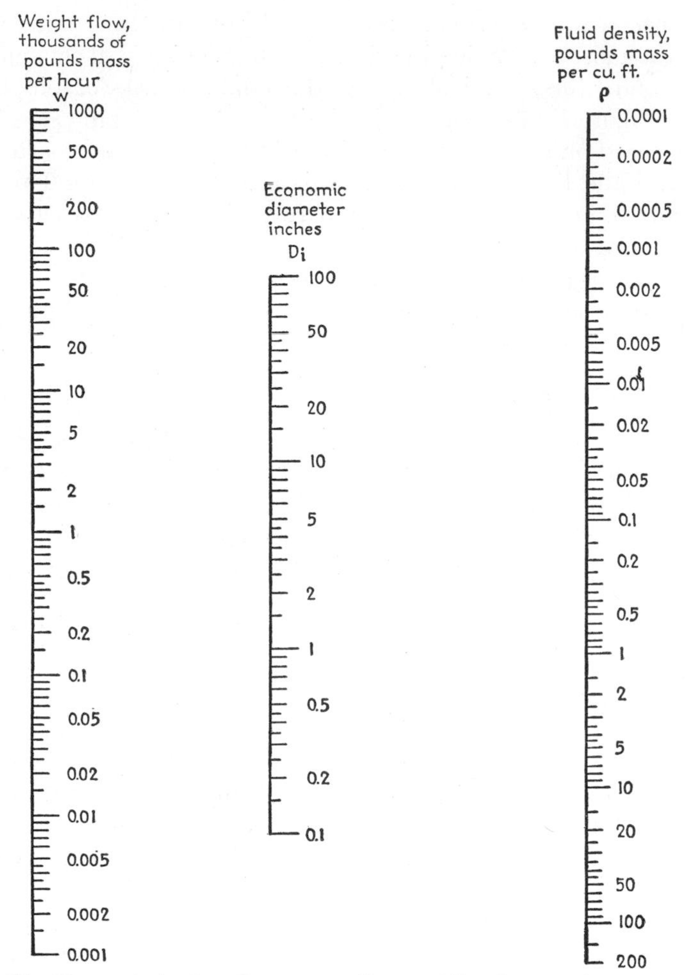

FIG. F-17.—Economical pipe diameters. (*Perry, "Chemical Engineers' Handbook," 3d ed., McGraw-Hill Book Company, Inc., New York,* 1950.)

Weymouth formula for high-pressure gas:

$$M = 28.0d^{2.667}\sqrt{\frac{p_1^2 - p_2^2}{SL_m}\frac{520}{T}} \tag{F-18}$$

where M = SCF/hr
d = inside pipe, diameter in.
S = specific gravity relative to air
L_m = length of pipe, miles
p_1 = initial pressure, psia
p_2 = final pressure, psia

T = absolute temperature, °R

Pressure drops in steam piping may readily be obtained by a simplified version of the Darcy formula developed by the Crane Co.

$$\Delta p = C_1 C_2 V \tag{F-19}$$

where Δp = pressure drop, psi per 100 ft of pipe
V = specific volume of steam (see steam Tables E-13 to E-15)
C_1 = discharge factor from Fig. F-18
C_2 = size factor from Table F-6

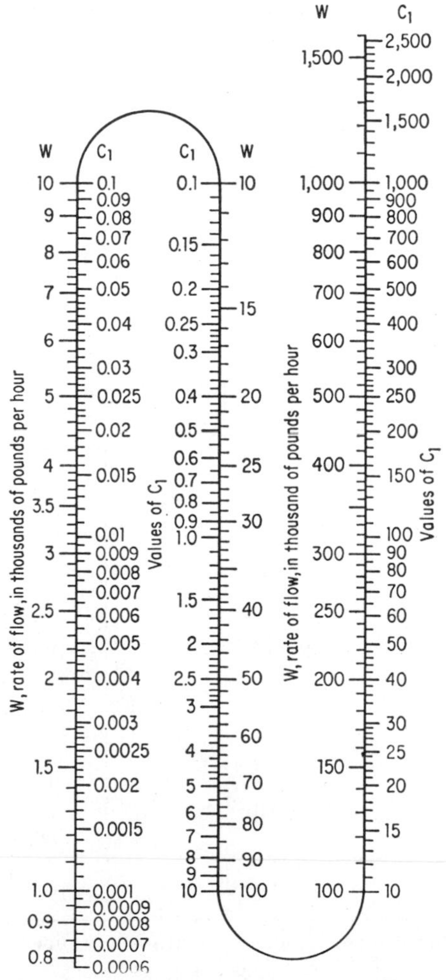

FIG. F-18.—Discharge factors C_1 for use in Eq. (F-19). *(Reprinted from Tech. Paper 410, Crane Co., Chicago, Ill., 1957, by permission.)*

TABLE F-6.—SIZE FACTORS (C_2) FOR USE IN EQ. (F-19)
(Reprinted from *Tech. Paper* 410, Crane Co., Chicago, Ill., 1957, by permission)

Nominal pipe size	Schedule No.*	Value of C_2
½	40*s*	93,500
	80*x*	186,100
	160	4,300,000
¾	40*s*	21,200
	80*x*	36,900
	160	100,100
1	40*s*	5,950
	80*x*	9,640
	160	22,500
1¼	40*s*	1,408
	80*x*	2,110
	160	3,490
1½	40*s*	627
	80*x*	904
	160	1,656
2	40*s*	169
	80*x*	236
	160	488
2½	40*s*	66.7
	80*x*	91.8
	160	146.3
3	40*s*	21.4
	80*x*	28.7
	160	48.3
4	40*s*	5.17
	80*x*	6.75
	120	8.94
	160	11.80
6	40*s*	0.610
	80*x*	0.798
6	120	1.015
	160	1.376
8	20	0.133
	30	0.135
	40*s*	0.146
	60	0.163
	80*x*	0.185
	100	0.211
	120	0.252
	160	0.333
10	20	0.0397
	30	0.0421
	40*s*	0.0447
	60*x*	0.0514
	80	0.0569
	100	0.0661
12	20	0.0157
	30	0.0168
	40	0.0180
	60	0.0206
	80	0.0231
	160	0.0423
14	10	0.00949
	20	0.00996
	30*s*	0.01046
	80	0.01416
	100	0.01657
	120	0.01898
	140	0.0218
	160	0.0252
16	10	0.00463
	20	0.00421
	30*s*	0.00504
	40*x*	0.00549
	160	0.01244

* *s* indicates standard pipe; *x* indicates extra-heavy pipe.

4. Pressure Drops in Valves and Fittings

Like pressure drops through orifices and nozzles, much of the pressure drop through valves and fittings is due to velocity changes. Friction is more important, but equations of form similar to those used in orifice calculations are applicable. The usual form is

$$h = K \frac{u^2}{2g} \tag{F-20}$$

where h = pressure loss in head fluid, ft
K = experimental coefficient (number of velocity heads)
u = average velocity in pipe leading to valve (or fitting), fps
g = 32.17, acceleration due to gravity, ft/sec²

Values of the resistance coefficients are given in Table F-7 and Figs. F-19 and F-20. These coefficients are reasonably accurate, provided the flow is turbulent. Specific data on particular valves can be secured from vendors. Figure F-21 is a correlation of resistance coefficients with equivalent lengths and length-to-diameter ratios.

The "equivalent" length method, though less accurate than the foregoing, is very convenient and is widely used. In this method, the

TABLE F-7.—REPRESENTATIVE "K" VALUES FOR VARIOUS VALVES AND FITTINGS
K is experimental coefficient for use in Eq. (F-20)
(Reprinted from *Tech. Paper* 409, Crane Co., Chicago Ill., by permission)

Type	K	Authority
Globe valve	10.0	Crane tests
Angle valve	5.0	Crane tests
Swing check valve (fully open)	2.5	Crane tests
Close return bend	2.2	
Standard tee	1.8	Giesecke & Badgett
Standard elbow	0.9	Giesecke & Badgett
Medium-sweep elbow	0.75	Crane tests
Long-sweep elbow	0.60	*Univ. Texas Bull.* 2712
45-deg elbow	0.42	*Univ. Texas Bull.* 2712
Gate valve (fully open)	0.19	*Univ. Wisconsin Bull.* 252
¼ closed	1.15	*Univ. Wisconsin Bull.* 252
½ closed	5.6	*Univ. Wisconsin Bull.* 252
¾ closed	24.0	*Univ. Wisconsin Bull.* 252
Borda entrance	0.83	Daugherty, "Hydraulics"
Sudden enlargement:		
$d_1/d_2 = 1/4$	0.92	Daugherty, "Hydraulics"
$d_1/d_2 = 1/2$	0.56	Daugherty, "Hydraulics"
$d_1/d_2 = 3/4$	0.19	Daugherty, "Hydraulics"
Ordinary entrance	0.5	Daugherty, "Hydraulics"
Sudden contraction:		
$d_2/d_1 = 1/4$	0.42	Daugherty, "Hydraulics"
$d_2/d_1 = 1/2$	0.33	Daugherty, "Hydraulics"
$d_2/d_1 = 3/4$	0.19	Daugherty, "Hydraulics"

valve or fitting is said to be equivalent to so many feet or so many "pipe diameters" of pipe. This length is added to the pipe length, and the entire pressure drop is computed at once. Both methods have advantages. Values for equivalent lengths are shown in Fig. F-21 and Table F-8.

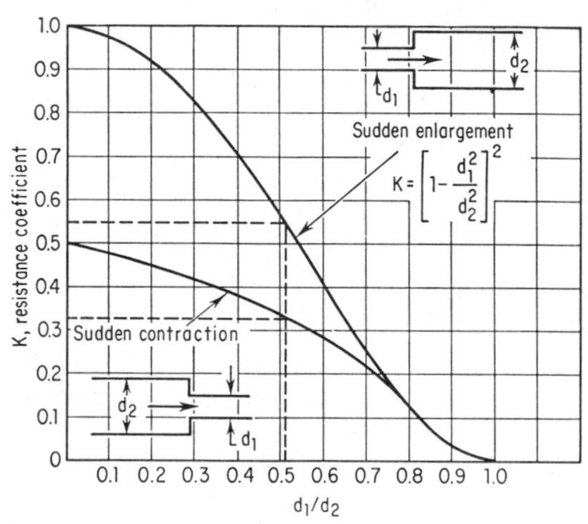

NOTE: The values for the resistance coefficient, K, are based on velocity in the small pipe. To determine K values in terms of the greater diameter, multiply the chart values by $(d_2/d_1)^4$.

FIG. F-19.—Flow resistance from sudden enlargements and contractions. (*Reprinted from Tech. Paper* 410, *Crane Co., Chicago, Ill.*, 1957, *by permission.*)

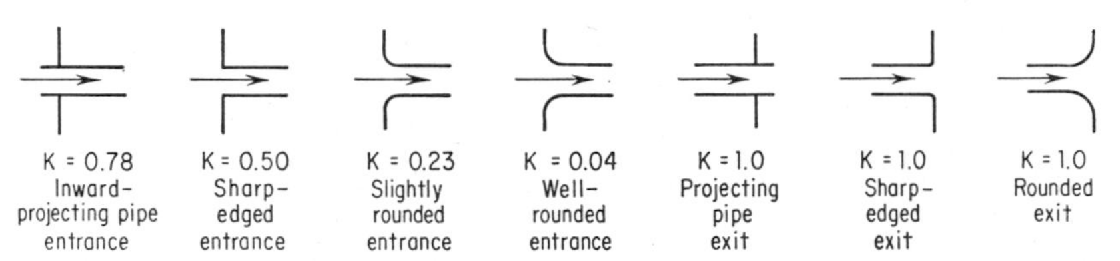

Problem.—Determine the total resistance coefficient for a pipe one diameter long having a sharp-edged entrance and a sharp-edged exit.

Solution.—The resistance of pipe one diameter long is small and can be neglected ($K = fL/D$).

From the diagrams, note:

Resistance for a sharp-edged entrance = 0.5
Resistance for a sharp-edged exit = 1.0

Then,

Total resistance K for the pipe = 1.5

FIG. F-20.—Flow resistance from pipe entrances and exits. (*Reprinted from Tech. Paper* 410, *Crane Co., Chicago, Ill.*, 1957, *by permission.*)

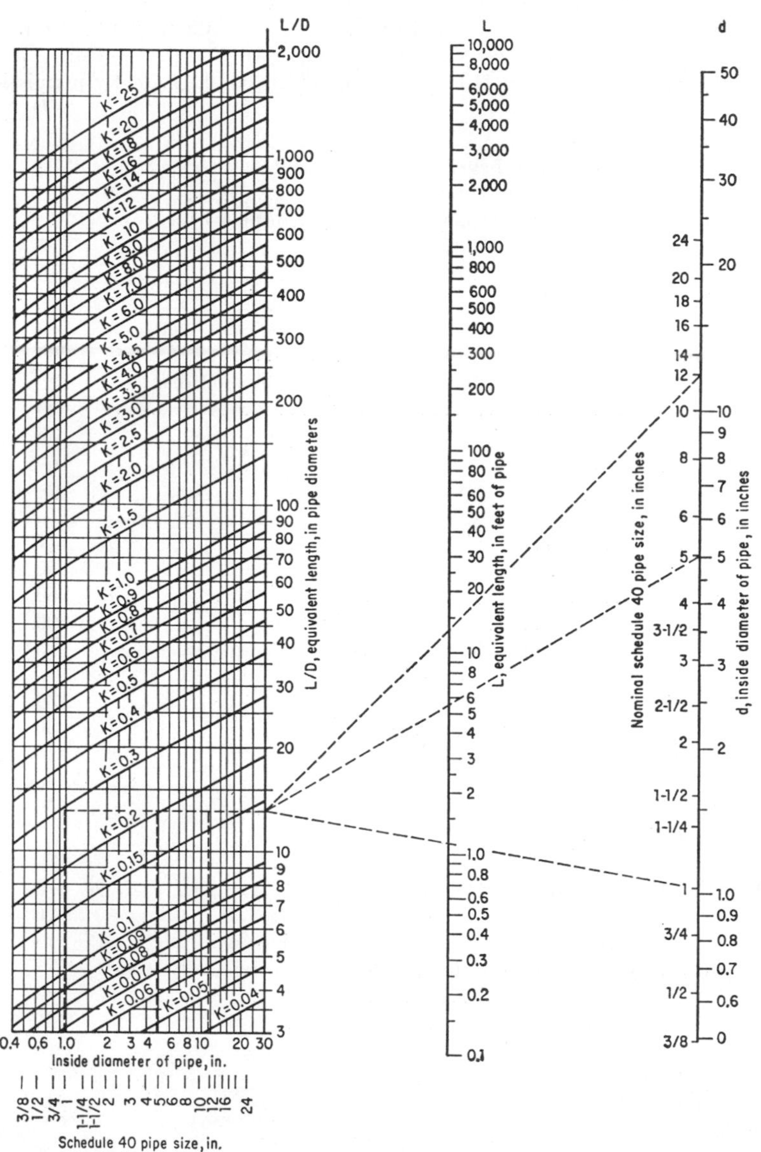

Problem—Find the equivalent length in pipe diameters and feet of schedule 40 pipe, and the resistance factor K for 1-, 5-, and 12-in. fully opened gate valves.

Solution

Valve size, in.	1	5	12	Refer to
Equivalent length, pipe diameters	13	13	13	Table F-8.
Equivalent length, ft of sched. 40 pipe	1.1	5.5	13	Dashed lines
Resist. factor K, based on sched. 40 pipe	0.30	0.20	0.17	on chart

FIG. F-21.—Resistance coefficient K, equivalent length L and L/D. (*Reprinted from Tech. Paper* 410, *Crane Co., Chicago, Ill.*, 1957, *by permission.*)

TABLE F-8.—REPRESENTATIVE EQUIVALENT LENGTHS (L/D) IN PIPE DIAMETERS OF VARIOUS VALVES AND FITTINGS

(Reprinted from *Tech. Paper* 410, Crane Co., Chicago, Ill., 1957, by permission)

	L/D
Globe valves, conventional:	
With no obstruction in flat, bevel, or plug type seat; fully open	340
With wing or pin-guided disk, fully open	450
Globe valves, Y pattern (no obstruction, flat, bevel, or plug-type seat):	
With stem 60° from run of pipeline, fully open	175
With wing or pin-guided disk, fully open	145
Angle valves, conventional:	
With no obstruction in flat, bevel, or plug-type seat; fully open	145
With wing or pin-guided disk, fully open	200
Gate valves, conventional wedge disk, double disk, or plug disk:	
Fully open	13
Three-quarters open	35
One-half open	160
One-quarter open	900
Gate valves, pulp stock:	
Fully open	17
Three-quarters open	50
One-half open	260
One-quarter open	1,200
Gate valves, conduit pipeline, fully open	3*
Check valves:	
Conventional swing, 0.5,† fully open	135
Clearway swing, 0.5,† fully open	50
Globe lift or stop, 2.0,† fully open	As for globe
Angle lift or stop, 2.0,† fully open	As for angle
In-line ball, 2.5 vert. and 0.25 horiz.,† fully open	150
Foot valves with strainer:	
With poppet lift-type disk, 0.3,† fully open	420
With leather-hinged disk, 0.4,† fully open	75
Butterfly valves (6 in. and larger), fully open	20
Cocks, straight-through, rectangular port, 100% pipe area, fully open	18
Cocks, three-way, rectangular port area, 80% of pipe area (fully open):	
Flow straight through	44
Flow through branch	140
Fittings:	
90° standard elbow	30
45° standard elbow	16
90° long-radius elbow	20
90° street elbow	50
45° street elbow	26
Square corner elbow	57
Standard tee:	
With flow through run	20
With flow through branch	60
Close pattern return bend	50

* Equivalent length is equal to the length between flange faces or welding ends.

† Minimum calculated pressure drop across valve to lift disk.

5. Piping Systems

The pressure loss for flow through a complete line is the entrance loss; the summation of drops through all pipe lengths, fittings, and valves; and exit losses. Rather than make the summation in such a tedious fashion, the calculation is more conveniently made by either of two methods: summation of velocity heads or summation of equivalent lengths.

The first of these is the more accurate and should be used for critical calculations, *i.e.*, those in which only a limited head is available or is costly to secure. To apply this method:

1. Make one calculation for each pipe size in line.
2. Compute average velocity and Reynolds number.
3. Find f from Figs. F-13 to F-15.
4. For pipe, find n from Eq. (F-13).
5. Sum up velocity heads for fittings and valves, remembering that a K of 1 represents one velocity head.
6. Add 4 and 5 plus an additional velocity head unless exit velocity is utilized as entrance to downstream lines.
7. Compute pressure drop due to one velocity head.
8. Multiply results of 6 and 7 to secure total.

The numerical values of velocity heads may be computed from the following equations:

$$h = \frac{u^2}{2g} = \frac{u^2}{64.34} \qquad \text{(F-21)}$$

$$\Delta p = \frac{u^2}{2g}\frac{\rho}{144} = \frac{u^2\rho}{9{,}264} \qquad \text{(F-22)}$$

where h = head loss due to friction, ft of flowing liquid
Δp = pressure loss due to friction, psi
u = average velocity, fps
g = acceleration due to gravity, 32.17 ft/sec^2
ρ = density of flowing liquid, lb/cu ft

If top accuracy is not needed or the friction factor is about 0.065 or less, the more rapid equivalent-length method can be used. The line is divided into groups if there is more than one pipe size. For each group, add equivalent lengths of valves and fittings to pipe length and compute the total system drop. Total pressure drop is the summation of the system drop and pressure change due to difference in static head between entrance and exit. This method is useful with Fig. F-16 to estimate liquid pressure drops. It is in largest error for low-pressure gases.

TABLE F-9.—STEEL-PIPE DIMENSIONS, CAPACITIES, AND WEIGHTS*
(Perry, "Chemical Engineers' Handbook," 3d ed., McGraw-Hill Book Company, Inc., New York, 1950)

Nominal pipe size, in.	Outside diam., in.	Schedule No.	Wall thick-ness, in.	Inside diam., in.	Cross-sectional area metal, sq in.	Inside sectional area, sq ft	Circumference, ft, or surface, sq ft/ft, of length		Capacity at 1 ft/sec velocity		Weight of pipe per ft, lb
							Outside	Inside	U.S. gpm	Lb/hr water	
1/8	0.405	40†	0.068	0.269	0.072	0.00040	0.106	0.0705	0.179	89.5	0.25
		80‡	0.095	0.215	0.093	0.00025	0.106	0.0563	0.112	56.0	0.32
1/4	0.540	40†	0.088	0.364	0.125	0.00072	0.141	0.0954	0.323	161.5	0.43
		80‡	0.119	0.302	0.157	0.00050	0.141	0.0792	0.224	112.0	0.54
3/8	0.675	40†	0.091	0.493	0.167	0.00133	0.177	0.1293	0.596	298.0	0.57
		80‡	0.126	0.423	0.217	0.00098	0.177	0.1110	0.440	220.0	0.74
1/2	0.840	40†	0.109	0.622	0.250	0.00211	0.220	0.1630	0.945	472.5	0.85
		80‡	0.147	0.546	0.320	0.00163	0.220	0.1430	0.730	365.0	1.09
		160	0.187	0.466	0.384	0.00118	0.220	0.1220	0.529	264.5	1.31
3/4	1.050	40†	0.113	0.824	0.333	0.00371	0.275	0.2158	1.665	832.5	1.13
		80‡	0.154	0.742	0.433	0.00300	0.275	0.1942	1.345	672.5	1.48
		160	0.218	0.614	0.570	0.00206	0.275	0.1610	0.924	462.0	1.94
1	1.315	40†	0.133	1.049	0.494	0.00600	0.344	0.2745	2.690	1,345.0	1.68
		80‡	0.179	0.957	0.639	0.00499	0.344	0.2505	2.240	1,120.0	2.17
		160	0.250	0.815	0.837	0.00362	0.344	0.2135	1.625	812.5	2.85

1¼	1.660	40†	0.140	1.380	0.669	0.01040	0.435	0.362	4.57	2,285.0	2.28
		80‡	0.191	1.278	0.881	0.00891	0.435	0.335	3.99	1,995.0	3.00
		160	0.250	1.160	1.107	0.00734	0.435	0.304	3.29	1,645.0	3.77
1½	1.900	40†	0.145	1.610	0.799	0.01414	0.498	0.422	6.34	3,170.0	2.72
		80‡	0.200	1.500	1.068	0.01225	0.498	0.393	5.49	2,745.0	3.64
		160	0.281	1.338	1.429	0.00976	0.498	0.350	4.38	2,190.0	4.86
2	2.375	40†	0.154	2.067	1.075	0.02330	0.622	0.542	10.45	5,225.0	3.66
		80‡	0.218	1.939	1.477	0.02050	0.622	0.508	9.20	4,600.0	5.03
		160	0.343	1.689	2.190	0.01556	0.622	0.442	6.97	3,485.0	7.45
2½	2.875	40†	0.203	2.469	1.704	0.03322	0.753	0.647	14.92	7,460.0	5.80
		80‡	0.276	2.323	2.254	0.02942	0.753	0.609	13.20	6,600.0	7.67
		160	0.375	2.125	2.945	0.02463	0.753	0.557	11.07	5,535.0	10.0
3	3.500	40†	0.216	3.068	2.228	0.05130	0.917	0.804	23.00	11,500.0	7.58
		80‡	0.300	2.900	3.016	0.04587	0.917	0.760	20.55	10,275.0	10.3
		160	0.437	2.626	4.205	0.03761	0.917	0.688	16.90	8,450.0	14.3
3½	4.000	40†	0.226	3.548	2.680	0.06870	1.047	0.930	30.80	15,400.0	9.11
		80‡	0.318	3.364	3.678	0.06170	1.047	0.882	27.70	13,850.0	12.5
4	4.500	40†	0.237	4.026	3.173	0.08840	1.178	1.055	39.6	19,800.0	10.8
		80‡	0.337	3.826	4.407	0.07986	1.178	1.002	35.8	17,900.0	15.0
		120	0.437	3.626	5.578	0.07170	1.178	0.950	32.2	16,100.0	19.0
		160	0.531	3.438	6.621	0.06447	1.178	0.901	28.9	14,450.0	22.6

* Based on A.S.A. Standards B36.10-1939.

† Designates former "standard" sizes.

‡ Former "extra strong."

TABLE F-9.—STEEL-PIPE DIMENSIONS, CAPACITIES, AND WEIGHTS* (*Continued*)

Nominal pipe size, in.	Outside diam., in.	Schedule No.	Wall thick-ness, in.	Inside diam., in.	Cross-sectional area metal, sq in.	Inside sectional area, sq ft	Circumference, ft, or surface, sq ft/ft, of length		Capacity at 1 ft/sec velocity		Weight of pipe per ft, lb
							Outside	Inside	U.S. gpm	Lb/hr water	
5	5.563	40†	0.258	5.047	4.304	0.1390	1.456	1.322	62.3	31,150.0	14.7
		80‡	0.375	4.813	6.112	0.1263	1.456	1.263	57.7	28,850.0	20.8
		120	0.500	4.563	7.953	0.1136	1.456	1.197	51.0	25,500.0	27.1
		160	0.625	4.313	9.696	0.1015	1.456	1.132	45.5	22,750.0	33.0
6	6.625	40†	0.280	6.065	5.584	0.2006	1.734	1.590	90.0	45,000.0	19.0
		80‡	0.432	5.761	8.405	0.1810	1.734	1.510	81.1	40,550.0	28.6
		120	0.562	5.501	10.71	0.1650	1.734	1.445	73.9	36,950.0	36.4
		160	0.718	5.189	13.32	0.1469	1.734	1.360	65.8	32,900.0	45.3
8	8.625	20	0.250	8.125	6.570	0.3601	2.258	2.130	161.5	80,750.0	22.4
		30†	0.277	8.071	7.260	0.3553	2.258	2.115	159.4	79,700.0	24.7
		40†	0.322	7.981	8.396	0.3474	2.258	2.090	155.7	77,850.0	28.6
		60	0.406	7.813	10.48	0.3329	2.258	2.050	149.4	74,700.0	35.7
		80‡	0.500	7.625	12.76	0.3171	2.258	2.000	142.3	71,150.0	43.4
		100	0.593	7.439	14.96	0.3018	2.258	1.947	135.3	67,650.0	50.9
		120	0.718	7.189	17.84	0.2819	2.258	1.883	126.5	63,250.0	60.7
		140	0.812	7.001	19.93	0.2673	2.258	1.835	120.0	60,000.0	67.8
		160	0.906	6.813	21.97	0.2532	2.258	1.787	113.5	56,750.0	74.7
10	10.75	20	0.250	10.250	8.24	0.5731	2.814	2.685	257.0	128,500.0	28.1
		30†	0.307	10.136	10.07	0.5603	2.814	2.655	252.0	126,000.0	34.3
		40†	0.365	10.020	11.90	0.5475	2.814	2.620	246.0	123,000.0	40.5

		60‡	0.500	9.750	16.10	0.5185	2.814	2.550	233.0	116,500.0	54.8
		80	0.593	9.564	18.92	0.4989	2.814	2.503	224.0	112,000.0	64.4
		100	0.718	9.314	22.63	0.4732	2.814	2.440	212.0	106,000.0	77.0
		120	0 843	9.064	26.24	0.4481	2.814	2.373	201.0	100,500.0	89.2
		140	1.000	8.750	30.63	0.4176	2.814	2.290	188.0	93,750.0	105.0
		160	1.125	8.500	34.02	0.3941	2.814	2.230	177.0	88,500.0	116.0
12	12.75	20	0.250	12.250	9.82	0.8185	3.338	3.21	367.0	183,500.0	33.4
		30†	0 330	12.090	12.87	0.7972	3.338	3.17	358.0	179,000.0	43.8
		40	0.406	11.938	15.77	0.7773	3.338	3.13	349.0	174,500.0	53.6
		60	0.562	11.626	21.52	0.7372	3.338	3.05	331.0	165,500.0	73.2
		80	0.687	11.376	26.03	0.7058	3.338	2.98	317.0	158,500.0	88.6
		100	0.843	11.064	31.53	0.6677	3.338	2.90	299.0	149,500.0	108.0
		120	1.000	10.750	36.91	0.6303	3.338	2.82	283.0	141,500.0	126.0
		140	1.125	10.500	41.08	0.6013	3.338	2.75	270.0	135,000.0	140.0
		160	1.312	10.126	47.14	0.5592	3.338	2.66	251.0	125,500.0	161.0
14	14.0	10	0.250	13.500	10.80	0.9940	3.665	3.54	446.0	223,000.0	36.8
		20	0.312	13.376	13.42	0.9750	3.665	3.51	438.0	219,000.0	45.7
		30	0.375	13.250	16.05	0.9575	3.665	3.47	430.0	215,000.00	54.6
		40	0.437	13.126	18.61	0.9397	3.665	3.44	422.0	211,000.0	63.3
		60	0.593	12.814	24.98	0.8956	3.665	3.36	402.0	201,000.0	85.0
		80	0.750	12.500	31.22	0.8522	3.665	3.28	382.0	191,000.0	107.0
		100	0.937	12.126	38.45	0.8020	3.665	3.18	360.0	180,000.0	131.0
		120	1.062	11.876	43.17	0.7693	3.665	3.11	345.0	172,500.0	147.0
		140	1.250	11.500	50.07	0.7213	3.665	3.01	324.0	162,000.0	171.0
		160	1.406	11.188	55.63	0.6827	3.665	2.93	306.0	153,000.0	190.0

* Based on A.S.A. Standards B36.10 – 1939.
† Designates former "standard" sizes.
‡ Former "extra strong."

TABLE F-9.—STEEL-PIPE DIMENSIONS, CAPACITIES, AND WEIGHTS* (*Concluded*)

Nominal pipe size, in.	Outside diam., in.	Schedule No.	Wall thickness, in.	Inside diam., in.	Cross-sectional area metal, sq in.	Inside sectional area, sq ft	Circumference, ft, or surface, sq ft/ft, of length		Capacity at 1 ft/sec velocity		Weight of pipe per ft, lb
							Outside	Inside	U.S. gpm	Lb/hr water	
16	16.0	10	0.250	15.500	12.37	1.3104	4.189	4.06	587.0	293,500.0	42.1
		20	0.312	15.376	15.38	1.2895	4.189	4.03	578.0	289,000.0	52.3
		30	0.375	15.250	18.41	1.2680	4.189	4.00	568.0	284,000.0	62.6
		40	0.500	15.000	24.35	1.2272	4.189	3.93	550.0	275,000.0	82.8
		60	0.656	14.688	31.62	1.1766	4.189	3.85	528.0	264,000.0	108.0
		80	0.843	14.314	40.14	1.1175	4.189	3.76	500.0	250,000.0	137.0
		100	1.031	13.938	48.48	1.0596	4.189	3.65	474.0	237,000.0	165.0
		120	1.218	13.564	56.56	1.0035	4.189	3.56	450.0	225,000.0	193.0
		140	1.437	13.126	65.74	0.9397	4.189	3.44	422.0	211,000.0	224.0
		160	1.562	12.876	70.85	0.9043	4.189	3.37	405.0	202,500.0	241.0
18	18.0	10	0.250	17.50	13.94	1.6703	4.712	4.59	748.0	374,000.0	47.4
		20	0.312	17.376	17.34	1.6468	4.712	4.55	738.0	369,000.0	59.0
		30	0.437	17.126	24.11	1.5993	4.712	4.49	717.0	358,500.0	82.0
		40	0.562	16.876	30.79	1.5533	4.712	4.42	697.0	348,500.0	105.0
		60	0.718	16.564	38.98	1.4964	4.712	4.34	670.0	335,000.0	133.0
		80	0.937	16.126	50.23	1.4183	4.712	4.23	635.0	317,500.0	171.0
		100	1.156	15.688	61.17	1.3423	4.712	4.11	602.0	301,000.0	208.0
		120	1.343	15.314	70.28	1.2791	4.712	4.02	573.0	286,500.0	239.0
		140	1.562	14.876	80.66	1.2070	4.712	3.90	540.0	270,000.0	275.0
		160	1.750	14.500	89.34	1.1467	4.712	3.80	514.0	257,000.0	304.0

20	20.0	10	0.250	19.500	15.51	2.0740	5.236	5.11	930.0	465,000.0	52.8
		20	0.375	19.250	23.12	2.0211	5.236	5.05	902.0	451,000.0	78.6
		30	0.500	19.000	30.63	1.9689	5.236	4.98	883.0	441,500.0	105.0
		40	0.593	18.814	36.15	1.9305	5.236	4.94	866.0	433,000.0	123.0
		60	0.812	18.376	48.95	1.8417	5.236	4.81	826.0	413,000.0	167.0
		80	1.031	17.938	61.44	1.7550	5.236	4.70	787.0	393,500.0	209.0
		100	1.250	17.500	73.63	1.6703	5.236	4.59	750.0	375,000.0	251.0
		120	1.500	17.000	87.18	1.5762	5.236	4.46	707.0	353,500.0	297.0
		140	1.750	16.500	100.3	1.4849	5.236	4.32	665.0	332,500.0	342.0
		160	1.937	16.126	109.9	1.4183	5.236	4.22	635.0	317,500.0	374.0
24	24.0	10	0.250	23.500	18.65	3.012	6.283	6.16	1,350.0	675,000.0	63.5
		20	0.375	23.250	27.83	2.948	6.283	6.09	1,325.0	662,500.0	94.7
		30	0.562	22.876	41.39	2.854	6.283	6.00	1,280.0	640,000.0	141.0
		40	0.687	22.626	50.31	2.792	6.283	5.94	1,254.0	627,000.0	171.0
		60	0.937	22.126	67.89	2.670	6.283	5.80	1,200.0	600,000.0	231.0
		80	1.218	21.564	87.17	2.536	6.283	5.65	1,136.0	568,000.0	297.0
		100	1.500	21.000	106.0	2.405	6.283	5.50	1,080.0	540,000.0	361.0
		120	1.750	20.500	122.3	2.292	6.283	5.37	1,030.0	515,000.0	416.0
		140	2.062	19.876	142.1	2.155	6.283	5.21	965.0	482,500.0	484.0
		160	2.312	19.376	157.5	2.048	6.283	5.08	918.0	409,000.0	536.0
30	30.0	10	0.312	29.376	29.10	4.707	7.854	7.69	2,110.0	1,055,000.0	99.0
		20	0.500	29.000	46.34	4.587	7.854	7.60	2,055.0	1,027,500.0	158.0
		30	0.625	28.750	57.68	4.508	7.854	7.53	2,020.0	1,010,000.0	197.0

* Based on A.S.A. Standards B36.10 – 1939.
† Designates former "standard" sizes.
‡ Former "extra strong."

TABLE F-10.—CONDENSER AND HEAT-EXCHANGER TUBE DIMENSIONS
(Perry, "Chemical Engineers' Handbook, 3d ed., McGraw-Hill Book Company, Inc., New York, 1950)

Outside diam., in.	Wall thickness		Inside diam., in.	Cross-sectional area metal, sq in.	Inside sectional area, sq ft	Circumference, ft, or surface, sq ft/ft of length		Velocity, fps for 1 U.S. gpm	Capacity at 1 fps velocity		Weight per ft, lb†
	B.W.G. and Stubs' gauge*	In.				Outside	Inside		U.S. gpm	Lb/hr water	
¼	14	0.083	0.084	0.0435	0.000039	0.0654	0.0219	57.14	0.0175	8.75	0.161
	16	0.065	0.120	0.0377	0.000079	0.0654	0.0314	28.20	0.0355	17.73	0.140
	18	0.049	0.152	0.0309	0.000126	0.0654	0.0397	17.68	0.0566	28.30	0.115
	20	0.035	0.180	0.0236	0.000177	0.0654	0.0471	12.59	0.0794	39.70	0.0876
	22	0.028	0.194	0.0195	0.000205	0.0654	0.0507	10.869	0.0920	46.00	0.0724
	24	0.022	0.206	0.0157	0.000231	0.0654	0.0539	9.645	0.1037	51.85	0.0584
⅜	14	0.083	0.209	0.0761	0.000238	0.0981	0.0547	9.362	0.1068	53.40	0.282
	16	0.065	0.245	0.0633	0.000327	0.0981	0.0641	6.814	0.1468	73.40	0.235
	18	0.049	0.277	0.0501	0.000418	0.0981	0.0725	5.330	0.1876	93.80	0.186
	20	0.035	0.305	0.0373	0.000507	0.0981	0.0798	4.395	0.2275	113.8	0.139
	22	0.028	0.319	0.0305	0.000555	0.0981	0.0835	4.015	0.2494	124.7	0.113
	24	0.022	0.331	0.0243	0.000597	0.0981	0.0866	3.732	0.2679	134.0	0.0904
½	12	0.109	0.282	0.1338	0.000433	0.1309	0.0748	5.142	0.1945	97.25	0.493
	14	0.083	0.334	0.1087	0.000608	0.1309	0.0874	3.662	0.2730	136.5	0.403
	16	0.065	0.370	0.0888	0.000747	0.1309	0.0969	2.981	0.3352	167.5	0.329
	18	0.049	0.402	0.0694	0.000882	0.1309	0.1052	2.530	0.3952	197.6	0.258
	20	0.035	0.430	0.0511	0.001009	0.1309	0.1125	2.209	0.4528	226.4	0.190
⅝	10	0.134	0.357	0.2067	0.000695	0.1636	0.0935	3.206	0.3119	156.0	0.769
	11	0.120	0.385	0.1904	0.000808	0.1636	0.1008	2.758	0.3626	181.3	0.708
	12	0.109	0.407	0.1767	0.000903	0.1636	0.1066	2.468	0.4053	202.7	0.657
	13	0.095	0.435	0.1582	0.00103	0.1636	0.1139	2.163	0.4623	231.2	0.588
	14	0.083	0.459	0.1460	0.00115	0.1636	0.1202	1.938	0.5161	258.1	0.526
	15	0.072	0.481	0.1250	0.00126	0.1636	0.1259	1.768	0.5655	258.9	0.465
	16	0.065	0.495	0.1143	0.00134	0.1636	0.1296	1.663	0.6014	300.7	0.425
	17	0.058	0.509	0.1033	0.00141	0.1636	0.1333	1.580	0.6328	316.4	0.384
	18	0.049	0.527	0.0887	0.00151	0.1636	0.1380	1.476	0.6777	338.9	0.330
	19	0.042	0.541	0.0596	0.00160	0.1636	0.1469	1.393	0.7181	359.1	0.286
¾	10	0.134	0.482	0.2593	0.00127	0.1963	0.1262	1.754	0.5700	235.0	0.965
	11	0.120	0.510	0.2375	0.00142	0.1963	0.1335	1.569	0.6373	318.7	0.884
	12	0.109	0.532	0.2195	0.00154	0.1963	0.1393	1.447	0.6912	345.6	0.817
	13	0.095	0.560	0.1955	0.00171	0.1963	0.1466	1.303	0.7674	383.7	0.727
	14	0.083	0.584	0.1739	0.00186	0.1963	0.1529	1.198	0.8348	417.4	0.647
	15	0.072	0.606	0.1534	0.00200	0.1963	0.1587	1.114	0.8976	448.8	0.571
	16	0.065	0.620	0.1398	0.00210	0.1963	0.1623	1.061	0.9425	471.3	0.520
	17	0.058	0.634	0.1261	0.00219	0.1963	0.1660	1.017	0.9829	491.5	0.469
	18	0.049	0.652	0.1079	0.00232	0.1963	0.1707	0.962	1.041	520.5	0.401
	19	0.042	0.666	0.0934	0.00242	0.1963	0.1744	0.920	1.086	543.0	0.348
⅞	9	0.148	0.579	0.3380	0.00183	0.2291	0.1516	1.218	0.8213	410.7	1.26
	10	0.134	0.607	0.3119	0.00201	0.2291	0.1589	1.109	0.9021	451.1	1.16

* B.W.G. = Birmingham wire gauge, commonly used for ferrous tubing; it is identical with Stubs' = Stubs' iron-wire gauge.

† In brass, sp gr = 8.56; sp gr of steel = 7.8.

TABLE F-10.—CONDENSER AND HEAT-EXCHANGER TUBE DIMENSIONS (*Continued*)

Outside diam., in.	Wall thickness		Inside diam., in.	Cross-sectional area metal, sq in.	Inside sectional area, sq ft	Circumference, ft, or surface, sq ft/ft of length		Velocity, fps for 1 U.S. gpm	Capacity at 1 fps velocity		Weight per ft, lb†
	B.W.G. and Stubs' gauge*	In.				Outside	Inside		U.S. gpm	Lb/hr water	
⅞	11	0.120	0.635	0.2846	0.00220	0.2291	0.1662	1.012	0.9874	493.7	1.06
	12	0.109	0.657	0.2623	0.00235	0.2291	0.1720	0.948	1.055	527.5	0.976
	13	0.095	0.685	0.2328	0.00256	0.2291	0.1793	0.870	1.149	574.5	0.866
	14	0.083	0.709	0.2065	0.00274	0.2291	0.1856	0.813	1.230	615.0	0.768
	15	0.072	0.731	0.1816	0.00291	0.2291	0.1914	0.766	1.306	653.0	0.676
	16	0.065	0.745	0.1654	0.00303	0.2291	0.1950	0.735	1.360	680.0	0.615
	17	0.058	0.759	0.1488	0.00314	0.2291	0.1987	0.709	1.409	704.5	0.554
	18	0.049	0.777	0.1271	0.00329	0.2291	0.2034	0.678	1.477	738.5	0.473
	19	0.042	0.791	0.1099	0.00341	0.2291	0.2071	0.654	1.530	751.5	0.409
1	7	0.180	0.640	0.4637	0.00223	0.2618	0.1676	0.999	1.001	500.5	1.73
	8	0.165	0.670	0.4328	0.00245	0.2618	0.1754	0.909	1.100	505.0	1.61
	9	0.148	0.704	0.3962	0.00270	0.2618	0.1843	0.826	1.212	606.0	1.47
	10	0.134	0.732	0.3654	0.00292	0.2618	0.1916	0.763	1.310	655.0	1.36
	11	0.120	0.760	0.3318	0.00315	0.2618	0.1990	0.707	1.414	707.0	1.23
	12	0.109	0.782	0.3051	0.00334	0.2618	0.2048	0.667	1.499	750.0	1.14
	13	0.095	0.810	0.2701	0.00358	0.2618	0.2121	0.622	1.607	803.5	1.00
	14	0.083	0.834	0.2391	0.00379	0.2618	0.2183	0.588	1.701	850.5	0.890
	15	0.072	0.856	0.2099	0.00400	0.2618	0.2241	0.557	1.795	897.5	0.781
	16	0.065	0.870	0.1909	0.00413	0.2618	0.2277	0.538	1.854	927.0	0.710
	17	0.058	0.884	0.1716	0.00426	0.2618	0.2314	0.523	1.912	956.0	0.639
	18	0.049	0.902	0.1463	0.00444	0.2618	0.2361	0.501	1.993	996.5	0.545
	19	0.042	0.916	0.1264	0.00458	0.2618	0.2398	0.486	2.056	1028.0	0.470
1⅛	7	0.180	0.765	0.5355	0.00319	0.2945	0.2003	0.698	1.432	716.0	1.99
	8	0.165	0.795	0.4979	0.00345	0.2945	0.2081	0.646	1.548	774.0	1.85
	9	0.148	0.829	0.4546	0.00375	0.2945	0.2170	0.594	1.683	841.5	1.69
	10	0.134	0.857	0.4175	0.00401	0.2945	0.2244	0.556	1.800	900	1.55
	11	0.120	0.885	0.3792	0.00427	0.2945	0.2317	0.521	1.916	958	1.41
	12	0.109	0.907	0.3479	0.00449	0.2945	0.2375	0.496	2.015	1008	1.29
	13	0.095	0.935	0.3074	0.00477	0.2945	0.2448	0.467	2.141	1071	1.14
	14	0.083	0.959	0.2717	0.00502	0.2945	0.2511	0.443	2.253	1127	1.01
	15	0.072	0.981	0.2381	0.00525	0.2945	0.2568	0.424	2.356	1178	0.886
	16	0.065	0.995	0.2165	0.00540	0.2945	0.2605	0.412	2.424	1212	0.805
	17	0.058	1.009	0.1944	0.00555	0.2945	0.2642	0.401	2.491	1246	0.723
	18	0.049	1.029	0.1624	0.00575	0.2945	0.2694	0.387	2.581	1291	0.616
	19	0.042	1.041	0.1429	0.00591	0.2945	0.2725	0.377	2.652	1326	0.532
1¼	7	0.180	0.890	0.6051	0.00432	0.3271	0.2330	0.516	1.939	969.5	2.25
	8	0.165	0.920	0.5624	0.00462	0.3271	0.2409	0.482	2.073	1037	2.09
	9	0.148	0.954	0.5124	0.00496	0.3271	0.2498	0.449	2.226	1113	1.91
	10	0.134	0.982	0.4698	0.00526	0.3271	0.2572	0.424	2.361	1181	1.75
	11	0.120	1.010	0.4260	0.00556	0.3271	0.2644	0.401	2.495	1248	1.58
	12	0.109	1.032	0.3907	0.00581	0.3271	0.2701	0.384	2.608	1304	1.45
	13	0.095	1.060	0.3447	0.00613	0.3271	0.2775	0.363	2.751	1376	1.28

* B.W.G. = Birmingham wire gauge, commonly used for ferrous tubing; it is identical with Stubs' = Stubs' iron-wire gauge.

† In brass, sp gr = 8.56; sp gr of steel = 7.8.

TABLE F-10.—CONDENSER AND HEAT-EXCHANGER TUBE DIMENSIONS (*Continued*)

Outside diam., in.	Wall thickness		Inside diam., in.	Cross-sectional area metal, sq in.	Inside sectional area, sq ft	Circumference, ft, or surface, sq ft/ft of length		Velocity, fps for 1 U.S. gpm	Capacity at 1 fps velocity		Weight per ft, lb†
	B.W.G. and Stubs' gauge*	In.				Outside	Inside		U.S. gpm	Lb/hr water	
1¼	14	0.083	1.084	0.3042	0.00641	0.3271	0.2839	0.348	2.877	1439	1.13
	15	0.072	1.106	0.2665	0.00667	0.3271	0.2896	0.334	2.993	1497	0.991
	16	0.065	1.120	0.2419	0.00684	0.3271	0.2932	0.326	3.070	1535	0.900
	17	0.058	1.134	0.2172	0.00701	0.3271	0.2969	0.318	3.146	1573	0.808
	18	0.049	1.152	0.1848	0.00724	0.3271	0.3015	0.308	3.249	1625	0.688
	19	0.042	1.166	0.1590	0.00742	0.3271	0.3053	0.300	3.330	1665	0.593
1⅜	7	0.180	1.015	0.6758	0.00562	0.3620	0.2657	0.397	2.522	1261	2.51
	8	0.165	1.045	0.6272	0.00596	0.3620	0.2736	0.374	2.675	1338	2.33
	9	0.148	1.079	0.5705	0.00635	0.3620	0.2825	0.351	2.850	1425	2.12
	10	0.134	1.107	0.5224	0.00668	0.3620	0.2898	0.334	2.998	1499	1.94
	11	0.120	1.135	0.4731	0.00703	0.3620	0.2971	0.317	3.155	1578	1.76
	12	0.109	1.157	0.4335	0.00730	0.3620	0.3029	0.305	3.276	1638	1.61
	13	0.095	1.185	0.3820	0.00766	0.3620	0.3102	0.291	3.438	1719	1.42
	14	0.083	1.209	0.3369	0.00797	0.3620	0.3165	0.280	3.577	1789	1.25
	15	0.072	1.231	0.2947	0.00827	0.3620	0.3223	0.269	3.712	1856	1.10
	16	0.065	1.245	0.2675	0.00845	0.3620	0.3259	0.264	3.792	1896	0.995
	17	0.058	1.259	0.2399	0.00865	0.3620	0.3296	0.258	3.882	1941	0.893
	18	0.049	1.277	0.2041	0.00889	0.3620	0.3343	0.251	3.990	1995	0.759
	19	0.042	1.291	0.1759	0.00909	0.3620	0.3380	0.245	4.080	2040	0.654
1½	7	0.180	1.140	0.7464	0.00709	0.3925	0.2985	0.314	3.182	1591	2.78
	8	0.165	1.170	0.6920	0.00747	0.3925	0.3063	0.298	3.353	1677	2.57
	9	0.148	1.204	0.6286	0.00791	0.3925	0.3152	0.282	3.550	1775	2.34
	10	0.134	1.232	0.5750	0.00828	0.3925	0.3225	0.269	3.716	1858	2.14
	11	0.120	1.260	0.5202	0.00866	0.3925	0.3299	0.257	3.887	1944	1.94
	12	0.109	1.282	0.4763	0.00896	0.3925	0.3356	0.249	4.021	2011	1.77
	13	0.095	1.310	0.4193	0.00936	0.3925	0.3430	0.238	4.201	2101	1.56
	14	0.083	1.334	0.3694	0.00971	0.3925	0.3492	0.229	4.358	2176	1.37
	15	0.072	1.358	0.3187	0.0100	0.3925	0.3555	0.223	4.488	2244	1.20
	16	0.065	1.370	0.2930	0.0102	0.3925	0.3587	0.218	4.578	2289	1.09
	17	0.058	1.384	0.2627	0.0104	0.3925	0.3623	0.214	4.668	2334	0.978
	18	0.049	1.402	0.2234	0.0107	0.3925	0.3670	0.208	4.802	2401	0.831
	19	0.042	1.416	0.1923	0.0109	0.3925	0.3707	0.204	4.892	2446	0.716
1⅝	7	0.180	1.265	0.8171	0.00873	0.4254	0.3312	0.255	3.918	1959	3.04
	8	0.165	1.295	0.8354	0.00915	0.4254	0.3390	0.243	4.107	2054	2.82
	9	0.148	1.329	0.6868	0.00963	0.4254	0.3479	0.231	4.322	2161	2.55
	10	0.134	1.357	0.6198	0.0100	0.4254	0.3553	0.223	4.488	2244	2.34
	11	0.120	1.385	0.5674	0.0105	0.4254	0.3626	0.212	4.712	2356	2.11
	12	0.109	1.407	0.5191	0.0108	0.4254	0.3684	0.206	4.847	2424	1.93
	13	0.095	1.435	0.4566	0.0112	0.4254	0.3757	0.199	5.027	2514	1.70
	14	0.083	1.459	0.4020	0.0116	0.4254	0.3820	0.192	5.206	2603	1.50
	15	0.072	1.481	0.3512	0.0120	0.4254	0.3877	0.186	5.386	2693	1.31

* B.W.G. = Birmingham wire gauge, commonly used for ferrous tubing; it is identical with Stubs' = Stubs' iron-wire gauge.

† In brass, sp gr = 8.56; sp gr of steel = 7.8.

TABLE F-10.—CONDENSER AND HEAT-EXCHANGER TUBE DIMENSIONS (*Concluded*)

Outside diam., in.	Wall thickness		Inside diam., in.	Cross-sectional area metal, sq in.	Inside sectional area, sq ft	Circumference, ft, or surface, sq ft/ft of length		Velocity, fps for 1 U.S. gpm	Capacity at 1 fps velocity		Weight per ft, lb†
	B.W.G. and Stubs' gauge*	In.				Outside	Inside		U.S. gpm	Lb/hr water	
1⅝	16	0.065	1.495	0.3186	0.0122	0.4254	0.3914	0.183	5.475	2738	1.19
	17	0.058	1.509	0.2855	0.0124	0.4254	0.3951	0.180	5.565	2783	1.06
	18	0.049	1.527	0.2426	0.0127	0.4254	0.3998	0.175	5.700	2850	0.903
	19	0.042	1.541	0.2300	0.0130	0.4254	0.4034	0.171	5.834	2967	0.777
1¾	7	0.180	1.390	0.8878	0.0105	0.4582	0.3639	0.212	4.712	2356	3.30
	8	0.165	1.420	0.8216	0.0110	0.4582	0.3718	0.203	4.937	2469	3.06
	9	0.148	1.454	0.7449	0.0115	0.4582	0.3807	0.194	5.161	2581	2.77
	10	0.134	1.482	0.6803	0.0120	0.4582	0.3880	0.186	5.386	2693	2.53
	11	0.120	1.510	0.6145	0.0124	0.4582	0.3953	0.180	5.565	2783	2.29
	12	0.109	1.532	0.5620	0.0128	0.4582	0.4011	0.174	5.745	2873	2.09
	13	0.095	1.560	0.4939	0.0133	0.4582	0.4084	0.168	5.969	2985	1.84
	14	0.083	1.584	0.4346	0.0137	0.4582	0.4147	0.163	6.149	3075	1.62
	15	0.072	1.606	0.3796	0.0141	0.4582	0.4205	0.158	6.328	3169	1.41
	16	0.065	1.620	0.3441	0.0143	0.4582	0.4241	0.156	6.418	3209	1.28
	17	0.058	1.634	0.3083	0.0146	0.4582	0.4278	0.153	6.552	3276	1.15
	18	0.049	1.652	0.2619	0.0149	0.4582	0.4325	0.150	6.687	3344	0.974
	19	0.042	1.666	0.2253	0.0151	0.4582	0.4362	0.148	6.777	3389	0.838
1⅞	7	0.180	1.515	0.9585	0.0125	0.4909	0.3966	0.178	5.610	2805	3.57
	8	0.165	1.545	0.8864	0.0130	0.4909	0.4045	0.171	5.834	2917	3.30
	9	0.148	1.579	0.8030	0.0136	0.4909	0.4134	0.164	6.104	3052	2.99
	10	0.134	1.607	0.7329	0.0141	0.4909	0.4207	0.158	6.328	3164	2.73
	11	0.120	1.635	0.6616	0.0146	0.4909	0.4280	0.153	6.552	3276	2.46
	12	0.109	1.657	0.6048	0.0150	0.4909	0.4338	0.149	6.732	3366	2.25
	13	0.095	1.685	0.5312	0.0155	0.4909	0.4411	0.144	6.956	3478	1.98
	14	0.083	1.709	0.4594	0.0159	0.4909	0.4474	0.140	7.136	3568	1.74
	15	0.072	1.731	0.4078	0.0163	0.4909	0.4532	0.137	7.315	3658	1.52
	16	0.065	1.745	0.3695	0.0166	0.4909	0.4568	0.134	7.450	3725	1.38
	17	0.058	1.759	0.3310	0.0169	0.4909	0.4605	0.132	7.585	3793	1.23
	18	0.049	1.777	0.2811	0.0172	0.4909	0.4652	0.130	7.719	3860	1.05
	19	0.042	1.791	0.2418	0.0175	0.4909	0.4689	0.127	7.854	3927	0.900
2	7	0.180	1.640	1.0289	0.0147	0.5233	0.4294	0.152	6.597	3299	3.83
	8	0.165	1.670	0.9511	0.0152	0.5233	0.4372	0.147	6.822	3411	3.54
	9	0.148	1.704	0.8608	0.0158	0.5233	0.4461	0.141	7.091	3546	3.20
	10	0.134	1.732	0.7855	0.0164	0.5233	0.4534	0.136	7.360	3680	2.92
	11	0.120	1.760	0.7084	0.0169	0.5233	0.4608	0.132	7.585	3793	2.64
	12	0.109	1.782	0.6475	0.0173	0.5233	0.4665	0.129	7.764	3882	2.41
	13	0.095	1.810	0.5686	0.0179	0.5233	0.4739	0.125	8.034	4017	2.12
	14	0.083	1.834	0.4998	0.0183	0.5233	0.4801	0.122	8.213	4107	1.86
	15	0.072	1.856	0.4359	0.0188	0.5233	0.4859	0.118	8.437	4219	1.62
	16	0.065	1.870	0.3951	0.0191	0.5233	0.4896	0.117	8.572	4286	1.47
	17	0.058	1.884	0.3542	0.0194	0.5233	0.4932	0.115	8.707	4354	1.32
	18	0.049	1.902	0.3000	0.0197	0.5233	0.4979	0.113	8.841	4421	1.12
	19	0.042	1.916	0.2584	0.0200	0.5233	0.5016	0.111	8.976	4488	0.961

* B.W.G. = Birmingham wire gauge, commonly used for ferrous tubing; it is identical with Stubs' = Stubs' iron-wire gauge.

† In brass, sp gr = 8.56; sp gr of steel = 7.8.

CHAPTER G

HEAT TRANSFER

Heat transfer is important to all industry. Since it is of special importance to chemical plants, it would be desirable for all process engineers to be experts on heat transfer. The subject is so complex and comprehensive that all but a few engineers are well advised to depend on those more expert for the final design of critical heat-transfer equipment. This chapter presents data for estimating the simpler types of heat transfer. The accompanying discussion of theory is extremely brief and, alone, would not give the reader an adequate grasp of the subject. A few of the many excellent texts on the subject are listed below.

Kern, "Process Heat Transfer," McGraw-Hill Book Company, Inc., New York, 1950.

Bosworth, "Heat Transfer Phenomena: The Flow of Heat in Physical Systems," John Wiley & Sons, Inc., New York, 1952.

Brown and Marco, "Introduction to Heat Transfer," 3d ed., McGraw-Hill Book Company, Inc., New York, 1958.

McAdams, "Heat Transmission," 3d ed., McGraw-Hill Book Company, Inc., New York, 1954.

Hottel, H. C., "Notes on Radiant Heat Transmission by Non-absorbing Media," Technology Store, Cambridge, Mass., 1951.

Jakob, Max, "Heat Transfer," 2 vols., John Wiley & Sons, Inc., New York, 1948, 1957.

Wilkes, G. B., "Heat Insulation," John Wiley & Sons, Inc., New York, 1950.

Fishenden and Saunders, "An Introduction to Heat Transfer," Oxford University Press, New York, 1950.

Eckert and Drake, "Heat and Mass Transfer," 2d ed., McGraw-Hill Book Company, Inc., New York, 1959.

Jakob and Hawkins, "Elements of Heat Transfer," 3d ed., John Wiley & Sons, Inc., New York, 1957.

Knudsen and Katz, "Fluid Dynamics and Heat Transfer," McGraw-Hill Book Company, Inc., New York, 1958.

Kays and London, "Compact Heat Exchangers," McGraw-Hill Book Company, Inc., New York, 1958.

1. General Theory of Heat Flow

a. Types of Heat Transfer.—There are three principal mechanisms. *Radiation* is one that is familiar to anybody who has sat in the sun or stood in front of an open-hearth steel furnace. The heat is carried by radiation of visible and invisible light rays. Even cold objects emit some radiant energy which may be received by any object that is still colder.

Thermal conduction occurs wherever temperature inequalities exist. If a steel bar is heated on one end, the other end soon becomes warm.

Convection, the third method of thermal transfer, results from physical movement of matter. For example, the movement of hot air from a register carries heat into a cold room, and the heat reaches the far corners by the physical mixing of cold and warm air. Boiling liquids, condensing vapors, freezing liquids, and melting solids also involve movement of material and are classed as convection.

b. General Methods of Attack.—Several mechanisms proceed concurrently in most industrial heat transfer. For example, in a water heater, heat travels from the burning gases to the tube walls by radiation and convection, through the tube metal by conduction, and, finally, to the water by convection. If all the intermediate temperatures were known, the transfer of heat could be estimated from any one of the three steps. However, usually the intermediate conditions are unknown and all types of transfer must ultimately be treated together. This would be extremely difficult were it not that procedures have been developed that permit the calculations to be performed in steps.

The keys for understanding this division of operations are the concepts of thermal resistance and mean temperature difference. Thermal resistance is analogous to electrical resistance. It permits the definition of the over-all heat-transfer coefficient. The mean temperature difference has much the significance implied by the name and is a mathematical expedient of great utility. The procedures used vary with the nature of the particular problem and the preference of the worker.

In general, material and heat balances are required to define a heat-exchange problem and trial-and-error procedures are needed for evaluation of the problem. From the nature of the materials involved and their flow ranges, the type of equipment most desirable can usually be chosen and tentatively sized using the solution to the problem. Such a trial-and-error solution allows local velocities and heat-transfer coefficients to be evaluated and then combined into either the final solution or the basis for a new trial-and-error solution of the problem.

This chapter treats the component parts before it discusses the over-all transfer. The latter section of the chapter discusses methods for choosing tentative solutions to trial-and-error problems with a view to minimizing the trial errors.

2. Radiation

a. Thermal Radiation.—The laws governing the emission of radiant energy were derived by considering a "Hohlraum," a closed space filled only with the radiation in equilibrium with the walls. From the laws of thermodynamics and quantum mechanics the quantity (and also quality distribution of wave lengths) may be derived.

$$\frac{Q}{A} = 0.173\left(\frac{T}{100}\right)^4$$

where Q = Btu/hr (radiant energy *leaving* surface)
A = area of surface
T = absolute temperature, °R

If the surface is isolated and can receive no radiant energy from any source, the above equation gives the *maximum amount* that can be emitted and is termed the "black-body radiation." To secure the actual emission, the black-body radiation must be multiplied by an experimental factor called "emissivity."

In general, the engineer is not interested in isolated surfaces, but rather in the transfer of heat to or from one surface to others. In this case radiant energy is both emitted and absorbed. The net transfer is given by the following equation:

$$Q_1 = 0.173A_1F\left[\left(\frac{T_1}{100}\right)^4 - \left(\frac{T_2}{100}\right)^4\right] \quad \text{(G-1)}$$
$$= 0.173A_1F\left(\frac{T_1}{100}\right)^4 - 0.173A_1F\left(\frac{T_2}{100}\right)^4$$

where Q_1 = net radiant energy leaving surface 1, Btu/hr
A_1 = area of surface 1, sq ft
T_1 = temperature of surface 1, °R
T_2 = temperature of surface 2, °R
F = factor depending on emissivities and geometrical arrangement (see Table G-1)

This equation is based on the assumptions that the medium between the surfaces does not absorb any of the radiant energy and that the emissivity and absorptivity are the same. The factor F is evaluated from the surface emissivities and the laws of optics. Table G-1 permits computation for a few simple cases; values of emissivities are given in Table G-2. Figure G-1 will be found useful in calculations based on Eq. (G-1).

TABLE G-1.—FACTORS FOR USE IN EQ. (G-1)

ϵ_1 = emissivity of surface 1

ϵ_2 = emissivity of surface 2

Arrangement of Surfaces	Emissivity Factor, F
Surface 1 small compared to surface 2 (for example, heat loss of equipment to surroundings)	ϵ_1
Two parallel planes of equal area when length and breadth are large compared to distance between them	$\dfrac{1}{\dfrac{1}{\epsilon_1} + \dfrac{1}{\epsilon_2} - 1}$
Surface 1 is cylinder (radius = r_1) inside larger concentric cylinder (radius = r_2) when length is large compared to diameter	$\dfrac{1}{\dfrac{1}{\epsilon_1} + \dfrac{r_1}{r_2}\left(\dfrac{1}{\epsilon_2} - 1\right)}$
Surface 1 is sphere (radius = r_1) inside of concentric sphere (radius = r_2)	$\dfrac{1}{\dfrac{1}{\epsilon_1} + \left(\dfrac{r_1}{r_2}\right)^2\left(\dfrac{1}{\epsilon_2} - 1\right)}$

NOTE: Factors based on surface separated by transparent medium.

TABLE G-2.—EMISSIVITIES OF SURFACES

Surface	Temperature, °F	Emissivity
Dull brass	120–660	0.22
Oxidized copper	77	0.78
Rolled sheet steel	70	0.66
Oxidized iron	67	0.74
Cast plate, smooth	73	0.80
Brick:		
Building	....	0.80–0.95
Refractory	....	0.75–0.90
Aluminum paint	....	0.3–0.6
Oil paint	212	0.92–0.96
Roofing paper	69	0.91

Very frequently radiant transfer is a small fraction of the total and may be neglected. Where radiation is a factor in heat exchangers, the radiant heat loss may be expressed as Btu/(hr)(sq ft)(°F) and added to the convection coefficient. This is done by guessing the unknown temperature, computing the radiation on that basis, and dividing the rate by the temperature difference. This expedient is mathematically unsound but is a justifiable approximation, provided the assumed temperature is reasonably close to that found by a final computation.[1]

[1] See McAdams, "Heat Transmission," 2d ed., McGraw-Hill Book Company, Inc., New York, 1942, for a complete discussion.

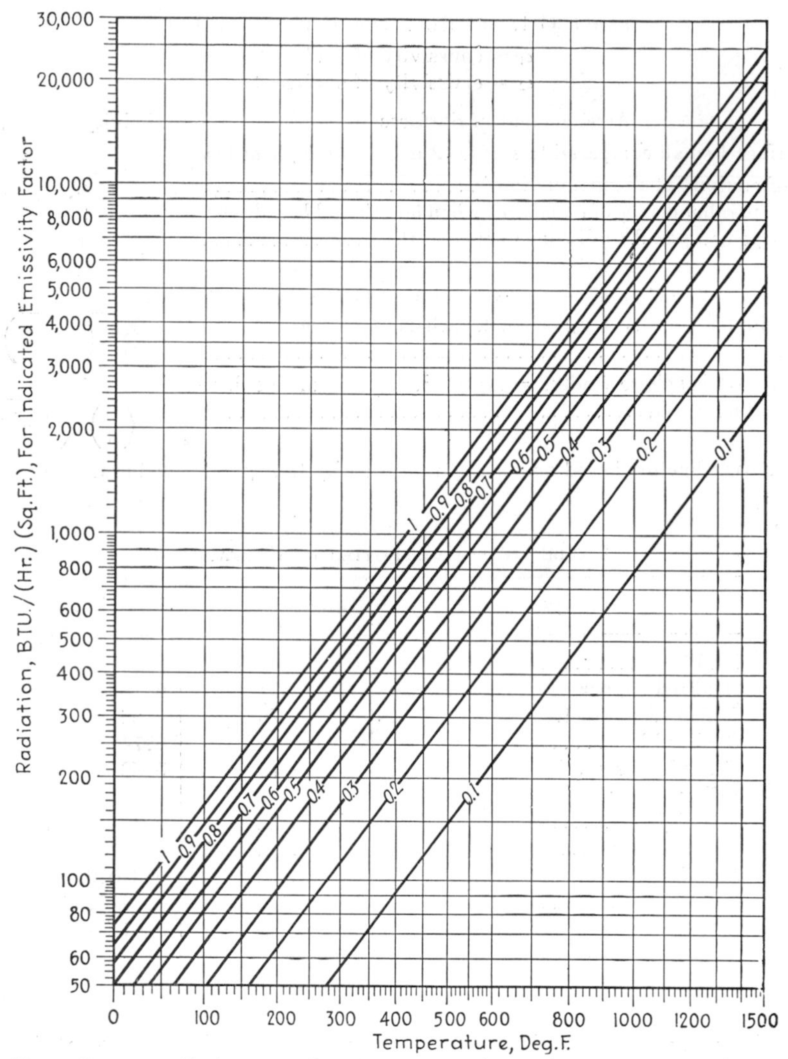

NOTE: Loss = radiation at surface temperature less radiation at surrounding temperature [see Eq. (G-1)].

FIG. G-1.—Chart for computing radiation losses.

b. Solar Radiation.—During the day, open equipment is heated by sunlight. Figure G-2 gives data on solar radiation and may be used for the estimation of the maximum heating effect of the sun on clear days. The fraction of this absorbed depends on the absorptivity which may be taken as equal to the emissivity for approximate calculations. Table G-3 gives experimental values of the rise in temperature produced on different surfaces exposed to the sun.

TABLE G-3.—DAILY MEAN RISE IN TEMPERATURE OF TEST PANELS EXPOSED TO THE SUN
(*Nat. Bur. Standards, Building Materials and Structures Report* BMS64)

	Aug. 2, 1939	Aug. 3, 1939	Aug. 1, 1939	July 31, 1939	Aug. 7, 1939
Panel inclination from horizontal, deg.	90	90	60	45	30
	°F	°F	°F	°F	°F
Black (lampblack)	20.9	21.0	37.4	46.3	48.5
Galvanized iron	16.1	15.3	28.1	32.0	37.7
Roofing shingle:					
Aluminum	19.4	20.2	34.1	40.7	41.6
Green	19.5	20.7	33.3	41.3	43.4
Red	21.5	23.1	37.2	44.8	46.0
Aluminum foil	9.8	8.3	15.0	17.3	19.7
White road-marking paint	12.3	12.1	19.7	22.9	24.7
Aluminum paint	14.6	14.5	24.4	29.0	29.3
Glossy white paint	8.9	7.9	12.1	13.0	15.5
Flat white paint	9.1	8.3	13.2	15.6	17.2
Ivory paint	10.2	9.3	14.9	16.8	19.2
Canary-yellow paint	10.9	10.4	16.7	19.2	21.6
Pearl-gray paint	13.3	13.7	20.3	24.3	25.6
Silver-gray paint	13.9	14.2	20.3	24.6	26.3
Light-lead paint	15.1	15.2	22.9	27.4	29.7
Slate paint	16.8	17.1	26.7	32.4	35.4
Medium-green paint (trim color)	20.4	20.5	35.3	42.7	46.3

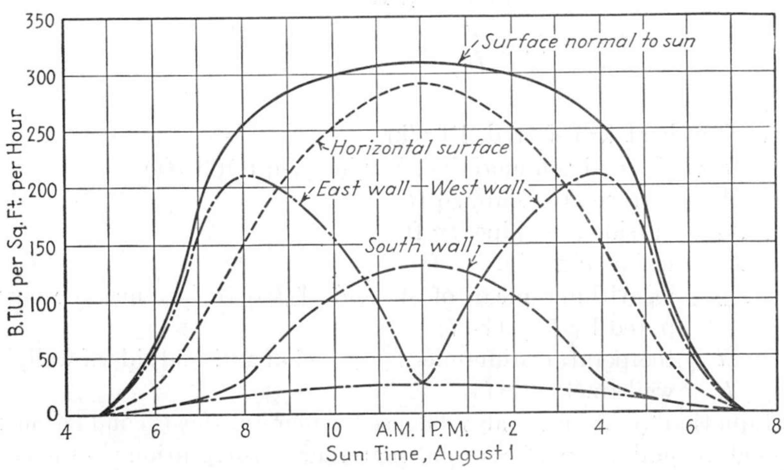

FIG. G-2.—Solar intensity normal to sun on horizontal surface and on walls for Aug. 1 at 40° north latitude. (*Reprinted from "Heating, Ventilating, Air Conditioning Guide," p.* 144, 1942.)

3. Conduction

a. Steady Conduction.—If two faces of a solid are maintained at different fixed temperatures, heat will flow through the metal at a steady rate. Most industrial heat-exchange equipment operates under conditions of steady heat conduction; that is, local temperatures and heat-flow rates change but slowly from one hour to the next. The relations for steady conduction reduce to the following very simple basic equation:

$$\frac{dQ}{dA} = -k\,\frac{dt}{dx} \qquad \text{(G-2)}$$

where Q = heat flow, Btu/hr
A = area at right angles to direction of heat flow, sq ft
t = temperature, °F
x = position relative to any fixed plane, and measured in the direction of heat flow, ft
k = thermal conductivity, Btu/(hr)(sq ft)(°F/ft)

Equation (G-2) may readily be integrated over the volume of the body to obtain the over-all transfer through it. For the two most common cases it becomes

$$Q = -\,kA\,\frac{\Delta t}{L} \qquad \text{(slab of large area)} \qquad \text{(G-3)}$$

$$Q = -k\,\frac{A_0 - A_i}{\ln\,(A_0/A_i)}\,\frac{\Delta t}{L} \qquad \text{(long pipe)} \qquad \text{(G-4)}$$

$$= -kA_m\,\frac{\Delta t}{L}$$

where Q = heat transferred, Btu/hr
k = thermal conductivity, Btu/(hr)(sq ft)(°F/ft)
A = surface area slab, sq ft
A_0 = outside area pipe, sq ft
A_i = inside area pipe, sq ft
A_m = logarithmic mean of A_0 and A_i, sq ft (A_m may be computed by Fig. G-17)
Δt = temperature difference between hot and cold side of wall, °F
L = wall thickness, ft

Equation (G-2) may also be used when the heat conduction is through several layers of different materials. Integrations in the case of slabs and pipes are given as follows:

$$Q = A\Delta t \frac{1}{\dfrac{L_1}{k_1} + \dfrac{L_2}{k_2} + \dfrac{L_3}{k_3} + \cdots} \quad \text{(G-5)}$$

$$Q = \Delta t \frac{1}{\dfrac{L_1}{k_1 A_1} + \dfrac{L_2}{k_2 A_2} + \dfrac{L_3}{k_3 A_3} + \cdots} \quad \text{(G-6)}$$

where Q = heat transferred, Btu/hr
Δt = temperature difference between hottest and coldest surface
L_1, L_2, L_3 = thickness of layers 1, 2, 3, ft
k_1, k_2, k_3 = thermal conductivities, Btu/(hr)(sq ft)(°F/ft) of layers
A = slab area, sq ft
A_1, A_2, A_3 = logarithmic mean area of pipe components of layers 1, 2, 3, sq ft

b. Thermal Conductivity Data.—Table G-4 gives data on metals and Table G-5 gives conductivities of insulating materials.

Table G-4.—Thermal Conductivities of Metals
Values given in Btu/(hr)(sq ft)(°F/ft)
Values given in parentheses are extrapolated more than 50°F

Metal	Temperature, °F									
	−300	−100	0	100	200	300	400	600	800	1000
Aluminum	164	127	116	117	118	121	124	134	147	
Brass (70% Cu)	...	.....	55	58	60	62	63	66	67	
Cast iron	...	.....	(33)	(32)	31	30	29	28	26	(23)
Cast high-silicon iron	...	.....	.....	(30)						
Copper (pure)	290	230	225	222	219	217	215	211	209	206
Iron (99.9%)	...	.....	(42)	(40)	38	37	35	32	28	24
Lead	...	.....	21	20	19	18	18			
Nickel	...	.....	37	35	34	33	33	32		
Monel	8.7	11.6	12.1	15						
Steel:										
0.1% C	...	.....	(36)	35	33	30	28	26	24	23
0.8% C (quenched)	...	.....	.....	(24)	24	24	23	23	21	
18-8 Cr-Ni (low C)	4.8	7.2	7.9	8.6	9.3	9.8	10.2	11.0	11.8	(12.6)
0.35% C, 1% W	...	.....	(23)	(22)	22	21	21	19		
12% Cr	...	.....	.....	(13.8)	14.0	14.2	14.3	14.6	14.9	(15.3)
Wrought iron	...	.....	(36)	(35)	33	31	31	28	25	(23)

Note: Additional data on a few other metals are given in Table D-2.

References:

Perry, "Chemical Engineers' Handbook," 2d ed., McGraw-Hill Book Company, Inc., New York, 1941; Schack, Goldschmidt, and Partridge, "Industrial Heat Transfer," John Wiley & Sons, Inc., New York, 1933; Shelton, *Natl. Bur. Standards, J. Research*, **12,** 441 (1934); "Heat Transfer Through Metallic Walls," International Nickel Co., New York; Chelton and Mann, "Cryogenic Data Book," National Bureau of Standards, Boulder, Colo., 1956.

TABLE G-5.—SOME PHYSICAL AND THERMAL CHARACTERISTICS OF COMMERCIAL INSULANTS
(Saginor, *Chem. & Met. Eng.*, January, 1941)

Material	Commercial forms available	Upper limiting temp., °F	Density, lb/cu ft	Thermal conductivity, Btu/(hr)(sq ft)(°F/in.), at mean temperature, °F											
				100	200	300	400	500	600	800	1000	1200	1500	2000	2500
Aluminum foil	Built-up sectional and block	1000	0.2	0.28	0.37	0.39	0.46	0.52	0.58						
Asbestos	Loose fiber	800	36	0.16	1.23	1.40	1.45	1.48	1.50	1.56					
Asbestos finishing cements	Loose dry	1000	48–60	....	1.50										
Asbestos air cell	Sectional and block														
	4 plies per in.	300	12	0.53	0.65	0.77	0.89								
	8 plies per in.	300	19	0.48	0.56	0.63	0.71								
Asbestos felts (sponge felt)	Sectional and block														
	40 laminations per in.	700	30	0.40	0.44	0.49	0.53	0.59							
	20 laminations per in.	700	20	0.55	0.61	0.67	0.73	0.79							
Asbestos millboard	Sheets	1000	54	0.90	0.95	1.00	1.20	1.40	1.50						
Asbestos sprayed	Loose fiber and liquid binder	800	7	0.36	0.41	0.46	0.53	0.63							
Concrete—insulating type	Loose granular	1800	31	1.72	1.80	1.87	1.94	2.01	2.08	2.23	2.37	2.51	2.73	3.43	
Concrete—insulation refr'y	Powder	2200	65	....	....	2.78	2.82	2.86	2.90	2.98	3.09	3.20	3.44	3.84	
Cork	Sectional and block	200	11	0.30											
Diatomaceous earth	Fine powder	1600	15–17	0.38	0.41	0.44	0.48	0.51	0.55	0.62	0.69	0.76	0.86		
	Coarse powder	1600	22	0.45	0.48	0.52	0.55	0.59	0.62	0.70	0.77	0.84	0.95		
High temp. alumina and diatomac. silica mixtures	Sectional and block	1900–2000	24–30	....	0.59	0.61	0.64	0.68	0.70	0.76	0.82	0.88			
Hair felt	Blanket	200	13	0.30											
Silica brick	Brick	3100–3300	130–150	6.20	6.50	6.80	7.00	7.30	7.60	8.20	8.80	9.40	10.20	11.70	13.20
Firebrick	Brick	2250–3200	110–112	5.80	6.00	6.20	6.50	6.70	6.90	7.40	7.80	8.30	9.00	10.20	11.30
Red brick	Brick	500–1000	100–120	4.50	4.70	5.00	5.20	5.50	5.80	6.30	6.80	7.30	8.10		
Insulating brick	Brick	2500	39–53	....	1.57	1.73	1.84	2.01	2.14	2.50	2.53	2.83			

TABLE G-5.—SOME PHYSICAL AND THERMAL CHARACTERISTICS OF COMMERCIAL INSULANTS (*Concluded*)

Material	Commercial forms available	Upper limiting temp., °F	Density, lb/cu ft	Thermal conductivity, Btu/(hr)(sq ft)(°F/in.), at mean temperature, °F											
				100	200	300	400	500	600	800	1000	1200	1500	2000	2500
Insulating brick	Brick	2000	28–38	....	1.12	1.15	1.19	1.23	1.27	1.37	1.48	1.62			
Insulating brick	Brick	1600	30	....	0.78	0.81	0.85	0.89	0.93	1.00	1.09	1.16			
Kapok	Blanket	212	1	0.25											
Magnesia—85 %	Sectional and block	600	17	0.43	0.47	0.51	0.55								
Magnesia cements	Loose dry	600	21	0.48	0.54	0.60	0.66								
Mineral wools (untreated):*															
Glass wool	Loose, flexible		3	0.26	0.32	0.41	0.61								
	Sectional and blanket	1000	4	0.24	0.30	0.38	0.52								
			8	0.22	0.26	0.30	0.36								
Rock wool	Loose, flexible		5	0.29	0.37	0.47	0.64								
	Sectional and blanket	1000	10	0.28	0.33	0.39	0.46								
			15	0.29	0.35	0.40	0.47								
			20	0.34	0.38	0.43	0.49								
Slag wool	Loose, flexible														
	Sectional and blanket	1000	Generally same as rock wool												
Mineral-wool-type cements	Loose dry	1500	24–30	0.57	0.61	0.66	0.77	0.88	1.02	1.15	1.44				
Mineral wool (waterproofed low temperature type)	Board	225	18	0.36											
	Sectional and block	165	15	0.30											
	Blanket	225	16	0.29											
Vermiculite	Loose, sectional, and block	1600	17	0.57	0.64	0.69	0.73	0.76	0.80	0.89	0.98				
Vermiculite-type cements	Loose dry	1800	16	0.74	0.79	0.84	0.89	0.94	1.04	1.14	1.25				
Wool felt	Sectional or block	212	24	0.36											

NOTE: Conductivity figures represent average value of commercial makes of various manufacture. Conductivity of cements, loose fills, and flexible insulants varies with applied density.

* Some manufacturers recommend 1200°F limit for mineral wools.

c. Heating and Cooling of Solids.—The basic equation for the general case of heat flow by conduction is

$$\frac{\rho C_p}{k}\frac{\partial t}{\partial \theta} = \frac{\partial^2 t}{\partial x^2} + \frac{\partial^2 t}{\partial y^2} + \frac{\partial^2 t}{\partial z^2} \tag{G-7}$$

where t = local temperature, °F, at time θ
ρ = density, lb/cu ft
C_p = specific heat, Btu/(lb)(°F)
k = thermal conductivity, Btu/(hr)(sq ft)(°F/ft)
x, y, z = Cartesian coordinates

This equation can be handled by comparatively complex methods. Solutions have been published for a number of cases,[1] and solutions for two cases (spheres and infinite slabs, *i.e.*, slabs whose length and breadth are large compared with the thickness) are approximated by Fig. G-3. This figure is based on three dimensionless groups:

Conductivity ratio, $m = k/rh$
Relative time, $k\theta/(\rho c r^2)$
Temperature approach, $(t_1 - t)/(t_1 - t_0)$

where k = thermal conductivity, Btu/(hr)(sq ft)(°F/ft)
r = radius of sphere, half thickness of slab, ft
h = film coefficient of heat transfer, Btu/(hr)(sq ft)(°F)
θ = time, hr
c = specific heat, Btu/(lb)(°F)
ρ = density, lb/cu ft
t_0, t_1, t = temperature of surrounding fluid, initial solid temperature, final mean temperature, °F

The film coefficient, which is discussed later in the chapter, expresses the rate of transfer of heat to the solid. The "time" factor includes physical properties and a dimension that are unrelated to time itself but which are needed to make the correlation a general one. The temperature-approach factor is almost self-explanatory and is closely related to the "transfer unit" concept.

To use Fig. G-3, the film coefficient should be estimated separately and the conductivity ratio computed. Either the relative time or the temperature approach can be computed, and the other read from the chart. The final desired result then can be obtained by algebraic solution for temperature at specified time or time for specified temperature.

[1] Heisler, *Trans. Am. Soc. Mech. Engrs.*, **49**, 227 (1947); cf. Schneider, "Conduction Heat Transfer," Addison-Wesley Publishing Company, Reading, Mass., and other standard texts.

If the film coefficient is low and the thermal conductivity is high, the conductivity ratio may be above 2 and out of the range covered by the chart. The equations given below give approximate solutions

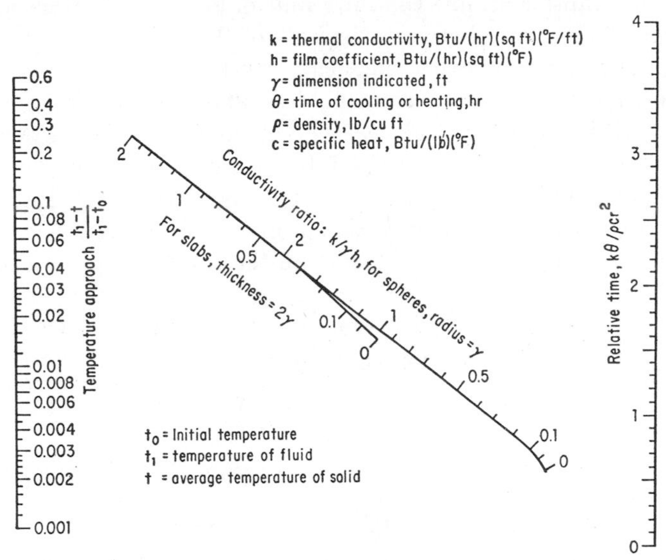

FIG. G-3.—Nomograph for heating or cooling of solids.

covering the range for conductivity factors m larger than 0.5 and are exact for the limiting case of very large conductivity ratios.

$$\frac{t_1 - t}{t_1 - t_0} = e^{-ak/c\rho r^2} \tag{G-8}$$

where a = a factor dependent on shape and conductivity ratio, m

$$= \frac{3m}{m + 0.21} \text{ for spheres}$$

$$= \frac{2m}{m + 0.26} \text{ for long cylinders}$$

$$= \frac{m}{m + 0.35} \text{ for large area slabs}$$

The other symbols are as defined in the preceding discussion.

4. Forced Convection

a. Pressure Drop and Heat Transfer.—The most convenient way of stating convective heat-transfer rates is by the film coefficient which is the heat transfer in unit time per unit of boundary surface and unit temperature difference between the bulk of the fluid and the boundary surface. As might be expected, heat transferred (or convected) to a solid from a fluid flowing by it would depend on the physical properties of the fluid, its velocity, and the type of flow—laminar or turbulent. It is not surprising, then, that the Reynolds number which determines the type of flow is also a prime factor in heat transfer. Indeed, one of the many tools used in correlation of heat-transfer data has been the Colburn[1] analogy between heat transfer and friction (also mass transfer).

The Sieder-Tate[2] modification of the Colburn method leads to definition of a j factor as follows:

$$j = (\mathrm{St})(\mathrm{Pr})^{2/3}\left(\frac{\mu_s}{\mu}\right)^{0.14} \qquad \text{(G-9)}$$

where $\mathrm{St} = h/(C_pG)$ = Stanton number
$\mathrm{Pr} = C_p\mu/k$ = Prandtl number
h = film coefficient of heat transfer
C_p = specific heat at constant pressure
G = mass velocity
k = thermal conductivity of fluid
μ, μ_s = fluid viscosity at bulk, surface temperature

The original analogy presumed j factors to be proportional to friction factors for the particular flow pattern involved. A broader assumption that j factors depend only on the Reynolds number for turbulent flows and on both the Reynolds number and the ratio of flow length to the diameter for the streamline and transition flows has led to widely used correlations of great value for engineering calculations. For streamline flow, under some conditions, both the liquid movement and the heat-transfer coefficients are increased by the natural convection induced by thermal gradients and resultant density changes. It is possible to make corrections for these.[3] Detailed mathematical analyses at turbulent conditions have shown that the complete relationship between heat transfer and fluid flow is more complex.

[1] Colburn, *Trans. Am. Inst. Chem. Engrs.*, **29**, 522 (1930).

[2] Sieder and Tate, *Ind. Eng. Chem.*, **28**, 1429 (1936).

[3] For example, see pp. 233–237 of McAdams, "Heat Transmission," 3d ed., McGraw-Hill Book Company, Inc., New York, 1954.

Experimental work confirms the more advanced relations for at least one case from which the error of the simpler relationships can be evaluated.[1] On this basis, the indicated errors are moderate for Prandtl numbers between 0.5 and 20 covering the bulk of engineering problems.

b. Coefficients inside Conduits.—Table G-6 presents a cross reference to recommended *general* relationships covering heat transfer and pressure drops of fluids inside conduits. The basic equations are given first in dimensionless form and then repeated in dimensional equations. *Simplified* relationships applicable to a number of special cases are given later (starting on page 236) together with suggestions and data for estimating physical properties of fluids for heat-transfer calculations.

TABLE G-6.—GENERAL EQUATIONS AND CHARTS APPLICABLE TO FLOW INSIDE CONDUITS

	Circular	Rectangular	Double pipe annuli	Finned tube annuli
Laminar flow:				
Range* R	0–2,100	0–300	0–300	0–300
Friction† f	$64/R$	....	$64\psi/R$	$64/R$
Heat transfer	Eqs. (G-10), (G-11)	Eqs. (G-10), (G-11)	Eqs. (G-10), (G-11)	Eqs. (G-10), (G-11)
"Transition":				
Range R	1,000–20,000	100–100,000	100–100,000	100–100,000
Friction f	Figs. F-13 to F-15	Fig. G-5	Figs. G-5, G-6	Fig. G-5
Heat transfer	Fig. G-4	Fig. G-5	Fig. G-5	Fig. G-5
Fully turbulent:				
Range R	10,000 up	10,000 up	10,000 up §	10,000 up §
Friction f	Figs. F-13 to F-15	Fig. G-5	Figs. G-5, G-6	Fig. G-5
Heat transfer	Eq. (G-12) Figs. G-7,‡ G-22‡	Eq. (G-13)	Eq. (G-13)	Eq. (G-13)

* Reynolds number R defined by Eq. (F-12).
† Friction factor f defined by Eq. (F-13); factor ψ given by Fig. G-6.
‡ Alternate treatment to Eq. (G-12), recommended for Prandtl numbers above 20.
§ Approximate range only, see Fig. G-5.

It will be noted that two equations are recommended for turbulent flow in circular tubes—the Sieder and Tate[2] Eq. (G-11) for its simplicity and convenience and the Friend and Metzner[3] Eq. (G-14) for its greater accuracy at high Prandtl numbers, particularly for those above 20. Neither the latter equation nor its graphical representation (Fig. G-7) makes any correction for high thermal gradients. The

[1] The calculated values are below experimental values at high Prandtl numbers. The reader may judge the difference by comparing Eqs. (G-12) and (G-14); cf. Friend and Metzner, *J. Am. Inst. Chem. Eng.*, **4**, 393 (1958).

[2] *Ind. Eng. Chem.*, **28**, 1429 (1930).

[3] *J. Am. Inst. Chem. Engrs.*, **4**, 393 (1958).

expedient of evaluating the Prandtl number at the film temperature (average between surface and bulk temperatures) is recommended as a corrective measure for estimating coefficients when the temperature differences are large.

Correlations covering fluid friction and heat transfer inside conduits follow in sequence: dimensionless basic equations, graphical correlations, dimensional forms of basic equations and groups, and simplified relations restricted to special cases. Any *consistent* set of units may be used in the dimensionless groups and equations that follow:

Reynolds number:

$$R = \frac{D_e G}{\mu} \tag{F-12}$$

Friction losses:

$$\frac{z}{L} = \frac{u^2}{2gD_e} f \tag{F-13}$$

j factor:

$$j = \mathrm{St}(\mathrm{Pr})^{2/3} \left(\frac{\mu_s}{\mu}\right)^{0.14} \tag{G-9}$$

$$= \frac{h}{C_p G} \left(\frac{C_p \mu}{k}\right)^{2/3} \left(\frac{\mu_s}{\mu}\right)^{0.14}$$

For heat transfer, streamline flow:[1]

$$j = 1.86 R^{-2/3} \left(\frac{\pi L}{P}\right)^{1/3} = 1.86 R^{-2/3} \left(\frac{L}{D}\right)^{1/3} \tag{G-10}$$

or (preferred form for tubes):

$$\frac{hD}{k} = 2.0 \left(\frac{WC_p}{kL}\right)^{1/3} \left(\frac{\mu}{\mu_s}\right)^{1/3} \tag{G-11}$$

and for fully turbulent flow:

$$j = 0.023 R^{-0.2} \qquad \text{(circular tube)} \tag{G-12}$$

$$j = 0.020 R^{-0.2} \qquad \text{(other shapes)} \tag{G-13}$$

Alternate equation,[2] turbulent flow inside tubes:

$$\mathrm{St} = \frac{f}{9.6 + 33.4 f^{1/2} (\mathrm{Pr} - 1)/\mathrm{Pr}^{1/3}} \tag{G-14}$$

[1] Neglecting natural convection.

[2] Friend and Metzner, *loc. cit.*

where D = inside diameter of circular tube
D_e = equivalent diameter
C_p = specific heat at constant pressure
G = mass velocity
g = acceleration due to gravity
f = friction factor (four times Fanning friction factor) (see Fig. F-14 for values in circular tubes)
j = heat-transfer analogue to f, defined by Eq. (G-9)
k = thermal conductivity
L = straight tube length
R = Reynolds number, defined by Eq. (F-12)
W = flow through tube or annulus, lb/hr (between two fins for finned annulus)
μ, μ_s = viscosity at bulk, surface temperature
u = average velocity
h = film coefficient of heat transfer
z = friction loss as column of flowing fluid
P = wetted perimeter
Pr = $C_p\mu/k$, Prandtl number (see Table G-9, Fig. G-10)
St = $h/(C_pG)$, Stanton number

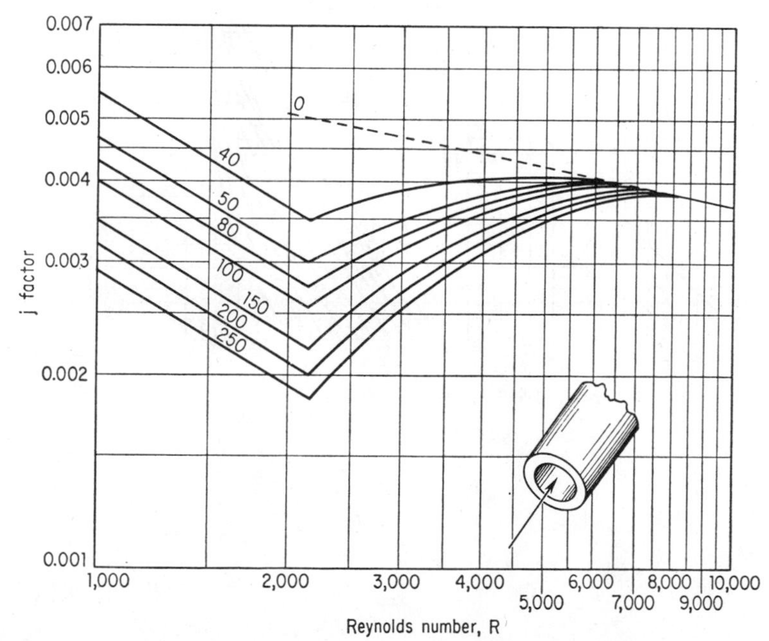

FIG. G-4.—Correlation of heat transfer inside circular tubes. [*Based on correlation of Sieder and Tate, Ind. Eng. Chem.*, **28,** 1429 (1936).]

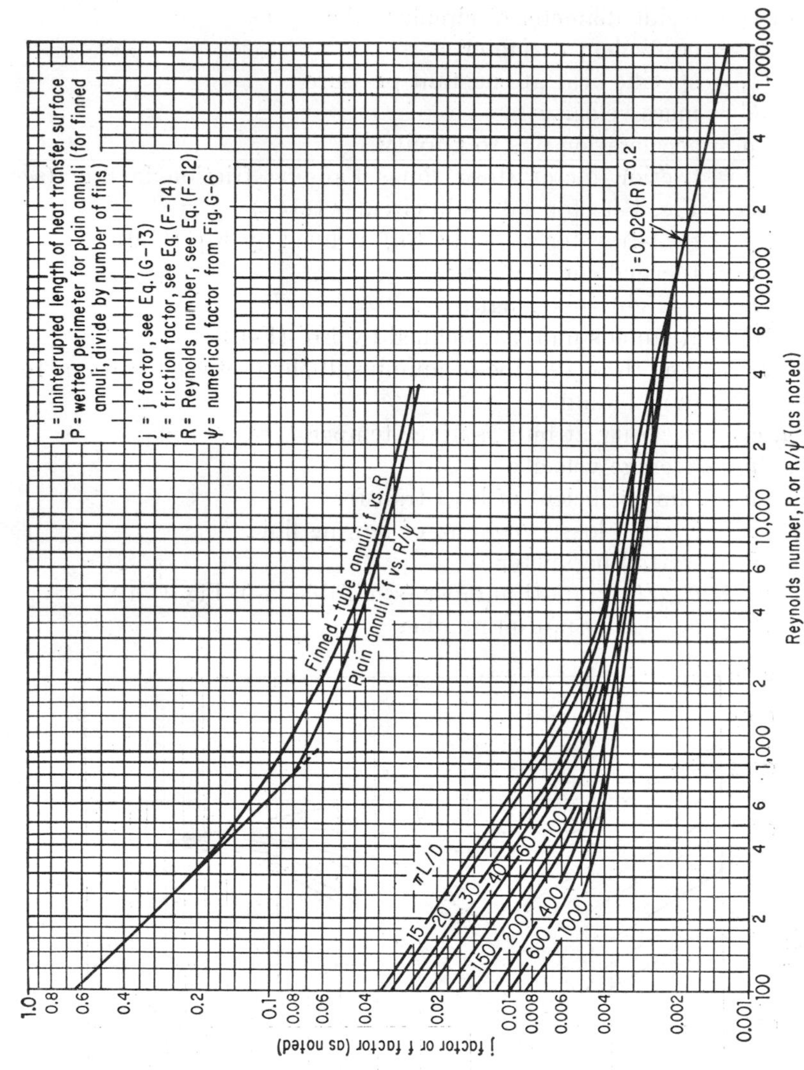

FIG. G-5.—Friction factors f and j factors for heat transfer of fluids in plain and finned tube annuli. See Fig. G-6 for values of "shape" factor ψ.

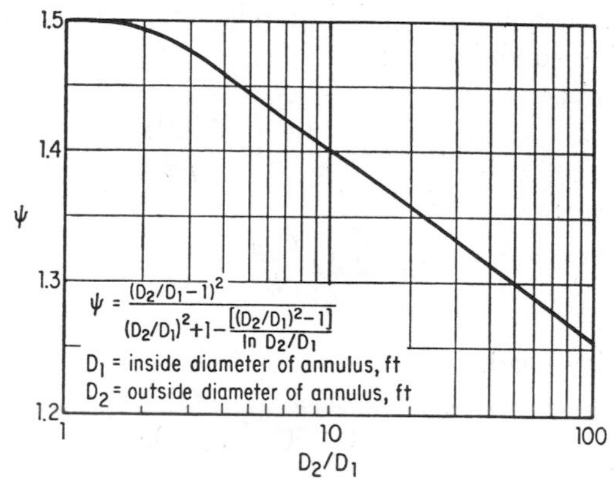

FIG. G-6.—Shape factor used for estimating friction factors in concentric annuli by Fig. G-5.

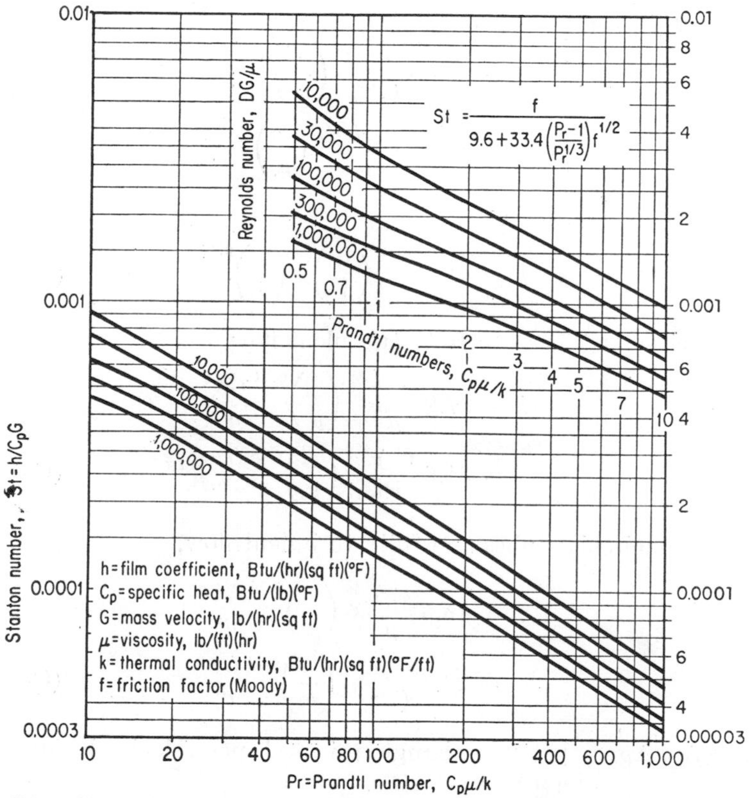

FIG. G-7.—Chart showing Friend and Metzner correlation of heat-transfer data for turbulent flow inside circular tubes. [*J. Am. Inst. Chem. Engrs.*, **4**, 392 (1958).]

Dimensional equations applying to heat transfer and pressure drops for fluids inside conduits are:

Reynolds number:

$$R = \frac{124 d_e G}{\mu} = \frac{6.32 W}{\mu d} \tag{F-12}$$

Prandtl number:

$$Pr = 2.42 \frac{C_p \mu}{k} \tag{G-15}$$

Pressure drop:

One velocity head,

$$\Delta p_v = \frac{u^2 \rho}{9{,}250} = \frac{G^2}{9{,}250 \rho} \tag{F-22}$$

In conduit,

$$\Delta p = \Delta p_v \frac{L}{D_e} f$$

Film coefficients:

From j factors,

$$h = 3{,}600 \frac{C_p G}{(\mathrm{Pr})^{2/3}} \left(\frac{\mu}{\mu_s}\right)^{0.14} j$$

$$= 1{,}990 \frac{C_p^{1/3} k^{2/3} G}{\mu^{2/3}} \left(\frac{\mu}{\mu_s}\right)^{0.14} j \tag{G-9a}$$

For streamline flow,

$$h = \frac{24 k^{2/3} C_p^{1/3}}{d_e} \left(\frac{W}{L}\right)^{1/3} \left(\frac{\mu}{\mu_s}\right)^{0.14} \tag{G-11a}$$

For circular tubes and fully turbulent flow,

$$h = 17.9 \frac{k^{2/3} C_p^{1/3}}{\mu^{7/15}} \left(\frac{\mu}{\mu_s}\right)^{0.14} \frac{G^{0.8}}{d^{0.2}}$$

$$= 31.6 \frac{C_p \mu^{0.2}}{(\mathrm{Pr})^{2/3}} \left(\frac{\mu}{\mu_s}\right)^{0.14} \frac{G^{0.8}}{d^{0.2}} \tag{G-12a}$$

For noncircular tubes and fully turbulent flow,

$$h = 15.5 \frac{k^{2/3} C_p^{1/3}}{\mu^{7/15}} \left(\frac{\mu}{\mu_s}\right)^{0.14} \frac{G^{0.8}}{d_e^{0.2}}$$

$$= 27.4 \frac{C_p \mu^{0.2}}{(\mathrm{Pr})^{2/3}} \left(\frac{\mu}{\mu_s}\right)^{0.14} \frac{G^{0.8}}{d_e^{0.2}} \tag{G-13a}$$

where Δp, Δp_v = pressure change, psi total, per velocity head
u = average velocity, fps
d, D = diameter of circular tube, in., ft
d_e, D_e = equivalent diameter, in., ft
f = friction factor, dimensionless

C_p = specific heat at constant pressure, Btu/(lb)(°F)
G = mass velocity, lb/(sq ft)(sec)
h = film coefficient of heat transfer, Btu/(hr)(sq ft)(°F)
k = thermal conductivity, Btu/(hr)(sq ft)(°F/ft)
L = length of conduit, ft
Pr = Prandtl number
ρ = fluid density, lb/(cu ft)
μ, μ_s = fluid viscosity, centipoises; at bulk, surface temperature
W = flow rate, lb/hr
Q = flow rate, gpm

For a number of cases, simplified relations may be used. Figure G-8 and Eq. (G-16) are based on a Prandtl number of 0.78 and a

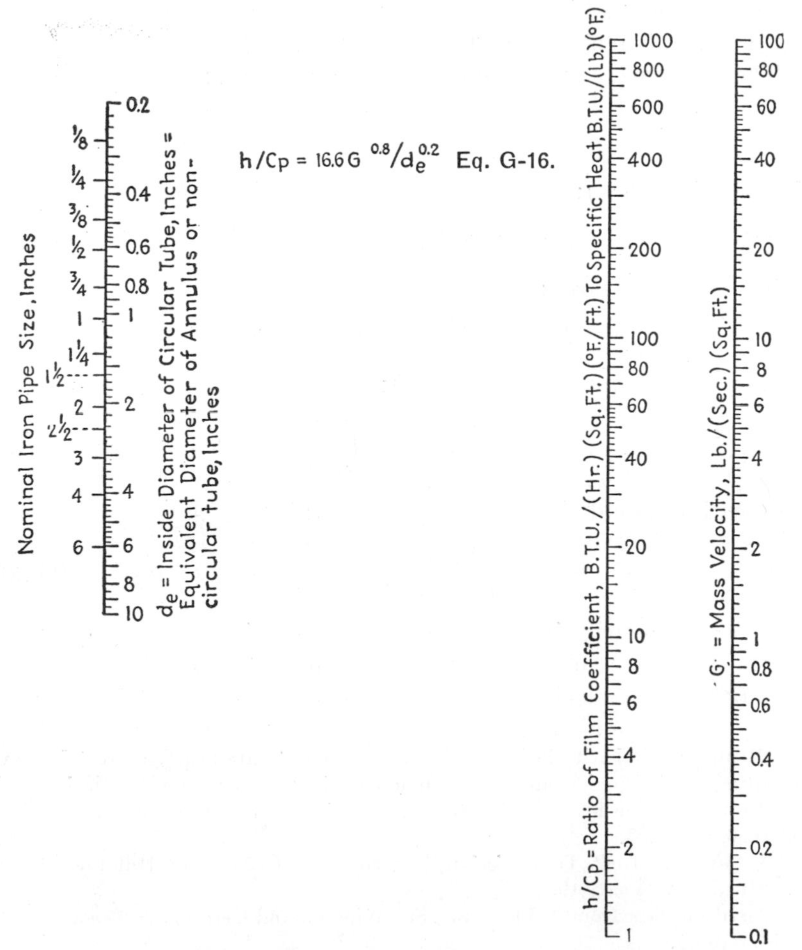

FIG. G-8.—Film coefficients for gases in turbulent flow inside tubes.

viscosity of 0.019 centipoise, conditions that are fairly representative for conservative calculation. Use of Eq. (G-17) in conjunction with Table G-7 is recommended for air.[1]

In a number of equations, the term $k^{2/3}C_p^{1/3}$ appears. Values of this combination for liquid hydrocarbons are given in Table G-8 and are the basis of Fig. G-9 for hydrocarbon film coefficients.

In the absence of specific data, the thermal conductivity of alcohols and organic acids may be taken as 0.1; other organic compounds as 0.08 Btu/(hr)(sq ft)(°F/ft) at room temperature (based on averages of data for 40 compounds). Values of Prandtl numbers are given in Table G-9 (gases) and Fig. G-10 (liquids).

Figure G-11 in combination with Table G-10 gives a rapid method for estimating film coefficients for a number of liquids. To use these, the base factors in Table G-10 are multiplied by the corrective factors from Fig. G-11 to secure the film coefficient.

The simplified equations are summarized below:

Gases[2] (turbulent flow):

$$h = 16.6C_pG^{0.8}/d^{0.2} \tag{G-16}$$

Air (turbulent flow):

$$h = AG^{0.8}/d^{0.2} \tag{G-17}$$

Water (turbulent flow):

$$h = 5.6(1 + 0.011t)G^{0.8}/d^{0.2} \tag{G-18}$$

Hydrocarbon liquids[3] (laminar flow):

$$h = \frac{2.9}{sd}\left(\frac{W}{L}\right)^{1/3}\left(\frac{\mu}{\mu_s}\right)^{0.14} \tag{G-19}$$

Hydrocarbon liquids (fully turbulent flow):

$$h = \frac{31.2}{\mu^{7/15}d}\left(\frac{Q}{d}\right)^{0.8}\left(\frac{\mu}{\mu_s}\right)^{0.14} \tag{G-20}$$

where h = film coefficient of heat transfer, Btu/(hr)(sq ft)(°F)

C_p = specific heat at constant pressure, Btu/(lb)(°F)

G = mass velocity, lb/(sec)(sq ft)

[1] Because of similarity in properties, the same data are nearly correct for oxygen, nitrogen, nitric oxide, and carbon monoxide if h, as calculated for air, is multiplied by the specific heat relative to air. These factors are, respectively, 0.905, 1.03, 0.96, and 1.03.

[2] McAdams, "Heat Transmission," 3d ed., p. 226, McGraw-Hill Book Company, Inc., New York, 1954.

[3] Based on data from Table G-8. See Winston and Clarke, *Petroleum Refiner*, **51,** 147 (1955).

d = inside diameter of tube, in.
μ, μ_s = viscosity at bulk, surface temperature
s = specific gravity relative to water (60/60°F)
Q = flow rate per tube, gpm
W/L = ratio of flow rate per tube, lb/hr, to channel length, ft
t = temperature, °F
A = constant from Table G-7

TABLE G-7.—PROPERTIES OF AIR FOR HEAT-TRANSFER CALCULATIONS
(Data compiled by authors)

Pressure, psia	Temperature, °F	Thermal conductivity k, Btu / (hr)(sq ft)(°F/ft)	Viscosity μ, lb/(ft)(hr)	Specific heat, Btu / (lb)(°F)	Prandtl number, Pr	$\mu^{0.2}$	Constant A of Eq. (G-17)*
14.7	−300	0.0048	0.0152	0.252	0.80	0.427	3.3
	−200	0.0077	0.0242	0.240	0.76	0.475	3.5
	0	0.0131	0.0393	0.240	0.72	0.523	4.0
	200	0.0178	0.0517	0.242	0.70	0.552	4.4
	500	0.0245	0.0672	0.247	0.68	0.582	4.8
	1000	0.0335	0.088	0.262	0.69	0.615	5.3
	2000	0.048	0.121	0.290	0.73	0.655	5.8
	3000	0.058	0.145	0.30	0.75	0.68	6.2
100	−200	0.0080	0.0260	0.260	0.84	0.482	3.6
	0	0.0133	0.0400	0.243	0.73	0.525	4.1
	200	0.0180	0.0520	0.242	0.70	0.554	4.4
	500	0.0246	0.0675	0.248	0.69	0.583	4.8
	1000	0.0336	0.090	0.263	0.70	0.618	5.3
500	−200	0.0105	0.031	0.450	1.32	0.499	5.0
	0	0.0140	0.0416	0.260	0.77	0.530	4.2
	200	0.0186	0.0539	0.248	0.72	0.558	4.5
	500	0.0248	0.070	0.252	0.71	0.588	4.8
1,000	0	0.0155	0.0450	0.282	0.82	0.538	4.4
	200	0.0196	0.0563	0.257	0.74	0.562	4.6
	500	0.0251	0.071	0.254	0.72	0.589	4.8
2,000	0	0.0190	0.0540	0.320	0.91	0.556	5.0
	200	0.0213	0.0605	0.268	0.77	0.571	4.7
	500	0.0259	0.072	0.258	0.72	0.591	4.9
3,000	0	0.0224	0.065	0.343	0.99	0.579	5.3
	200	0.0229	0.067	0.278	0.82	0.582	4.8
	500	0.0267	0.074	0.263	0.72	0.595	5.1

* Numerical value = $26.5 C_p \mu^{0.2}/\mathrm{Pr}^{0.6}$, derived in same fashion as Eq. (G-16).

TABLE G-8.—COMBINED THERMAL CONDUCTIVITY, SPECIFIC HEAT FACTORS OF PETROLEUM PRODUCTS FOR USE IN HEAT-TRANSFER CALCULATIONS IN EQS. (G-9*a*), (G-11*a*), (G-12*a*), (G-13*a*), (G-25*a*), (G-26*a*), AND (G-27*a*)

(Based on data of Cragoe, *Nat. Bur. Standards, Misc. Pub.* 97)

Density of hydrocarbon		Temperature range, °F	Factor $k^{2/3}C_p^{1/3}$ where k = thermal conductivity, Btu/(hr)(sq ft)(°F/ft); C_p = specific heat, Btu/(lb)(°F)
°A.P.I.	Specific gravity (60/60)		
10	1.0000	0–800	0.12
20	0.9340	0–800	0.13
30	0.8762	0–800	0.14
40	0.8251	0–600	0.14–0.15
50	0.7796	0–400	0.15–0.16
60	0.7389	0–400	0.16

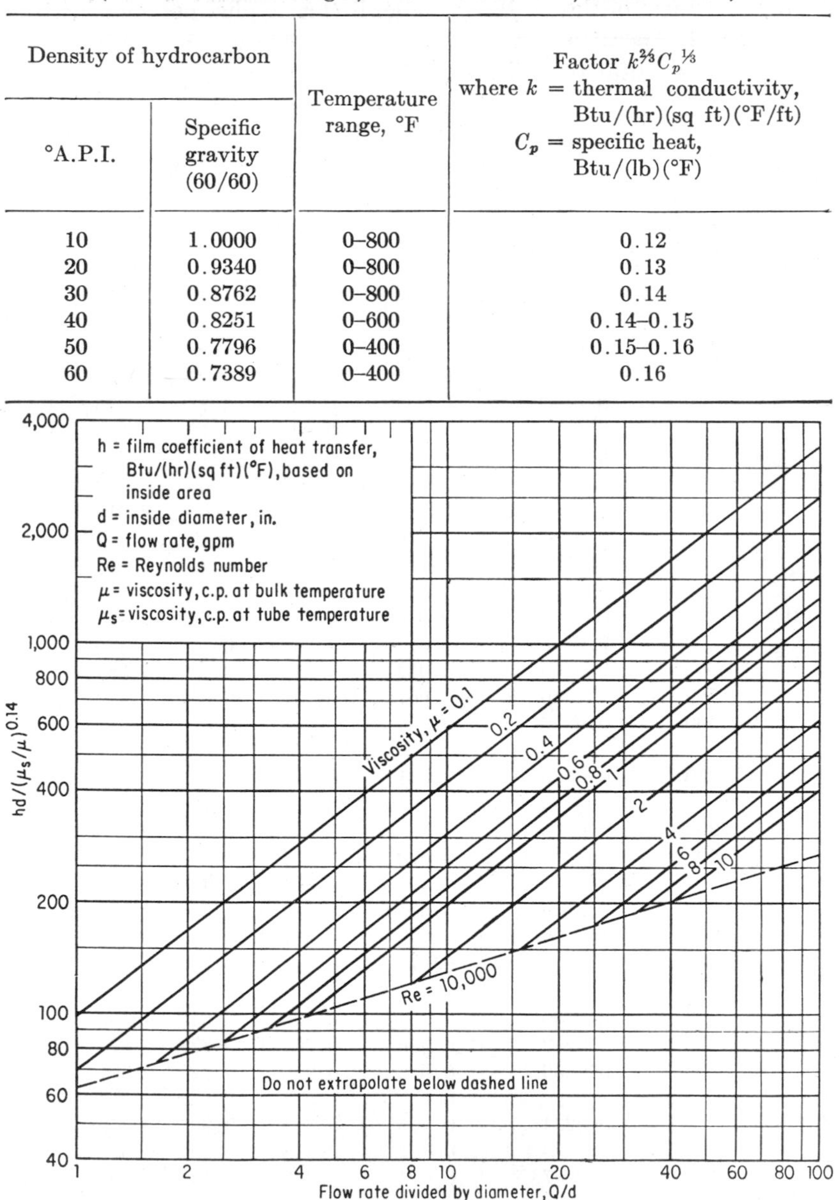

FIG. G-9.—Film coefficients for hydrocarbons inside circular tubes. (*Winston and Clarke, Petroleum Refiner, May,* 1956, *p.* 227.)

TABLE G-9.—PRANDTL NUMBERS OF COMMON CASES
(Data compiled by authors)

Substance	Temperature, °F				
	−300	0	500	1000	2000
	Prandtl number				
Ammonia	....	0.90	0.89		
Argon	0.70	0.68	0.66	0.66	0.67
Carbon monoxide	0.80	0.75	0.72	0.74	0.79
Carbon dioxide	....	0.79	0.69	0.68	0.5
Hydrogen	0.71	0.71	0.67	0.66	0.66
Methane	0.81	0.74	0.72	0.6	0.4
Nitrogen	0.79	0.73	0.68	0.70	0.74
Oxygen	0.83	0.72	0.70	0.71	0.76
Water vapor	....	....	1.02	0.99	0.98

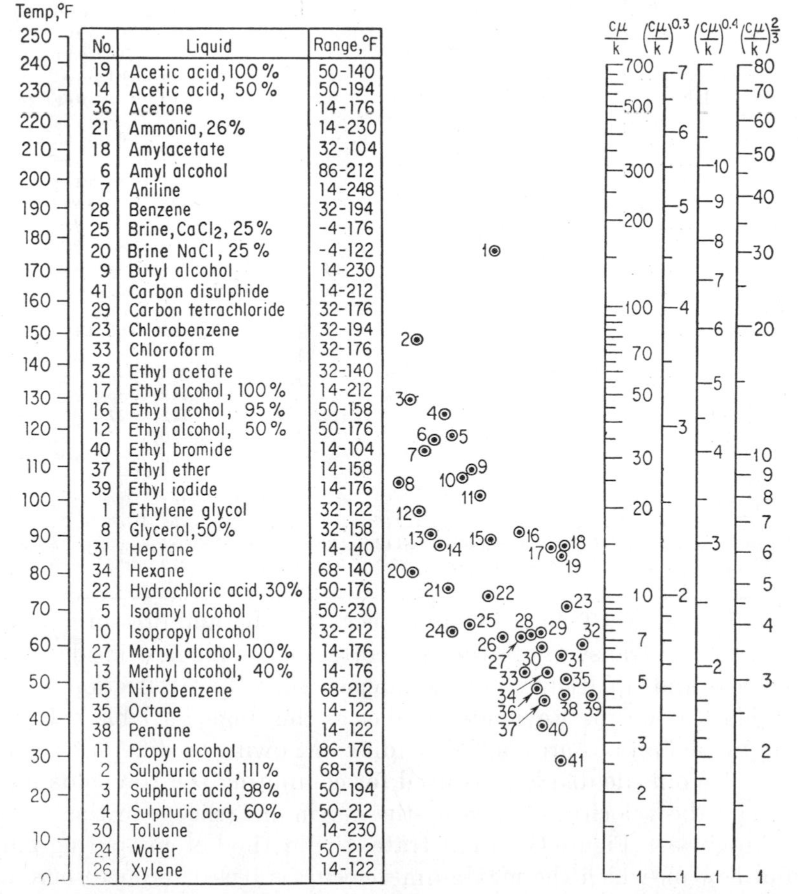

No.	Liquid	Range, °F
19	Acetic acid, 100%	50-140
14	Acetic acid, 50%	50-194
36	Acetone	14-176
21	Ammonia, 26%	14-230
18	Amylacetate	32-104
6	Amyl alcohol	86-212
7	Aniline	14-248
28	Benzene	32-194
25	Brine, $CaCl_2$, 25%	-4-176
20	Brine NaCl, 25%	-4-122
9	Butyl alcohol	14-230
41	Carbon disulphide	14-212
29	Carbon tetrachloride	32-176
23	Chlorobenzene	32-194
33	Chloroform	32-176
32	Ethyl acetate	32-140
17	Ethyl alcohol, 100%	14-212
16	Ethyl alcohol, 95%	50-158
12	Ethyl alcohol, 50%	50-176
40	Ethyl bromide	14-104
37	Ethyl ether	14-158
39	Ethyl iodide	14-176
1	Ethylene glycol	32-122
8	Glycerol, 50%	32-158
31	Heptane	14-140
34	Hexane	68-140
22	Hydrochloric acid, 30%	50-176
5	Isoamyl alcohol	50-230
10	Isopropyl alcohol	32-212
27	Methyl alcohol, 100%	14-176
13	Methyl alcohol, 40%	14-176
15	Nitrobenzene	68-212
35	Octane	14-122
38	Pentane	14-122
11	Propyl alcohol	86-176
2	Sulphuric acid, 111%	68-176
3	Sulphuric acid, 98%	50-194
4	Sulphuric acid, 60%	50-212
30	Toluene	14-230
24	Water	50-212
26	Xylene	14-122

FIG. G-10.—Prandtl numbers $c\mu/k$ for liquids. (*McAdams, "Heat Transmission," 3d ed., McGraw-Hill Book Company, Inc., New York,* 1954). NOTE: To obtain $c\mu/k$ at a specified temperature for a given liquid, as usual a straight line is drawn from the specified temperature through the numbered point for the liquid to the $c\mu/k$ scale; to obtain $c\mu/k$ raised to one of the three fractional powers shown, a horizontal alignment is made from the $c\mu/k$ scale.

Table G-10.—Base Factors for Heat-transfer Coefficients Inside Tubes

	Average temperature of the liquid, °F											
	Liquids heated						Liquids cooled					
	0	50	100	150	200	250	0	50	100	150	200	250
Acetic acid (100 %)	.....	.....	97.2	101	105	109	.....	.....	72.2	75.0	78.7	81.6
Acetic acid (50 %)	.....	117	156	180	203	228	.....	85.2	122	153	179	201
Acetone	104	122	134	137	139	142	88.4	105	118	121	124	126
Ammonia	350	425	507	599	690	790	314	380	507	650	797	938
Amyl acetate	65.0	66.2	67.7	71.8	78.5	86.0	48.1	50.7	52.3	53.3	54.4	55.5
Amyl alcohol (*iso*)	21.6	35.8	52.7	73.3	96.0	118	.....	22.7	36.0	53.0	72.3	91.8
Aniline	.....	43.8	58.4	76.5	99.2	123	.....	31.6	45.2	63.3	86.9	110
Benzene	.....	75.6	94.5	108	121	134	.....	61.7	79.1	93.5	107	120
Brine, $CaCl_2$ (25 %)	139	190	257	332	420	517	106	152	217	312	397	510
Butyl alcohol (*n*)	31.2	45.5	62.4	83.0	107	133	19.2	31.4	45.6	64.0	84.9	107
Carbon disulphfide	114	119	125	129	132	133	103	110	116	121	126	128
Carbon tetrachloride	57.4	69.2	78.6	82.6	85.8	88.2	40.4	55.8	67.3	72.0	76.0	78.7
Chlorobenzene	64.6	73.3	78.8	80.5	82.0	82.8	54.5	57.5	60.7	63.8	65.0	65.6
Ethyl acetate	126	126	125	123	122	121	83.3	84.1	84.1	84.1	83.3	82.4
Ethyl alcohol (100 %)	58.0	73.6	92.3	112	132	151	41.2	54.8	71.1	89.0	108	128
Ethyl alcohol (40 %)	61.5	104	162	228	292	389	.....	70.4	121	176	230	315
Ethyl bromide	97.8	104	110	114	119	122	84.9	92.9	99.9	106	112	116
Ethylene glycol (50 %)	71.4	105	158	222	299	380	44.2	71.7	120	183	261	
Ethyl ether	100	115	123	130	137	144	86.0	99.0	109	118	126	135
Glycerol (50 %)	59.0	90.5	131	182	242	302	34.9	59.5	94.5	134	179	222
Heptane	81.4	87.0	94.7	102	112	122	66.8	74.2	81.5	88.8	97.0	104
Hexane	85.8	93.8	102	109	114	117	70.7	78.5	87.2	95.2	102	108
Methyl alcohol (100 %)	83.0	110	126	138	149	160	65.4	88.5	105	118	132	146
Methyl alcohol (90 %)	86.0	114	136	154	172	188	68.5	92.0	112	128	142	156
Methyl alcohol (40 %)	64.0	110	164	213	264	312	38.2	80.0	127	177	236	292
Octane (*n*)	72.0	79.0	85.9	92.0	97.0	102	56.0	63.8	70.9	77.2	82.2	88.0
Pentane (*n*)	103	105	110	115	118	121	82.8	89.6	96.4	101	106	110
Propyl alcohol (*iso*)	25.7	49.3	71.5	94.5	117	139	.....	32.4	51.5	71.7	92.7	116
Sulphur dioxide	167	171	175	180	182	194	150	155	161	166	174	178
Sulphuric acid (60 %)	.....	65.9	79.4	94.5	110	129	.....	35.3	54.4	66.9	74.0	77.7
Toluene	77.3	86.9	96.6	104	112	119	61.8	71.5	81.4	90.3	97.6	103
Water	.....	225	322	408	392	508	.....	153	273	355	427	483

c. Flow across Tubes.—Data on heat transfer and pressure drop of fluids flowing transverse to tubes is correlated as for flow inside tubes, by charts and equations showing a dependence of both friction factors and j factors on the Reynolds number as the single variable. In this case, the definitions are somewhat different owing to different geometry. The outside diameter is used for computing the Reynolds number, and the velocity is taken as the maximum (except in the case of single tubes). Figure G-14 illustrates the method of computing minimum flow on which the maximum velocity is based. Since there are very many different tube configurations and variable tube spacings, it is not surprising that more than one correlation is required to cover even the arrangements of principal interest.

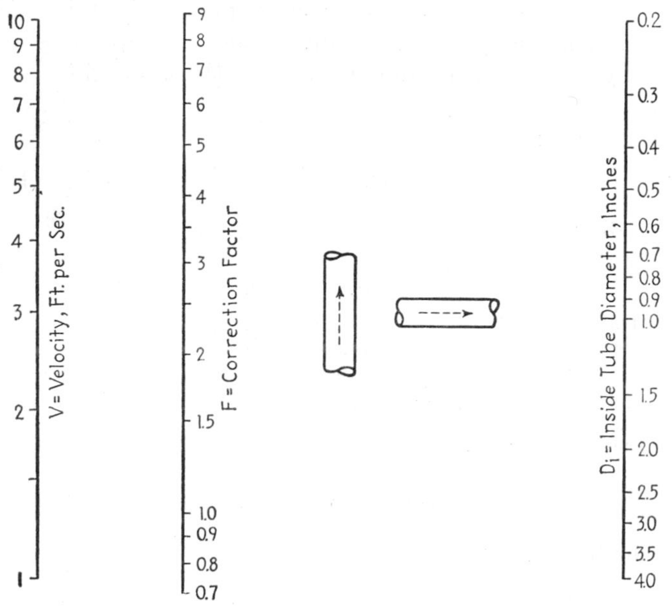

Fig. G-11.—Case 1, correction factors for liquids heated or cooled inside tubes. To obtain film coefficient h in Btu/(hr)(sq ft)(°F), multiply base factors from Table G-10 by factor F from this chart. (*Stoever, Chem. Met. Eng., May,* 1944.)

The friction factor for flow across tubes is defined by Eq. (F-12*a*), and its relation to the Reynolds number given by Fig. G-12 and by Eqs. (G-21) to (G-24) for several tube bundle arrangements. Correlation of heat transfer by j factors is given by Fig. G-13 and Eqs. (G-25) to (G-27).

The basis of the recommended correlations for both friction and j factors at low Reynolds numbers is the recent work of Bergelin, Leighton, Lafferty, and Pigford.[1] Small differences in heat transfer for the three staggered tube arrangements tested were neglected in the recommended relations. For friction factors, the data were extended to high Reynolds numbers by assuming a two-tenths power relationship to the Reynolds numbers consistent with earlier data.[2,3] For Reynolds numbers between 1,000 and 10,000, the Delaware work exhibits higher j factors than correlations of older data, such as those by McAdams[4] and Grimison.[5] In effect, the recommendations given here are conservative.

[1] Univ. of Delaware, *Eng. Expt. Sta. Bull.* 4, 1958.

[2] Perry, "Chemical Engineers' Handbook," 3d ed., McGraw-Hill Book Company, Inc., New York, 1950.

[3] Grimison, *Trans. Am. Soc. Mech. Engrs.*, **57**, 583 (1937).

[4] McAdams, "Heat Transmission," 3d ed., McGraw-Hill Book Company, Inc., New York, 1954.

[5] Grimison, *Trans. Am. Soc. Mech. Engrs.*, **59**, 583 (1937); **60**, 381 (1938).

A summary of dimensionless equations applying for fully turbulent flow for friction and j factors is given below. Any consistent set of units may be used in these equations or the dimensionless groups therein.

Reynolds number:

$$R = \frac{DG}{\mu} \tag{F-12a}$$

Friction factor:

$$f_c = \frac{\Delta p g}{2G^2 n_r} = \frac{zg}{2u^2 n_r} \tag{G-21}$$

Triangular staggered:

$$P/D = 1.25 \text{ to } 1.5$$

$$f_c = \frac{0.78}{(P/D)^{1/2}} R^{-0.2} \tag{G-22}$$

Staggered square banks:

$$\frac{P}{D} = 1.25 \qquad f_c = 0.63R^{-0.2} \tag{G-23}$$

In-line tube banks:

$$\frac{P}{D} = 1.25 \text{ to } 1.5 \qquad f_c = \frac{0.2}{P/D - 1} R^{-0.2} \tag{G-24}$$

j factor:

$$j = \frac{h}{C_p G} \left(\frac{C_p \mu}{k}\right)^{2/3} \left(\frac{\mu_s}{\mu}\right)^{0.14} \tag{G-9}$$

Staggered tubes:

$$j = 0.33F_n R^{-0.4} \tag{G-25}$$

In-line (square) tubes:

$$j = 0.30F_n R^{-0.4} \tag{G-26}$$

Single tubes:

$$j = 0.26R^{-0.4} \tag{G-27}$$

Dimensional equations for pressure drops and heat transfer of fluids flowing across tubes are:

Reynolds number:

$$R = \frac{124dG}{\mu} = \frac{124u\rho d}{\mu} \tag{F-12}$$

Prandtl number:

$$\text{Pr} = 2.42 \frac{C_p \mu}{k} \tag{G-15a}$$

Pressure drop:

One velocity head,

$$\Delta p_v = \frac{u^2 \rho}{9{,}250} = \frac{G^2}{9{,}250\rho} \tag{F-15}$$

Total drop,

$$\Delta p = n_r \Delta p_v f_c \tag{G-28}$$

Film coefficients:

From j factors,

$$h = 3{,}600 \frac{C_p G}{(\mathrm{Pr})^{2/3}} \left(\frac{\mu}{\mu_s}\right)^{0.14} j$$

$$= 1{,}990 \frac{C_p^{1/3} k^{2/3} G}{\mu^{2/3}} \left(\frac{\mu}{\mu_s}\right)^{0.14} j \tag{G-9a}$$

Turbulent flow, staggered tubes,

$$h = 95 F_n \frac{k^{2/3} C_p^{1/3}}{\mu^{4/15}} \left(\frac{\mu}{\mu_s}\right)^{0.14} \frac{G^{0.6}}{d^{0.4}} \tag{G-25a}$$

Turbulent flow, in-line tubes,

$$h = 86 F_n \frac{k^{2/3} C_p^{1/3}}{\mu^{4/15}} \left(\frac{\mu}{\mu_s}\right)^{0.14} \frac{G^{0.6}}{d^{0.4}} \tag{G-26a}$$

Turbulent flow, single tubes,

$$h = 75 \frac{k^{2/3} C_p^{1/3}}{\mu^{4/15}} \left(\frac{\mu}{\mu_s}\right)^{0.14} \frac{G^{0.6}}{d^{0.4}} \tag{G-27a}$$

where d = inside tube diameter, in.
- P = tube pitch (center-to-center spacing), ft
- D = outside tube diameter, ft
- C_p = specific heat at constant pressure
- f_c = friction factor for cross flow
- F_n = factor depending on number of rows of tubes traversed: 10 or more, $F_n = 1.00$; 7, $F_n = 0.97$; 4, $F_n = 0.90$; 2, $F_n = 0.78$
- g = acceleration due to gravity
- G = mass velocity, lb/(sec)(sq ft), maximum except for single tubes (see Fig. G-14)
- h = film coefficient of heat transfer
- k = thermal conductivity, Btu/(hr)(sq ft)(°F/ft)
- n_r = number of rows of tubes traversed
- Δp = pressure drop, psi
- Δp_v = pressure drop equivalent to one velocity head, psi
- Pr = Prandtl number, dimensionless
- R = Reynolds number, dimensionless
- u = linear velocity, fps
- z = head loss as a fluid column
- μ, μ_s = viscosity, centipoise at bulk, surface temperature
- ρ = fluid density, lb/cu ft

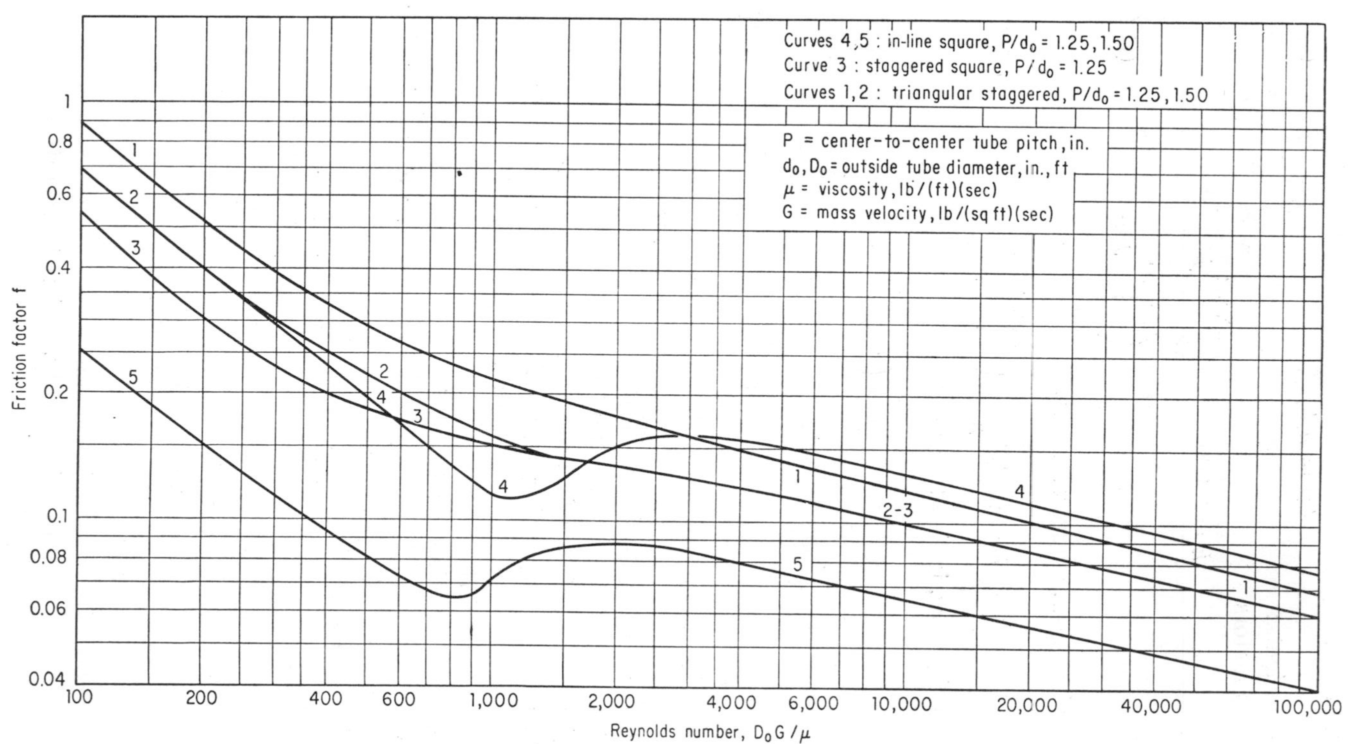

FIG. G-12.—Friction factors for flow across tube banks. *(Prepared largely from data of Bergelin, Leighton, Lafferty, and Pigford, Univ. Delaware Expt. Sta. Bull.* 4, *April*, 1958.)

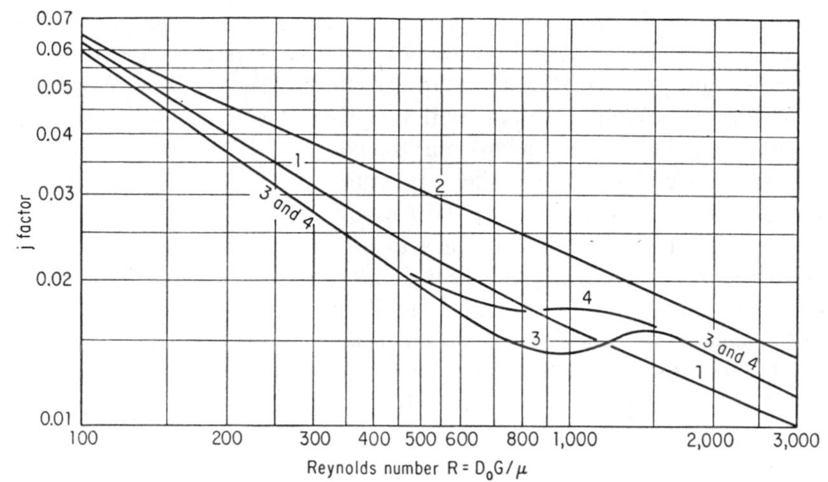

Single tubes, curve 1
Staggered tubes:
Triangular, $P/D_o = 1.25$ and 1.50 } curve 2
Square, $P/D_o = 1.25$ } curve 2
In-line square, $P/D_o = 1.25$ and 1.50, curves 3, 4

$$j = \frac{h}{C_pG}\left(\frac{C_p}{k}\right)^{2/3}\left(\frac{\mu_s}{\mu}\right)^{0.14}$$

where h = film coefficient, Btu/(hr)(sq ft)(°F)
C_p = specific heat, Btu/(hr)(sq ft)(°F)
G = mass velocity, lb/(hr)(sq ft)
k = thermal conductivity, Btu/(hr)(sq ft)(°F/ft)
μ, μ_s = viscosity, lb/(ft)(hr), at bulk, surface temperatures
P/D_o = ratio of center-to-center tube pitch to outside tube diameter

FIG. G-13.—Chart for j factors for fluid flow transverse to single tubes or banks of tubes having 10 or more rows. (*Based on data of Bergelin, Leighton, Lafferty, and Pigford, Univ. Delaware Expt. Sta. Bull.* 4, *April*, 1958; *and McAdams, "Heat Transmission," 3d ed., McGraw-Hill Book Company, Inc., New York*, 1954.)

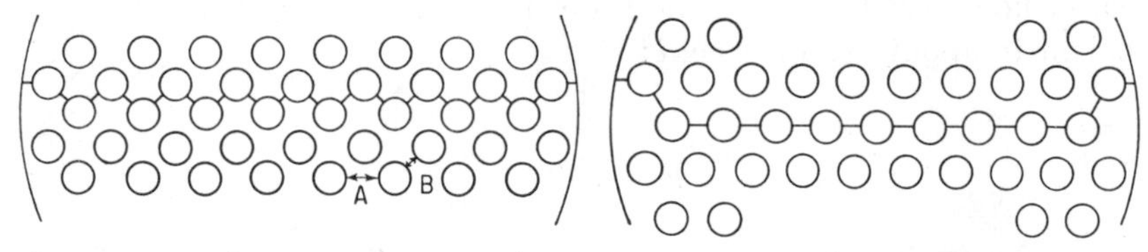

FIG. G-14.—Illustration of basis for computing maximum flow across tube banks.

d. Flow in Shells.—In most commercial tube-and-shell exchangers the flow patterns cannot be described as either parallel or transverse to the tubes, but rather as a mixed flow. In the rare cases that baffles are omitted, the flow will be principally parallel to the tube bundle and the film coefficient may be computed as for other conduits.

A number of different baffle arrangements are employed to promote cross flow. For the common half-moon type, the flow through the window is largely parallel to the tube. Between windows and at the center of the exchanger, the flow is largely transverse. The normal construction techniques require a small, but definite, clearance between baffle holes and tubes and between the outside of the baffle and the shell. The fluid will, to some extent, flow through the openings left by these clearances. Such baffle by-passing, or leakage, reduces the extent to which the desired cross flow is achieved.

For a first approximation to the film coefficient in this complex situation, baffle leakage may be neglected and the film coefficient taken as 60 per cent of the calculated value for true transverse flow at the center line. Even this approximation should not be made unless the baffle window is between 25 and 40 per cent and the heat exchanger in question is well designed, *i.e.*, has low clearances between baffle holes and tubes and the tube bundle nearly fills the shell *so that the by-passing of the tube bundle is minimized.*

Pressure drops computed on the basis of no leakages and true transverse flow will likewise be high but will suffice as a first approximation. Frequently the pressure drop on the shell side is not a limiting factor.

For open ("trombone," or "trickle") coolers used in cooling fluids flowing through single horizontal tubes or tube bundles by trickling water flow over the pipes, the following equation[1] gives the film coefficient to about ± 25 per cent:

$$h = 940 \sqrt{\frac{m}{d_o L}}$$

where m = gallons of water per minute flowing over a tube
L = tube length, ft
d_o = outside tube diameter, in.
h = film coefficient, Btu/(hr)(sq ft)(°F)

[1] McAdams, "Heat Transmission," 3d ed., McGraw-Hill Book Company, Inc., New York, 1954.

5. Condensing Vapors

The rate of transfer of heat to cold surfaces depends on whether the condensation is of the film type or dropwise type. The names are descriptive of the mechanisms. Dropwise condensation of steam frequently occurs on polished surfaces and is promoted by addition of oil or other organic substances which prevent complete wetting of the surface by water. Although dropwise coefficients are higher, film-type condensation is more common in practice and will alone be considered here. The original equations for rate of condensation were derived by Nusselt in 1916 under the following assumptions:

1. Heat-transfer rate is determined solely by the rate of transfer through condensed liquid film
2. This condensed liquid film is continuous
3. Flow of film results from the effect of gravity alone
4. Flow of film is streamline

The recommended equations given below[1] are based largely on the original derivation. These equations are valid unless the liquid flow is so heavy that turbulent flow results. If this occurs, higher heat transfer results, so that the equations are safe. Turbulence occurs when the Reynolds number exceeds the critical value for flow in condensate film. The critical condition, expressed in terms of condensation rates, may be computed by the following equations:

For vertical tubes $$\frac{W}{N_v d\mu} = 300$$

For horizontal tubes $$\frac{W}{N_h L\mu} = 2{,}500$$

where W = condensate, lb/hr
d = diameter of vertical tube, in.
μ = viscosity, centipoises
N_v = number of vertical tubes
N_h = number of horizontal rows of horizontal tubes
L = length of all tubes in one horizontal row, ft

Outside vertical tubes:

$$h = 129\left(\frac{k^3\rho^2\Delta H}{\mu L\Delta t}\right)^{1/4} \qquad \text{(G-29)}$$

[1] McAdams, *op. cit.*

Outside a vertical tier of horizontal tubes:

$$h = 154 \left(\frac{k^3 \rho^2 \Delta H}{\mu N d_o \Delta t} \right)^{1/4} \tag{G-30}$$

where h = film coefficient, Btu/(hr)(sq ft)(°F)
k = thermal conductivity of condensed liquid (at temperature of cold surface), Btu/(hr)(sq ft)(°F/ft)
ρ = density of liquid at cold surface temperature, lb/cu ft
ΔH = latent heat of vaporization at condenser pressure, Btu/lb
μ = viscosity of liquid at cold surface temperature, centipoise
L = length of vertical tubes, ft
N = number of rows of tubes in a vertical tier
d_o = outside pipe diameter, in.
Δt = difference in temperature between vapor and cold surface

Figure G-15 and Table G-11 permit the estimation of film coefficients for the condensation of 18 vapors on horizontal tubes. Figure

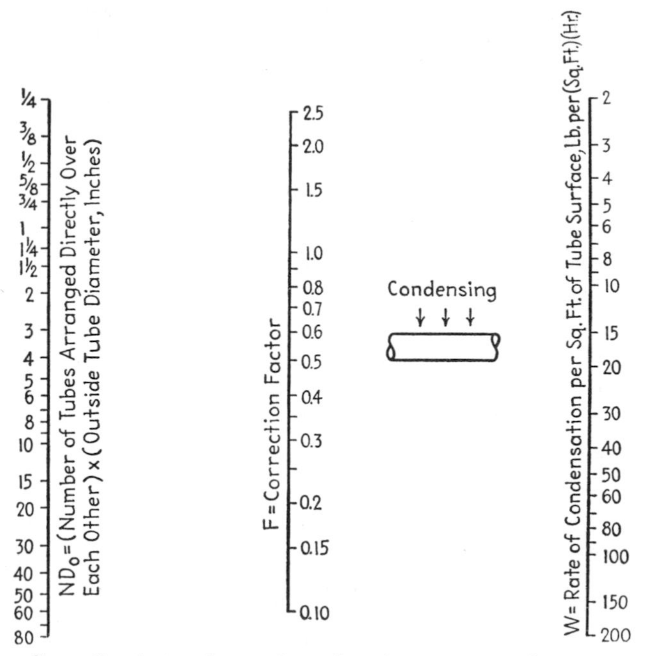

FIG. G-15.—Correction factors for condensation of pure saturated vapors on horizontal tubes. (*Stoever, Chem. & Met. Eng., May,* 1944.)

G-16 gives the film coefficients for 17 vapors on a horizontal tube 1.9 in. outside diameter. Rates of condensation of vapor mixtures are usually intermediate between the rates of the component vapors; to be safe, the lowest film coefficient of any component may be assumed for the mixture.

TABLE G-11.—BASE FACTORS FOR CONDENSATION OF PURE SATURATED VAPORS ON HORIZONTAL TUBES
(Stoever, *Chem. & Met. Eng.*, May, 1944, p. 98)

	Temperature of condensate film (assume equal to tube wall)					
	50°F	100°F	150°F	200°F	250°F	300°F
Acetic acid	.....	511	495	470	424	373
Acetone	772	789	805	805	795	780
Ammonia	2,768	3,145	3,459	3,711	*3,875*	*3,965*
Aniline	*275*	405	544	685	830	977
Benzene	554	609	658	706	755	798
Carbon disulphide	924	933	933	924	905	868
Carbon tetrachloride	551	580	569	*482*		
Chloroform	735	791	847	*895*	*950*	*997*
Ethyl acetate	*702*	772	835	889	936	990
Ethyl alcohol	*495*	556	618	678	745	807
Ethyl ether	620	646	665	678	691	705
Heptane	*488*	537	580	607	628	645
Hexane	*525*	552	576	592	608	614
Methyl alcohol	695	772	850	920	972	103
Octane	*482*	513	538	554	575	585
Propyl alcohol (*iso*)	*284*	400	488	548	*596*	*632*
Steam	1,830	2,440	3,020	3,590	4,120	4,660
Sulphur dioxide	1,260	1,200	1,115	1,010	*900*	*780*

For vapor mixtures that form immiscible condensates, a partial dropwise condensation probably results. Data of Baker and coworkers[1] indicate that film coefficients over 400 Btu/(hr)(sq ft) (°F) are usually obtainable for condensation of steam and organic vapors to form a two-phase condensate.

[1] Baker and Tsao, *Trans. Am. Inst. Chem. Engrs.*, **36**, 517 (1940). Baker and Mueller, *Ind. Eng. Chem.*, **29**, 1067 (1937). (Vapors studied: benzene, toluene, heptane, chlorobenzene, and trichloroethylene.)

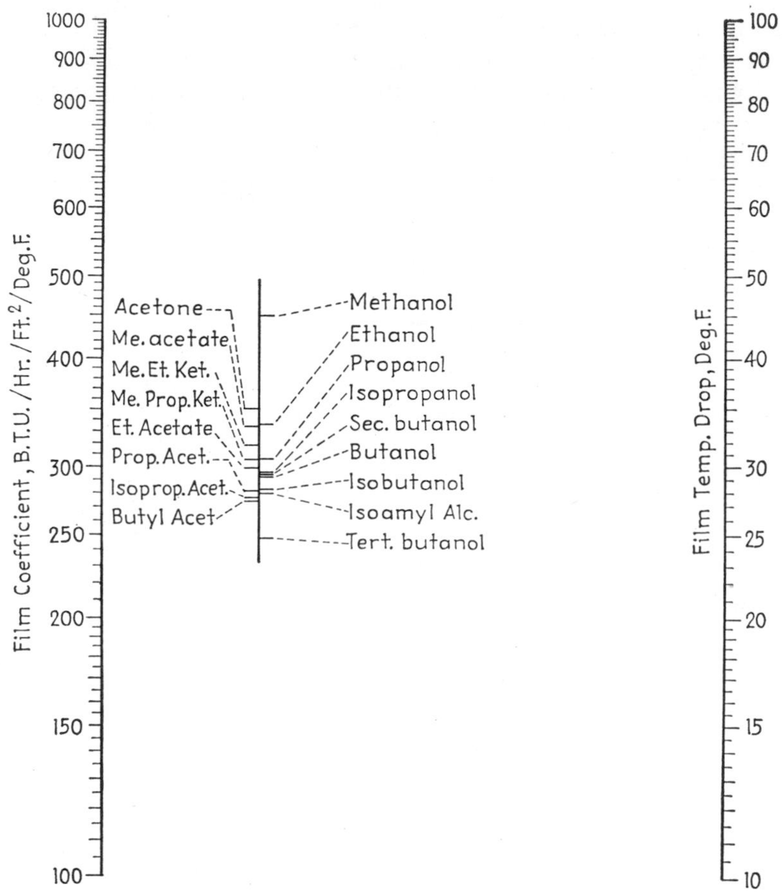

FIG. G-16.—Film coefficients of condensing alcohols, ketones, and esters on 1.9-in. tube. [*Reprinted from Othmer and Berman, Ind. Eng. Chem.*, **35**, 1068 (1943), *by permission of the American Chemical Society.*]

6. Boiling Liquids

a. Types of Boiling.—The rate of evaporation of a liquid depends on the turbulence produced by boiling. At quite low rates, little turbulence results and the rate of heat transfer is little more than would be expected from natural convection. At reasonable rates, the boiling does produce turbulence and in effect promotes itself to fairly high steady-state boiling. This steady state, termed "nucleate boiling," is quite sensitive to the temperature difference and other factors.

As the temperature difference is increased, the boiling rate reaches a maximum beyond which the vapor cannot escape and the surface becomes partially vapor bound, causing the evaporation rate to drop with increased temperature difference. This relatively unstable film-boiling condition becomes stable film boiling at somewhat higher temperature differences wherein all the surface is completely vapor bound.

b. Stable Film Boiling.—This type of boiling has been theoretically analyzed,[1] and the recommended equation for boiling outside of tubes is

$$h = 0.62 \left[\frac{k^3 \rho_v (\rho_l - \rho_v)}{D \mu_v \Delta t} \right]^{1/4} \qquad \text{(G-31)}$$

where h = film coefficient, Btu/(hr)(sq ft)/(°F)
k = thermal conductivity of vapor, Btu/(hr)(sq ft)(°F)/(ft)
ρ_v, ρ_l = vapor, liquid density, lb/cu ft
D = outside tube diameter, ft
Δt = temperature difference, °F
μ_v = vapor viscosity, centipoises

All vapor properties are evaluated at the average temperature (one-half saturated liquid plus surface).

The preceding equation does not include radiation, but a small contribution of radiation can be separately computed and added to secure the total heat transfer. If the radiation effect is large, the two are not additive.

Film-boiling rates are greatly augmented by forced circulation, so that considerably higher heat-transfer rates can readily be achieved.

c. Nucleate Boiling.—Results of experiments on nucleate boiling are complex. The rates are sensitive to the nature and cleanliness of the heated surface, to the physical properties of the liquid and its vapor, and to the applied temperature difference.

Maximum boiling rates have been correlated in terms of reduced pressure[2] and in terms of physical properties.[3,4] The recommended equation[4] is as follows:

$$W = 110 \rho_l^{2/3} \rho_v^{1/3} \qquad \text{(G-32)}$$

where W = maximum weight of liquid evaporated, lb/hr
ρ_l, ρ_v = liquid, vapor density, lb/cu ft

Values computed by this equation represent a reasonable approximation. The observed rates with different surfaces vary somewhat. Values given by Eq. (G-32) are usually obtainable and frequently exceeded.

[1] Bromley, *Trans. Am. Inst. Chem. Engrs.*, **46**, 221 (1950). Equation (G-31) differs slightly from the original derivative and is quoted from McAdams, *op. cit.*

[2] Cichelli and Bonilla, *Trans. Am. Inst. Chem. Engrs.*, **41**, 455 (1945).

[3] Addoms Thesis in Chemical Engineering, Massachusetts Institute of Technology, 1948.

[4] Clarke, unpublished data.

The temperature differences at which the maximum boiling rates occur are generally between 30 and 100°F for boiling at atmospheric pressure, but smaller temperature differences, low as 2 or 3°F, are observed at higher pressures. As a guide, typical ranges of dimensionless groups involving the temperature difference are as follows:

$$\text{St} = \frac{C_p \Delta t}{L} \qquad \text{between 0.03 and 0.30}$$

$$= \frac{C_p \Delta t}{L(\text{Pr})^{0.6}} \qquad \text{between 0.02 and 0.10}$$

$$= \frac{C_p \Delta t}{L(\text{Pr})} = \frac{k \Delta t}{L\mu} \qquad \text{between 0.045 and 0.07}$$

where St = Stanton number
Pr = Prandtl number
C_p = specific heat of liquid
Δt = temperature difference
L = latent heat of vaporization
k = thermal conductivity of liquid
μ = viscosity of liquid

Below the maximum boiling rate, the heat-transfer rate is usually taken as a power of temperature difference:

$$\frac{h}{h_m} = \left(\frac{\Delta t}{\Delta t_m}\right)^n \tag{G-33}$$

where h, h_m = film coefficient at Δt, Δt_m
Δt = temperature difference less than maximum
Δt_m = temperature difference at maximum boiling rate
n = constant exponent

The exponent in Eq. (G-33) is usually between 2.0 and 2.4, though some experimental data at higher pressures suggest very much lower values.

The preceding information permits only a first approximation of nucleate boiling coefficients. Experimental data should be sought whenever possible.

Gilmour[1] gives a correlation of nucleate boiling that is helpful though not very convenient or entirely satisfactory, since small discrepancies with the correlation can lead to larger errors in the calculated coefficient.

Since boiling coefficients are high, conservative design frequently

[1] *Chem. Eng. Prog.*, **54** (10), 77 (1958).

represents small cost differentials and therefore an entirely acceptable solution.

Forced or induced circulation increases boiling coefficients. In this case the coefficient can be computed on the basis of transfer without boiling with complete assurance that the computed coefficient will be very safely conservative. Use of the so-called "thermal syphon" reboilers to induce circulation is very common. The difference between the densities of the liquid and the boiling mass provides a substantial head, and very large circulating rates can be achieved.

7. Over-all Heat Transfer

a. Heat-transfer Equipment.—The shell and tube exchangers are in very common usage for industrial heat transfer. Their design involves a careful balance between a large number of factors,[1] the most important of which is securing the maximum velocities with available pressure drops. On the tube side this is effected by relation of tube diameter to total length. If a long path is essential to secure the velocity and area in reasonable sized tubes, several passes or multiple units may be required. The mean temperature difference is usually reduced by more than one pass in a single shell. On the shell side, the velocity may be adjusted by tube and baffle spacing, or occasionally by more than one shell pass.

Appreciation of these factors is important to design of the process prior to equipment specification. When the temperature changes on both fluids are large, the problem of securing proper velocities is likely to be more severe than if the temperature changes are small. In some cases it is economic to use more cooling water with a smaller temperature rise. This favors increased water velocities and multipass arrangements.

Double pipe exchangers are the easiest to design and are frequently employed where a small amount of heat is to be transferred, as in the case of a pilot plant. Where the film coefficient on one side is much greater than on the other, increased area on the side of low rate is an advantage. Finned tubes are frequently used to accomplish this as in the case of air heaters. Fins may be either around the tube for cross flow, or along the length for parallel flow. In the oil industry, open or "trombone" coolers are fairly common. In these, cooling is effected by spraying water over a tube bundle. Such coolers are inexpensive

[1] Baylock, *Trans. Am. Inst. Chem. Engrs.*, **40,** 593 (1944) gives an excellent brief discussion.

but require a larger area. In addition, there are submerged coils, steam jackets, and many other devices for heat transfer.[1]

Cooling water towers and contact coolers are used for heat transfer but present an entirely different type of problem and are treated briefly in Chapter N on Water.

Tube bundles are usually made by expanding the tubes into tube sheets. Table G-12 gives minimum center-to-center spacing for this type of construction.

TABLE G-12.—MINIMUM CENTER-TO-CENTER SPACING FOR TUBES EXPANDED IN TUBE SHEET

[Blaylock, *Trans. Am. Inst. Chem. Engrs.* **40,** 593 (1944)]

Outside tube diameter, in.	5/8	3/4	1	1 1/4
Triangular spacing, in.	13/16	15/16	1 1/4	1 9/16
Rectangular spacing, in.	7/8	1	1 1/4	1 9/16

The diameter of the shell depends on the number and arrangement of tubes in addition to the center-to-center spacing. Table G-13 gives data on various numbers of tubes arranged in a hexagonal pattern (either around a central tube, or starting with three tubes grouped around the center). Table G-14 gives data on rectangular spacing.

TABLE G-13.—DIAMETER RATIOS FOR TUBES ARRANGED HEXAGONALLY IN SHELLS

(Deutsch, *Chem. & Met. Eng.*, August, 1940, p. 538)

No. of tubes	Ratio: $\frac{\text{shell diameter}}{\text{center-to-center spacing}}$	No. of tubes	Ratio: $\frac{\text{shell diameter}}{\text{center-to-center spacing}}$
3	2.15	61	9.00
7	3.00	63	9.09
12	4.05	69	9.33
13	4.46	73	9.77
19	5.00	85	10.16
27	6.03	90	10.86
31	6.28	96	11.06
37	7.00	102	11.26
42	7.43	109	11.58
48	8.02	114	12.01
55	8.21	121	12.14

[1] Schwarz, *Chem. & Met. Eng.*, May, 1944, p. 120.

TABLE G-14.—DIAMETER RATIOS FOR TUBES ARRANGED RECTANGULARLY IN SHELLS

No. of Tubes	Ratio: $\frac{\text{Shell diameter}}{\text{Center-to-center spacing}}$
4	2.41
5	3.00
8	4.00
12	4.44
13	5.00
16	5.3
22	6.2
29	7.05
37	7.6

For a large number of tubes, the shell diameter may be approximately estimated by the equations below:
Rectangular arrangement:

$$D = 1.25d\sqrt{n} \tag{G-34}$$

Hexagonal arrangement:

$$D = 1.15d\sqrt{n} \tag{G-35}$$

where D = shell diameter, in.
d = center-to-center tube spacing, in.
n = number of tubes

It will be noted that the hexagonal arrangement of tubes is the more economical of space and, therefore, affords higher shell velocities. However, the rectangular array is frequently used because cleaning of the tubes is considerably easier.

b. Mean Temperature Difference.—The over-all transfer of heat is usually computed as the product of three quantities: the surface area, the mean temperature difference, and the over-all coefficient of heat transfer. This obviates the need for integration over the entire surface. Actually the mean temperature difference is derived by such an integration under the assumptions that the over-all coefficient of heat transfer is constant and that any variation in temperature of the fluids is linear with respect to the heat absorbed or lost. These conditions are seldom accurately fulfilled, but the error is not often serious. The results of this integration for single-pass exchangers (either cocurrent or countercurrent) may be most conveniently expressed as the integrated average effect of the temperature gradient between the two fluids. This turns out to be the logarithmic mean temperature difference, which is given by Eq. (G-36) or in graphical form by Fig. G-17.

$$\Delta t_m = \frac{\Delta t_1 - \Delta t_2}{2.303 \log \dfrac{\Delta t_1}{\Delta t_2}} \tag{G-36}$$

where Δt_m = mean temperature difference
Δt_1 = largest temperature difference
Δt_2 = smallest temperature difference

The same integration may be performed for multipass exchangers, but the results are algebraically complicated and are usually expressed in graphical form as a factor for correction of the logarithmic mean temperature difference. Figure G-18 is such a chart. Where the tube and shell passes are the same in number and flow relations, no correction factor is needed. Also if one side or the other of the exchanger is at constant temperature (as in the case of condensing steam, etc.), no correction factor need be used for the number of passes.

A word of caution should, perhaps, be added. When one of the two fluids is partly, but not totally, evaporated or condensed, the true mean temperature difference may be quite different from what would be computed by the methods just discussed. In this case, the smallest temperature difference may occur even in the middle of the exchanger. Again, the coefficients of heat transfer may change significantly, so that logarithmic mean temperature is not valid. The over-all heat transfer of any exchanger can be computed by integration or approximated by computing by sections and adding the results. An approximation useful for variable coefficients is to compute the heat transfer from the equation

$$q = A \frac{U_1 \Delta t_2 - U_2 \Delta t_1}{2.303 \log (U_1 \Delta t_2 / U_2 \Delta t_1)}$$

where q = heat transferred, Btu/hr
A = heat exchange area, sq ft
U_1, U_2 = over-all coefficient of heat transfer, ends 1 and 2, Btu/(hr)(sq ft)(°F)
$\Delta t_1, \Delta t_2$ = temperature difference, ends 1 and 2, °F

It is important to note that each product contains Δt at one end and U at the other.

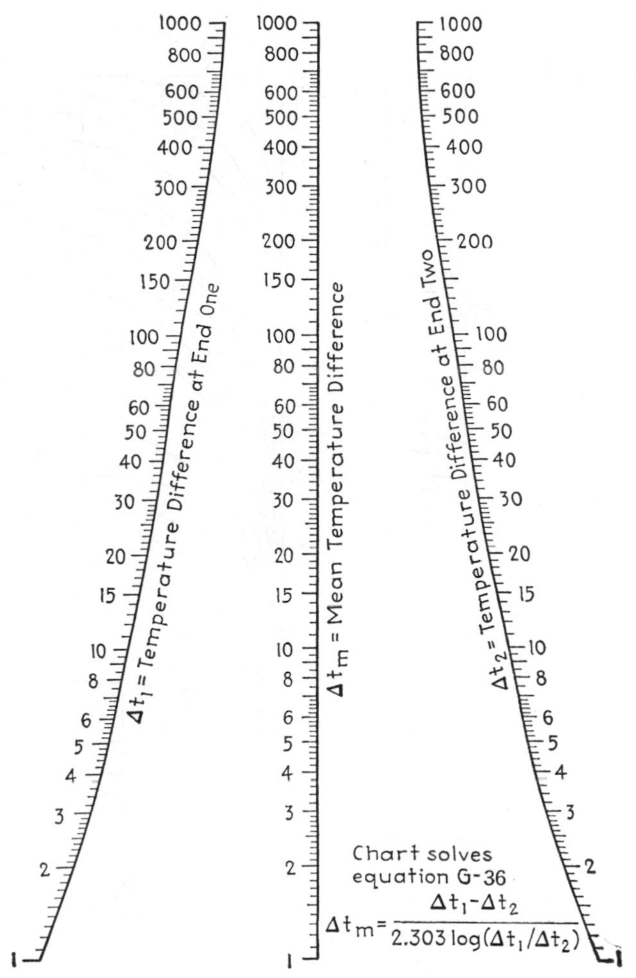

FIG. G-17.—Logarithmic temperature differences.

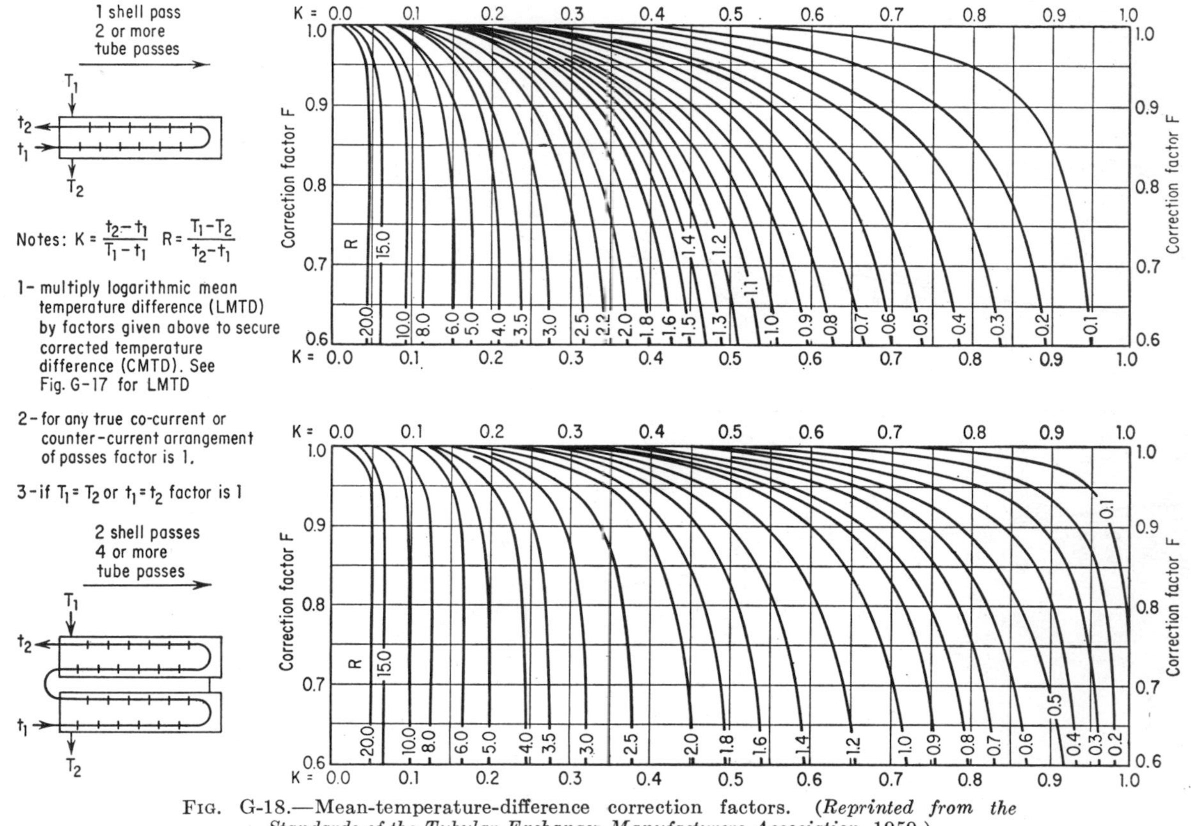

FIG. G-18.—Mean-temperature-difference correction factors. (*Reprinted from the Standards of the Tubular Exchanger Manufacturers Association*, 1959.)

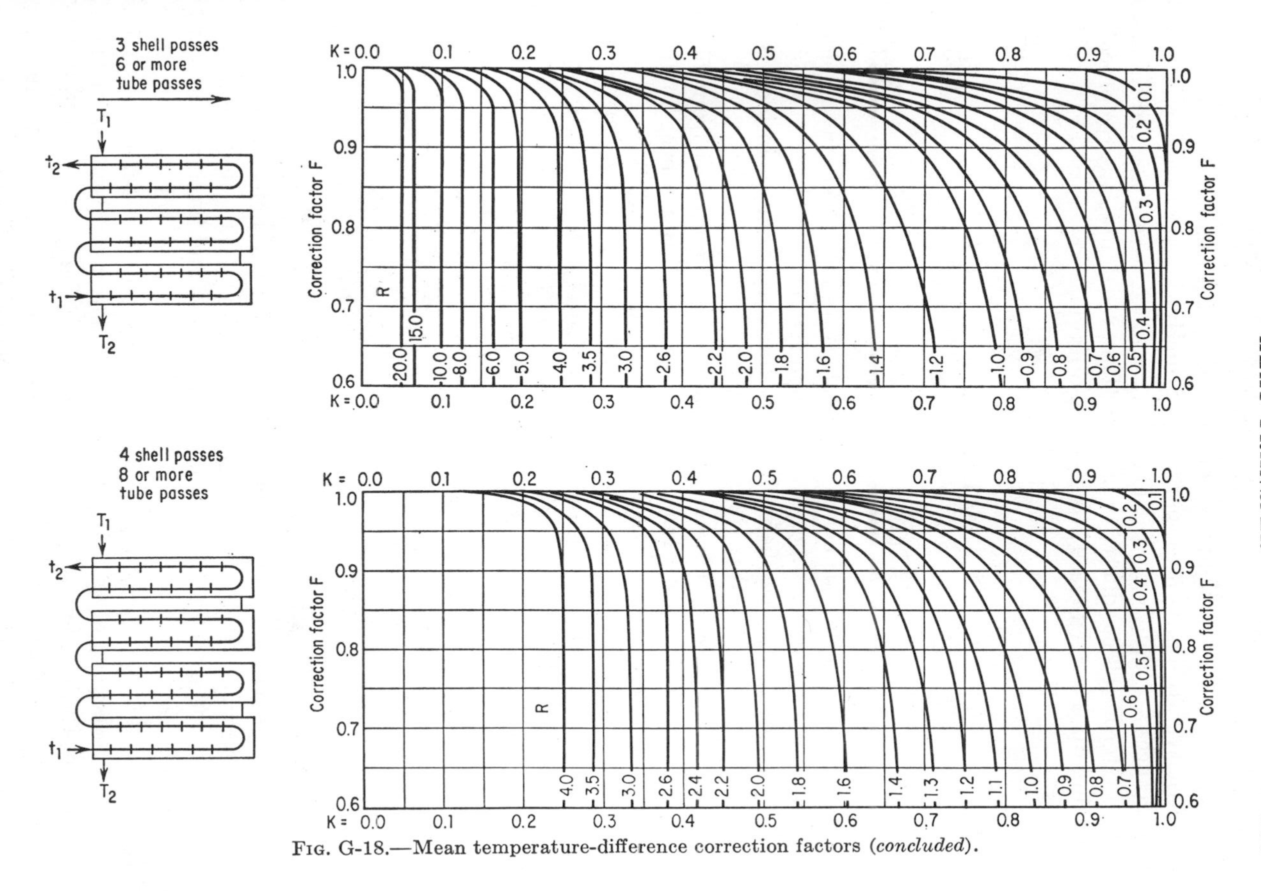

FIG. G-18.—Mean temperature-difference correction factors (*concluded*).

c. Over-all Coefficients.—The flow of heat is analogous to that of electricity. In both cases, parallel conductivities are additive, series resistances are additive. In heat flow, however, we are used to thinking in terms of conductivity rather than resistance, its reciprocal. Adding up the resistances in series, we obtain

$$\frac{1}{U_o} = \frac{1}{h_o} + \frac{d_o}{d_i}\frac{1}{h_i} + f_o + \frac{d_o}{d_i}f_i + \frac{d_o}{d_m}\frac{L}{k} \qquad \text{(G-37)}$$

$$\frac{1}{U_i} = \frac{1}{h_i} + \frac{d_i}{d_o}\frac{1}{h_o} + f_i + \frac{d_i}{d_o}f_o + \frac{d_i}{d_m}\frac{L}{k} \qquad \text{(G-38)}$$

$$\frac{1}{U_f} = \frac{1}{h_i} + \frac{1}{h_o} \qquad \text{(G-39)}$$

where U_o = over-all coefficient of heat transfer based on outside surface of tubes, Btu/(hr)(sq ft)(°F)

U_i = over-all coefficient of heat transfer based on inside area of tubes, Btu/(hr)(sq ft)(°F)

U_f = over-all coefficient of heat transfer computed from film coefficients alone, neglecting metal resistance, fouling resistance, and area ratio, Btu/(hr)(sq ft)(°F)

h_o = film coefficient, outside of tube, Btu/(hr)(sq ft)(°F)

h_i = film coefficient, inside of tube, Btu/(hr)(sq ft)(°F)

f_o = fouling factor outside of tube (see Table G-17)

f_i = fouling factor inside of tube (see Table G-17)

d_o = outside diameter of tube, in.

d_i = inside diameter of tube, in.

d_m = logarithmic mean of d_o and d_i, in.

$$\left(\text{usually factors } \frac{d_i}{d_m} \text{ or } \frac{d_o}{d_m} \text{ may be neglected}\right)$$

L = thickness of tube wall, ft

k = thermal conductivity of tube wall, Btu/(hr)(sq ft)(°F/ft)

For preliminary calculations the simpler Eq. (G-39) may be used; for close design, the inclusion of metal resistance and fouling factors is desirable. Figure G-19 affords a graphical aid to solution of Eq. (G-37), and Tables G-15 and G-16 give data on diameter ratios and thermal resistance of tube walls. These equations presume that the various resistances involved are independent of the temperature differences. In some cases, such as condensing vapors and boiling liquids, the film coefficient depends on the temperature difference. Consequently these temperatures must be assumed and checked by subsequent calculation. For steady operation, the heat passes through the successive layers at the same rate. Expressing this algebraically,

$$U_o\Delta t_m = h_o\Delta t_o = h_i\frac{d_o}{d_i}\Delta t_i, \text{ etc.} \qquad \text{(G-40)}$$

where U_o = over-all coefficient based on outside tube area, Btu/(hr)(sq ft)(°F)
Δt_m = mean temperature difference, °F
h_o = film coefficient outside of tube, Btu/(hr)(sq ft)(°F)
h_i = film coefficient inside of tube, Btu/(hr)(sq ft)(°F)
d_i, d_o = diameter of tube, in.; inside, outside
Δt_o = mean temperature difference between shell fluid and outside of wall, °F
Δt_i = mean temperature difference between tube fluid and inside of wall, °F

Equation (G-40) permits the estimation of the mean temperature differences and, therefore, a check on the assumptions originally made in computing the film coefficients.

For extended surface exchangers, such as those using finned tubes, the over-all coefficient is computed in an analogous fashion such as that described by the equation below:

$$\frac{1}{U_A} = \frac{1}{E_A}\left(\frac{1}{h_A} + f_A\right) + \frac{A}{BE_B}\left(\frac{1}{h_B} + f_B\right) + \frac{A}{C}\frac{L}{k} \qquad \text{(G-41)}$$

where U_A = over-all coefficient referred to side A, Btu/(hr)(sq ft)(°F)
h_A, h_B = film coefficient, side A, B, Btu/(hr)(sq ft)(°F)
f_A, f_B = fouling factor, side A, B (see Table G-17)
E_A, E_B = fin efficiency factor (see Fig. G-20)
L = length, ft, of thermal path between sides A and B through (common) area C
k = thermal conductivity, Btu/(hr)(sq ft)(°F/ft)
A, B, C = surface area side A, side B, and common to both sides, sq ft

It will be noted that a new term, "fin efficiency," is introduced by this equation. Use of this term is a method for including the thermal resistance of the metal in the fins. Figure G-20 gives data in fin efficiencies for several shapes of fins. These efficiencies apply only to the finned area. The over-all factor is

$$E_A = \frac{A_n + A_f E_f}{A} \qquad \text{(G-42)}$$

where E_A = fin efficiency factor, side A
E_f = fin efficiency
A_n, A_f, A = nonfinned, finned, and total area for side A

Table G-18 gives dimensional data on longitudinal finned-tube exchangers.

Tables G-19 to G-23 give typical over-all coefficients of heat transfer.

TABLE G-15.—DIAMETER RATIOS FOR COMMON TUBING

Condenser tubes							Steel pipe			
OD, in.	B.W.G. No.						Nom. size, in.	Schedule No.		
	10	12	14	16	18	20		40	80	160
	d_o/d_i = ratio OD to ID							d_o/d_i		
3/8	....	....	1.79	1.53	1.35	1.23	1/4	1.48	1.79	
1/2	....	1.77	1.50	1.35	1.24	1.16	3/8	1.37	1.60	
5/8	1.75	1.54	1.36	1.26	1.18	1.13	1/2	1.35	1.54	1.80
3/4	1.55	1.41	1.28	1.21	1.15	1.10	3/4	1.28	1.42	1.71
7/8	1.44	1.33	1.23	1.17	1.12	1.09	1	1.26	1.38	1.61
1	1.37	1.28	1.20	1.15	1.11	1.08	1 1/2	1.18	1.27	1.42
1 1/4	1.27	1.21	1.15	1.12	1.08	1.06	2	1.15	1.23	1.41
1 1/2	1.22	1.17	1.12	1.09	1.07	1.05	3	1.14	1.21	1.34
2	1.15	1.12	1.09	1.07	1.05	1.04	4	1.12	1.18	1.31

TABLE G-16.—THERMAL RESISTANCE OF TUBE WALLS

k = thermal conductivity, Btu/(hr)(sq ft)(°F/ft); values given are average values 0 to 250°F

d_o/d_m = ratio of outside diameter to log mean of inside and outside diameter (values used for condenser tubes here based on 3/4-in.-OD tubes)

L = thickness of tube wall, ft

Tube or pipe wall		Condenser tube, B.W.G.					Schedule 40 pipe		
		10	12	14	16	18	1/2 in.	1 in.	2 in.
Material	k	Values of $1{,}000 r_w$, $r_w = \frac{d_o}{d_m}\frac{L}{k}$							
Copper	223	0.06	0.05	0.03	0.03	0.02	0.05	0.06	0.06
Aluminum	131	0.10	0.08	0.06	0.05	0.03	0.08	0.10	0.11
Red brass	92	0.15	0.12	0.08	0.07	0.05	0.12	0.13	0.15
Yellow brass	69	0.20	0.15	0.11	0.09	0.06	0.15	0.18	0.20
Admiralty	63	0.22	0.17	0.12	0.10	0.07	0.17	0.20	0.22
Steel, 0.2% C	38	0.36	0.28	0.20	0.16	0.11	0.28	0.33	0.36
Nickel	35	0.39	0.31	0.22	0.17	0.13	0.30	0.35	0.39
Lead	20	0.68	0.53	0.39	0.30	0.22	0.53	0.61	0.69
Silicon bronze	18.8	0.73	0.57	0.41	0.31	0.23	0.56	0.65	0.73
304–316 stainless and inconel	8.8	1.56	1.22	0.88	0.68	0.50	1.20	1.38	1.56

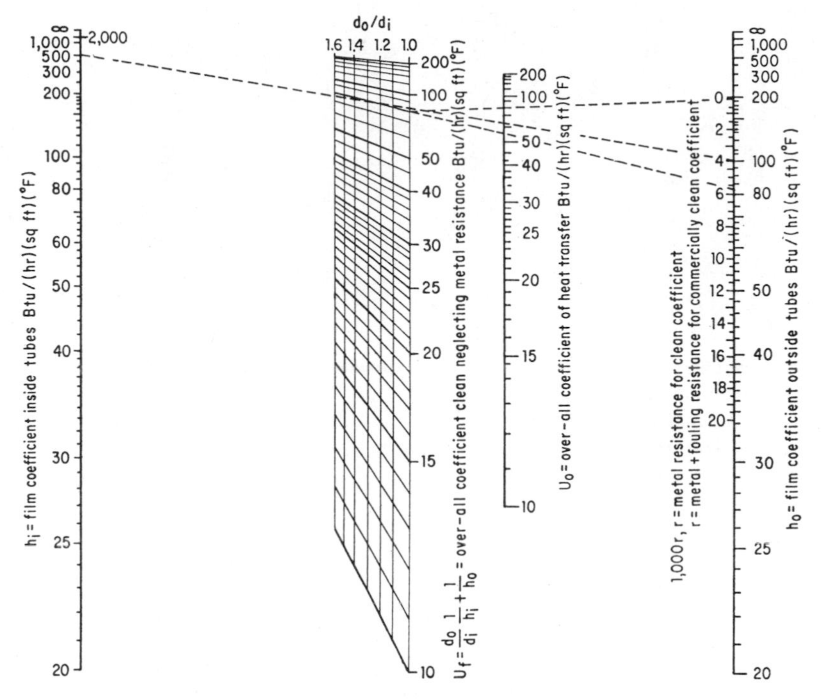

ILLUSTRATIVE EXAMPLE

Problem.—Condensing acetone by untreated cooling-tower water (in tubes). Tubes are ¾-in. 16 B.W.G. steel. $h_i = 500$, $h_o = 100$ (already determined).

Diameter Ratio.—$d_o/d_i = 1.21$ from Table G-15.

Clean Coefficient, Neglecting Metal Resistance.—Connect h_i with h_o and read $U_f = 80$ Btu/(lb)(hr)(sq ft) at grid for $d_o/d_i = 1.21$.

Metal Resistance.—$1{,}000r_w = 0.16$, say 0.2, from Table G-16.

For Clean Coefficient.—Connect r_w and U_f, then $U_o = 79$.

Fouling Resistance.—$1{,}000f_o = 0.5$, $1{,}000f_i = 5$ (choose high and neglect d_o/d_i).

Total Resistance.—$1{,}000r = 0.2 + 0.5 + 5 = 5.7$.

Commercially Clean Coefficient.—Connect r and U_f, then $U_o = 55$.

FIG. G-19.—Over-all coefficients of heat transfer.

TABLE G-17.—FOULING RESISTANCES
(Based on Standards of Tubular Exchanger Manufacturers Association, 4th ed., 1959)
[See Eqs. (G-37) and (G-38)]

Fuel oil	0.005
Crude oil	0.002–0.006
Clean circulating oil	0.001
Machinery and transformer oil	0.001
Quenching oil	0.004
Vegetable oils	0.003
Kerosene	0.002
Manufactured gas	0.01
Organic vapors	0.0005
Solvent vapors	0.001
Clean steam	0.0005
Exhaust steam (oil bearing)	0.001
Refrigerating vapors (condensing from reciprocating compressors)	0.002
Air, compressed	0.002
Natural gas	0.001
Organic liquids	0.001
Refrigerating liquids	0.001
Brine (cooling)	0.001
Sea water:	
0–240°F	0.0005
240–400°F	0.001
Brackish water	0.001–0.003
Cooling-tower water:	
Treated, 0–240°F	0.001
Treated, 240–400°F	0.002
Untreated	0.003
City or well water:	
0–240°F	0.001
240–400°F	0.002
River:	
Minimum	0.001–0.002
Mississippi	0.002–0.003
Delaware, Schuylkill	0.002–0.003
East River, New York harbor	0.002–0.003
Hard (over 15 grains/gal)	0.003–0.005
Distilled	0.0005
Treated boiler feed	0.001
Engine jacket	0.001
Monoethanolamine solution	0.002
Caustic solution	0.002
Molten heat transfer salts	0.0005

TABLE G-18.—DIMENSIONAL DATA, LONGITUDINAL FINNED-TUBE EXCHANGERS
(By permission of Brown Fintube Co. and Griscom-Russell. List does not include all sizes manufactured)

Mfr.*	Type	Inner pipe, ips	Flow area, sq in.	Outer pipe, ips	No. fins	Fin thickness, in.	Fin height, in.	O.S./I.S.	Hydraulic Eq. D, in.	Net flow area, sq in.	$\frac{P}{\text{ft}}$	$\left(\frac{P}{\pi L}\right)^{1/3}$ L = 20 ft	$\frac{P_1}{P_1 + P_2}$
BFT	CA	½	0.32	1½	12	0.035	0.5	11.7	0.389	1.811	0.1128	0.121	0.296
G-R	A&B	¾	0.533	2½	18	0.024	0.5	8.22	0.511	3.71	0.123	0.125	0.314
BFT	EC12	1	0.86	2	12	0.035	0.5	6.16	0.434	2.345	0.134	0.128	0.362
BFT	EC16	1	0.86	2	16	0.035	0.5	7.8	0.354	2.271	0.183	0.143	0.295
BFT	EC20	1	0.86	2	20	0.035	0.5	9.4	0.295	2.193	0.114	0.122	0.244
BFT	JC12	1	0.86	3	12	0.035	1	11.0	0.671	6.173	0.239	0.156	0.291
BFT	JC16	1	0.86	3	16	0.035	1	14.3	0.540	6.028	0.333	0.174	0.233
BFT	JC20	1	0.86	3	20	0.035	1	17.6	0.446	5.883	0.21	0.149	0.193
G-R	C&D	1½	2.038	3	24	0.024	0.5	5.93	0.43	4.27	0.122	0.125	0.298
BFT	017-4	1½	2.038	3	24	0.035	0.5	5.93	0.415	4.103	0.122	0.125	0.291
BFT	017-5	1½	2.038	3	28	0.035	0.5	6.72	0.37	4.03	0.116	0.123	0.258
G-R	C2S	1½	2.038	3	32	0.024	0.5	7.53	0.351	4.16	0.112	0.121	0.238
BFT	017-7	1½	2.038	3	36	0.035	0.5	8.3	0.301	3.88	0.109	0.12	0.207
G-R	C6S	1½	2.038	3	16	0.05	0.5	4.36	0.526	4.17	0.14	0.13	0.378
BFT	LH24	1½	2.038	3½	24	0.035	0.75	8.29	0.48	6.391	0.168	0.139	0.24
BFT	LH28	1½	2.038	3½	28	0.035	0.75	9.48	0.425	6.281	0.162	0.137	0.211
G-R	L&M	1½	2.038	3½	16	0.05	0.735	5.82	0.637	6.47	0.187	0.144	0.325
BFT	217-4	1½	2.038	4	24	0.035	1	10.65	0.541	9.025	0.215	0.15	0.211
BFT	217-5	1½	2.038	4	28	0.035	1	12.25	0.475	8.88	0.208	0.149	0.186
BFT	017-4	1½	2.038	3	24	0.05	0.5	5.93	0.395	3.916	0.121	0.124	0.28
BFT	017-5	1½	2.038	3	28	0.05	0.5	6.75	0.349	3.808	0.116	0.122	0.247
BFT	LH24	1½	2.038	3½	24	0.05	0.75	8.29	0.459	6.109	0.168	0.139	0.233
BFT	LH28	1½	2.038	3½	28	0.05	0.75	9.48	0.403	5.951	0.162	0.137	0.203
BFT	217-4	1½	2.038	4	24	0.05	1	10.65	0.52	8.655	0.215	0.15	0.205
BFT	217-5	1½	2.038	4	28	0.05	1	12.25	0.451	8.445	0.208	0.149	0.179

* BFT = Brown Fintube Co.; G-R = Griscom-Russell; P_1 = pipe perimeter; P_2 = perimeter of fins; $P = P_1 + P_2$

TABLE G-19.—OVER-ALL HEAT-TRANSFER COEFFICIENTS FOR EVAPORATION OF WATER BY SUBMERGED COPPER COIL (CLAASSEN)
(McAdams, "Heat Transmission," 2d ed., McGraw-Hill Book Company, Inc., New York, 1942)
U = Btu/(hr)(sq ft)(°F)

	Δt_o, °F							
	20	30	40	50	60	70	80	90
U for 212°F	390	490	560					
U for 187°F	...	360	440	520	600			
U for 158°F	...	...	...	...	510	600	660	720

TABLE G-20.—OVER-ALL HEAT-TRANSFER COEFFICIENTS FOR EVAPORATION OF WATER BY A SUBMERGED COIL AT ATMOSPHERIC PRESSURE (CLAASSEN)
(McAdams, "Heat Transmission," 2d ed., McGraw-Hill Book Company, Inc., New York, 1942)
U = Btu/(hr)(sq ft)(°F)

	Weight, % solids							
	0	10	20	30	40	50	60	70
U, salt solutions, Δt_o = 18°F	420	430	440					
U, molasses, Δt_o = 12°F	360	350	340	320	290	250	210	170

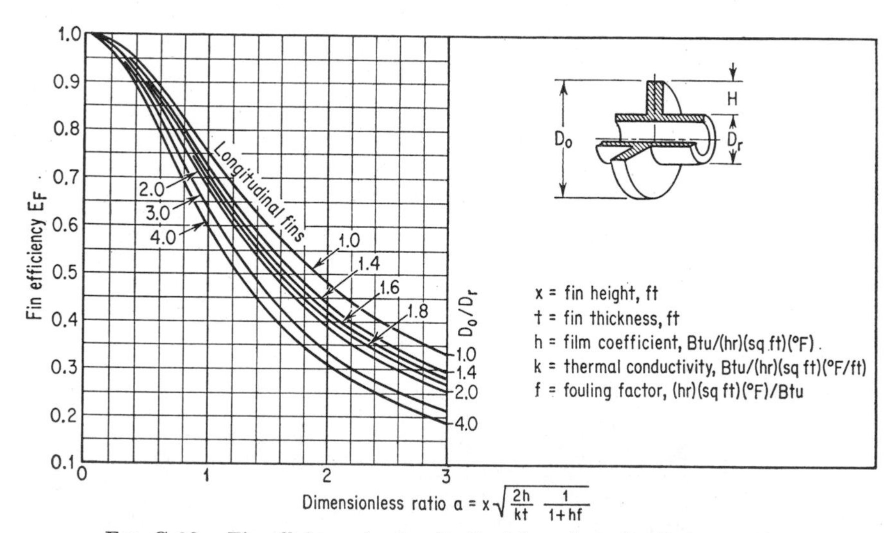

FIG. G-20.—Fin efficiency for longitudinal fins of constant cross section.

TABLE G-21.—ORDINARY RANGES OF OVER-ALL COEFFICIENTS
(Under special conditions higher or lower values may be realized.)
(Perry, "Chemical Engineers' Handbook," 3d ed., McGraw-Hill Book Company, Inc., New York, 1950)
U = Btu/(hr)(sq ft) (°F)

Type of heat exchanger	State of controlling resistance		Typical fluid	Typical apparatus
	Free convection, U	Forced convection, U		
Liquid to liquid......	25– 60	150–300	Water	Liquid-to-liquid heat exchangers
Liquid to liquid......	5– 10	20– 50	Oil	
Liquid to gas*........	1– 3	2– 10	....	Hot-water radiators
Liquid to boiling liquid	20– 60	50–150	Water	Brine coolers
Liquid to boiling liquid	5– 20	25– 60	Oil	
Gas* to liquid........	1– 3	2– 10	....	Air coolers, economizers
Gas* to gas..........	0.6– 2	2– 6	....	Steam superheaters
Gas* to boiling liquid.	1– 3	2– 10	....	Steam boilers
Condensing vapor to liquid............	50–200	150–800	Steam to water	Liquid heaters and condensers
Condensing vapor to liquid............	10– 30	20– 60	Steam to oil	
Condensing vapor to liquid	40– 80	60–150	Organic vapor to water	
Condensing vapor to liquid............	......	15–300	Steam-gas mixture	
Condensing vapor to gas*.............	1– 2	6– 16	....	Steam pipes in air, air heaters
Condensing vapor to boiling liquid.......	40–100	....	....	Scale-forming evaporators
Condensing vapor to boiling liquid.......	300–800	....	Steam to water	
Condensing vapor to boiling liquid.......	50–150	....	Steam to oil	

* At atmospheric pressure.

TABLE G-22.—OVER-ALL COEFFICIENTS OF HEAT TRANSFER FOR CONDENSING STEAM TO BOILING WATER AT 167°F IN A SUBMERGED-TUBE EVAPORATOR
(McAdams, "Heat Transmission," 2d ed., McGraw-Hill Book Company, Inc., New York, 1942)
U = Btu/(hr)(sq ft)(°F)

	Δt_o, over-all Δt, °F				
	18	27	36	45	54
U_o, rusty iron	280	300	325	350	375
U_o, clean iron	300	385	460	535	610
U_o, slightly dirty copper	580	780	950	1120	
U_o, polished copper	820	1110	1470	1810	2120

TABLE G-23.—OVER-ALL COEFFICIENTS OF HEAT TRANSFER FROM CONDENSING STEAM TO BOILING WATER IN A BASKET EVAPORATOR. 31 STEEL TUBES 1 IN. OUTSIDE DIAMETER AND 4 FT LONG
(McAdams, "Heat Transmission," 2d ed., McGraw-Hill Book Company, Inc., New York, 1942)
U = Btu/(hr)(sq ft)(°F)

	Δt_o, °F					
	10	20	30	40	50	60
U for 140°F	120	205	270	320	350	370
U for 167°F	220	320	400	440	480	500
U for 212°F	320	440	500	520		

8. Exchanger Design

The accuracy of heat-transfer calculations may be fairly high under favorable conditions. Usually accuracy suffers somewhat from uncertainties in basic physical data, in equations, and in approximations used. When such is the case, factors of safety should be correspondingly generous.

Methods used in heat-exchanger design usually involve trial-and-error procedures, and the most appropriate method of handling problems will vary from case to case. A formal outline of one general procedure follows:

1. Define the problem. This involves computing the material and heat balances and establishing terminal temperature differences and allowable pressure drops (where they are limiting). Establish fouling factors that are applicable.

2. Select the type of heat-exchange equipment to be used.

3. Compute the log mean temperature difference and the corrected mean if a multipass arrangement is to be used.

4. Make a tentative (or trial) design of the heat exchanger based on assumed over-all coefficients, number of transfer units, or other appropriate method. For example, if a tube-and-shell exchanger is proposed, choose the number, size, and spacing of the tubes and the baffle type and spacing.

5. Set appropriate fouling factors. It may be noted that these are, in a sense, judgment factors to assure an adequate period of sustained operation between scheduled shutdowns for cleaning and maintenance.

6. Compute film coefficients and metal resistances for each step in the over-all transfer.

7. From the individual coefficients, compute the over-all coefficient of heat transfer.

8. Compute the total heat-transfer area required, and compare with tentative design.

9. Compare the result of 7 with assumed design, and determine the nature of discrepancies. From this comparison a new design can be made and checked out until the assumed design is proved adequate.

Various expedients can be used to approximate a proper design, thereby reducing the extent of trial and error. Gauging the problem by reference to typical over-all coefficients is very useful. Many problems can be simplified by use of the "transfer unit" concept. For heat transfer between a metal surface and a fluid traversing it, the number of transfer units involved is given by the equation below:

$$n_T = \frac{t_h - t_c}{\Delta t'_m} = \frac{h_A}{U_A} \frac{t_h - t_c}{\Delta t_m} \tag{G-43}$$

where n_T = number of transfer units, side A

t_h, t_c = temperature at hot, cold ends, °F

$\Delta t'_m$ = mean temperature difference between wall and fluid

Δt_m = over-all mean temperature difference

h_A = film coefficient of heat transfer side A (tubes, shell, annulus, etc.) of exchanger, Btu/(hr)(sq ft)(°F)

U_A = over-all coefficient of heat transfer, Btu/(hr)(sq ft)(°F)

For straight heating and cooling of fluids (*i.e.*, no phase change involved), a relationship between the number of transfer units and the Stanton number can readily be established by evaluating the latter in terms of temperatures, specific heat, and area. Resulting equations follow:

In general:

$$n_T = \frac{A_H}{A_F}\,(\mathrm{St}) \tag{G-44}$$

In tubes:

$$n_T = \frac{4L}{D}\,(\mathrm{St}) \tag{G-45}$$

Flow across tubes:

$$n_T = \frac{\pi}{P/D_o - 1}\,n_r n_b(\mathrm{St}) \tag{G-46}$$

where n_T = number of transfer units
St = Stanton number
A_H = heat-transfer area
A_F = flow area used in computing mass velocity
L = tube length
D, D_o = inside, outside tube diameter
P = center-to-center tube pitch
n_r = total number of tubes in path of flow between baffles
n_b = number of baffle passes

Correlations of heat transfer (such as those by j factors) basically state that the Stanton number is fixed by the Prandtl and the Reynolds numbers. Moreover, the friction factor is fixed by the Reynolds number. The combination of these two functional dependencies permits correlation of pressure drops expressed as velocity heads with thermal-transfer units. Figures G-21 and G-22, constructed on this basis, are useful for many problems of heat transfer, particularly those involving large temperature changes. On the other hand, they are not particularly useful when small temperatures are involved and are not applicable for boiling or condensing. Transfer units may be used for *single-pass* arrangements when *streamline* flow prevails. The most convenient methods for streamline flow differ from the relations just discussed.

To use Fig. G-22 for flow inside tubes, compute n_T from Eq. (G-43) and read L/D and n_v from the chart at the best guess as to expected Reynolds number. If pressure drop is limiting, compute the maximum velocity from the following:

$$G = \sqrt{\frac{9{,}250\rho\Delta p}{n_v}} = 96\sqrt{\frac{\rho\Delta p}{n_v}} \tag{G-47}$$

$$u = 96\sqrt{\frac{\Delta p}{\rho n_v}} \tag{G-48}$$

where G = mass velocity, lb/(sec)(sq ft)
ρ = fluid density, lb/cu ft
Δp = pressure drop, psi
n_v = number of velocity heads
u = linear velocity, fps

The Reynolds number can be computed from this velocity, and the calculation repeated for greater accuracy. This is seldom necessary except for viscous fluids. In this case the calculation may show that achievement of turbulent flow cannot be realized within the allowable pressure drop. This difficulty often can be circumvented by putting the viscous liquid in the shell or by using "cut-and-twist" type of longitudinal finned tubes to give many short sections in which it is advantageous for high transfer rates in laminar flow.

Use of the chart for shells can also be an advantage for simplifying design calculations. Once the tube side of an exchanger is laid out, the number of tubes traversed between baffles is set and the chart may be used to estimate the number of baffles required. If the pressure drop estimated for the proposed layout is too large, a more open bundle (larger tube pitch or square pitch instead of triangular) is indicated.

One of the first decisions to make in design of tube-and-shell exchangers is which fluid will use the tube and which the shell side. This is often decided by special factors such as the need for frequent mechanical cleaning, which is easier to perform on the tube side. Usually condensing liquids are put on the shell. Design of the shell side for very high pressures is more costly than on the tube side. Many other examples of special consideration could be given. In the absence of such factors, the greater flow is usually assigned to the shell because it is generally easier to provide a large flow area on the shell side.

The transfer-unit method directly implies that the *number of velocity heads per transfer unit* is established by the Prandtl and Reynolds numbers. This ratio is a measure of the efficiency of utilizing the pressure drop for heat transfer. Thus, a comparison of the velocity head to transfer-unit ratios for different types of equipment is one basis of gauging their suitability for design where pressure drops are important. The authors have made such a comparison and observe that flow through tubes has the highest "efficiencies" at high Reynolds numbers but is not so efficient as flow across tubes at Reynolds numbers below 10,000. The performance of the different tube arrangements for cross flow does not differ widely. This circumstance favors

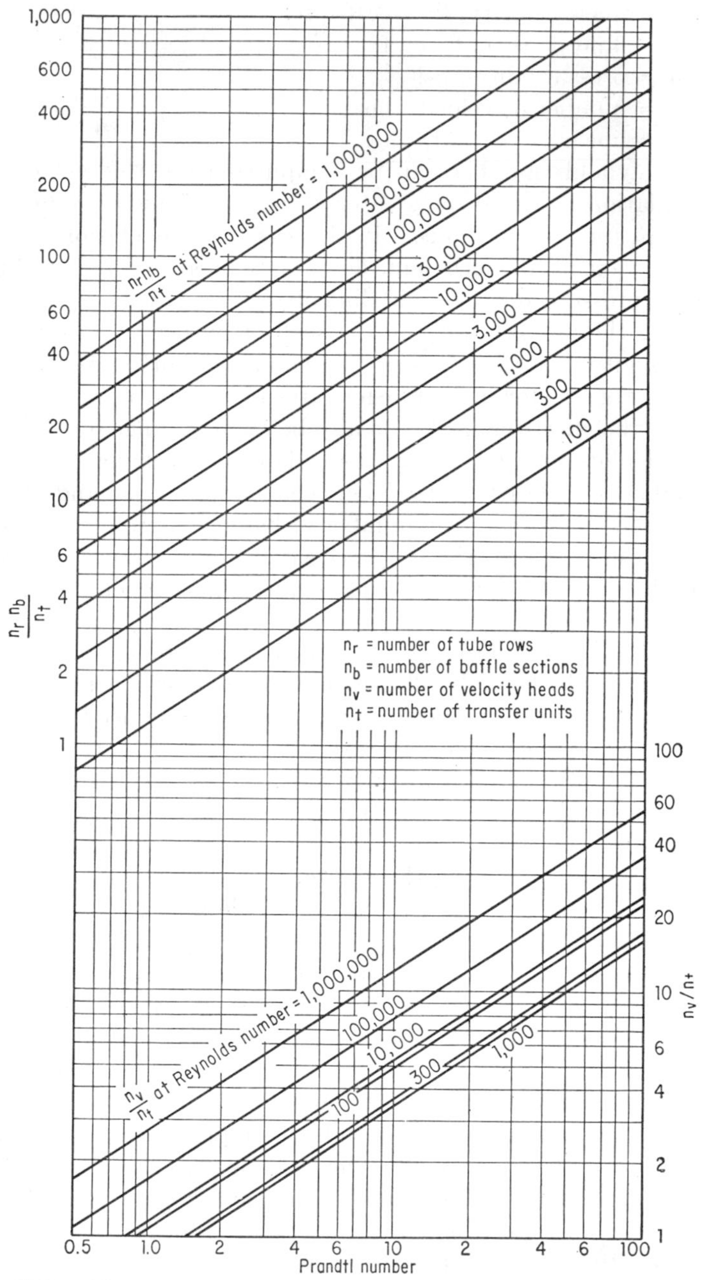

FIG. G-21.—Chart for transfer-unit method of design for flow transverse to tube banks. NOTE: Applicable for true cross flow. For estimation of commercial shells, divide values obtained by 0.6.

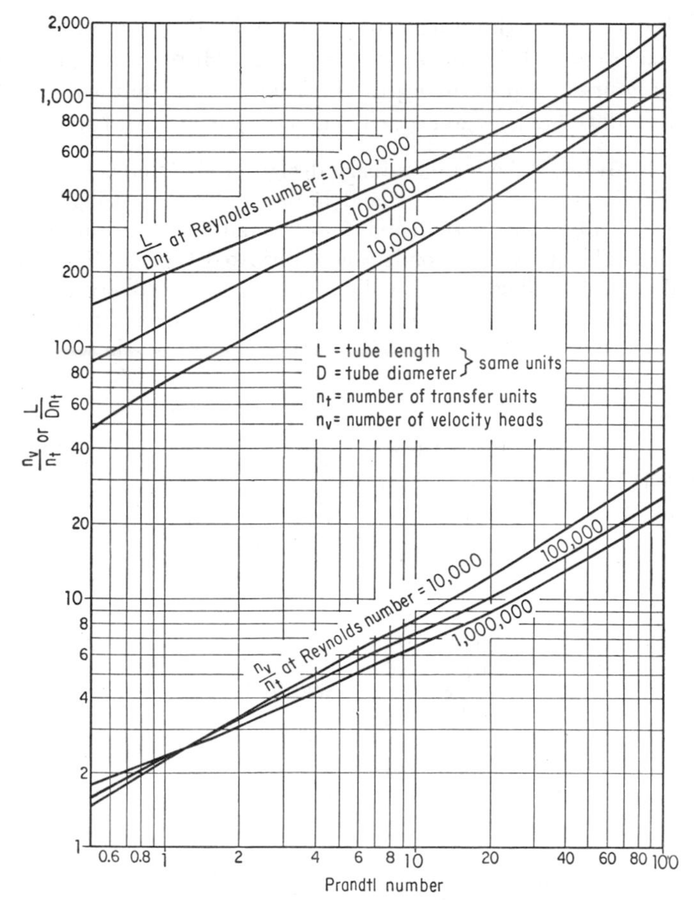

Fig. G-22.—Chart for transfer-unit method of design for flow inside tubes. (Based on data from Figs. F-14 and G-7.)

use of the compact triangular tube layout because of its economy in size of exchanger shells and tube sheets. More open arrangement should, of course, be used if required for ease in mechanical cleaning or to accommodate very large flows (large compared with tube-side flow).

Example G-1. Make a preliminary design of an aftercooler which cools 1,000 SCF/min of dry air at 260°F and 25 psia to 90°F by using 80°F cooling water. Allow 10°F rise on the cooling water and 0.50 psi drop on the compressed air. The air compressor is a nonlubricated type, so that the air can be considered nonfouling. The cooling water is of relatively good quality, and a fouling factor of 0.001 is adequate. The exchanger should be all-steel construction, and the tubes used should, preferably, be chosen to conform with sizes stocked at the plant:

⅝-in. 12 and 16 B.W.G. in 12 and 16 ft lengths
¾-in. 12 B.W.G. in 12 and 16 ft lengths
1-in. 12 B.W.G. in 12, 16, and 20 ft lengths
1½-in. 8 and 12 B.W.G. in 16 and 20 ft lengths

SOLUTION:

A. Heat Balance:

1. Weight flow:

   1 lb = 13.07 SCF (Table H-10)

   $$1{,}000 \text{ SCF/min} = \frac{1{,}000(60)}{13.07} = 4{,}590 \text{ lb/hr}$$

2. Heat transferred:

   $C_p = 0.241$ Btu/lb (Table G-7)

   $Q = 0.241(260 - 90)4590 = 188{,}000$ Btu/hr

3. Cooling water required:

   1 lb/hr = 1 × 10 = 10 Btu/hr

   $$\text{lb/hr} = \frac{188{,}000}{10} = 18{,}800$$

   $$\text{gpm} = \frac{18{,}800}{(8.32)(60)} = 37.6$$

B. Tentative Decisions:

1. Use tube side for air because of the low pressure drop available and the low viscosity of air permitting attainment of reasonable Reynolds numbers at low pressure drops
2. Take the ratio of the over-all film coefficient (based on the tube side) to the air coefficient as 1.3. This is based on the relatively low coefficients characterizing low-pressure gases and the high uses obtainable using water
3. Use the compact triangular tube bundle configuration, as pressure drop on the water side is not considered to be a problem

C. Mean Temperature Difference:
  1. Temperature difference at cold end = 90 − 80 = 10°F
  2. Temperature difference at warm end = 260 − 90 = 170°F
  3. Mean temperature difference = 55°F (by Fig. G-17)

D. Tube Side:
  1. Transfer units:

$$n_T = \frac{h_A}{U_A}\frac{t_h - t_c}{t_m} \tag{G-43}$$

$$= 1.3\,\frac{260 - 90}{55} = 3.1$$

  2. Prandtl number = 0.71 (Table G-7)
  3. Assumed Reynolds number = 100,000
  4. Application of Fig. G-22:

$$\frac{L}{Dn_t} = 110 \qquad \frac{L}{D} = 110(3.1) = 342$$

$$\frac{n_v}{n_t} = 1.9 \qquad n_v = 1.9(3.1) = 5.9$$

  5. Tube selection (data from Table F-10):

Tube	ID, in.	Flow area, sq ft	Calculated length, ft
⅝ × 16 B.W.G.	0.495	0.00134	342(0.495)/12 = 14.1
¾ × 12 B.W.G.	0.532	0.00154	342(0.532)/12 = 15.1
1 × 12 B.W.G.	0.782	0.00334	342(0.782)/12 = 22.3

Try ⅝-in. tubes with 12- or 16-ft lengths.

  6. Allowable velocity:
Velocity head for tube friction only amounts to 6 in round numbers; allow 4 more for tube entrance, exit, and exchanger nozzles, making a total of 10
Specific volume of air at average temperature of 185°F and pressure of 24.8 psia by proportion:

$$v = 13.07\,\frac{14.73}{24.8}\,\frac{460 + 185}{460 + 60} = 9.62$$

$$G = 96\sqrt{\frac{\rho\Delta p}{n_v}} \quad \text{[Eq. (G-47)]} \quad = 96\sqrt{\frac{0.50}{9.62(10)}} = 6.93 \text{ lb/(sec)(sq ft)}$$

  7. Number of tubes:
Capacity of one tube = 6.93(0.00134)3,600 = 33.4 lb/hr

$$\text{Minimum number of tubes} = \frac{4{,}590}{33.4} = 137$$

8. Check on Reynolds number:
   Viscosity = 0.049 lb/(ft)(hr) (Table G-7)
   $$R = \frac{3{,}600}{12} \times \frac{6.93(0.495)}{0.049} = 20{,}800$$
9. Recheck by Fig. G-22:
   $$\frac{L}{D} = 85(3.1) = 264$$
   $$n_v = 1.8 \times 3.1 = 5.6$$
   $$\text{Tube length} = \frac{264(0.495)}{12} = 10.8 \text{ ft}$$
10. Film coefficient:
   $h = 4.1G^{0.8}/d^{0.2}$ [Table G-7 and Eq. (G-17)]
   $= 4.1(6.93)^{0.8}/(0.495)^{0.2} = 22.2$ Btu/(hr)(sq ft)(°F)

E. Shell Side:

1. Tube layout should be sketched to determine flow areas and number of tube rows. For preliminary consideration, these can be approximated as below
2. Tube pitch—use $^{13}\!/_{16}$ in. as closest practical arrangement (see Table G-12)
3. Approximate shell diameter = $(^{13}\!/_{16})(1.15\sqrt{137}) = 10.8$ in. [Eq. (G-35)]: use 11.5 in.
4. Approximate number of tubes in center row $= \dfrac{10.8}{(^{13}\!/_{16})} = 13$
5. Approximate number of staggered rows:
   $$\text{Distance between rows} = \frac{(^{13}\!/_{16})(\sqrt{3})}{2} = 0.705 \text{ in.}$$
   $$\text{Number} = \frac{10.8}{0.705} = 15$$
6. Assume baffle spacing of 18 in., 1.5 ft
7. Flow area $= 1.5\left[\dfrac{11.5 - 13(^5\!/_8)}{12}\right] = 0.30$ sq ft
8. Mass velocity of water:
   $$G = \frac{18{,}800}{0.3} = 62{,}300 \text{ lb/(hr)(sq ft)}$$
   $$= 17.4 \text{ lb/(sec)(sq ft)}$$
9. Reynolds number:
   Viscosity at 85°F = 0.85 centipoise (Fig. F-1)
   $$R = 124\frac{dG}{\mu} \qquad \text{[in units used, Eq. (F-12)]}$$
   $$= \frac{124(^5\!/_8)(17.4)}{0.85} = 1{,}600$$

10. Film coefficient:

    j factor = 0.018 (Fig. G-13)

    Prandtl number = 5.7 (Fig. G-10)

    $$h = \frac{j(C_pG)}{(Pr)^{2/3}} = \frac{(0.018)(62{,}300)}{(5.7)^{2/3}} = 353 \qquad \text{[Eq. (G-9}a\text{)]}$$

    This is the coefficient expected for true cross flow. Multiply by 0.6 for shell-side coefficient

    $h_o = 0.6(353) = 211$

F. Over-all Heat Transfer:

1. Data from Tables G-15 and G-16:

   Diameter ratio, $\dfrac{d_o}{d_i} = 1.28$

   Metal resistance $= \dfrac{0.28}{1{,}000}$

2. Over-all coefficient:

   $\dfrac{d_o}{d_i}\dfrac{1}{h_i} = \dfrac{1.28}{22.2}$ 0.0576

   $\dfrac{1}{h_o} = \dfrac{1}{211}$ 0.0047

   $\dfrac{d_o}{d_n}\dfrac{L}{k}$ 0.0003

   f_o 0.0010

   $\dfrac{1}{U_o}$ 0.0636

   $U_o = 15.7$

3. Required heat-transfer area:

   $$HTA = \frac{Q}{U_o\Delta t} = \frac{188{,}000}{(15.7)(55)} = 218 \text{ sq ft}$$

4. Minimum tube length:

   Perimeter of 137 tubes = 137(0.1636) = 22.4 sq ft

   Length = $^{218}/_{22.4}$ = 9.7 (say 10 ft)

5. Area supplied by standard 12-ft length = 22.4(12) = 267 sq ft

G. Conclusion:

The use of one hundred thirty-seven ⅝-in. tubes, 12 ft long, in a shell of 11½ in. diameter is an acceptable preliminary design. Ten-foot tubes could be used if deviation of plant standards is acceptable. For final design, a detailed tube layout should be made, and such layouts usually show that addition of a few more tubes to fill up the shell is desirable.

H. Alternate Designs:

The above is not the only acceptable design. Since the L/D ratio is nearly independent of the tube size (it is affected only by small changes in Reynolds number and the somewhat greater change in shell coefficient), we can readily estimate the effect of tube size with considerable confidence.

For example, the tube length required would become 10(0.782)/0.495, or 15.8 ft for a 1-in. by 12 B.W.G. tube.

We can now readily estimate the effect of reversing the tube and shell sides. From Fig. G-21, at R = 40,000, n_v/n_t = 1.1 for true cross flow, or 1.1/0.6 = 1.8 for a well-designed shell side. This leads to about eight velocity heads including the shell nozzles and to an allowable mass velocity of 7.7 lb/(sec)(sq ft) and a Reynolds number of 52,000.

The flow area at the bundle center becomes 4,590/(3,600 × 7.7) = 0.167, and the required baffle spacing becomes 18(0.167)/0.300 or 10 in. instead of 18 used for water. Referring again to Fig. G-21, a value of

$$n_r n_b/n_t = 12/0.6 = 20$$

is obtained.

Since there are 15 rows per baffle pass, the number of baffled spaces n_b becomes 20(3.1)/15 = 4.2; *i.e.*, five baffled spaces each 10 inches long, or a total length of 4 ft, would be required. The baffle spacing of 10 in. is close, so that baffle leakage could be too severe to sustain a coefficient as high as 60 per cent of that for true cross flow. Moreover, it can be seen that the Reynolds number of water in the tubes would be low, so that its coefficient would become more important. Nonetheless, it would still appear that use of the shell side for air flow would lead to a good design. Fewer tubes of greater length should be used to improve the water coefficient and to widen the baffle spacing on the shell side.

Use of larger size tubes, when the air is used in the shell, can readily be shown to be disadvantageous. Tubes with low integral fins having three times the area of plain tubes are available from one manufacturer and these could be used to advantage.

The preceding example illustrates some of the uncertainties of heat-exchanger design and methods used in the resolution of the problem. The example covers a tube-and-shell exchanger, which is by far the most common type. The advantages of other types of exchangers and regenerators should not be overlooked, however.

9. Heat Losses by Open Equipment

Bare surfaces lose heat by radiation and convection. Radiation has already been treated earlier in the chapter. The heat loss by natural convection in absence of any wind, *i.e.*, to still air, may be estimated by the equations given below.[1]

Top-surface horizontal plates...... $h = 0.38(\Delta t)^{0.25}$
Bottom-surface horizontal plates... $h = 0.2(\Delta t)^{0.25}$
Vertical plates (over 1 ft high)..... $h = 0.27(\Delta t)^{0.25}$
Piping.......................... $h = 0.27(\Delta t/D_o)^{0.25}$

[1] McAdams, *op. cit.*

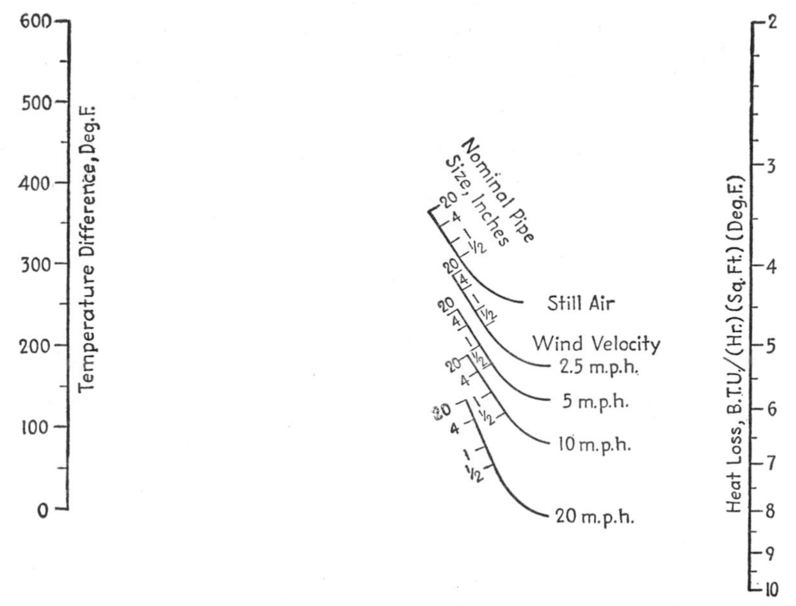

FIG. G-23.—Heat losses from bare iron pipe. *(Based on data from "Heat Insulation Handbook," Ehret Magnesia Manufacturing Company, Valley Forge, Pa.,* 1942.)

where h = film coefficient, Btu/(hr)(sq ft)(°F)
Δt = temperature difference, °F
D_o = outside diameter, ft

It is more convenient for most purposes to lump the radiant and convection transfers together. Figure G-23 gives such data on losses by bare pipe. It should be remembered that rates in rainy weather will be much larger. Similar values for insulated pipe are given in Fig. G-24. The loss from large flat surfaces is about 0.20 Btu/(hr)(sq ft)(°F) when insulated with 2 in. of rock wool or about 0.24 Btu/(hr)(sq ft)(°F) when insulated with 2 in. of 85 per cent magnesia.

Figure G-25 permits estimation of heat losses from furnace walls. This chart is based on still air at 80°F. Wind increases these somewhat and lowers the temperature of the cold faces considerably.

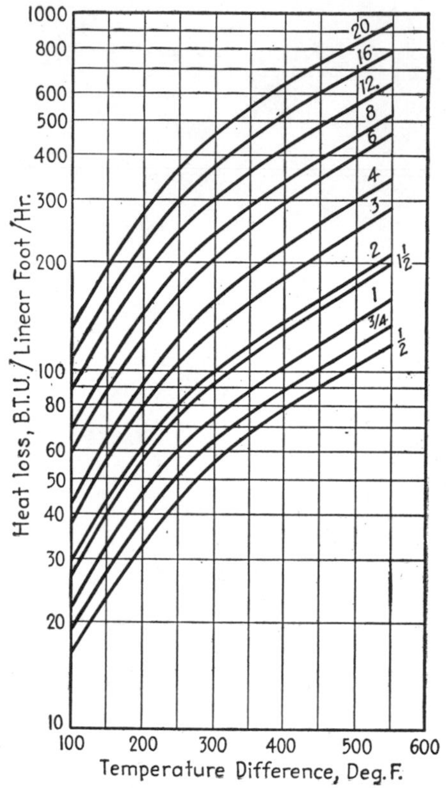

FIG. G-24.—Heat losses from pipe insulated with standard 85 per cent magnesia.

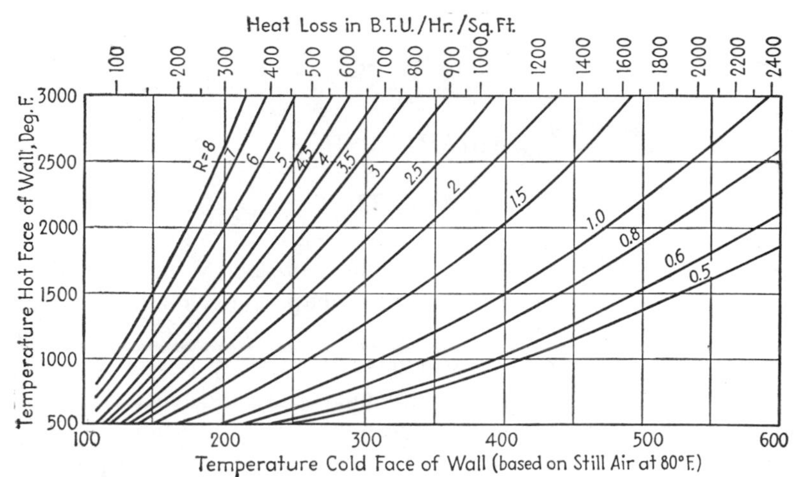

Temperature at first joint from hot face = hot face temperature $- \left(\frac{\text{Btu loss} \times L_1}{K_1}\right)$

Temperature at second joint from hot face = first joint temperature $- \left(\frac{\text{Btu loss} \times L_2}{K_2}\right)$

For a one-component wall $R = \frac{L}{K}$ where L = thickness of wall, in.

K = conductivity (Btu/sq ft/°F difference in temperature across 1 in. thickness)

For a wall of n components, $R = \frac{L_1}{K_1} + \frac{L_2}{K_2} + \cdots + \frac{L_n}{K_n}$

where L_1, L_2, etc. = thickness of each component

K_1, K_2, etc. = conductivity of each component

FIG. G-25.—Heat losses through furnace walls to still air. *(Copyright by Universal Atlas Cement Div., U.S. Steel Corp. Reproduced by their permission.)*

CHAPTER H

COMBUSTION

By all odds the most important source of energy for processing is the combustion of fuels with air. This chapter gives data on air and specific fuels. The last section is intended to centralize data common to all fuels, though some of the charts therein are based on specific fuels. Although combustion equipment is treated very briefly, an effort has been made to mention factors of importance in the choice of fuels and for estimating utility requirements of combustion processes.

Several of the many excellent texts on fuel and combustion are listed below. Information is frequently given in works primarily devoted to other subjects, notably thermodynamics and power.

Haslam and Russell, "Fuels and Their Combustion," McGraw-Hill Book Company, Inc., New York, 1926.

"Combustion," 3d ed., American Gas Association, New York, 1932.

Trinks, "Industrial Furnaces," Vol. I, 4th ed., 1951; Vol. II, 3d ed., 1955, John Wiley & Sons, Inc., New York.

Dunstan, "Science of Petroleum," 4 vols., Oxford Press, England, 1938.

Griswold, "Fuels, Combustion and Furnaces," McGraw-Hill Book Company, Inc., New York, 1946.

National Research Council, "The Chemistry of Coal Utilization," 2 vols., John Wiley & Sons, Inc., 1945.

1. Composition of Air

The composition of air is, fortunately, very constant. The only appreciable variation arises from local contamination, and even this is seldom sufficient to affect combustion processes. The composition of pure dry air is as follows:

	Mole per Cent
Oxygen (O_2)	21.0
Nitrogen (N_2)	78.0
Argon (A)	0.94
Carbon dioxide (CO_2)	0.03
Other (Ne Kr, etc.)	0.03

Since argon and other rare gases are normally reported as nitrogen in gas analysis and are, like nitrogen, inert toward combustion, air may be regarded as 21 mole per cent oxygen and 79 mole per cent "atmospheric nitrogen." On a weight basis air is 23.2 per cent oxygen, and 4.31 lb of dry air are required to supply 1 lb of oxygen.

The average molecular weight of air is 28.97. "Atmospheric nitrogen" has an average molecular weight of 28.17 or about 0.6 per cent greater than that of pure nitrogen. Other physical properties of "atmospheric nitrogen" differ only slightly from pure nitrogen; for example, the specific heats differ about 0.2 per cent.

2. Coal and Coke

a. Testing and Classification.—Because the properties of coal from different localities are extremely variable, various tests are necessary to specify the behavior of coal to be used for fuel or producing coke. These tests are largely empirical and their utility depends on a careful standardization of procedure. Below are listed publications of A.S.T.M. on standard methods:

D 121-30	Definitions of Terms Relating to Coal and Coke
D 141-48	Method of Shatter Test for Coke
D 166-24	Specifications for Gas and Coking Coals
D 167-24	Method of Test for Volume of Cell Space of Lump Coke
D 197-30	Method of Test for Fineness of Powdered Coal
D 271-48	Methods of Laboratory Sampling and Analysis of Coal and Coke
D 291-29	Method of Test for Cubic Foot Weight of Crushed Bituminous Coal
D 292-29	Method of Test for Cubic Foot Weight of Coke
D 293-50	Method of Test for Sieve Analysis of Coke
D 294-50	Method of Tumbler Test for Coke
D 310-34	Method of Test for Size of Anthracite
D 311-30	Method of Test for Sieve Analysis of Crushed Bituminous Coal
D 346-35	Method of Sampling Coke for Analysis
D 388-38	Specifications for Classification of Coals by Rank
D 389-37	Specifications for Classification of Coals by Grade
D 407-44	Definition of the Terms Gross Calorific Value and Net Calorific Value of Fuels
D 409-51	Method of Test for Grindability of Coal by the Hardgrove-machine
D 410-38	Method of Test for Screen Analysis of Coal
D 431-44	Method for Designating the Size of Coal from Its Screen Analysis
D 440-49	Method of Drop Shatter Test for Coal
D 441-45	Method of Tumbler Test for Coal
D 492-48	Method of Sampling Coals Classed According to Ash Content
D 493-39	Definitions for Varieties of Bituminous and Subbituminous Coals
D 547-41	Index of Dustiness of Coal and Coke
D 720-57	Free-swelling Index of Coal
D 980-53	Sampling for Volatiles (Smoke Ordinances)
D 1412-56T	Equilibrium Moisture of Coal (Tentative)
E 11-39	Specifications for Sieves for Testing Purposes (Wire-cloth Sieves, Round-hole and Square-hole Screens or Sieves)

TABLE H-1.—CLASSIFICATION OF COALS BY RANK*
(U.S. Bureau of Mines Bulletin 446, 1942)

Class	Group	Limits of fixed carbon or Btu, mineral-matter-free basis	Requisite physical properties
I. Anthracitic.........	1. Meta-anthracite	Dry F.C., 98 % or more (dry V.M., 2 % or less)	
	2. Anthracite	Dry F.C., 92 % or more and less than 98 % (dry V.M. 8 % or less and more than 2 %)	
	3. Semianthracite	Dry F.C., 86 % or more and less than 92 % (dry V.M., 14 % or less and more than 8 %)	Nonagglomerating†
II. Bituminous‡.......	1. Low-volatile bituminous coal	Dry F.C., 78 % or more and less than 86 % (dry V.M., 22 % or less and more than 14 %)	
	2. Medium-volatile bituminous coal	Dry F.C., 69 % or more and less than 78 % (dry V.M., 31 % or less and more than 22 %)	
	3. High-volatile A bituminous coal	Dry F.C., less than 69 % (dry V.M., more than 31 %); and moist § Btu 14,000 ‖ or more	
	4. High-volatile B bituminous coal	Moist § Btu, 13,000 or more and less than 14,000 ‖	
	5. High-volatile C bituminous coal	Moist Btu, 11,000 or more and less than 13,000 ‖	
III. Subbituminous....	1. Subbituminous A coal	Moist Btu, 11,000 or more and less than 13,000 ‖	Both weathering and nonagglomerating¶
	2. Subbituminous B coal	Moist Btu, 9500 or more and less than 11,000 ‖	
	3. Subbituminous C coal	Moist Btu, 8300 or more and less than 9500 ‖	
IV. Lignitic...........	1. Lignite	Moist Btu, less than 8300	Consolidated
	2. Brown coal	Moist Btu, less than 8300	Unconsolidated

F.C. = fixed carbon

Btu = British thermal units.

V.M. = volatile matter.

* This classification does not include a few coals which have unusual physical and chemical properties and which come within the limits of fixed carbon or Btu of the high-volatile bituminous and subbituminous ranks. All these coals either contain less than 48 per cent dry, mineral-matter-free fixed carbon or have more than 15,500 moist, mineral-matter-free Btu.

† If agglomerating, classify in low-volatile group of the bituminous class.

‡ It is recognized that there may be noncaking varieties in each group of the bituminous class.

§ Moist Btu refers to coal containing its natural bed moisture but not including visible water on the surface of the coal.

‖ Coals having 69 per cent or more fixed carbon on the dry, mineral-matter-free basis shall be classified according to fixed carbon, regardless of Btu.

¶ There are three varieties of coal in the high-volatile C bituminous coal group: variety 1, agglomerating and nonweathering; variety 2, agglomerating and weathering: variety 3, nonagglomerating and nonweathering.

The properties most important for use as fuel are as listed below:

1. Heating value.
2. Proximate analysis:
   Moisture (loss at 105°C for 1 hr)
   Volatile matter (additional loss on ignition at 950°C)
   Ash (residue after burning)
   Fixed carbon (remainder)
3. Fusing point of ash.
4. Ultimate analysis, % C, H, N, O, and S.

There are numerous systems of classifying coal all of which have advantages and defects. The most widely used is the U.S. Bureau of Mines classification which has been adopted by the A.S.T.M. and the A.S.A. This classification, which is based on the proximate analysis, is presented in Table H-1.

b. Typical Analyses of Coals.—Coal analyses of the United States are presented in Table H-2. These data are only illustrative unless the reader is interested in one of these few coals. Fortunately thousands of coal analyses are published by the U.S. Bureau of Mines and similar organizations, so that specific data may usually be obtained on any particular coal. The analyses of some typical cokes are given in Table H-3. The ultimate analysis is given on the ash- and moisture-free basis in accordance with custom. The advantage of this is to exclude the hydrogen and oxygen present in the ash and water from combustion calculations. This distinction, though not precise, affords a sound workable method.

c. Heating Value of Coal.—The heating value of coal determines the amount required to supply a given quantity of heat. Though best determined directly, gross heating values may be predicted with reasonable accuracy from the ultimate analysis by the modified Dulong formula (or by the equivalent factors from Table H-4).

$$\text{Btu/lb} = 155.44\text{C} + 621\ (\text{H} - \tfrac{1}{8}\text{O}) + 40.5\text{S}$$

where C = carbon %
H = hydrogen %
O = oxygen %
S = sulphur %

If the ultimate analysis is expressed in the ash- and moisture-free basis, the calculated heating value will be on the same basis and the heating value based on the coal as received may be obtained by multiplying by the following factor:

$$\frac{100}{100 - \%\ \text{ash} - \%\ \text{moisture}}$$

TABLE H-2.—ANALYSES OF TYPICAL UNITED STATES COALS

Coal		As-received basis							Ash- and moisture-free basis				
No.	Source: bed or mine	Moisture: Loss by air drying, %	Moisture: Total, %	Volatile matter, %	Fixed carbon, %	Ash, %	Gross heating value, Btu/lb	Softening point of ash, °F	% C	% H	% O	% N	% S
1	Buck Mountain..	2.4	5.2	2.8	85.3	6.7	13,090	....	95.0	1.9	2.0	0.7	0.4
2	Pocahontas......	1.0	1.6	17.4	74.5	6.5	14,400	2780	91.1	4.5	2.4	1.3	0.7
3	Lower Freeport..	1.1	2.0	24.9	66.7	6.4	14,310	2840	88.6	5.3	3.8	1.5	0.8
4	Black Creek.....	1.2	3.1	36.0	58.3	2.6	14,190	2390	84.0	5.7	7.2	1.9	1.2
5	Pittsburgh......	0.5	2.0	41.3	48.7	8.0	13,320	2090	80.6	5.7	7.5	1.3	4.9
6	No. 6 Illinois....	....	7.9	32.1	47.7	12.3	11,540	2360	82.2	5.3	9.8	1.7	1.0
7	Eureka Navy....	2.4	4.4	35.5	47.9	12.2	12,310	....	81.9	6.3	8.9	2.1	0.8
8	No. 1 Wyoming..	4.3	10.8	37.7	47.8	3.7	11,750	....	78.0	5.5	13.8	1.5	1.2
9	Newcastle.......	4.4	10.9	36.1	37.0	16.0	9,910	....	75.9	6.5	15.3	1.8	0.5
10	Black Diamond..	5.8	11.1	37.3	44.8	6.8	11,600	(2470)	79.5	5.8	12.5	1.4	0.8
11	Colstrip.........	....	25.1	28.5	38.8	7.6	......	....	76.7	5.2	16.2	1.1	0.8
12	Knife River.....	26.5	35.6	26.2	32.6	5.6	7,130	....	72.0	4.7	21.1	1.0	1.2

DETAILED DESCRIPTION OF COALS AND REFERENCES
(All references are U.S. Bureau of Mines Technical Papers)

1. Buck Mountain bed, bottom bench, coal from Highland No. 5 Mine, Luzern County, Pa. *Tech. Paper* 512.
2. Pocahontas No. 3 bed coal from Buckeye No. 3 Mine, Stephanson, Wyoming County, W.Va. *Tech. Paper* 604.
3. Lower Freeport bed coal from Indiana County, Pa. *Tech. Paper* 621.
4. Black Creek bed coal from Empire Mine, Walker County, Ala. *Tech. Paper* 531.
5. Pittsburgh bed coal from Pittsburgh Terminal No. 9 Mine, Washington County, Pa. *Tech. Paper* 571.
6. No. 6 bed coal from Orient No. 2 Mine, West Frankport, Franklin County, Ill. *Tech. Paper* 524.
7. Navy No. 6 Lower Bench coal from Eureka Navy Mine, Cumberland, King County, Wash. *Tech. Paper* 512.
8. No. 1 bed coal from Rock Springs Mine No. 4, Rock Springs, Sweetwater County, Wyo. *Tech. Paper* 512.
9. Upper Bagley bed coal from Newcastle Mine, King County, Wash. *Tech. Paper* 512.
10. Black Diamond bed coal from Gallup Southwestern Mine, Gallup, McKinley, County, N.M. *Tech. Paper* 569.
11. Rosebud bed coal from Colstrip Mine, Rosebud County, Mont. *Tech. Paper* 622.
12. Lignite from Knife River Mine, Beulah, Mercer County, N.D. *Tech. Paper* 512.

The heating value of coal may also be predicted from the proximate analysis alone, by using Fig. H-1.

d. Combustion Properties of Coal Constituents.—Table H-4 gives the combustion properties of the elements commonly found in coal. This table is intended to facilitate calculations based on the ultimate analysis of coal. Use of this table is illustrated in Examples H-1, H-3, and H-4 given later in this chapter. The calculation of these factors is illustrated by Example H-5.

TABLE H-3.—ANALYSES OF COKES
(Marks, "Mechanical Engineers' Handbook," 5th ed., McGraw-Hill Book Company, Inc., New York, 1951)

Kind of process	As-received basis									
	Proximate, %				Ultimate, %					High heat value, Btu/lb
	Moisture	Volatile matter	Fixed carbon	Ash	Hydrogen	Carbon	Nitrogen	Oxygen	Sulphur	
Gas-works coke:										
Horizontal retorts	0.8	1.4	88.0	9.8	0.7	86.8	1.1	0.9	0.7	12,820
Vertical retorts	1.3	2.5	86.3	9.9	1.1	85.4	1.4	1.5	0.7	12,770
Narrow coke ovens	0.7	2.0	85.3	12.0	0.6	84.6	1.2	0.9	0.7	12,550
By-product coke	0.8	1.4	87.1	10.7	0.8	85.0	1.3	1.2	1.0	12,690
Beehive coke	0.5	1.8	86.0	11.7	0.8	84.4	1.2	0.9	1.0	12,527
Low-temperature coke	0.0	9.8	83.5	6.7	3.2	84.2	1.9	3.2	0.8	14,030
Low-temperature coke	2.8	15.1	72.1	10.0	3.5	74.5	1.6	8.6	1.8	12,600
Pitch coke	0.3	1.1	97.6	1.0	0.6	96.6	0.7	0.6	0.5	14,097
Petroleum coke	1.1	7.0	90.7	1.2	3.3	90.8	0.8	3.1	0.8	15,060

TABLE H-4.—COMBUSTION PROPERTIES OF COAL AND COKE CONSTITUENTS

Element	Carbon	Hydrogen	Oxygen	Nitrogen	Sulphur
Formula	C	H	O	N	S
Atomic weight	12.010	1.008	16.000	14.008	32.06
Free element at 60°F, 30 in. Hg	Solid, graphite	Gas, H_2	Gas, O_2	Gas, N_2	Solid, rhombic
Gross heating value, Btu/lb:					
1. As free element	14,100	61,042	0	0	3,983
2. Apparent in coal (Dulong)	15,544	62,100	−7,762	0	4,050
Theoretical air required:					
1. Weight, lb/lb combustible	11.48	34.26	−4.31	0	4.30
2. Volume, SCF/lb combustible	150.0	447.8	−56.3	0	56.2
Primary products of combustion with oxygen:					
1. Formula	CO_2	H_2O	*	N_2	SO_2
2. Weight, lb/lb combustible	3.668	8.936	....	1.000	1.998
3. Volume, SCF/lb combustible	31.38	188.7	....	13.52	11.69
Nitrogen† in products of combustion with theoretical air:					
1. Weight, lb/lb combustible	8.82	26.31	−3.31	1.00	3.30
2. Volume, SCF/lb combustible	118.5	353.6	−44.5	13.5	44.4

* Oxygen included in products of combustion of the other elements.
† Includes other inert gases reported as nitrogen in standard gas analyses.

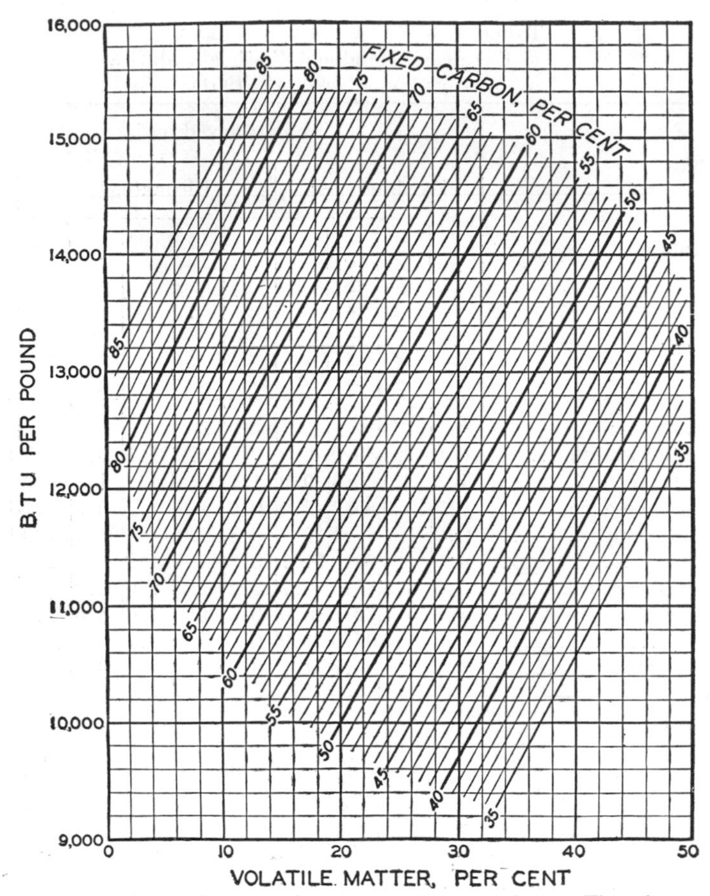

FIG. H-1.—Heating value of coals from proximate analysis. This chart will give approximate results (±300 Btu/lb) for eastern coals containing less than 10 per cent moisture and normal amounts of ash. (*Based on results published in Analyses of Coals, U.S. Bur. Mines Bull.* 22, *reprinted from Graf, "Gas Engineers' Handbook," McGraw-Hill Book Company, Inc., New York,* 1934.)

e. Types of Coal Stokers.—Except for small units, coal is seldom hand-fired. Table H-5 summarizes information on stokers. The coking properties, the quantity and fusing points of ash, and the amount of volatile matter determine the relative advantages of the different types of stoking equipment.

High ash content is obviously undesirable because of the cost of disposal and because some of the heating value is inevitably lost as unburned coke carried by the ash. Ash content of fines is usually higher than that of the larger sizes from the same mine. Frequently

TABLE H-5.—SUMMARY OF INFORMATION ON STOKERS
(Perry, "Chemical Engineers' Handbook," 3d ed., McGraw-Hill Book Company, Inc., New York, 1950)

	Chain grates	Overfeed stokers	Underfeed stokers
Description	Fuel from hopper is carried horizontally into furnace on a continuous web. The web is cooled coming out through the ash pit and returns to coal hopper	Grate extends from hopper into furnace at a rather steep incline, although coal flows down only if a rocking or plunger movement is imparted to grate bars	Raw coal is pushed up through the fuel bed and cinder falls off onto cinder plate. Gases are distilled in an oxidizing atmosphere. Capable of high overloads
Fuel used	1. Coke breeze, steam sizes of anthracite, or 2. High-volatile midwestern coals. In general, any noncoking, clinkering coal may be used, but preferably no mixtures	All coals can be used. Good all-round stoker. Is mainly used on midwestern fuels. Coking coals can be used. Will burn refuse fuels	High-volatile coals, coking coals, and slack or fines may be burned. Ash must not be easily fusible
Draft	Natural: 0.25 to 0.60 in. water Forced: 1- to 2-in. water pressure with coke 1- to 4-in. water pressure with Illinois and similar coals	Natural: 0.25 to 0.60 in. water Forced: 1- to 3-in. water pressure	All forced draft Normal: 2- to 4-in. water pressure in wind box Maximum: 5- to 7-in. water pressure in wind box
Rate of combustion, lb per sq ft per hr	Average: 30–35 10 lb per 0.1 in. water, for natural drafts	Average: 25–35	Average: 30–40 10 lb per 1-in. water pressure
Means of regulation	1. Height of coal grate 2. Speed of grate 3. Amount and distribution of air	1. Rate of plunger feeder 2. Rate of ash removal 3. Amount and distribution of air	1. Rate of feed 2. Amount and distribution of blast
Miscellaneous	Watch for live coals going over end of grate	Fireman needed for ash removal at times and fire must be cleaned	Air is admitted over clinker plate to burn out cinder. Fuel bed from 12 to 24 in. deep

the ash content may be reduced by washing. Clinker formation is favored by high ash content and by low fusing point. Ashes fusing in the range of 1900 to 2200°F are considered low fusing; those above 2600°F are considered high fusing. The fusing point alone is not a clear-cut criterion of tendency to clinker. In some of the larger furnaces the fuel bed is kept at a high temperature and the ash is tapped off as molten slag. This type of operation is very advantageous for handling coals with low-fusing ash.

The burning of coal takes place in two zones: on the fire bed itself (primary combustion) and in the flame above (secondary combustion). This arises for two reasons: (1) a large portion of the coal is volatilized on heating to the temperature of the fire bed; (2) the primary combustion is seldom complete. In practice, therefore, considerable space is provided above the fire bed for secondary combustion and some of the air is admitted directly to the firebox. The combustion volume required obviously depends on the amount of the coal volatilized. The volatile matter determination of the proximate analysis is a semiquantitative measure of the fraction of coal volatilized and is, therefore, a rough measure of the secondary combustion requirements for volume and air.

f. Powdered Coal.—The use of coal in powdered form is very advantageous for securing efficient combustion with low-cost operation and maintenance. Further, a furnace may readily be designed for alternate use of powdered coal, fuel oil, or gas. Since the ash is partly blown out of the stack, provision for ash removal may be required, particularly in urban communities. Fineness of grinding is important and is conveniently determined by a screen analysis using, say 50-, 100-, and 200-mesh screens. Blake and Purdy[1] give typical screen analyses, as shown in the accompanying table.

	Low-volatile coal, %	High-volatile coal, %
Through:		
50-mesh	99.5	99
100-mesh	95	90
200-mesh	80	65

The power required for pulverizing coal depends on the dryness and the properties of the particular coal. This may be estimated by one of the A.S.T.M. methods listed earlier. Dry coal requires considerably less power but is considerably dustier. Figure H-2 gives the power required for pulverizing a Pennsylvania bituminous coal.

g. Storage of Coal.—Four things may occur to stored coal: (1) spalling from alternate wetting and drying, (2) loss of heating value owing to oxidation, (3) impairment of coking properties, and (4) spontaneous combustion.

[1] Perry, "Chemical Engineers' Handbook," 3d ed., McGraw-Hill Book Company, Inc., New York, 1950.

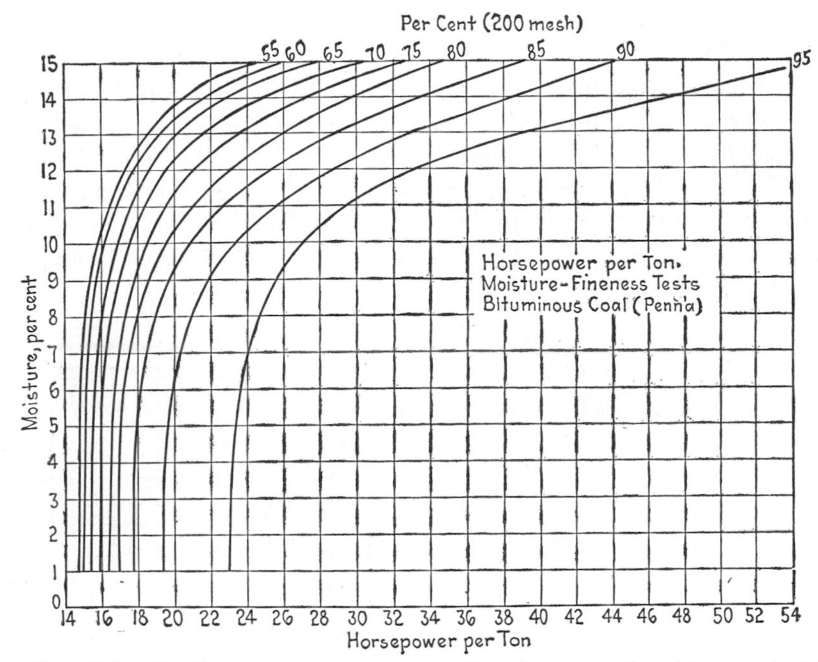

FIG. H-2.—Relation among power consumption, fineness, and moisture content in pulverizing Pennsylvania bituminous coal. (*Haslam and Russell, "Fuels and Their Combustion," McGraw-Hill Book Company, Inc., New York,* 1926.)

Spalling may be serious in the case of low-rank coals but is less important in the case of bituminous or anthracite coals. The loss of heating value is generally small, in the order of 1 per cent per year. Since some gain in weight accompanies the oxidation, the pile loss is somewhat less than indicated by the heating value per pound. Loss in coking properties is not usually serious and is no disadvantage for coal to be used as fuel.

Spontaneous combustion deserves careful consideration.[1] Coal temperature should be watched and the coal moved if the temperature goes over 120°F. Avoiding the segregation of sizes and the storing of larger sizes is helpful. Avoiding too high piles is also a good precaution. Wetting the coal does *not* retard oxidation.

Storage of coal above ground is limited in some localities by law.

[1] For a brief discussion see comments by Fieldner; Marks, "Mechanical Engineers' Handbook," 5th ed., McGraw-Hill Book Company, Inc., New York, 1951.

3. Fuel Oil

a. Fuel-oil Specification.—Table H-6 gives specifications of five grades of oil. It will be noted that the former Grade 4 has now been omitted from the list. Grades 5 and 6 are frequently referred to as Bunker B and Bunker C. The heavier oils are cheaper and more popular for large units, whereas the lighter grades are usually employed for small units because of their greater convenience. Table H-7 gives data on the composition and properties of several oils.

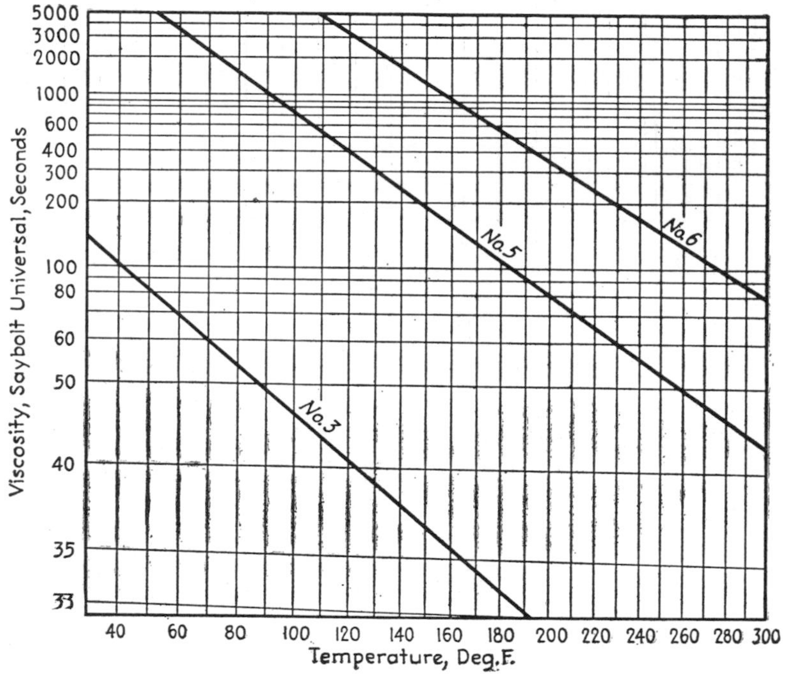

FIG. H-3.—Approximate maximum viscosity of three grades of fuel oils.

The flash and fire points are of importance in determining conditions for safe handling. An oil below the flash point is safe since insufficient vapors are produced for possible ignition. The carbon residue and distillation data are of importance if the burner is designed to vaporize a considerable portion of the oil.

The viscosity of the oil is important to the process engineer for sizing pipe lines and determining the temperature to which the oil must be heated for burner use. The most favorable viscosity is, of course, determined by the particular burner; however, Trinks[1] recommends 300 Saybolt Universal as a general working figure. The effect of temperature on viscosity differs considerably for different oils; nevertheless, the relation is qualitatively the same so that the temperature required for a given viscosity may be estimated to within 10 or 20°F by use of Fig. H-3.

As in the case of coal, an ultimate analysis may be made, but this usually serves no useful purpose because fuel oils differ only slightly in composition. Table H-7 gives the analyses of several oils.

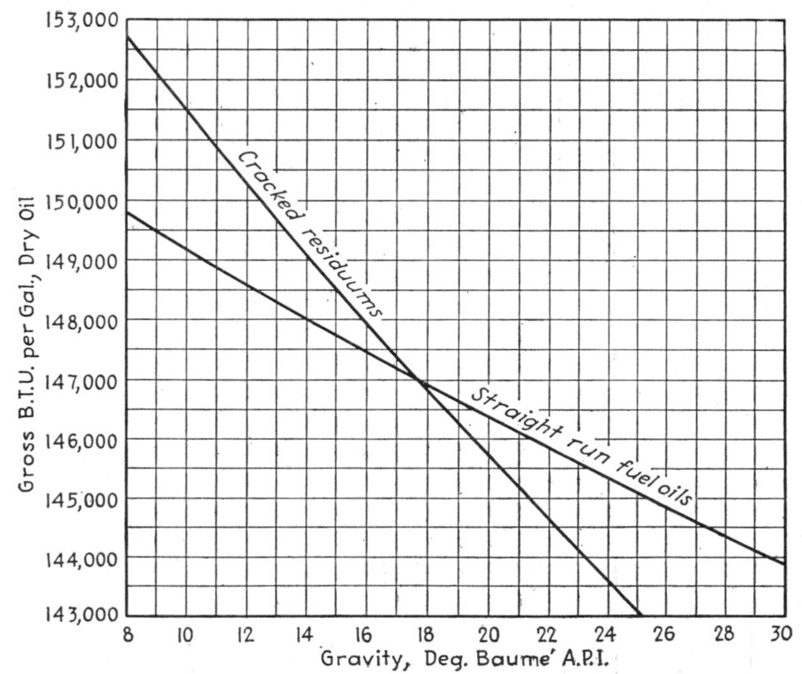

Fig. H-4.—Relation between heating value of dry fuel oil and A.P.I. gravity. (*"Combustion," 3d ed., American Gas Association, New York*, 1932.)

b. Heating Values of Fuel Oils.—These range between 130,000 and 153,000 Btu/gal (gross). The light oils generally have the lower heating values per gallon but higher per pound; conversely, the heavier oils have larger heating values per gallon though less per pound. The heating value of a fuel oil should be directly determined. However, in the absence of specific data approximate values may be obtained by use of Fig. H-4.

[1] Trinks, "Industrial Furnaces," vol. I, 4th ed., 1951; vol. II, 3d ed., 1955, John Wiley & Sons, Inc., New York.

TABLE H-6.—DETAILED REQUIREMENTS FOR FUEL OILS*
[*Nat. Bur. Standards Com. Standard* CS12-40 (1940)]

Grade†		Flash point, °F		Pour point, °F	Water and sediment, %	Carbon residue, %	Ash, %	Distillation temperatures, °F				Viscosity sec			
								10% point	90% point		End point	Saybolt Universal at 100°F		Saybolt Furol at 122°F	
No.	Description of fuel oil	Min	Max	Max	Max	Max	Max	Max	Max	Min	Max	Max	Min	Max	Min
1	Distillate oil for use in burners requiring a volatile fuel	100 or legal	165	0	Trace	0.05 on 10% residuum§	....	410	...	.....	560‖				
2	Distillate oil for use in burners requiring a moderately volatile fuel	110 or legal	190	10‡	0.05	0.25 on 10% residuum¶	....	440	600						
3	Distillate oil for use in burners requiring a low-viscosity fuel	110 or legal	230	20‡	0.10	0 15 straight	....	...	675	600**	....	45			
5	Oil for use in burners requiring a medium-viscosity fuel	130 or legal	...	...	1.00	....	0.10	...	...	.....	....	..	50	40	
6	Oil for use in burners equipped with preheaters, permitting a high-viscosity fuel	150	...	...	2.00††	....	....	...	...	.....	....	..	..	300	45

* Recognizing the necessity for low-sulphur fuel oils used in connection with heat-treatment, nonferrous metal, glass, and ceramic furnaces, and other special uses, a sulphur requirement may be specified in accordance with the following table:

Grade of Fuel Oil, No.	Sulphur (max.), %
1	0.5
2	0.5
3	0.75
5	No limit
6	No limit

Other sulphur limits may be specified only by mutual agreement between the buyer and seller.

† It is the intent of these classifications that failure to meet any requirement of a given grade does not automatically place an oil in the next lower grade unless in fact it meets all requirements of the lower grade.

‡ Lower or higher pour points may be specified whenever required by conditions of storage or use. However, these specifications shall not require a pour point lower than 0°F under any conditions.

§ For use in other than sleeve-type blue-flame burners, carbon residue on 10 per cent residuum may be increased to a maximum of 0.12 per cent. This limit may be specified by mutual agreement between the buyer and seller.

‖ The maximum end point may be increased to 590°F when used in burners other than sleeve-type blue-flame burners.

¶ To meet certain burner requirements, the carbon-residue limit may be reduced to 0.15 per cent on 10 per cent residuum.

** The minimum distillation temperature of 600°F for 90 per cent may be waived if A.P.I. gravity is 26 or lower.

†† Water by distillation, plus sediment by extraction. Sum, maximum 2.0 per cent. The maximum sediment by extraction shall not exceed 0.50 per cent. A deduction in quantity shall be made for all water and sediment in excess of 1.0 per cent.

TABLE H-7.—COMPOSITION AND PROPERTIES OF VARIOUS OILS
("Combustion," 3d ed., American Gas Association, New York, 1932)

	A.P.I gravity, °Bé	Specific gravity at 60°F	Lb/gal	Per cent bottom sediment (centrifuge method)	Pour point, °F	Viscosity		Analysis, weight basis, %					Heating value (bomb calorimeter)		
						furol sec	At °F	Water	Sulphur	Carbon	Hydrogen	Undetermined	Gross Btu/lb	Net* Btu/lb	Gross Btu/gal
Domestic fuel oils															
Commercial gasoline	55.5	0.7567	6.304	....	..	...	...	....	....	84.3	15.7	....	21,000	19,506	132,384
Commercial kerosene	41.8	0.8165	6.802	....	..	...	...	....	0.02	84.7	15.3	....	20,000	18,545	136,040
Gas oil—Kansas	35.0	0.8498	7.080	....	..	...	...	....	0.18	....	.....	....	19,748	......	139,816
Gas oil—Oklahoma	32.0	0.8654	7.210	....	..	...	...	....	0.21	....	.....	....	19,474	......	140,407
Straight-run fuel oils															
California	26.2	0.8973	7.475	0.10	42	14	77	0.89	1.19	85.8	12 09	0.04	19,053	18,001	142,421
Oklahoma Mid-Continent	25.3	0.9024	7.518	0.05	32	60	77	....	0.46	86 4	12.38	0.76	19,353	18,173	145,496
Mixed Kansas Mid-Continent	25.5	0.9013	7.509	0.10	38	23	77	....	0.46	86.5	12.38	0.76	19,182	18,002	144,031
Mixed Kansas Mid-Continent	25.1	0.9036	7.528	0.10	38	16	122	0.33	0.39	86.2	12.39	0.69	19,256	18,076	144,956
Archer & Wichita Cos., Texas	25.5	0.9013	7.509	0.05	32	42	77	0.11	0.60	....	.....	....	19,277	18,093	144,758
Low-level cracked residuums															
California	16.5	0.9561	7.965	0.10	Below 0	13	77	0.05	1.14	87.5	10.17	1.14	18,319	17,351	145,911
Oklahoma Mid-Continent	14.3	0.9705	8.085	0.00	Below 0	57	77	0.05	0.70	87.6	10.27	1.38	18,454	17,479	149,200
Mixed Kansas Mid-Continent	22.0	0.9218	7.680	0.20	Below 0	14	77	0.05	0.42	87.4	11.10	1.03	18,778	17,723	144,215
Mixed Kansas Mid-Continent	8.1	1.0140	8.448	0.80	Below 0	24	122	0.25	0.67	88.5	9.07	1.51	18,077	17,212	152,714
Archer & Wichita Cos., Texas	18.6	0.9427	7.854	0.80	Below 0	20	77	....	0.79	....	.....	....	18,732	17,781	147,121
Flashed cracked residuums															
California	8.9	1.0070	8.389	0.30	21	174	77	0.05	1.20	88.3	9.50	0.95	18,084	17,179	151,707
Oklahoma Mid-Continent	10.7	0.9951	8.290	0.20	43	124	122	....	0.77	88.5	9.92	0.81	18,293	17,349	151,649
Mixed Kansas Mid-Continent	8.6	1.0110	8.423	1.00	25	52	122	....	0.56	88.9	9.80	0.74	18,277	17,345	153,947
Mixed Kansas Mid-Continent	9.2	1.0030	8.356	2.60	42	159	122	0.40	0.68	88 4	9.95	0.81	18,274	17,324	152,697

* Gross Btu/lb observed $-[(\text{lb H} \times 9) + \text{lb } H_2O \text{ present}]\ 1057 =$ net Btu/lb

c. Types of Oil Burners.—Efficient performance of a burner depends on its ability to atomize the oil and mix thoroughly with the air stream before the ignition zone. This atomization may be effected by high- or low-pressure air, by steam, or by mechanical means. Table H-8 gives data on air atomization. Haslam and Russell[1] state that between 1/16 and 1 lb of steam is required to atomize 1 lb of fuel. Oil may also be atomized by pressuring through liquid spray nozzles. In general 75 to 150 psi head pressure is required. Rotary burners atomize oil by a combination of air flow and centrifugal force of a rotating cylinder. These burners are very flexible. Although they may be made very efficient, usually their efficiency is low and large excesses of air (100 to 200 per cent) may be required.

TABLE H-8.—BURNER PERFORMANCE WITH DIFFERENT TYPES OF AIR ATOMIZATION
(Haslam and Russell, "Fuels and Their Combustion," McGraw-Hill Book Company, Inc., New York, 1926)

Type	Velocity of atomizing air, fps	Per cent of air for combustion used for atomization
Low-pressure (turboblower) air atomization, 10 oz per sq in.	300	60
Low-pressure (positive-pressure pump) air atomization, 2 psi	500	50
High-pressure air atomization, 50 psi	2,500	12

d. Combustion Properties of Constituents of Fuel Oil.—The data given in Table H-4 for coal and coke may be used, except that the heating value of oils computed by this method will, however, generally be about 600 Btu/lb below the actual value.

4. Gaseous Fuels

a. Analyses.—Gaseous fuels are usually analyzed by either the traditional absorption and combustion analyses or by low-temperature distillations.

Carbon dioxide, oxygen, illuminants, hydrogen, carbon monoxide, methane, ethane, and nitrogen are the quantities usually determined by absorption and combustion analyses. These analyses are reason-

[1] "Fuels and Their Combustion," McGraw-Hill Book Company, Inc., New York, 1926.

ably specific except for the hydrocarbons. The illuminants include acetylene, aromatics, higher olefins, and isobutane in addition to ethylene, the principal constituent. Some of these components may be determined separately. The methane and ethane reported will be fictitious if other saturated hydrocarbons are present, that is, 15 per cent CH_4, 5 per cent C_2H_6, and 1.0 per cent C_3H_8 would be reported as 14 per cent CH_4 and 7 per cent C_2H_6. For combustion purposes this is not at all serious as the calculated heating values and products of combustion will be correct. Any sulphur dioxide or hydrogen sulphide present will be reported as carbon dioxide, unless separately determined.

If considerably higher hydrocarbons are present, analysis by low-temperature distillation is advisable. This low-temperature analysis, however, does not distinguish among H_2, CH_4, CO, N_2, and O_2. Further, carbon dioxide and acetylene are usually removed before analysis. Consequently such constituents must be determined separately.

b. Units of Measurement.—Fuel gases are customarily measured as cubic feet. Because the amount of gas contained in a cubic foot varies with temperature and pressure, it is necessary to define the conditions of measurement. The most widely used standard is that of the American Gas Association which defines a standard cubic foot (SCF) as the quantity of gas contained in a cubic foot of space at 60°F and 30 in. of mercury (14.73 psi) absolute pressure. The gas volumes are usually based on actual physical measurements, though frequently perfect gas volumes are used for simplicity, as this avoids any difficulty in choice of values for components that are liquid at the standard conditions.

Gas sales are generally contracted on the basis of one of three tables as adopted by the American Gas Association,[1] the Natural Gasoline Association of America,[2] or the California Natural Gasoline Association.[3] Unfortunately these tables are not in agreement. The American Gas Association values are based on actual gas volumes for true gases, and perfect gas volumes for vapors of liquids. The California Natural Gasoline Association values are based on the perfect gas law (using a different value of the gas constant than is employed in this book). They give factors for gas-law deviation correction and

[1] Shnidman, "Gaseous Fuels," American Gas Association, New York, 1954.

[2] "Engineering Data Book," 7th ed., Natural Gasoline Supply Men's Association, Tulsa, 1957.

[3] Matteson and Hanna, *Proceedings* of the April, May, June, 1942, meetings, California Natural Gasoline Association; cf. *Oil Gas J.*

recommend procedures for applying these to metering[1] and correction of gas analyses.[2]

These differences, though the cause of considerable irritation, are in reality no great cause for concern, since the differences involved are less than the usual accuracies of gas analysis and measurement. The values given in this chapter are taken largely from these three tables or references quoted by them. No claim is made that the values chosen are the best available; however, an effort has been made to make the given data self-consistent. Values are given on both the perfect gas and the actual volume basis so that the reader may choose whichever standard appears more suitable.

Though this book uses only the American Gas Association standard conditions to define the cubic foot, it should be stated that many other standards of gas measurement are employed. The compressed gas industry uses 68°F and standard atmospheric pressure (14.70 psi) as the basis of gas measurement. Other standards are occasionally encountered: 70°F and 14.70 psi; 60°F and 14.40 psi; and 32°F and 14.70 psi.

c. Typical Fuel Gases.—Table H-9 gives analyses, properties, and combustion products of some commercial fuels.

d. Properties of Constituents of Gaseous Fuels.—Very fortunately nearly all properties of fuel gases are additive, and the properties of any fuel gas (at low pressure at least) may be readily computed from those of its constituents. As already mentioned under Units of Gas Measurement, there are two methods of doing this: by perfect gas volumes or by actual gas volumes. This choice is largely arbitrary and is left to the reader's judgment.

Table H-10 gives specific gravities, specific volumes, and heating values. Table H-11 gives combustion ratios for computing air and combustion properties. The basic units and references to source material for these tables already have been discussed in Section 4*b* on page 297.

[1] Tentative Standards for the Determination of Superexpansibility Factors in High Pressure Gas Measurement, *California Natural Gasoline Assocation*, Bull. TS-354 (1936).

[2] Tentative Standard Method for Analysis of Natural Gas and Gasoline by Fractional Distillation, *California Natural Gasoline Association*, Bull. TS-411 (1957).

TABLE H-9.—PROPERTIES OF TYPICAL COMMERCIAL GASES
("Combustion," 3d ed., American Gas Association, New York, 1932)

No.	Gas	Constituents of gas—% by volume							Illuminants		Sp gr	Cu ft air req. for comb. of cu ft of gas	Btu/cu ft gross	Btu/cu ft net	Products of combustion per cu ft of gas				Ultimate % CO_2	Btu net/cu ft of prod. of comb.	Flame temp., no excess air
		CO_2	O_2	N_2	CO	H_2	CH_4	C_2H_6	C_2H_4	C_6H_6					H_2O	CO_2	N_2	Total			
1	Natural gas (Birmingham)	....	....	5.0	....	....	90.0	5.0	....	....	0.60	9.41	1002	904	2.02	1.00	7.48	10.50	11.8	86.0	3565
2	Natural gas (Pittsburgh)	....	....	0.8	....	....	83.4	15.8	....	....	0.61	10.58	1129	1021	2.22	1.15	8.37	11.73	12.1	87.0	3562
3	Natural gas (So. California)	0.7	....	0.5	....	....	84.0	14.8	....	....	0.64	10.47	1116	1009	2.20	1.14	8.28	11.62	12.1	87.0	3550
4	Natural gas (Los Angeles)	6.5	....	....	....	....	77.5	16.0	....	....	0.70	10.05	1073	971	2.10	1.16	7.94	11.20	12.7	86.7	3550
5	Natural gas (Kansas City)	0.8	....	8.4	....	....	84.1	6.7	....	....	0.63	9.13	974	879	1.95	0.98	7.30	10.23	11.9	86.0	3535
6	Reformed natural gas	1.4	0.2	2.9	9.7	46.6	37.1	....	1.3	(C_3H_6 0.8)	0.41	5.22	599	536	1.30	0.53	4.16	5.99	11.3	89.6	3615
7	Mixed, natural, and water gas	4.4	2.1	4.7	25.5	35.1	23.1	4.7	0.2	0.2	0.61	4.43	525	477	1.01	0.64	3.55	5.20	15.3	91.7	3630
8	Coke-oven gas	2.2	0.8	8.1	6.3	46.5	32.1	....	3.5	0.5	0.44	4.99	574	514	1.25	0.51	4.02	5.78	11.2	87.0	3610
9	Coal gas (continuous verticals)	3.0	0.2	4.4	10.9	54.5	24.2	....	1.5	1.3	0.42	4.53	532	477	1.15	0.49	3.62	5.26	11.9	90.7	3645
10	Coal gas (inclined retorts)	1.7	0.8	8.1	7.3	49.5	29.2	....	0.4	3.0	0.47	5.23	599	540	1.23	0.57	4.21	6.01	11.9	89.9	3660
11	Coal gas (intermittent verticals)	1.7	0.5	8.2	6.9	49.7	29.9	....	3.0	0.1	0.41	4.64	540	482	1.21	0.45	3.75	5.41	10.7	89.0	3610
12	Coal gas (horizontal retorts)	2.4	0.75	11.35	7.35	47.95	27.15	....	1.32	1.73	0.47	4.68	542	486	1.15	0.50	3.81	5.46	11.6	89.0	3600
13	Mixed coke-oven and carbureted water gas	3.4	0.3	12.0	17.4	36.8	24.9	....	3.7	1.5	0.58	4.71	545	495	1.04	0.62	3.85	5.51	13.9	90.0	3630
14	Mixed coal, coke-oven, and carbureted water gas	1.8	1.6	13.6	9.0	42.6	28.0	....	2.4	1.0	0.50	4.52	528	475	1.11	0.50	3.71	5.32	11.8	89.3	3640
15	Carbureted water gas	3.0	0.5	2.9	34.0	40.5	10.2	....	6.1	2.8	0.63	4.60	550	508	0.87	0.76	3.66	5.29	17.2	96.2	3725
16	Carbureted water gas	4.3	0.7	6.5	32.0	34.0	15.5	....	4.7	2.3	0.67	4.51	534	493	0.75	0.86	3.63	5.24	17.1	94.2	3700
17	Carbureted water gas (low gravity)	2.8	1.0	5.1	21.0	47.5	15.0	....	5.2	2.4	0.54	4.61	549	501	0.98	0.64	3.70	5.31	14.7	94.3	3690
18	Water gas (coke)	5.4	0.7	8.3	37.0	47.3	1.3	....	....	....	0.57	2.10	287	262	0.53	0.44	1.74	2.71	20.1	96.6	3670
19	Water gas (bituminous)	5.5	0.9	27.6	28.2	32.5	4.6	....	0.4	0.3	0.70	2.01	261	239	0.47	0.41	1.86	2.74	18.0	87.2	3510
20	Oil gas (Pacific coast)	4.7	0.3	3.6	12.7	48.6	26.3	....	2.7	1.1	0.47	4.73	551	496	1.15	0.56	3.77	5.48	12.9	90.5	3630
21	Producer gas (buckwheat anthracite)	8.0	0.1	50.0	23.2	17.7	1.0	....	....	.....	0.86	1.06	143	133	0.22	0.32	1.34	1.88	19.4	70.5	3040
22	Producer gas (bituminous)	4.5	0.6	50.9	27.0	14.0	3.0	....	....	....	0.86	1.23	163	153	0.23	0.35	1.48	2.06	18.9	74.6	3175
23	Producer gas (0.6 lb steam per lb coke)	6.4	....	52.8	27.1	13.3	0.4	....	....	....	0.88	1.00	135	128	0.17	0.34	1.32	1.82	20.5	70.3	3010
24	Blast-furnace gas	11.5	....	60.0	27.5	1.0	.....	....	....	....	1.02	0.68	92	92	0.02	0.39	1.14	1.54	25.5	59.5	2650
25	Commercial butane	(C_4H_{10} 93.0)			....	(C_3H_8 7.0)		....	....	....	1.95	30.47	3225	2977	4.93	3.93	24.07	32.93	14.0	90.5	3640
26	Commercial propane	(C_3H_8 100.0)			.....	.....	.....	....	....	.....	1.52	23.82	2572	2371	4.17	3.00	18.82	25.99	13.7	91.2	3660

TABLE H-10.—SPECIFIC GRAVITIES, SPECIFIC VOLUMES, AND HEATS OF COMBUSTION OF GASES AND VAPORS FOR COMBUSTION CALCULATIONS

Substance	Formula	Molecular weight	Specific gravity (air = 1)		Specific volume SCF/lb		Heat of combustion					
							Btu/lb		Btu/SCF of perfect gas		Btu/SCF of actual gas	
			Perfect gas	Actual gas	Perfect gas	Actual gas	Gross	Net	Gross	Net	Gross	Net
Hydrogen	H_2	2.016	0.0696	0.0695	187.85	187.98	61,070	51,600	325	275	325	275
Methane	CH_4	16.042	0.554	0.555	23.61	23.56	23,880	21,520	1011	911	1014	914
Acetylene	C_2H_2	26.036	0.899	0.906	14.55	14.43	21,430	20,710	1473	1423	1485	1435
Ethylene	C_2H_4	28.052	0.968	0.974	13.50	13.42	21,640	20,290	1603	1503	1613	1512
Ethane	C_2H_6	30.069	1.038	1.046	12.59	12.49	22,320	20,430	1772	1622	1787	1636
Propylene	C_3H_6	42.079	1.453	1.480	9.000	8.83	21,040	19,690	2338	2188	2383	2231
Propane	C_3H_8	44.095	1.522	1.546	8.588	8.45	21,660	19,940	2522	2322	2563	2360
1-3 Butadiene	C_4H_6	54.088	1.867	1.934	7.002	6.76	20,250	19,200	2892	2742	2996	2840
Isobutylene	C_4H_8	56.105	1.937	2.008	6.750	6.51	20,760	19,420	3076	2877	3189	2968
1-Butene	C_4H_8	56.105	1.937	2.001	6.750	6.53	20,840	19,490	3087	2887	3191	2970
2-Butene, *cis*	C_4H_8	56.105	1.937	2.001	6.750	6.53	20,800	19,440	3081	2880	3185	2960
2-Butene, *trans*	C_4H_8	56.105	1.937	2.001	6.750	6.53	20,770	19,410	3077	2876	3181	2955
Isobutane	C_4H_{10}	58.121	2.006	2.066	6.516	6.33	21,250	19,630	3261	3013	3357	3101
n-Butane	C_4H_{10}	58.121	2.006	2.070	6.516	6.31	21,310	19,680	3270	3020	3377	3119
Isopentane vapor	C_5H_{12}	72.147	2.491	2.60*	5.249	5.02*	21,050	19,480	4010	3710	4194*	3881*
n-Pentane vapor	C_5H_{12}	72.147	2.491	2.61*	5.249	5.00*	21,090	19,520	4018	3718	4218*	3904*
Benzene vapor	C_6H_6	78.108	2.697	2.78*	4.848	4.70*	18,020	17,290	3717	3566	3834*	3679*
n-Hexane vapor	C_6H_{14}	86.173	2.975	3.17*	4.395	4.12*	20,940	19,400	4765	4414	5082*	4709*
Toluene vapor	C_7H_8	92.134	3.181	3.31*	4.110	3.95*	18,250	17,430	4440	4240	4623*	4413*
n-Heptane vapor	C_7H_{16}	100.20	3.459	3.76*	3.780	3.47*	20,860	19,340	5518	5116	6011*	5573*
n-Octane vapor	C_8H_{18}	114.23	3.944	4.34*	3.316	3.01*	20,780	19,270	6266	5812	6904*	6902*
Naphthalene vapor	$C_{10}H_8$	128.16	4.421	4.86†	2.955	2.69†	17,300	16,710	5854	5654	6431†	6212†
Carbon monoxide	CO	28.010	0.967	0.968	13.52	13.50	4,350	4,350	321	321	321	321
Carbon dioxide	CO_2	44.010	1.519	1.528	8.605	8.555	0	0	0	0	0	0
Hydrogen sulphide	H_2S	34.076	1.176	1.189	11.113	10.99	7,100	6,545	639	589	646	596
Sulphur dioxide	SO_2	64.06	2.212	2.235	5.912	5.85	0	0	0	0	0	0
Water vapor	H_2O	18.016	0.622	0.628*	21.02	20.82*	0	0	0	0	0	0
Ammonia	NH_3	17.032	0.588	0.596	22.23	21.91	9,668	8,001	435	360	441	365
Oxygen	O_2	32.00	1.105	1.105	11.83	11.82	0	0	0	0	0	0
Nitrogen	N_2	28.016	0.967	0.967	13.52	13.52	0	0	0	0	0	0
Atmospheric nitrogen	....	28.17	0.972	0.972	13.44	13.44	0	0	0	0	0	0
Air	....	28.966	1.000	1.000	13.074	13.069	0	0	0	0	0	0

NOTES: One standard cubic foot (SCF) is quantity of gas contained in 1 cu ft at 60°F and 30 in. of mercury. One lb mole = 378.7 SCF of a perfect gas.

Actual gas values are based on experimental gas-density measurements if available, otherwise deviation from perfect gas law is estimated.

* Substance is liquid at 60°F and 30 in. of mercury; however, it can be present as a vapor component of a gas. Gas-law deviation is based on behavior in gas mixture. This table in no case assumes deviations over 10 per cent. Because of the arbitrary nature of this correction many chemists prefer to use perfect-gas-law values for vapors.

† Naphthalene is solid at 60°F and 30 in. of mercury.

5. Natural Gasoline and Liquefied Gases

a. Units of Measurement.—Sales are usually based on gallons even though a "gallon" of a liquefied gas does not lend itself to precise definition. For this reason sales are sometimes based on standard cubic feet (of vapor) as in the case of gases.

TABLE H-11.—COMBUSTION RATIOS OF THE LIGHT HYDROCARBONS AND OTHER FUEL COMPONENTS

BASIS: Perfect combustion with no excess air

Substance	Formula	SCF/SCF (perfect gas) or mole/mole combustible					Lb/lb of combustible				
		Required for combustion		Combustion products			Required for combustion		Combustion products		
		Oxygen	Air	CO_2	H_2O	N_2	Oxygen	Air	CO_2	H_2O	N_2
Hydrogen..........	H_2	0.5	2.382	0	1.0	1.882	7.937	34.344	0	8.937	26.407
Methane...........	CH_4	2.0	9.528	1.0	2.0	7.528	3.990	17.265	2.744	2.246	13.725
Acetylene..........	C_2H_2	2.5	11.911	2.0	1.0	9.411	3.073	13.927	3.381	0.692	10.224
Ethylene...........	C_2H_4	3.0	14.293	2.0	2.0	11.293	3.422	14.807	3.138	1.285	11.385
Ethane............	C_2H_6	3.5	16.675	2.0	3.0	13.175	3.725	16.119	2.927	1.798	12.394
Propylene..........	C_3H_6	4.5	21.439	3.0	3.0	16.939	3.422	14.807	3.138	1.285	11.385
Propane...........	C_3H_8	5.0	23.821	3.0	4.0	18.821	3.629	15.703	2.994	1.634	12.074
Butylenes..........	C_4H_8	6.0	28.585	4.0	4.0	22.585	3.422	14.807	3.138	1.285	11.385
Butanes............	C_4H_{10}	6.5	30.967	4.0	5.0	24.467	3.579	15.487	3.029	1.550	11.908
Pentanes...........	C_5H_{12}	8.0	38.114	5.0	6.0	30.114	3.548	15.353	3.050	1.498	11.805
Benzene...........	C_6H_6	7.5	35.732	6.0	3.0	28.232	3.073	13.927	3.381	0.692	10.224
Hexanes...........	C_6H_{14}	9.5	45.260	6.0	7.0	35.760	3.528	15.266	3.064	1.464	11.738
Toluene............	C_7H_8	9.0	42.878	7.0	4.0	33.878	3.126	13.527	3.344	0.782	10.401
Heptanes..........	C_7H_{16}	11.0	52.41	7.0	8.0	41.41	3.513	15.20	3.076	1.438	11.69
Octanes............	C_8H_{18}	12.5	59.55	8.0	9.0	46.05	3.502	15.15	3.084	1.420	11.65
Naphthalene.......	$C_{10}H_8$	12.0	57.170	10.0	4.0	45.170	2.996	12.964	3.434	0.562	9.968
Carbon monoxide...	CO	0.5	2.382	1.0	0	1.882	0.571	2.471	1.571	0	1.900
Ammonia..........	NH_3	0.75	3.573	SO_2	1.5	3.323	1.409	6.097	SO_2	1.587	5.511
Hydrogen sulphide..	H_2S	1.5	7.146	1.0	1.0	5.646	1.409	6.097	1.880	0.529	4.688

When sales are on the gallon basis, the analysis is customarily expressed as liquid volume per cent computed on the basis of standard values of liquid densities assumed to be the same for a given component at 60°F, regardless of the pressure and composition of the mixture.

b. Properties of Component Hydrocarbons.—Data are given in Table H-12. The melting point, freezing point, liquid densities, and heats of vaporization at 1 atm pressure are based on the California Natural Gasoline Association table.[1] The heats of vaporization at 60°F are from various sources (or estimated), and the heats of com-

[1] Matteson and Hanna, April, May, June, 1942, meetings, California Natural Gasoline Association; *Proc. Oil Gas J.*

bustion are computed from those given in Table H-10. Vapor pressures were taken from the Natural Gasoline Association of America table.[1] Vapor equivalent to 1 gal was computed from the liquid densities and the specific volume (Table H-10).

As combustion ratios of a liquefied hydrocarbon are identical with those for its vapor, Table H-11 may be used without change.

TABLE H-12.—PROPERTIES OF LIGHT HYDROCARBONS

Substance	Formula	Molecular weight	Melting point, °F	Boiling point at 1 atm, °F	Liquid density, lb/gal	Heat of combustion, Btu/gal		Heat of vaporization, Btu/lb		Vapor pressure at 100°F, psi	Vapor from 1 gal	
						Gross	Net	At boiling point	At 60°F		SCF of perfect gas	SCF of actual gas
Methane*	CH_4	16.042	−296.5	−258.5	(2.5)*	(59,420)	(53,520)	245	(110)*	.....	59.02	58.90
Ethylene*	C_2H_4	28.052	−272.6	−154.7	(3.3)*	(70,920)	(66,460)	208	(150)*	.....	44.55	44.29
Ethane*	C_2H_6	30.069	−297.8	−128.2	(3.3)*	(73,160)	(66,920)	211	(150)*	.....	41.55	41.22
Propylene	C_3H_6	42.079	−300.8	−54.0	(4.2)	87,740	82,060	189	154	228	37.80	37.09
Propane	C_3H_8	44.095	−305.8	−43.8	4.239	91,180	83,900	183	153	189.5	36.40	35.82
Isobutylene	C_4H_8	56.105	−221.3	+19.4	5.040	103,820	97,070	172	162	63	34.02	32.81
1-Butene	C_4H_8	56.105	−319.0	20.4	5.001	103,420	96,670	174	164	66	33.76	32.56
2-Butene, *cis*	C_4H_8	56.105	−218.7	38.7	5.118	105,530	98,570	187	(182)	.....	34.55	33.42
2-Butene, *trans*	C_4H_8	56.105	−158.4	33.7	5.118	105,380	98,420	183	(178)	.....	34.55	33.42
Isobutane	C_4H_{10}	58.121	−229.0	10.0	4.697	99,110	91,500	158	146	73.5	30.61	29.73
n-Butane	C_4H_{10}	58.121	−218.2	31.1	4.863	102,850	94,980	166	159	52.0	31.69	30.69
Isopentane	C_5H_{12}	72.147	−257.1	82.4	5.201	108,700	100,430	146	151	20.3	27.30	26.11
n-Pentane	C_5H_{12}	72.147	−201.5	97.0	5.251	110,060	101,760	153	160	15.5	27.56	26.26
Benzene	C_6H_6	78.108	+41.9	176.2	7.344	131,750	125,660	170	184	.....	35.60	34.52
n-Hexane	C_6H_{14}	86.173	−139.9	155.7	5.526	114,830	106,320	146	....	6.1	24.29	22.76
Toluene	C_7H_8	92.134	−139.2	230.9	7.244	130,900	124,230	156	....	.....	29.77	28.61
n-Heptane	C_7H_{16}	100.2	−131.4	209.1	5.727	118,550	109,840	138	....	1.6	21.65	19.87
n-Octane	C_8H_{16}	114.2	−70.2	258.1	5.885	121,350	112,460	131	....	0.5	19.51	17.71

* Pure methane and ethylene do not exist as liquids at 60°F. Values given are intended to represent those of their solutions in hydrocarbon mixtures. Pure ethane does exist as a liquid of density of 3 lb/gal. The values given intend to represent properties in dilute solutions.

c. Vapor Pressure and Storage.—The vapor pressure of natural gasoline is customarily determined by the Reid apparatus. The resulting Reid pressure is closely related to the true vapor pressure, but not identical owing to inadequacies of the test. Figure H-5 gives

[1] "Engineering Data Book," 7th ed., Natural Gasoline Supply Men's Association, Tulsa, 1957.

the approximate relation to the true vapor pressure and also permits estimation of vapor pressures over a range in temperature.

Where the vapor pressure is not known, approximate values may be obtained from the vapor pressures of components and Raoult's law—or better from vaporization constants (see Chapter L).

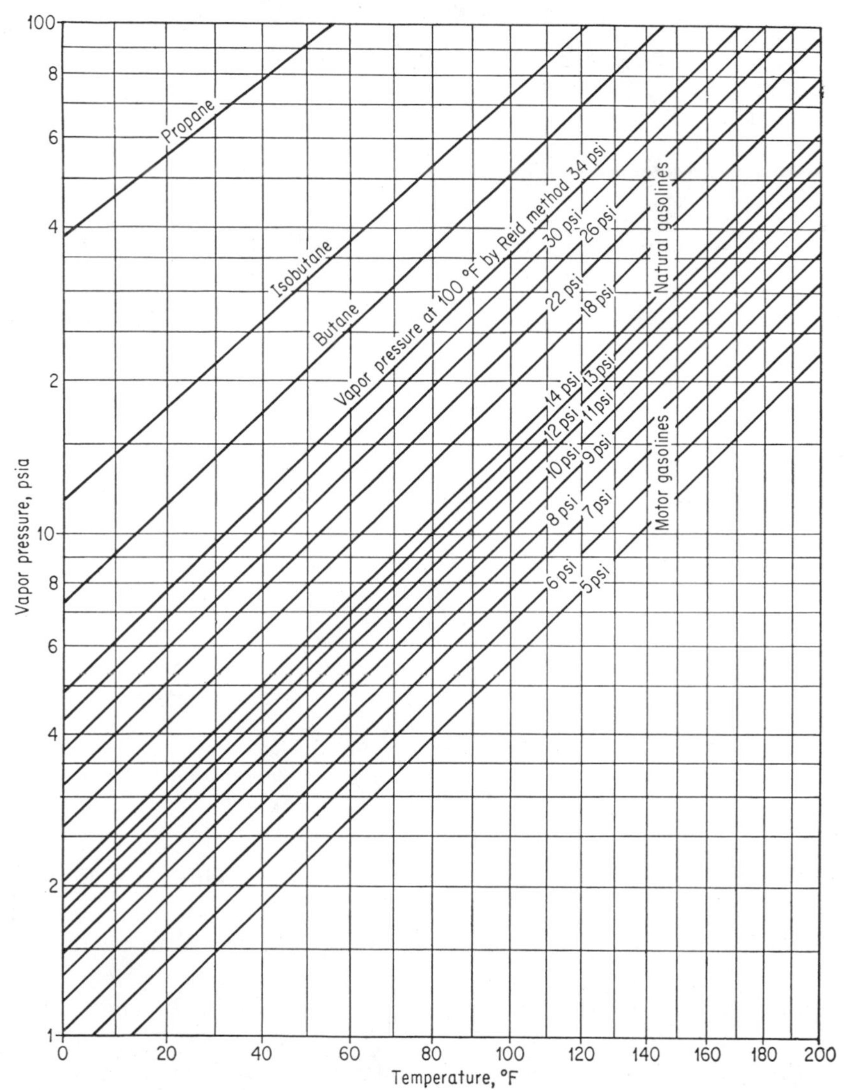

FIG. H-5.—Vapor pressures of motor and natural gasolines. (*Data of Chicago Bridge & Iron Co., "Engineering Data Book," 7th ed., Natural Gasoline Supply Men's Association, Tulsa,* 1957.)

The Natural Gasoline Supply Men's Association recommends a working storage pressure to prevent breathing losses through the vent as given by (assuming p less than Δ)

$$\Phi = P + (\Delta - p)\frac{T + 460}{t + 460}$$

where Φ = storage pressure, psig
P = vapor pressure of liquid at maximum surface temperature, psia
p = vapor pressure of liquid at minimum surface temperature, psia
Δ = pressure in tank at which vacuum vent opens, psia
T = maximum average temperature in air-vapor space, °F
t = minimum average temperature in air-vapor space, °F

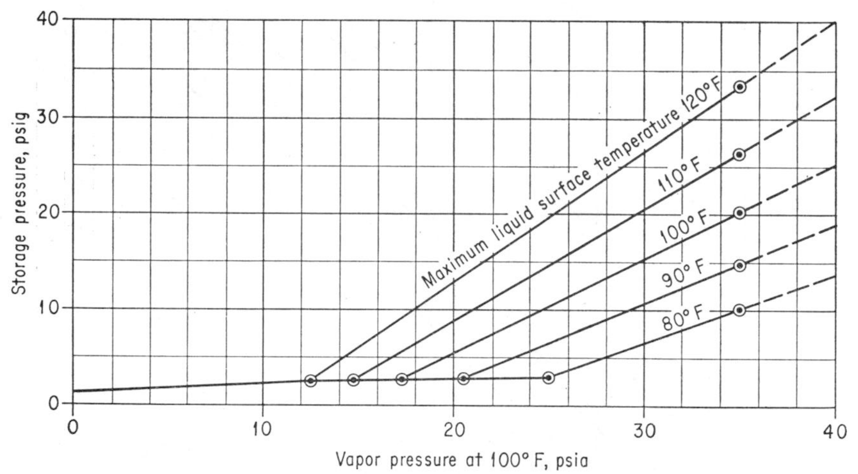

FIG. H-6.—Graph showing storage pressure required to prevent standing losses from volatile products. (*"Engineering Data Book," 7th ed., Natural Gasoline Supply Men's Association, Tulsa*, 1957.)

The storage pressure may be more simply determined by Fig. H-6 which is based on the above relation and the assumptions listed below:

1. Minimum liquid surface temperature 10°F less than maximum liquid surface temperature.
2. Maximum vapor space temperature 40°F greater than maximum liquid surface temperature.
3. Minimum vapor space temperature 15°F less than maximum liquid surface temperature.

These assumptions are extreme and often lead to high answers.

6. Combustion Data and Calculations for Fuels in General

a. Definitions of Heating Value.—There are two common heating values in use: net and gross. The difference between these values arises because one product of combustion, water, is produced as a vapor on burning the fuel but is condensed to liquid on recooling to room temperature.

The net heating value of a fuel is defined as the heat produced on combustion of the fuel at any given temperature (usual practice, 60°F) and assuming that the products of combustion are cooled back to the initial temperature, the water vapor being uncondensed.[1] The gross heating value of a fuel is defined as the total heat developed after the products are cooled back to the starting point, assuming that all of the water is condensed.[1]

In practice, neither heating value can be experimentally realized. A combustion calorimeter does cool the burned gases to nearly the original temperature, condensing most of the water and thus approximating closely the gross heating value as just defined. The differences due to uncondensed water vapor saturating the fixed gases and to solubility of gases in water are cause of concern to the thermochemist but need not concern us here. In actual furnaces the combustion products leave the stack without condensation of water vapor, so that the net heating value is closer to the heat realized by industrial use.

Since both heating values are essentially arbitrarily defined, it follows that choice between the two is largely one of taste. The net heating value is preferred by many European workers, whereas the use of gross heating values is more common to American practice and will be used for most purposes in this book. The units of heat used here are the British thermal unit (Btu) and the therm (100,000 Btu).

b. Excess Air and Furnace Losses.—Table H-13, pages 306–307, presents the heat distribution for several types of combustion equipment. It will be noticed that the stack loss is, in general, the most important item, though other losses frequently are of the same importance.

The stack loss is determined to some extent by the fuel employed, but to a greater extent by the temperature and the quantity of excess air. Figure H-7, or by the equation given with it, permits estimation of the quantity of excess air when carbon monoxide, carbon dioxide, sulphur dioxide, water vapor, oxygen, and nitrogen are the sole components of the stack gas, and the fuel is itself free from nitrogen. The

[1] "Combustion," 3d ed., American Gas Association, New York, 1932, *cf.* A.S.T.M., Standard, D 407-44, for more precise definitions.

TABLE H-13.—HEAT DISTRIBUTION
(Haslam and Russell, "Fuels and Their Combustion,"

	Boiler	Gas producer	Blue water gas set	Carbu-retted water gas set	Oil gas set	Coke oven
1. Heat input (per cent of total heat):						
(a) Potential heat in fuel	91.9	96.3	93.3	Coal 57.3 Oil 38.5	96.7	Coal 92.8 Gas 7.2
(b) Heat in auxiliaries (air and steam)	0.0	2.7	6.7	4.1	3.0	0.0
(c) Heat in raw materials	8.1	0.0	0.0	0.0	0.0	0.0
Total	100.0	100.0	100.0	100.0	100.0	100.0
2. Heat output.						
(a) Heat in product	71.2	72.7	49.9	60.2	42.8	87.5
(b) Heat absorbed in reactions	....	....	.....	....	....	....
(c) Furnace efficiency	71.2	Hot 90.2 Cold 72.7	49.9	60.2	42.8	87.5
(d) Heat losses, total	28.8	Hot 9.8 Cold 27.3	50.1	39.8	57.2	12.5
(1) Stack loss	21.0	Sensible heat in producer gas 17.5	39.2	19.2	7.2	2.3
(2) Cinder loss	3.9	1.7	7.0	4.0	0.0	....
(3) Radiation loss }	3.9	8.1	3.9	3.7	1.4	1.2
(4) Unaccounted-for losses }					1.3	2.5
(5) Cooling water loss	.....	....	....	....	....	....
(6) Miscellaneous Accounted-for losses	.....	....	.....	....	Potential heat in carbon 9.5 Potential heat in tar 3.1	Sensible heat in coke 4.1 Sensible heat in gas 2.2 Sensible heat in tar 0.2
Remarks	.....	....	.....	....	Potential heat in lampblack 26.2 Other losses 8.5	

relatively small nitrogen content of typical fuels (except, perhaps, producer gas) does not, in practice, impair the calculations. Other components are rarely present unless a deficiency of air is employed. Flue gases high in carbon monoxide may, however, also contain some unburned hydrogen and hydrocarbons resulting from local deficiencies of oxygen, particularly if the distribution of air to the burners or to various parts of coal bed is not uniform.

For a given fuel of constant known composition, the excess air may be computed from either the oxygen or carbon dioxide content of the

IN INDUSTRIAL FURNACES
McGraw-Hill Book Company, Inc., New York, 1926)

Blast furnace	Open-hearth furnace	Billet furnace: Non-continuous	Billet furnace: Continuous with recuperator	Rotary cement kiln	Rotary lime kiln	Shaft lime kiln, producer gas fired	Down-draft brick kiln	Tunnel kiln continuous	Tunnel kiln Dressler	Ring-type brick kiln	Blast-furnace stove
93.1	95.0	100.0	100.0	100.0	100.0	95.9	100.0	100.0	100.0	100.0	100.0
6.8	0.0	0.0	0.0	0.0	0.0	4.1	0.0	0.0	0.0	0.0	0.0
0.0	5.0	0.0	0.0	0.0	0.0	0.0	0.0	0.0	0.0	0.0	0.0
100.0	100.0	100.0	100.0	100.0	100.0	100.0	100.0	100.0	100 0	100.0	100.0
8.2	Steel 21.6 Slag 3.1	25.0	52.5	14.1	12.3	12.0	23.7	17.2	67.2	51.5	62.0
28.4	−7.9	......	.....	22.4	28.4	35.0	2.7	7.9	0.0	0.0	0.0
36.6	11.8	25.0	52.5	22.4	28.4	35.0	26.4	25.1	67.2	51.5	62.0
63.4	88.2	75.0	47.5	77.6	71.6	65.0	73.6	74.9	32.8	48.5	38.0
55.0 6.1	62.0	40.0	29.3	48.3	41.2	30.6	33.3	27.9	8.3	38.2	18.6
.....	....	2.0	.....	.....	.....	0.7	0.0	3.3			
0.3	21.2	27.0	16.1	15.2	7.9 10.2	12.4	40.3	43.7	5.4	10.3	10.1
2.0	....	4.0	2.1	.....	.....	4.7	.....	.....	......	.....	2.6
..... Potential heat in blast-furnace gas is not a real loss because later used	....	Heat in slag 2.0	.....	.....	.....	Loss in producer gas main 4.6 Over-all balance including producers and kiln	.....	.....	Heat in trucks 19.1	.....	6.7

stack gas. This result is more accurate if based on oxygen, but carbon dioxide is usually more convenient especially if a carbon dioxide recorder is available. Excess may be computed by the following formula:

$$E = 7900 \frac{b - a}{a(100 - b)}$$

where E = per cent excess air
a = per cent CO_2 (+ CO, if also present)
b = ultimate per cent CO_2 for perfect combustion

Actually this is only one method of estimating. Others are illustrated in Examples H-1 to H-3. Figures H-8 and H-9 permit rapid estimation of stack losses for different fuels as follows: bituminous coal, Fig. H-8; natural gas, Fig. H-9. These charts permit the direct computation of losses and excess air from the per cent of carbon dioxide in stack gas and stack temperature. The accuracy will, however, be improved if the excess air is estimated independently either from Fig. H-7 or by algebraic methods. This is particularly true if the composition of fuel differs from those on which the chart is based. For high accuracy or for fuels not covered by the charts, direct computation is required. Methods are illustrated in Examples H-2 and H-3.

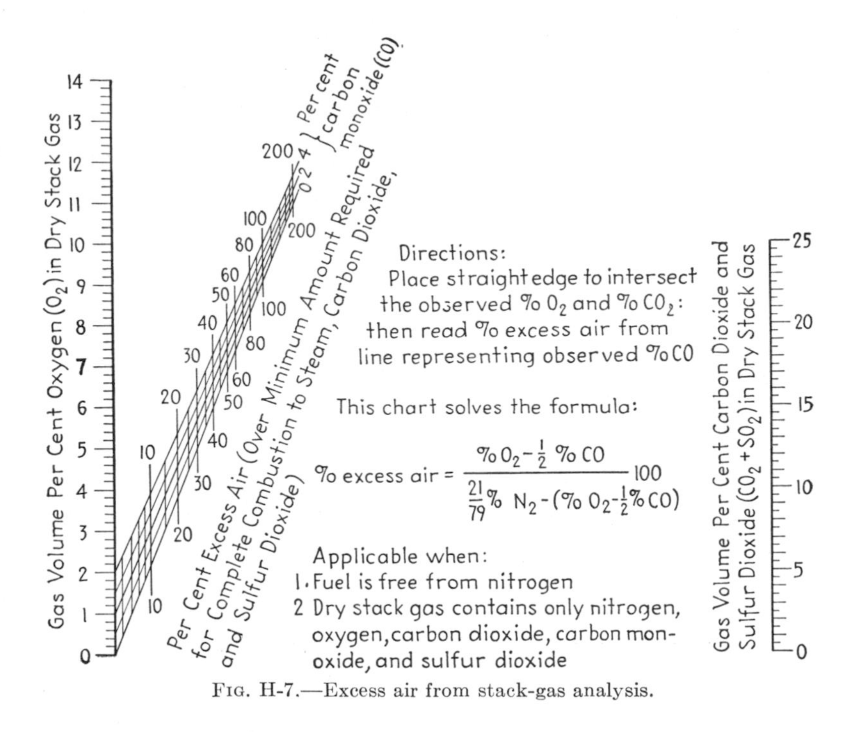

FIG. H-7.—Excess air from stack-gas analysis.

Theoretically, the maximum furnace efficiency is highest when a minimum of excess air is employed. Practical difficulties in securing close regulation of air may, however, dictate a large amount of air to ensure excess air at all times and throughout the furnace, particularly when the air flow is subject to the vagaries of natural draft. In forced-draft installation regulation is easier. Also gas burners can be regulated fairly closely even with natural draft, *provided* a draft break is employed between the furnace flue and the stack. Excess air is frequently employed to reduce the flame temperature (see discussion under that heading) and permit the use of cheaper refractories or avoid the need for water-cooled walls. In some metallurgical and ceramic furnaces the need for a controlled oxidizing or reducing atmosphere dictates the quantity of air employed. For one or more of these reasons excess air ranges from small to over 100 per cent excess air in practice. Unfortunately, no simple rule is available to cover the wide variety of installations that may be encountered.

In stoker-fed furnaces the cinder leaves hot, and still contains appreciable combustible material. For purposes of estimation, the cinder may be assumed to have a mean specific heat of 0.25 Btu/(°F)(lb), and the unburned material may be assumed to be carbon and to have a heating value of 14,000 Btu/lb. Estimating cinder loss is illustrated in Example H-3.

Furnace losses besides stack and cinder are largely heat leak by radiation and convection which may be estimated by standard methods of heat transfer (see Chapter G).

c. Flame Temperature.—The maximum temperature of a flame depends on the nature of fuel, the quantity of air used in its combustion, and, finally, the initial temperature of fuel and air. The flame temperature is useful for estimating firebox temperatures. At these high temperatures heat conduction through furnace walls is so rapid that the surface of the brick is less than that of the flame. A refractory capable of withstanding a temperature of 300°F below the flame temperature will be satisfactory. If the furnace walls are water-cooled, this difference may, of course, be considerably greater.

Figure H-10 gives flame temperatures of Pittsburgh natural gas with various quantities of air. Table H-9 gives the flame temperature of a number of commercial gases when burned with no excess air. Figure H-11 shows the effect of preheating air and fuel. It will be noticed that the temperatures do not vary markedly and that Fig. H-10 may be used for rough estimates of flame temperatures of a number of gaseous fuels.

Flame temperatures of liquid fuels would be expected to be lower than those of gaseous fuels owing to the heat used in vaporizing the fuel; however, because of the relatively higher combustion values of the heavy hydrocarbons, the difference may be small (see Example H-4). Flame temperatures of coal and coke are lower because of their oxygen content.

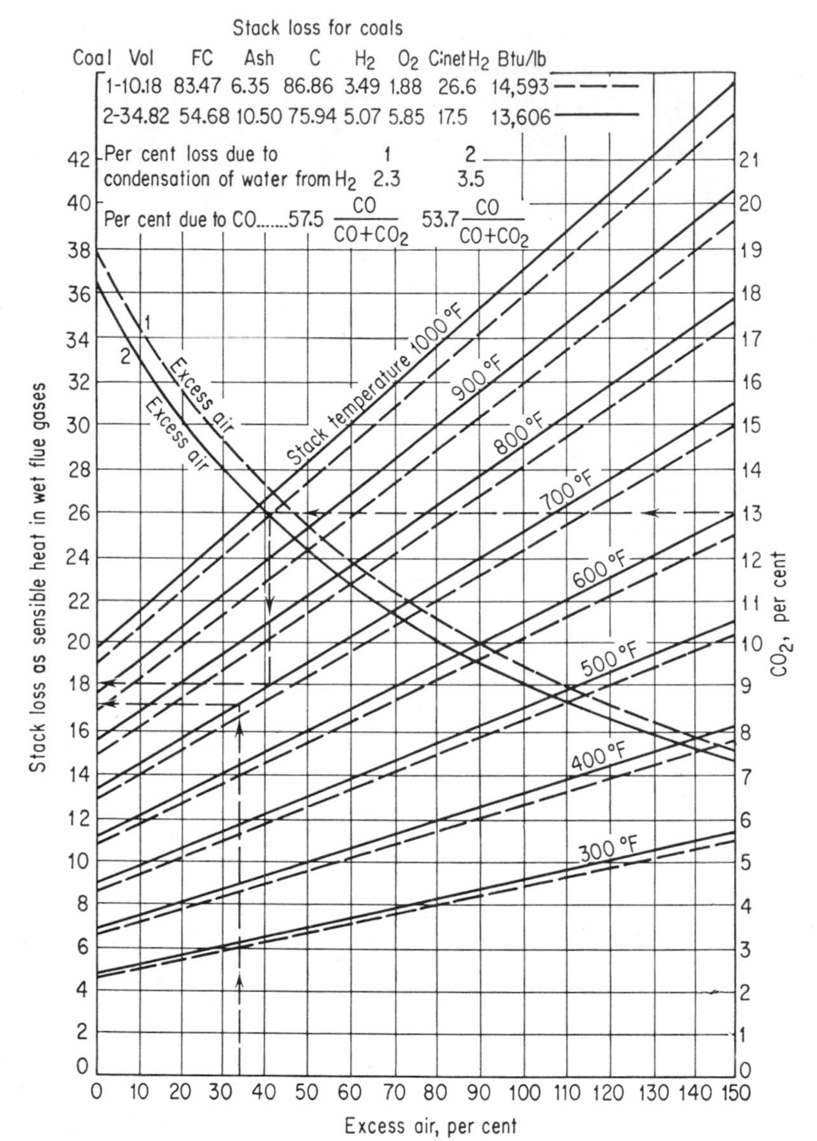

FIG. H-8.—Chart for graphical determination of per cent excess air and per cent stack loss in a furnace fired with a high-volatile or low-volatile bituminous coal. (*Haslam and Russell, "Fuels and Their Combustion," McGraw-Hill Book Company, Inc., New York*, 1926.)

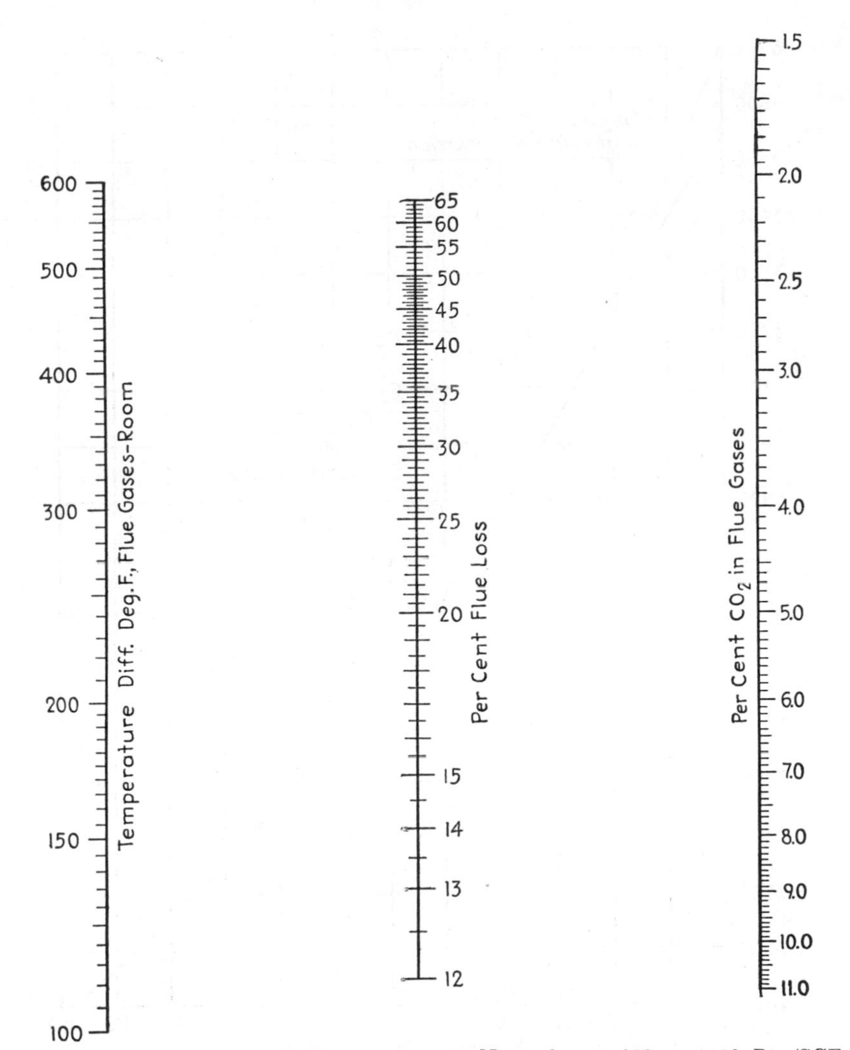

FIG. H-9.—Alignment chart for flue losses. Natural gas 1060 to 1190 Btu/SCF. (*"Fuel-Flue Gases," American Gas Association, New York,* 1940.)

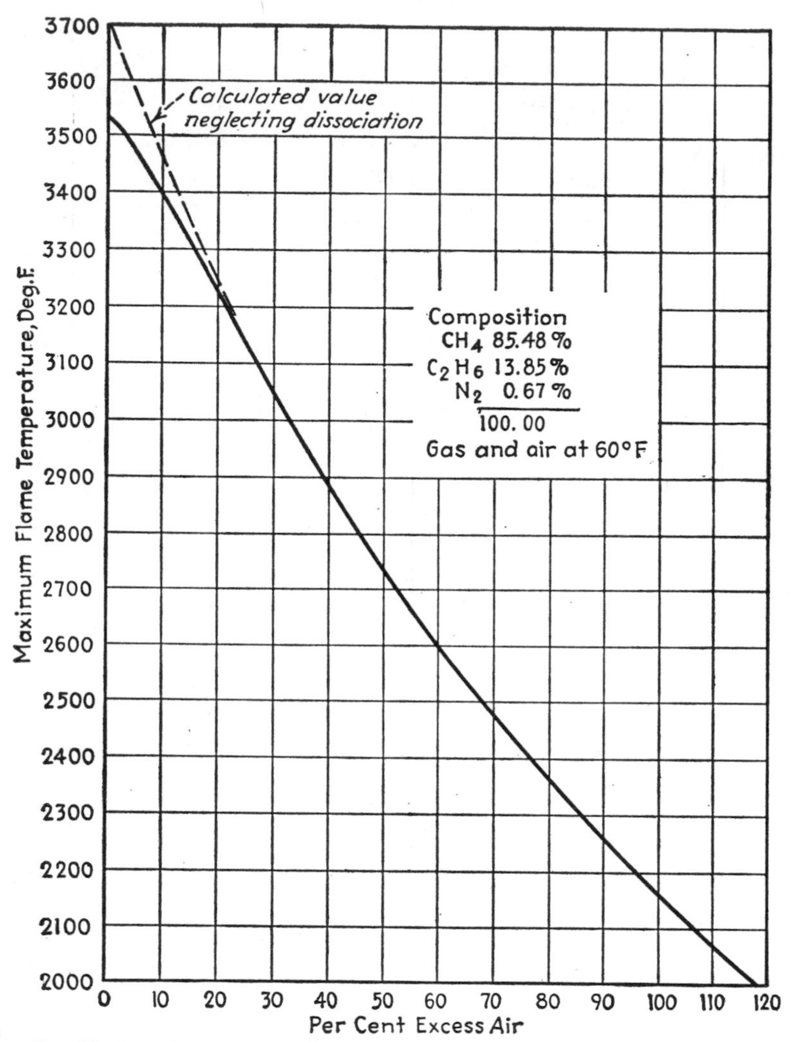

FIG. H-10.—Flame temperatures of Pittsburgh natural gas and air mixtures.

The calculation of flame temperatures is theoretical. However, calculated values have been closely approached (at least for fast-burning gases) by experimental temperatures which are but slightly lower because of small unavoidable heat losses. If no heat is lost, the heat content of the products of combustion must be equal to the heat of combustion plus the heat content of the fuel and air (or oxygen). At the high temperatures frequently involved, however, carbon dioxide and steam dissociate, oxides of nitrogen may be present, and even

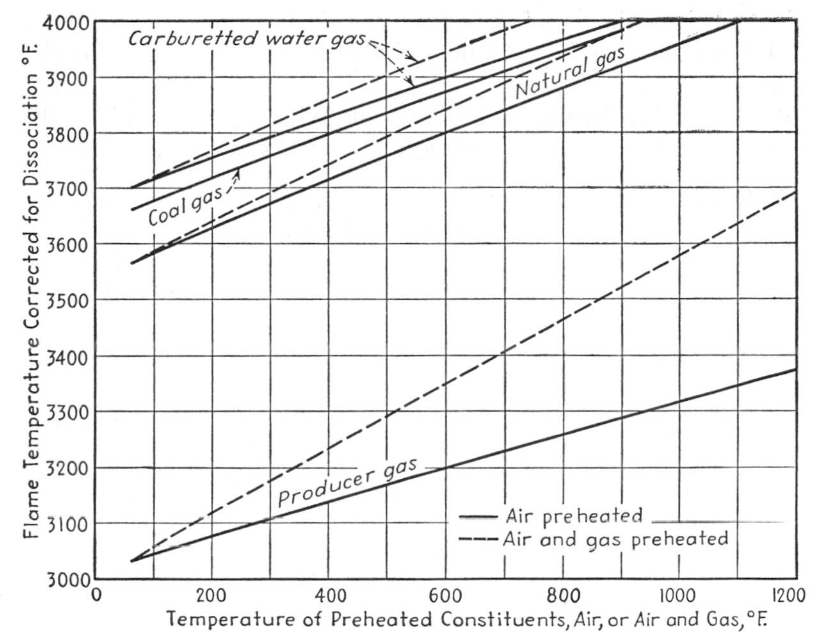

FIG. H-11.—Effect of preheat on flame temperature. ("*Combustion*," 3d *ed.*, *American Gas Association*, *New York*, 1932.)

atomic hydrogen and oxygen may be formed. Correction for the heats of these reactions must be made, and the computations are rendered very tedious so that details of the exact method will be omitted. Persons requiring data for exact calculations may find methods outlined in good texts on physical chemistry. An excellent and very readable treatment is given by the American Gas Association.[1] A very scholarly discussion based on good thermodynamic data has been given by Lewis and von Elbe.[2]

The error caused by neglect of dissociation is illustrated by the dotted line in Fig. H-10. It will be noted that, for no excess air, the discrepancy with the exact value is about 200°F and becomes negligible at 25 per cent excess air. This is due partly to the lower temperature and partly to repression of dissociation by the free oxygen contained in the excess air. Consequently it may be seen that if more than 20 per cent excess air is used, neglect of dissociation is justifiable for most process purposes, provided the temperatures are in the range of Fig. H-10. If the flame temperature is much higher because of the use of

[1] "Combustion," *op. cit.*

[2] *Phil. Mag*, **20,** 44 (1935); *J. Am. Chem. Soc.*, **57,** 612, 2737 (1934).

preheated air or the use of pure oxygen, this may not be true. Example H-4 gives one method of calculation of flame temperatures neglecting dissociation.

d. Illustrative Examples.—The underlying principles of combustion calculations are very simple and are outlined below.

1. Combustion reactions:
   $C + \frac{1}{2}O_2 = CO$ (incomplete combustion)
   $C + O_2 = CO_2$
   $2H + \frac{1}{2}O_2 = H_2O$
   $H_2 + \frac{1}{2}O_2 = H_2O$
   $S + O_2 = SO_2$
   $P + \frac{5}{4}O_2 = \frac{1}{2}P_2O_5$
   $C_cH_hO_oN_nS_s + (c + \frac{1}{4}h - \frac{1}{2}o + s)O_2 = cCO_2 + \frac{1}{2}hH_2O + \frac{1}{2}nN_2 + sSO_2$
2. Material balance:
   Carbon in = carbon out
   Hydrogen in = hydrogen out
   Etc.
3. Heat balance:
   Heat in = heat out
   Heat in includes sensible heat and heat of combustion
   Heat out includes sensible heat and heat of combustion of unburned material; also includes latent heat of water vapor if gross heat of combustion is used

These principles together with data on fuels (such as given in earlier sections of the chapter and in Chapter E) permit computation of heat losses, fuel and air required to supply needed heat, flame temperature heat losses, and similar problems. Methods of application are probably better shown by illustrations than by abstract discussion. Five examples follow.

Example H-1. Compute the combustion products and stack-gas analysis from one pound of Knife River Lignite when completely burned with 40% excess air. Composition of this coal may be taken from Table H-2.

SOLUTION:

1. Weight of ash- and moisture-free coal (lignite),

% moisture	35.6
% ash	5.6
Total	41.2

Ash- and moisture-free coal = 100 − 41.2 = 58.8% or 0.588 lb/lb

2. Air required (factors from Table H-4) to burn,

C: $0.588 \times 0.720 \times 150$	63.5 SCF/lb
H: $0.588 \times 0.047 \times 448$	12.4
O: $0.588 \times 0.211 \times (-56.3)$	−7.0
S: $0.588 \times 0.012 \times 56.2$	0.4
Subtotal	69.3
40% excess $= 0.40 \times 69.3$	27.7
Total	97.0 SCF/lb

3. Nitrogen in flue gas,

From air $= 0.79 \times 97.0$	76.6 SCF/lb
From fuel $= 0.588 \times 0.010 \times 13.5$	0.1
Total	76.7

4. Oxygen in flue gas,

Oxygen in excess air $= 0.21 \times 27.7$ 5.82

5. Carbon dioxide in flue gas,

$0.588 \times 0.720 \times 31.4$ 13.3

6. Sulphur dioxide in flue gas,

$0.588 \times 0.012 \times 11.7$ 0.08

7. Flue products exclusive of water vapor,

Total of items 3 to 6 .... 95.9

8. Composition of dry stack gas,

$\% (CO_2 + SO_2) = 100 \dfrac{13.3 + 0.08}{95.9}$	14.0
$\% O_2 = 100 \dfrac{5.82}{95.9}$	6.1
$\% N_2 = 100 \dfrac{76.7}{95.9}$	79.9
Total	100.0

9. Water vapor in flue gas,
From burning H,

$0.588 \times 0.047 \times 189$ .... 5.2 SCF/lb

From moisture in coal,

0.356×21.02	7.5
Total	12.7 SCF/lb

10. Total flue products,

Items 7 and 9 = 108.6 SCF/lb of coal

Example H-2. A furnace is fired with Los Angeles natural gas of composition given by Table H-9. The stack temperature is 620°F, and the stack gas is 9.1% CO_2, 0.0% CO, and 6.1% O_2. Find the per cent gross heating value lost as heat content of stack gas (stack loss).

SOLUTION:

1. Stack products from 1 SCF of fuel excluding excess air (data from Table H-9),

N_2	7.94 SCF/SCF
CO_2	1.16
Subtotal	9.10
H_2O	2.10
Total	11.20 SCF/SCF

2. Excess air computed from % CO_2,
Let x = SCF excess air/SCF of fuel.

$$\% \; CO_2 = 100 \frac{1.16}{9.10 + x} = 9.10$$

$$9.10 + x = \frac{116}{9.1} = 12.75$$

$$\text{Excess air} = 12.75 - 9.1 = 3.65 \text{ SCF/SCF of fuel}$$

$$\left(\% \text{ excess air} = 100 \frac{3.65}{10.05} = 36.3\% \right)$$

3. Excess air computed from % O_2,

Flue gas contains 6.1% O_2, or 6.1/0.21 = 29.0% air.

$$\text{Volume of excess air} = \frac{29}{100 - 29} 9.10 = 3.72 \text{ SCF/SCF of fuel}$$

($\%$ excess air $= 100 \dfrac{3.72}{10.05} = 37.0\%$, and this value agrees with that computed from % CO_2)

4. Total combustion products,

Use excess air as computed from % O_2.

$$\text{Total } N_2 = 7.94 + 0.79 \times 3.72 = 7.94 + 2.94 \ldots\ldots 10.88 \text{ SCF}$$

$$O_2 = 0.21 \times 3.72 \ldots\ldots 0.78 \text{ SCF}$$

(CO_2, H_2O from item 1)

5. Heat content of flue products,

Gas	SCF	Btu/SCF (from Fig. E-10)	Btu
N_2	10.88	10.4	113.1
O_2	0.78	10.8	8.4
CO_2	1.16	15.1	17.5
H_2O	2.10	12.3	25.8
Total sensible heat above 60°F.			164.8 Btu/SCF fuel
Latent heat,			
H_2O	2.10	50.36	105.6
Total heat content above 60°F.			270.4 Btu/SCF fuel

This is the stack loss.

6. % stack loss, $100 \dfrac{270.4}{1073} = 25.2\%$

Example H-3. Find the quantity of coal required for firing a calcining kiln. The heat required for the calcining process is 4,500,000 Btu/hr The coal to be used is very similar to the Lower Freeport coal listed in Table H-2. From similar type installations, using a different coal, it was found that (1) the stack analysis was 11.0% CO_2, 1.2% CO, and 6.6% O_2; (2) the stack temperature was 800°F; (3) the cinder contained 22% combustible; and (4) heat loss by radiation, convection, and miscellaneous factors was about 8% of the heating value of the fuel.

SOLUTION:

Method: Find heat supplied per pound of coal and from this the amount required.

1. Heating value of coal = 14,310 Btu/lb
2. Cinder formed from 1 lb coal,
   Ash content = 0.064 lb

$$\text{Total cinder formed} = \frac{0.064}{1 - 0.22} = 0.082 \text{ lb}$$

   Combustible left in cinder = $0.22 \times 0.082 = 0.0180$ lb
3. Cinder loss per pound of coal,

Sensible heat (at 1000°F) = $0.082 \times (1000 - 60) \times 0.25$ (taking specific heat as 0.25). . . . . 19 Btu/lb
Heat of combustion = $0.018 \times 14{,}000$. 252
Total loss. . . . . 271 Btu/lb

or

$$100 \frac{271}{14{,}310} = 1.9\%$$

4. Combustible material per pound of coal,

Ash	0.064 lb
Moisture	0.020
Total	0.084

Ash- and moisture-free coal = 1 − 0.084 = 0.916 lb

C = 0.916 × 0.886*	0.812
H = 0.916 × 0.053	0.048
O = 0.916 × 0.038	0.035
N = 0.916 × 0.015	0.014
S = 0.916 × 0.008	0.007
Total	0.916

* 0.886 = % C/100, etc.

Carbon actually burned,

$$0.812 - 0.018 = 0.794 \text{ lb/lb coal}$$

5. Products of perfect combustion per pound of coal excluding excess air,

CO_2 from C = 0.794 × 31.38		24.9 SCF/lb
H_2O from moisture = 0.020 × 21.02	0.42	
H_2O from H = 0.048 × 188.7	9.05	
Total water vapor		9.5
SO_2 from S = 0.007 × 11.69		0.1

Nitrogen from

C = 0.794 × 118.5	94.1	
H = 0.048 × 353.6	16.9	
N = 0.014 × 13.5	0.2	
S = 0.007 × 44.4	0.3	
Subtotal	111.5	
Less credit for oxygen in coal, −0.035 × 44.5	−1.5	
Total nitrogen		110.0
Total products of combustion		144.5

(Factors used above are from Table H-4, except specific volume of steam, which is from Table H-10.)

6. Correction for carbon monoxide,

$CO_2 + CO + SO_2 = 24.9 + 0 + 0.1 = 25.0$ SCF/lb coal (result of 5)

If CO in the proportion represented by 1.2% CO and 11.0% CO_2

$$\text{CO formed} = 25.0\,\frac{1.2}{11 + 1.2} \ldots\ldots\ldots\ldots\ldots \quad 2.4 \text{ SCF/lb coal}$$

$$\text{Leaving } CO_2 + SO_2 = 25 - 2.4 \ldots\ldots\ldots\ldots \quad 22.6 \text{ SCF/lb coal}$$

Nitrogen produced in combustion of CO

$$= 1.9 \text{ SCF/SCF of CO (Table H-11)}$$

or

$$1.9 \times 2.4 = 4.6 \text{ SCF/lb coal}$$

This nitrogen was included in calculation 5 but should be omitted as the CO was not actually burned.

Corrected flue products are then,

$$N_2 = 110.0 - 4.6 = 105.4 \text{ SCF/lb coal}$$
$$\text{Total} = 144.5 - 4.6 = 139.9$$
$$H_2O = 9.5$$

Exit stack gas, dry basis = 139.9 − 9.5 = 130.4 SCF/lb

7. Excess air,

Let y = SCF of excess air/lb of coal

$0.21y$ = SCF of excess oxygen/lb of coal

$$\%\ O_2 = 6.6 = 100\,\frac{0.21y}{130.4 + y}$$

$$861 + 6.6y = 21y$$
$$14.4y = 861$$
$$y = 59.8 \text{ SCF/lb coal}$$

Oxygen = 0.21 × 59.8	12.6 SCF/lb
Nitrogen = 0.79 × 59.8	47.2
Total	59.8 SCF/lb

8. Total products of combustion including excess air,

	SCF/lb	%
$CO_2 + SO_2$ (from calculation 6)	22.6	11.9
CO (from calculation 6)	2.4	1.3
O_2 (from calculation 7)	12.6	6.6
N_2 = 47.2 + 105.4 (from 6 and 7)	152.6	80.2
Subtotal	190.2	100.0
H_2O	9.5	
Total	199.7	

It will be noted that the % CO_2 and CO do not quite agree with the stack-gas analysis given. This is not surprising in view of inaccuracies of gas analyses and the difference between this coal and the one previously used.

9. Stack loss per pound of coal,
Sensible heats (from Fig. E-11) at 800°F.

	Btu/SCF	SCF	Btu
$CO_2 + SO_2$	20.6	22.6	466
CO	13.9	2.4	33
O_2	14.5	12.6	183
N_2	14.0	152.6	2,136
H_2O	16.4	9.5	156
Subtotal	.....	.....	2,974
Condensing water	50.4	9.5	479
Subtotal	.....	.....	3,453
Combustion of CO	321	2.4	771
Total	.....	.....	4,224

10. Total losses and heat supplied per pound of coal,

	Btu/lb coal	% of heating value
Cinder	271	1.9
Stack	4,224	29.6
Heat leak, etc	1,145	8.0
Total	5,640	39.5
Heating value of coal	14,310	
Net heat supplied	8,670	

11. Coal required,

$$\text{Heat required} = 4{,}500{,}000 \text{ Btu/hr}$$

$$\text{Coal} = \frac{4{,}500{,}000}{8670} = 520 \text{ lb/hr}$$

or

$$520 \frac{24}{2000} = 6.24 \text{ tons/24-hr day}$$

Example H-4. A furnace is being converted from coal to oil. The refractories in the firebox are considered safe for a flame temperature of 2600°F. Find the minimum of excess air required to assure a flame temperature of 2600°F or below. The oil to be used has gross heating value of 17,500 Btu/lb and may be taken as 90% carbon and 10% hydrogen (neglecting sulphur, oxygen, and nitrogen also present).

SOLUTION:

1. Products of perfect combustion of 1 lb of oil (factors from Table H-4),

CO_2 from C 0.90×31.38 . . . . 28.2 SCF/lb
H_2O from H 0.10×188.7 . . . . 18.9
N_2 from C 0.90×118.5 . . . . 106.6
N_2 from H 0.10×353.6 . . . . 35.4
Total N_2 . . . . 142.0
Total . . . . 189.1 SCF/lb

2. Heat content of these products at 2600°F,

Sensible heats from Fig. E-12

	SCF	Btu/SCF	Btu/lb oil
CO_2	28.2	84.0	2,368
N_2	142.0	52.0	7,384
H_2O sensible	18.8	65.4	1,230
H_2O condensing	18.8	50.4	948
Total	.....	....	11,930

3. Heat content of air at 2600°F,

O_2 0.21×54.8 . . . . 11.5
N_2 0.79×52.0 . . . . 41.1
Total . . . . 52.6 Btu/SCF

4. Excess air,

Heat of combustion . . . . 17,500 Btu/lb
Less heat of products 11,930
Residue 5,570

For heat balance, this residue must equal the heat content of excess air,

$$\text{Excess air, then} = \frac{5570}{52.6} = 106.0 \text{ SCF}$$

$$\text{Air for perfect combustion} = \frac{N_2}{0.79} = \frac{142}{0.79} = 180 \text{ SCF}$$

$$\% \text{ excess air} = 100\,{}^{106}\!/_{180} = 59\%$$

5. Composition of stack gas,

CO_2 from calculation 1 . . . . 28.2 SCF
O_2 from calculation 4 $= 0.21 \times 106$ 22.3
N_2 from calculation 4 $= 0.79 \times 106$ 83.7
N_2 from calculation 1 . . . . 142.0
Total 276.2

$$\% \ CO_2 = 100 \frac{28.2}{276.2} \cdots\cdots \quad 10.2$$

$$\% \ O_2 = 100 \frac{22.3}{276.2} \cdots\cdots \quad 8.1$$

$$\% \ N_2 = 100 \frac{83.7 + 142}{276.2} \cdots\cdots \quad 81.7$$

Total.. 100.0

Example H-5. Compute factors similar to those in Table H-4 for phosphorus.

SOLUTION:

1. Reaction,

	4P +	$5O_2$ =	$2P_2O_5$
Molecular weight..................	30.98	32.00	141.96
No. of moles......................	× 4	× 5	× 2
Pounds............................	123.92 +	160.00 =	283.92

2. Weight ratios:

$$\text{Lb } P_2O_5/\text{lb P} = \frac{283.92}{123.92} \cdots\cdots \quad 2.2912$$

$$\text{Lb } O_2/\text{lb P} = \frac{160}{123.92} \cdots\cdots \quad 1.2912$$

$P_2O_5 - O_2 = P$.......................... 1.000 (*check*)

3. Volume of oxygen on basis of perfect gas law,

123.92 lb phosphorous = 5 moles of oxygen

or

5 × 378.7 SCF (from definition of standard cubic foot)

$$\text{SCF oxygen/lb P} = \frac{5 \times 378.7}{123.92} = 15.280$$

4. Volume of air and nitrogen

Take air as 21% O_2, 79% N_2.

$$\text{Air} = \frac{15.280}{0.21} = 72.76$$

$$\text{Nitrogen} = 72.76 \times 0.79 = 57.48$$

Check: $O_2 + N_2 = 57.48 + 15.28 = 72.76$

5. Weight of air,

1 lb air = 13.074 SCF of perfect gas

$$\text{Weight air} = \frac{72.76}{13.074} = 5.56 \text{ lb/lb P}$$

Check: Air is 23.2 weight % O_2 and weight air $= \frac{1.291}{0.232} = 5.56$ lb/lb P.

6. Weight nitrogen:

Air − oxygen = 5.56 − 1.29 = 4.27 lb/lb P

CHAPTER I

POWER

Although the process engineer is rarely required to design a power plant, he is called on to state the process utility demands in a form digestible by the power-plant engineer. This may necessitate a revision of part or all of the process to secure a balance between the various forms of heat and power that favor a low over-all cost. Information here given is intended for these purposes rather than for final power-plant design. Power and its allied subjects have been well treated by many engineers. A few useful references are given below.

Barnard, Ellenwood, and Hirschfeld, "Elements of Heat-Power Engineering," John Wiley & Sons, Inc., New York, 1933.

Butterfield, Jennings, and Luce, "Steam and Gas Engineering," D. Van Nostrand Company, Inc., Princeton, N.J., 1947.

Norris, Therkelsen, and Trent, "Applied Thermodynamics," 3d ed., McGraw-Hill Book Company, Inc., New York, 1955.

Powell, "Water Conditioning for Industry," McGraw-Hill Book Company, Inc., New York, 1954.

Newell, "Diesel Engineering Handbook," 9th ed., Diesel Publications, New York, 1955.

Creager and Justin, "Hydro-electric Handbook," 2d ed., John Wiley & Sons, Inc., New York, 1950.

Barrows, "Water Power Engineering, 3d ed., McGraw-Hill Book Company, Inc., New York, 1943.

A.S.M.E. Power Test Codes, American Society of Mechanical Engineers, New York.

Keenan and Keyes, "Thermodynamic Properties of Steam," John Wiley & Sons, Inc., New York, 1936.

Allen and Bursley, "Heat Engines," 5th ed., McGraw-Hill Book Company, Inc., New York, 1941.

Griswold, "Fuels, Combustion and Furnaces," McGraw-Hill Book Company, Inc., New York, 1946.

Johnson and Auth, "Fuels and Combustion Handbook," McGraw-Hill Book Company, Inc., New York, 1951.

Solberg, Cramer, and Spalding, "Elementary Heat Power," 2d ed., John Wiley & Sons, Inc., New York, 1952.

1. Boilers

a. Boiler Efficiency.—The fraction of the heating value of the fuel burned which is transferred to the boiling water is the boiler efficiency. The remainder of the heating value represents the stack loss, unburned fuel, and heat losses to air by conduction and radiation. The boiler efficiency is almost universally based on the gross heating value of the fuel. This can be realized only if the water vapor in the flue gases is condensed, as it is in an efficient calorimeter used for determining the heating values. Definitions of heating values are given in Chapter H. Blake and Purdy[1] state that sustained efficiencies of 85 to 88 per cent (for a combined boiler unit comprising boilers, economizers, and air preheaters) are common in larger central-station units but that sustained efficiencies of 80 per cent are excellent for industrial boiler plants.

Values for efficiency range from 15 to 25% for small domestic boilers used for hot water supply; from 30 to 50% for coal boilers such as would be used for hot water supply for large buildings, apartment houses, hotels, etc.; 30 to 60% for domestic and small industrial heating boilers; 40 to 70% for small power boilers, not over 50 H.P.; from 50 to 70% for moderate size boilers, up to several hundred horsepower; and from 70 to 85% for large power boilers operating under high-load factor and moderate overloads.[2]

Natural gas or oil probably gives better efficiencies in average boiler-plant installations. However, the theoretical superiority of high-rank coal is actually realized in the most efficient units. The lower rank coals give poorer efficiencies than either bituminous coal or gas.

b. Fuel Requirements.—The over-all heat requirements of a boiler unit (including economizer, if used) may be readily computed from the thermodynamic properties of steam and water. Direct calculation gives the amount of heat received by the water, and this divided by the boiler efficiency is the total supplied by fuel. This is illustrated by Example I-1.

Example I-1. Find the net heat to be supplied for generating 150,000 lb/hr of steam generated at 285 psig (300 psia) and superheated to 800°F. Make a preliminary estimate of the coal to be burned. The available coal has a heating value of 13,600 Btu/lb. 100,000 lb/hr of condensate return is available; the remainder is made up from treated water.

[1] Perry, "Chemical Engineers' Handbook," 3d ed., McGraw-Hill Book Company, Inc., New York, 1950.

[2] From "Combustion," 3d ed., American Gas Association, New York, 1932.

SOLUTION:

1. Heating 100,000 lb/hr condensate to saturation temperature and pressure (data from Tables E-13 and E-14).

Heat content of water saturated at 300 psia.....	393.8 Btu/lb
Heat content of water at 170°F................	137.9
Net heat..................................	255.9

or

For 100,000 lb $\frac{100{,}000}{100{,}000} \times 255.9$.............. 255.9 therm/hr

2. Heating 50,000 lb/hr treated water to saturation temperature,

Heat content of water saturated at 300 psia.....	393.8 Btu/lb
Heat content of water at 70°F.................	38.0
Net heat..................................	355.8

or

For 50,000 lb, $\frac{50{,}000}{100{,}000} \times 355.8$.............. 177.9 therm/hr

3. Generation of 150,000 lb of steam from saturated water,

Heat content of steam at 300 psia and 800°F...	1420.6 Btu/lb (Table E-15)
Heat content of saturated water..............	393.8
Net heat................................	1026.8

or

For 150,000 lb, $\frac{150{,}000}{100{,}000} \times 1026.8$........... 1540.2 therm/hr

4. Total heat required,

Sum of above................................ 1974.0 therm/hr

5. Boiler horsepower $= \frac{1974}{0.33475}$

$= 6000$ (assuming no economizer employed)

6. Total heating value of fuel required

$\frac{1974}{0.70} = 2820$ therm/hr or 67,700 therm/day

(for an assumed boiler efficiency of 70%)

7. Estimated coal consumption,

$$\frac{67{,}700 \times 100{,}000}{13{,}600} = 492{,}000 \text{ lb/day}$$

or 246 tons/day.

8. For a 350-day operating year this amounts to

$$246 \times 350 = 86{,}100 \text{ tons/yr}$$

c. Boiler Rating.—Boilers are usually rated in boiler horsepower. One boiler horsepower is 33,475 Btu/hr. Early practice was to consider 10 sq ft of heat-transfer surface equivalent to a boiler horsepower. In recent years higher boiling rates are common, and operation at 200 or 300 per cent is usual. When operated at the highest rates, the boiler efficiency may be impaired slightly and the life of the boiler tubes shortened. Nevertheless Blake and Purdy[1] suggest 250 per cent as fair industrial practice. The A.S.M.E. Power Test Codes suggest a direct *maximum* rating as thousands of Btu per hour.

d. Boiler Feed Water.—The proper treatment of the boiler water is very important for both boiler performance and life. Calcium and magnesium sulphates and acid carbonates deposit boiler scale, which impairs efficiency and endangers tubes by possibility of overheating. Even though these constituents are removed by precipitation (by lime, other reagents, or zeolites), the water may still be hazardous because of the so-called "caustic embrittlement." This embrittlement is favored by sodium hydroxide, carbonates, and silicates but retarded by sodium sulphate and certain organic materials, such as quebracho.[2] In short, a careful analysis of the available water supply should be made and submitted to a qualified expert for recommendations regarding proper treatment. This is particularly true for higher-pressure boilers.

e. Dowtherm Boilers.—The use of Dowtherm (several types are supplied by the Dow Chemical Co.) as a medium of heat transfer has become common practice. The boilers are similar in design to steam boilers. (A small amount of fixed gas is liberated on use of Dowtherm so that a vent must be provided in the return lines.) Data on these boilers are readily obtainable from licensed manufacturers.

f. Boiler Auxiliaries.—The use of air preheaters in industrial plants is not very common unless powdered coal is employed as a fuel. To some extent air preheaters, stokers, and oil and gas burners are discussed under Combustion in Chapter H. Pumps and blowers are treated in Chapters J and K, respectively.

The recovery of heat from waste gas from steel mills, glass works, or exhaust from diesel or gas engines may be accomplished by waste-heat boilers, provided the gas is available at a sufficiently high temperature, say 1000°F or over. The exit temperature is usually reduced to 300 or 500°F. Forced draft is nearly always required to secure sufficiently good heat-transfer rates. Table I-1 gives data on waste-heat boilers operating on exhaust from diesel engines.

[1] Perry, *op. cit.*

[2] Schroeder, Berk, and O'Brien, various papers in the *Trans. Am. Soc. Mech. Engrs.* from 1934 on.

TABLE I-1.—WASTE-HEAT BOILER PERFORMANCE ON DIESEL ENGINE EXHAUST (Perry, Chemical Engineers' Handbook, 2d ed., McGraw-Hill Book Company, Inc., New York, 1941)

Lb of steam per hr, full load; pressure 5 psi			Sq ft of heating surface	
Hp	4-cycle	2-cycle	4-cycle	2-cycle
75	77	63	72	72
100	92	78	72	72
200	154	165	72	144
300	212	229	72	144
400	327	300	144	192
500	392	370	144	240
600	483	450	192	288
700	550	525	192	336
800	638	600	240	384
900	700	675	240	432
1000	780	750	288	480

Economizers for preheating the water by the flue gas are usually included in very large central-station units but very frequently omitted in small industrial units to obviate the need for induced draft. The equipment manufacturer should be consulted.

2. Steam Engines and Turbines

a. Types of Steam Engines and Turbines.—Steam engines are classified by the number of cylinders, the valve mechanism, speed, and exhaust pressure. The simple steam engine expands steam in one stage by a single piston. Most engines are of this type. Larger engines are usually compounded, particularly if condensing. A compound engine expands steam in two or more stages; the exhaust from the first-stage cylinder drives the second-stage cylinder. Tandem compound cylinders are in line, with a common piston rod; cross compound cylinders are side by side driving separate piston rods. The principal types of valve mechanisms are slide, Corliss, nonreleasing Corliss, and Unaflow. A condensing engine exhausts at low pressures, the vacuum being secured by condensing the steam and by removal of small amounts of fixed gases in the steam by steam ejectors or vacuum pumps. Condensing engines are rather uncommon. Noncondensing engines exhaust at atmospheric pressure or above so that the exhaust may be used to supply low-pressure process-steam requirements.

Like steam engines, turbines are classified as condensing or noncondensing. In addition there is the bleeder turbine which exhausts a portion of the steam between stages to obtain exhaust steam and various pressures. Turbines are classified further according to the number and type of stages. An impulse stage operates primarily from the force of steam velocity from the steam jets. Reaction stages operate chiefly from the direct pressure force of expanding steam.

The steam turbine has almost completely replaced the steam engine in large power plants, and even small electric generators are usually turbine-driven. For small units the steam engine is frequently attractive. Steam engines have a greater flexibility in regard to speed as certain "critical" velocities of turbines may have to be avoided. Bleeder turbines, on the other hand, afford the maximum flexibility as to exhaust because variable quantities of exhaust steam may be withdrawn at different pressures. For driving individual units, the choice of drive is usually determined by mechanical considerations such as the speed required and the type of motion. Turbines are ideal for driving generators, centrifugal compressors, etc. Reciprocating pumps and compressors are usually engine-driven.

The lubricating oil from a turbine or steam engine contaminates the exhaust. This oil (unless completely removed) tends to foul heat-transfer surfaces. This contamination is much less serious in the case of turbines and is sometimes a factor favoring their selection.

b. Efficiency and Steam Consumption.—The steam consumption depends not only on the inlet and exhaust conditions of the steam but also on the mechanical factors determining the relative efficiency. To simplify tabulations, a standard basis for comparing performance is desirable. The ideal steam engine (Rankine cycle) affords such a basis. This ideal engine has never been achieved but is rigorously defined by thermodynamic considerations of the maximum theoretical power developed by expansion of steam. The number of pounds of steam required to develop 1 hp-hr is called the "ideal" steam rate. Numerical values are given in Table I-2 and Fig. I-1. Steam rates of an actual engine may be obtained by dividing this ideal rate by its efficiency. This is essentially the definition of steam engine (or turbine) efficiency. Stated in other terms, the Rankine efficiency of an engine is the ratio of developed power to the power that would have been developed by the ideal engine operating under identical conditions.

Table I-3 gives *representative* data on efficiencies of steam turbines, and Fig. I-2 gives representative efficiencies of simple (single-stage) steam engines. The practical efficiencies of steam turbines are usually dictated by cost considerations. The several types of steam engines vary appreciably in efficiency, particularly with the compounding.

TABLE I-2.—THEORETICAL STEAM RATES, CONDENSING FOR ENGINES AND TURBINES*

(Reprinted from "Cameron Hydraulic Data," 12th ed., 1951, by permission of Ingersoll-Rand Co.)

Lb per hp-hr

Initial temp., °F	Exhaust pressure, in. Hg abs				
	3.0	2.5	2.0	1.5	1.0
	150 lb gauge 365.8°F saturated steam				
365.8	8.35	8.13	7.85	7.53	7.14
400	8.17	7.95	7.67	7.36	6.99
450	7.92	7.70	7.44	7.15	6.78
500	7.67	7.45	7.22	6.93	6.68
550	7.42	7.22	6.98	6.72	6.38
600	7.18	6.98	6.76	6.51	6.18
650	6.95	6.76	6.55	6.31	6.00
700	6.72	6.53	6.34	6.11	5.82
	175 lb gauge 377.4°F saturated steam				
377.4	8.11	7.89	7.64	7.34	6.97
400	8.00	7.78	7.52	7.23	6.87
450	7.74	7.53	7.29	7.01	6.66
500	7.50	7.30	7.07	6.80	6.46
550	7.25	7.07	6.84	6.59	6.27
600	7.02	6.84	6.63	6.39	6.08
650	6.79	6.62	6.43	6.19	5.90
700	6.58	6.41	6.23	6.00	5.72
	200 lb gauge 387.8°F saturated steam				
387.8	7.92	7.71	7.46	7.19	6.84
450	7.61	7.41	7.17	6.90	6.57
500	7.35	7.16	6.95	6.68	6.37
550	7.12	6.94	6.73	6.48	6.18
600	6.90	6.72	6.52	6.29	5.99
650	6.67	6.51	6.32	6.10	5.81
700	6.46	6.30	6.12	5.91	5.64
750	6.25	6.10	5.93	5.73	5.47
	250 lb gauge 406°F saturated steam				
406.0	7.62	7.43	7.20	6.95	6.62
450	7.39	7.20	6.99	6.74	6.43
500	7.15	6.97	6.76	6.52	6.22
550	6.92	6.75	6.55	6.32	6.04
600	6.69	6.53	6.35	6.13	5.85
650	6.49	6.33	6.15	5.94	5.68
700	6.28	6.13	5.96	5.76	5.51
750	6.08	5.94	5.77	5.58	5.34

* Calculated from Keenan and Keyes, "Theoretical Steam Rate Tables," Am. Soc. Mech. Engrs., New York, 1938.

TABLE I-2.—THEORETICAL STEAM RATES, CONDENSING FOR ENGINES AND TURBINES (*Concluded*)

Lb per hp-hr

Initial temp., °F	Exhaust pressure, in. Hg abs				
	3.0	2.5	2.0	1.5	1.0
	300 lb gauge 421.7°F saturated steam				
421.7	7.39	7.21	7.01	6.76	6.47
450	7.23	7.06	6.86	6.62	6.33
500	6.99	6.83	6.63	6.40	6.12
550	6.76	6.61	6.42	6.20	5.93
600	6.55	6.40	6.21	6.01	5.75
650	6.33	6.20	6.02	5.82	5.58
700	6.14	5.99	5.84	5.65	5.41
750	5.94	5.81	5.66	5.48	5.25
800	5.76	5.63	5.49	5.31	5.09
850	5.57	5.45	5.32	5.15	4.94
	400 lb gauge 448 1°F saturated steam				
448.1	7.07	6.91	6.72	6.50	6.23
500	6.78	6.63	6.46	6.25	5.98
550	6.63	6.40	6.23	6.04	5.79
600	6.34	6.19	6.03	5.84	5.60
650	6.13	5.99	5.84	5.66	5.43
700	5.94	5.81	5.67	5.49	5.26
750	5.75	5.63	5.49	5.32	5.11
800	5.57	5.45	5.32	5.16	4.96
850	5.40	5.28	5.16	5.01	4.82
	600 lb gauge 488.8°F saturated steam				
600	6.10	5.98	5.83	5.66	5.45
650	5.90	5.78	5.64	5.47	5.27
700	5.70	5.59	5.46	5.30	5.10
750	5.52	5.41	5.29	5.14	4.94
800	5.35	5.24	5.12	4.98	4.80
850	5.18	5.08	4.97	4.83	4.65
900	5.02	4.93	4.82	4.69	4.52
950	4.86	4.78	4.67	4.55	4.39
1000	4.71	4.63	4.53	4.42	4.27
	800 lb gauge 520.3°F saturated steam				
650	5.78	5.67	5.54	5.38	5.19
700	5.58	5.47	5.35	5.20	5.01
750	5.39	5.28	5.17	5.03	4.85
800	5.21	5.12	5.01	4.87	4.70
850	5.05	4.96	4.85	4.73	4.57
900	4.89	4.80	4.70	4.58	4.43
950	4.74	4.66	4.57	4.45	4.30
1000	4.60	4.53	4.43	4.32	4.18

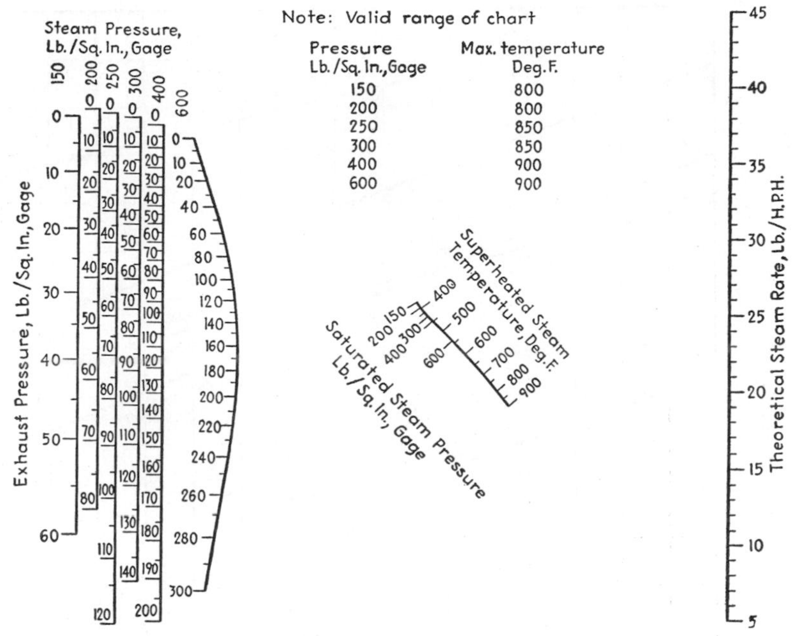

FIG. I-1.—Theoretical steam rates for steam engines and turbines, operating noncondensing.

TABLE I-3.—APPROXIMATE EFFICIENCIES OF IMPULSE STEAM TURBINES AT FULL LOAD

(Based on data from Perry, "Chemical Engineers' Handbook," 3d ed., 1950, and Marks, "Mechanical Engineers' Handbook," 5th ed., 1951, McGraw-Hill Book Company, Inc., New York)

Horsepower	Efficiency	
	Noncondensing	Condensing
150	0.47	
300	0.48	
450	0.61	
600	0.63	
700	0.64	0.68
1,000	0.65	0.70
1,500	0.67	0.70
2,000	0.67	0.70
2,800	0.68	0.70

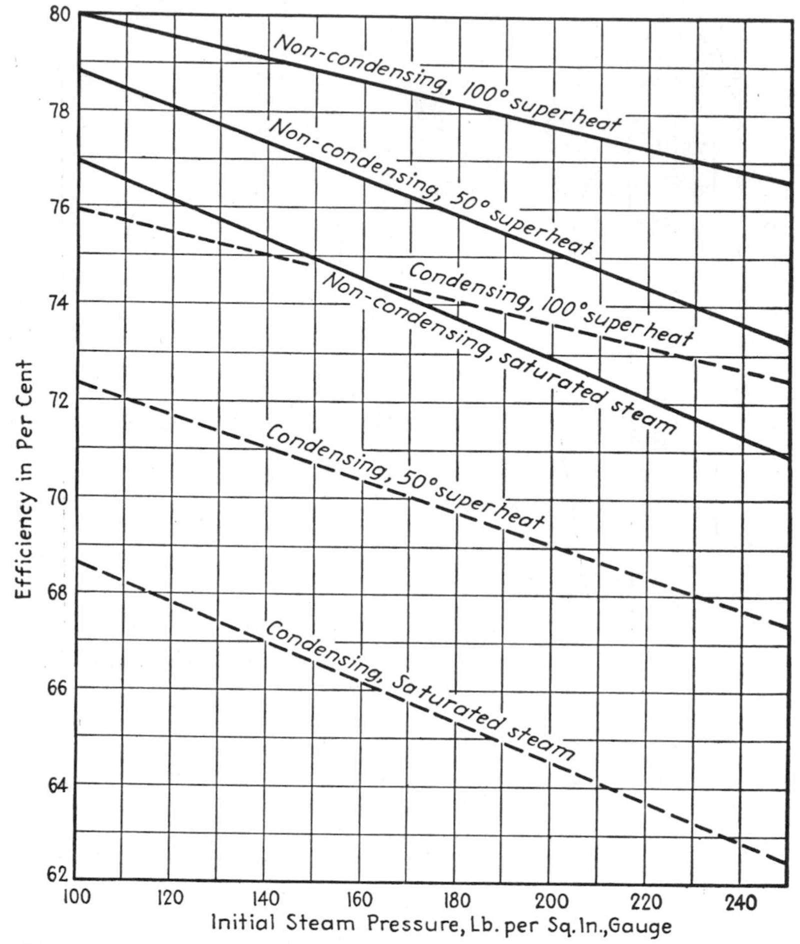

NOTE: Temperatures given are degrees Fahrenheit above saturation temperature.

FIG. I-2.—Approximate efficiencies, Uniflow steam engines. (*Adapted from Perry, "Chemical Engineers' Handbook," 2d ed., McGraw-Hill Book Company, Inc., New York,* 1941.)

c. Theory of the Ideal Steam Engine.—In the precise language of thermodynamics, the ideal engine expands steam by an adiabatic reversible process. Consequently,

$$S_2 = S_1$$
$$w = H_1 - H_2$$

where S_1 = entropy of steam, Btu/(lb)(°F)

S_2 = entropy of exhaust, Btu/(lb)(°F)

H_1, H_2 = heat content of steam, Btu/lb, at inlet, exhaust

w = work produced, Btu/lb

In other words, the entropy of the exhaust is precisely the same as that of the inlet steam, and the power developed is the mechanical equivalent of the heat abstracted from the steam. With this principle the theoretical steam rates for the ideal engine may be computed from the thermodynamic properties of steam. To compute the performance of a steam engine by use of a thermodynamic chart proceed as follows:

1. Find the point represented by the initial temperature and pressure of steam; read the heat content of steam, and record as h_1.

2. From this point follow the line of constant entropy until it crosses the pressure line corresponding to the exhaust pressure. Read the heat content and record as h_2.

3. The work produced is $h_2 - h_1$. This is in Btu per pound. The theoretical steam rate in pounds per horsepower-hour is $2544/(h_2 - h_1)$. The computation of the performance of an ideal steam engine from steam tables is illustrated by Example I-2.

Example I-2. Find the theoretical steam rate of an engine supplied with steam at 385 psig (400 psia) superheated to 600°F to be exhausted at 15.3 psig (30 psia). Finally compute the actual steam rate for an engine rated at 75 per cent efficiency.

SOLUTION:

1. Properties of inlet steam (400 psia, 600°F),

Heat content, $H_1 = 1306.9$ Btu/lb (Table E-15)
Entropy, $S_1 = 1.5894$ Btu/(lb)(°F)

2. Properties of steam saturated at 30 psia,

Heat content, $H = 1164.1$ Btu/lb (Table E-14)
Entropy, $S = 1.6993$ Btu/(lb)(°F)

3. Properties of water saturated at 30 psia,

Heat content, $h = 218.8$ Btu/lb
Entropy, $s = 0.3680$ Btu/(lb)(°F)

4. Quality of exhaust steam,
Let x = weight fraction of condensate in exhaust
$1 - x$ = weight fraction of live steam in exhaust (quality)

$$\text{Entropy of exhaust steam} = S_2 = S_1 = 1.5894 \text{ Btu/(lb)(°F)}$$

$$S_2 = S_1 = xs + (1 - x)S$$

or

$$S - S_1 = (S - s)x$$

and

$$x = \frac{S - S_1}{S - s} = \frac{1.6993 - 1.5894}{1.6993 - 0.3680} = \frac{0.1099}{1.3313}$$
$$= 0.0824$$
$$1 - x = 0.9176$$

5. Heat content of exhaust steam,

$$H_2 = (1 - x)H + xh$$
$$= H - xH + xh$$
$$= 1164.1 - 0.0824 \times 1164.1 + 0.0824 \times 218.8$$
$$= 1164.1 - 96.0 + 17.9 = 1086.0$$

6. Work produced per pound of steam,

$$H_1 - H_2 = 1306.9 - 1086.0 = 220.9 \text{ Btu/lb}$$

7. Theoretical steam rate,

$$\frac{2544}{220.9} = 11.5 \text{ lb steam/hp-hr developed}$$

8. Actual steam rate,

$$\frac{11.5}{0.75} = 15.4 \text{ lb steam/hp-hr}$$

d. Throttling Control.—The speed of a turbine or steam engine may be controlled by throttling the inlet steam to a lower pressure or by throttling the exhaust to a higher pressure. (In the case of the turbine the speed may also be controlled by the number of nozzles opened.) Whether this control is effected by a direct speed control or by the output of the driven equipment, the steam rate will be determined by the conditions prevailing at the engine inlet and exhaust rather than by the pressure in the supply and exhaust mains.

The relative effect of throttling the steam to a steam engine is illustrated by Example I-3.

Example I-3. A steam engine is supplied by a main carrying 200 psig saturated steam and exhausts to 15 psig. Find the steam rates per horsepower: (1) at line pressure and (2) if steam is throttled to 150 psig.

SOLUTION:

1. Steam at line pressure,

$$\text{Theoretical} = 16.9 \text{ lb/hp (Fig. I-1)}$$
$$\text{Efficiency} = 0.73 \text{ (Fig. I-2)}$$
$$\text{Actual rate} = \frac{16.9}{0.73} = 23.2 \text{ lb/hp-hr}$$

2. Properties of steam after throttling,
No work is performed and heat loss is negligible, hence, heat content of steam at reduced pressure will be unchanged. At 215 psia saturated steam has a heat content of 1199 Btu/lb (Table E-14), and a temperature of 388°F (by interpolation of Table E-15). Saturation temperature is 366°F.
3. Steam rate at throttled condition,

$$\text{Theoretical} = 19.5 \text{ lb/hp-hr}$$
$$\text{Efficiency} = 0.75$$
$$\text{Actual rate} = \frac{19.5}{0.75} = 26.0 \text{ lb/hp-hr}$$

e. Quality of Exhaust Steam.—The quality of steam exhausting from a steam engine or turbine may be computed readily from the properties of the steam, the power developed, and the heat lost. This is illustrated by Example I-4.

Example I-4. Compute the quality of steam exhaust from the engine considered in Example I-2, assuming a heat loss of 12 Btu/hp-hr.

SOLUTION:
1. Heat content of inlet steam = 1306.9 Btu/lb
2. Work produced for 1 lb steam,

$$\frac{1}{15.4} \text{ hp-hr, or } \frac{2544}{15.4} = 165.1 \text{ Btu/lb (result, Example I-2)}$$

3. Heat loss $= \dfrac{12}{15.4} = 0.8$ Btu/lb
4. Heat content of exhaust steam,

$$1306.9 - 165.1 - 0.8 = 1141.0 \text{ Btu/lb}$$

5. Heat content steam, water at exhaust pressure of 30 psia,

$$\text{Steam} = 1164.1 \text{ Btu/lb}$$
$$\text{Water} = 218.8 \text{ Btu/lb}$$

6. Quality,
Let x = weight fraction condensate in exhaust
$1 - x$ = weight fraction live steam in exhaust (quality)

$$1141.0 = 1164.1(1 - x) + 218.8x$$
$$-23.1 = -945.3x$$
$$x = \frac{23.1}{945.3} = 0.024$$
$$\text{Quality} = 1 - x = 0.976$$

3. Steam Balance

Steam is most efficiently used when all of the exhaust from turbines and engines is just sufficient to supply normal requirements for process steam. Once preliminary estimates of the power and steam requirements are available, a tentative steam balance is usually a valuable guide for rounding out process design to provide for a more efficient use of steam. Example I-5 illustrates such a preliminary steam balance.

Example I-5. Preliminary estimates on a new process place the average steam demand at 150,000 lb/hr and the average power requirement of steam and electric driven units at 8,000 hp. Assuming that the process steam is supplied at 35 psig, determine whether a steam balance is possible and make a tentative choice of boiler pressure and superheat.

SOLUTION:

1. Assumed average quality of exhaust = 97% live steam
2. Total steam to be generated,

$$\frac{150{,}000}{0.97} = 155{,}000 \text{ lb/hr}$$

3. Over-all steam rate of turbines and engines,

$$\frac{155{,}000}{8{,}000} = 19.4 \text{ lb/hp-hr}$$

4. Assumed over-all efficiency of power generation = 65%
5. Corresponding steam rate of ideal engine,

$$19.4 \times 0.65 = 12.6 \text{ lb/hp-hr}$$

6. Steam pressures and temperatures which will give this rate (from Fig. I-1),

Pressure, Psig	Steam Temp., °F
600	Saturated
400	660
300	780

CONCLUSION: 400 psig steam at 660°F appears most suitable for preliminary estimates.

4. Internal-combustion Engines

a. Gas Engines.—The approximate fuel consumption of gas engines is given in Table I-4. This table is for larger engines with pistons 10 in. in diameter or over. Smaller engines will require 5 to 15 per cent more fuel per horsepower. The engine jackets remove about 3500 to 4000 Btu/hp-hr. The jacket water should preferably be rather warm and should be fairly soft. Lubrication requirements depend on a variety of factors. About 1 gal for every 2,000 hp-hr should be ample allowance.[1]

[1] "Gas Engineers' Handbook," McGraw-Hill Book Company, Inc., New York, 1934.

TABLE I-4.—FUEL CONSUMPTION BY GAS ENGINES
(Graf, "Gas Engineers' Handbook," McGraw-Hill Book Company, Inc., New York, 1934)

Load	Fuel consumption, Btu/bhp-hr	
	4-cycle engines	2-cycle engines
5/4	11,000	15,000
4/4	10,000	13,000
3/4	11,000	16,000
2/4	12,500	17,000
1/4	17,000	23,000

b. Diesel Engines.—Data on performance of diesel engines are given in Table I-5. The jacket cooling water removes 2000 to 3500 Btu/hp-hr. Discharge temperatures above 180°F are not recommended. The lubricating oil requirement is about the same as for gas engines.

TABLE I-5.—MANUFACTURERS' FUEL GUARANTEES
(Perry, "Chemical Engineers' Handbook," 2d ed., McGraw-Hill Book Company, Inc., New York, 1941)

Make	Type*	Cycle	Size	Fuel consumption, lb/hp-hr		
				Full load	3/4 load	1/2 load
National Supply Co.	V, S	4	70–420	0.40	0.41	0.44
De La Vergne	V, A	4	100–300	0.42	0.435	0.52
Fulton	V, M, A	4	50–100	0.50		
Bush-Sulzer	V, A	2	500	0.425	0.44	0.52
Bush-Sulzer	V, S	2	1,800–2,400	0.39	0.40	0.43
American Locomotive	V, S	2	1,800	0.40	0.41	0.44
Nordberg	V, S	2	1,800–2,250	0.39	0.41	0.43
Hamilton M.A.N	V, S	2	2,250	0.39	0.40	0.43

* A = air injection; S = solid injection; V = vertical; M = marine.

c. Gasoline and Kerosene Engines.—Manufacturers' data quoted by Marks[1] range from 0.46 to 0.72 lb of fuel per horsepower-hour. In the absence of specific data, 0.6 and 0.7 lb of fuel per horsepower-hour may be assumed for gasoline and kerosene, respectively.

[1] Marks, "Mechanical Engineers' Handbook," 5th ed., McGraw-Hill Book Company, Inc., New York, 1951.

d. Waste-heat Recovery.—About one-third of the heat of combustion of fuel leaves engines with the exhaust. It is sometimes possible to utilize this heat. Table I-1 given earlier in the chapter gives data on waste-heat boilers operating on diesel exhaust.

5. Electrical Power

a. Generators.—The mechanical power required to drive an electric generator is

$$\mathrm{Hp} = \frac{\mathrm{kw}}{0.746 \times \mathrm{efficiency}}$$

Representative efficiencies are given in Table I-6. This table is based on three-phase alternating current. The efficiency of single-phase and

TABLE I-6.—PERFORMANCE DATA FOR ALTERNATORS, HORIZONTAL-COUPLED OR BELTED-TYPE ENGINE-DRIVEN GENERATORS
(Westinghouse Electric Corporation)
80% power factor, three-phase, 60-cycle, 240 to 2,400 volts

Kva	Poles	Speed, rpm	Kw excitation	Efficiency, %			Approx. net wt, lb
				½ load	¾ load	4/4 load	
25	4	1,800	0.5	81.5	85.7	87.6	800
93.8	8	900	2.0	87.0	89.5	90.9	2,900
250	12	600	5.0	90.0	91.3	92.2	6,000
500	18	400	10.0	91.7	92.6	93.2	8,000
1,000	24	300	15.0	92.6	93.4	93.9	15,000
3,125	48	150	40.0	93.4	94.2	94.6	52,000

Turbo-driven direct-connected type

Kva	Poles	Speed, rpm	Kw excitation @ 125 volts	Efficiency, %, at 80% power factor			Cu ft air per min
				½ load	¾ load	4/4 load	
2,500	2	3,600	25	95.1	96.0	96.2	5,000
6,250	2	3,600	37	95.5	96.4	96.8	12,000
9,375	2	3,600	45	96.0	96.8	97.1	17,000
12,500	2	3,600	60*	95.8	96.7	97.0	23,500
29,411†	2	3,600	130*	98.1†	98.3†	98.3†	
44,118†	2	3,600	155*	98.4†	98.5†	98.5†	

* Excitation for these ratings is usually at 250 volts.
† At 30 psig hydrogen pressure and 85% power factor.

d-c generators is not greatly different. Complete data are readily obtainable from the manufacturers.

b. Motors.—The electrical power consumption of a motor is

$$\text{Kw} = \frac{0.746 \times \text{hp}}{\text{efficiency}}$$

Average efficiencies are given in Table I-7. Efficiencies of most other types of motors are not greatly different. Variable-speed alternating motors, however, have relatively poor efficiencies, and their use should be avoided if other types of drive are available.

TABLE I-7.—AVERAGE EFFICIENCIES FOR VOLTAGES UP TO 550 IN THREE-PHASE MOTORS

Rating, hp	Induction motors			Unity power factor, synchronous motors		
	Full load	¾ load	½ load	Full load	¾ load	½ load
1–2	81	79	76			
3–5	85	84	82			
10–25	87.5	87	85			
25–50	89	88.5	86.5	90	88.5	85
75–100	91	90	88	92	91	88
150–200	92	91.5	89.5	93	92	89.5
Above 200	93.5	93	91	94.5	93.5	92

The above are averages for 1,200- and 1,800-rpm motors.

Low-speed motors: Efficiencies will be from 1 to 2 per cent lower; 2,200-volt motors, 100 hp and above, have approximately the same efficiency as low-voltage motors. On smaller sizes the 2,200-volt motors have efficiencies 1 to 2 per cent lower.

Slip-ring motors in the smaller sizes will have full-load efficiencies of 1 to 3 per cent lower.

c. Carrying Capacity of Copper Wire.—Loads are given in Table I-8.

d. General Electrical Data

Ohm's law:

$$I = \frac{E}{R}$$

where I = current, amp
E = applied voltage
R = resistance, ohms

TABLE I-8.—ALLOWABLE CURRENT-CARRYING CAPACITIES FOR COPPER CONDUCTORS

Not more than three conductors in raceway, cable, or direct burial (1956 National Electrical Code, Amended 1958, National Board of Fire Underwriters)

Size		Allowable amperes			Allowable amperes	
No. A.W.G.	Circular mils	Rubber insulation, types RH and RHW	Varnished cambric insulation, type V	Size, circular mils	Rubber insulation, types RH and RHW	Varnished cambric insulation, type V
14	4,107	15	25	250,000	255	270
12	6,530	20	30	300,000	285	300
10	10,380	30	40	350,000	310	325
8	16,510	45	50	400,000	335	360
6	26,250	65	70	500,000	380	405
4	41,740	85	90	600,000	420	455
3	52,640	100	105	700,000	460	490
2	66,370	115	120	750,000	475	500
1	83,690	130	140	800,000	490	515
0	105,500	150	155	900,000	520	555
00	133,100	175	185	1,000,000	545	585
000	167,800	200	210	1,250,000	590	645
0000	211,600	230	235	1,500,000	625	700
				1,750,000	650	735
				2,000,000	665	775

Applies to d-c circuits of any type and to a-c flow in noninductive circuits (such as heaters).

Power in direct and single-phase circuits:

$$P = fEI$$

where P = power, watts
f = power factor of circuit
= 1, for d-c and for any noninductive circuits or for synchronous motors
E = applied voltage
I = current, amp

Power in balanced 3-wire, 3-phase circuits:

$$\begin{aligned} p &= fEI \\ &= \sqrt{3}\, fE_L I_L \\ &= 3fE_p I_p \end{aligned}$$

where p = power, watts
f = power factor
E = circuit voltage
I = circuit current, amp
E_L = line voltage between any two wires
I_L = current in any one wire, amp
E_p = phase voltage
$= E_L$ for delta systems
$= \frac{E_L}{\sqrt{3}}$ for Y systems
I_p = phase current, amp
$= \frac{I_L}{\sqrt{3}}$ for delta systems
$= I_L$ for Y systems

Heat developed by resistance loads:

$$Q = 3.413EI = 3.413I^2R = 3.413E^2R$$

where Q = heat developed, Btu/hr
E = voltage
I = current, amp
R = resistance, ohms

Power consumed by electrolytic cells:

$$P = \frac{12.16nE}{\theta_e}$$

where P = power, kwhr/lb mole cell reaction
n = unit charge for 1 lb mole cell reaction, *i.e.*, number of equivalents per mole
E = applied voltage
= thermodynamic electromotive force divided by voltage efficiency
θ_e = current efficiency

CHAPTER J

PUMPS

Handling of liquids is a routine matter for the chemical engineer. This chapter gives performance data on several types of pumps and roughly outlines their fields of usefulness. The principal source of pump data for specific uses is the manufacturers who are very generous in supplying bulletins. Below are listed several texts on pumping liquids.

Kristal and Annett, "Pumps," 2d ed., McGraw-Hill Book Company, Inc., New York, 1953.

Church, "Centrifugal Pumps and Blowers," John Wiley & Sons, Inc., New York, 1944.

Hicks, "Pump Selection and Application," McGraw-Hill Book Company, Inc., New York, 1957.

Hicks, "Pump Operation and Maintenance," McGraw-Hill Book Company, Inc., New York, 1958.

1. General Data

a. Power Required for Pumping.—The work required to lift a liquid follows directly from the basic definition of work: force × distance. The force is the mass of liquid and the distance is the height raised, *i.e.*, the liquid head.

$$\text{Work (ft-lb)} = \text{weight (lb)} \times \text{head (ft)}$$

This is the useful work done; a greater amount must be supplied because of inevitable inefficiency. The over-all efficiency of a pump is defined as the ratio of work done to work supplied or, in terms of power, the ratio of water horsepower (theoretical power) to the brake horsepower applied to the pump drive. Introducing the efficiency and expressing work as power units (work per unit of time), the working equations given below may readily be obtained:

$$\begin{aligned} \text{Bhp} &= \frac{\text{lb/min} \times \text{ft head}}{33{,}000 \times \text{efficiency}} \\ &= \frac{\text{gpm} \times \text{psi differential pressure}}{1{,}713 \times \text{efficiency}} \\ &= \frac{\text{Gpm} \times \text{ft head} \times \text{sp gr}}{3{,}957 \times \text{efficiency}} \end{aligned}$$

Pumping efficiencies depend on the type of pump as well as on operating conditions; consequently, efficiencies are given under the type of pump. In computing the power required for pumping, the head and capacity must be those at the pump. The head must include pressure losses in lines and equipment, height through which the liquid is raised, pressure loss on control throttling in addition to the difference in pressures between the receiving vessel and the pump suction. Strictly speaking, the velocity head should also be included, but fortunately it may be neglected in all but a few cases, such as pumps operating at very low heads. The liquid pumped must include any by-passed to suction for control purposes. Treatment of these factors is illustrated by Example J-1.

Example J-1. 100 gpm of alcohol is to be supplied to a solvent recovery tower operated at 100 psig. As the process requires very close control, a dual control is employed, the pump discharge is maintained at constant pressure by a back-pressure controller by-passing liquid to the suction, and the flow into the system is finally controlled by a flow controller. The alcohol comes from a storage tank 1 ft above grade at atmospheric pressure. The liquid enters the tower at 49 ft above grade. The line has been sized for a total pressure drop from tank to tower of 10 psi. Estimate the power required using a turbine pump.

SOLUTION:

1. Head in pounds per square inch, differential,
net pressure head supplied,

Tower pressure − tank pressure.................... 100 psi

Liquid head = 49 − 1 = 48 ft alcohol,

or

$$\frac{\text{Specific gravity (60/60)}}{2.31} \times \text{head} = \frac{0.79}{2.31} \times 48 \ldots\ldots\ldots \quad 16$$

Line drop.................................... 10

Drop through control valve.................... 5

Total.................................... 131 psi

or, in feet of alcohol

$$\frac{131 \times 2.31}{0.79} = 383$$

2. Capacity,

Delivered to tower.................................... 100 gpm

Assume 10 per cent of this amount throttled back to suction...... 10

Total.................................... 110 gpm

3. Pump efficiency,

From Fig. J-18, 40%, (slightly above typical is justified because flow is in upper portion of the range)

4. Horsepower required,

$$\frac{\text{Gpm} \times \text{psi}}{1{,}713 \times \text{efficiency}} = \frac{110 \times 131}{1{,}730 \times 0.40} = 21 \text{ hp}$$

b. Suction Lifts.—A liquid cannot exist at pressure below its vapor pressure. Even if a liquid has no appreciable vapor pressure, it is unstable under tension. Consequently, the theoretical minimum *absolute* pressure at pump suction is the vapor pressure of the liquid. The maximum *suction lift* is then the *barometric pressure* less the *vapor pressure* when both of these pressures are expressed as *height of liquid.* This may be done by the relation:

$$\text{Liquid head, ft} = \frac{2.31 \times \text{psi}}{\text{specific gravity (60/60)}}$$

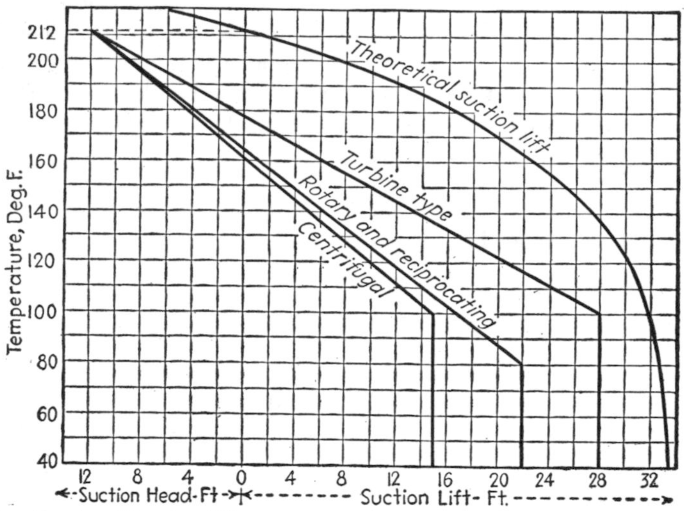

Fig. J-1.—Suction lifts for different types of pumps handling water. (*Kristal and Annett, "Pumps," McGraw-Hill Book Company, Inc., New York,* 1940.)

The suction lifts *practical* for *actual* pumps are considerably below the theoretical value, and it is by far the best practice to operate pumps with a slight positive pressure (flooded suction), whenever possible. However, pumps properly primed will lift liquids to some extent.

Figure J-1 gives data on lifts practical for pumping water and Fig. J-2 for other liquids. These figures are based on a barometric pressure of 14.70 psi. At high altitudes barometric pressures are lower (see Fig. C-1), and the lifts must be reduced accordingly. It will be noted from both of these figures that positive suction pressures are necessary for liquids of high vapor pressure. For boiler-feed pumps returning condensate under pressure and at temperature above the atmospheric boiling point, the Standards of the Hydraulic Institute recommends a pressure at suction of 14 ft of liquid head above the vapor pressure including an allowance for variation of temperature of the condensate.

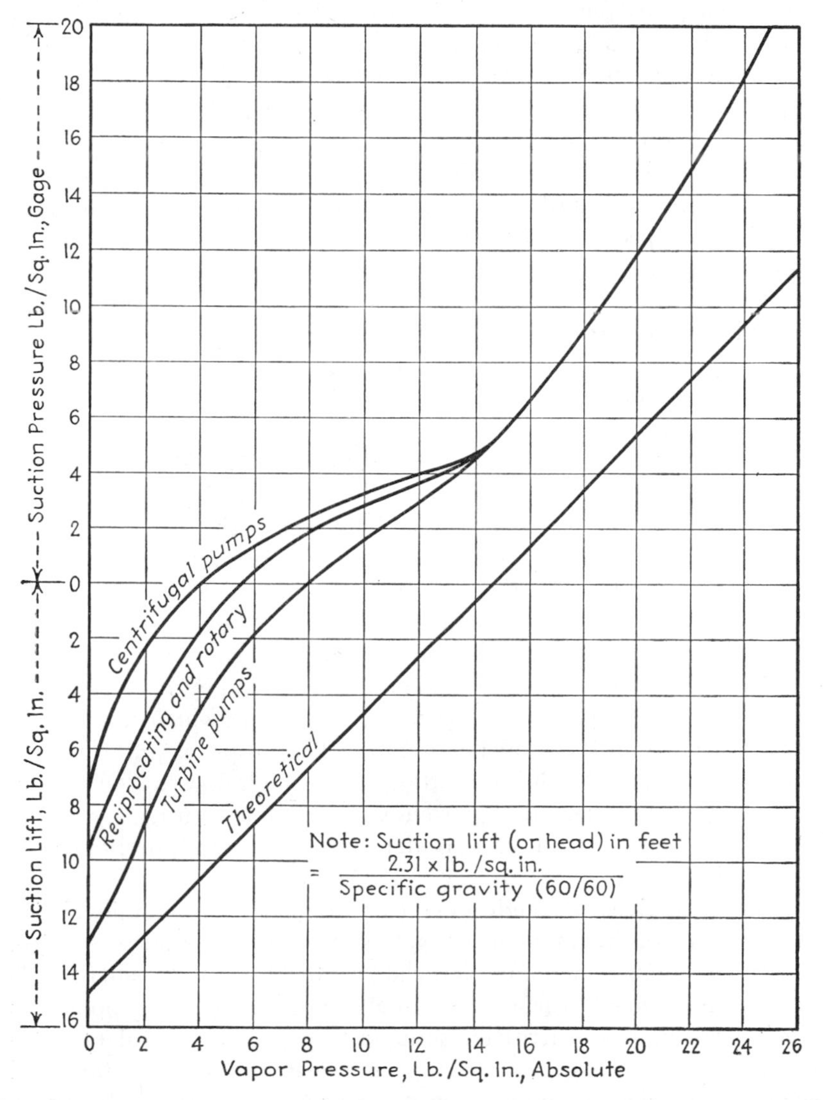

FIG. J-2.—Maximum suction lifts and minimum suction pressures for pumping volatile liquids.

This allowance amounts approximately to taking the vapor pressure about 10°F above the normal storage temperature. Presumably this same rule is applicable to other liquids; however, a greater allowance might well be desirable for liquids stored below the surrounding temperature (*e.g.*, liquefied natural gases).

The *above discussion is intended to apply to the use of standard pumps. Actually, lower suction heads can frequently be used by careful selection of pumps in cooperation with the manufacturer.*

c. Types of Pumps.—Although centrifugal and reciprocating pumps are the standard mainstays, other types have important applications. Reciprocating pumps are the oldest type and are extremely versatile since they handle almost any liquid and some slurries over any range of pressure and capacity. Their positive action favors control of rate by speed control. The type of motion is inconvenient for motor and turbine drives but is very advantageous for engine drives. In addition to the standard piston type, diaphragm pumps are available for handling corrosive and erosive materials.

Centrifugal pumps of volute and diffusion types are very popular. Although they cover a wide range of capacities, the smaller pumps are less efficient. Single-stage pumps afford only limited heads, but high pressures are easily obtained by multistage units. Rotary motion makes these pumps well suited to motor and turbine drives. The characteristics permit capacity control by either throttling or speed control, the former being a very attractive feature for electric-motor drives. Centrifugals are adaptable for pumping thick slurries.

Centrifugal pumps are classified further according to the shape of the impeller, which may be either single or double suction and may be designed to give a forward motion as well as outward, as in the Francis type, or still more in the axial or mixed flow type which is just one stage removed from the propeller pump. The propeller pump is well described by its name and is advantageous for high flows at low heads. These pumps supplement the "garden variety" of centrifugals, and the distinction between them is of secondary importance for purely process considerations.

Figure J-3 gives a rough idea of the relative economic choice between reciprocating and centrifugal pumps for boiler-water service.

Turbine pumps (herein distinguished from diffuser type centrifugals sometimes also called "turbine" pumps) are an important supplement to reciprocating and centrifugal pumps. Their most useful range is up to 150 gpm. They develop fairly high heads by a single stage and give efficiencies that compare favorably with low-capacity centrifugals. The turbine pump is well suited for high suction lifts.

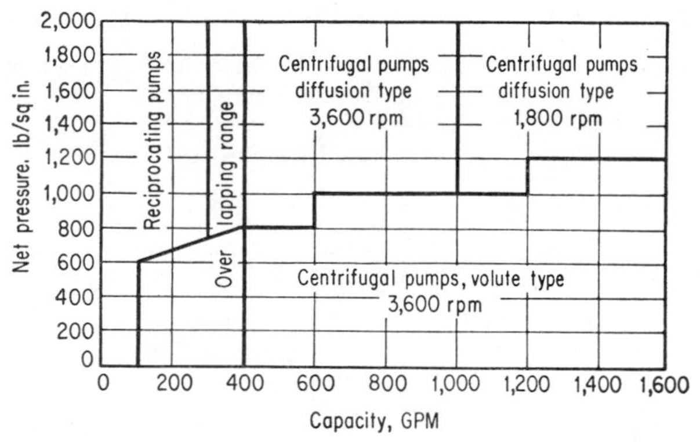

FIG. J-3.—Suggested economic operating range for reciprocating and centrifugal boiler-feed pump. (*Kristal and Annett, "Pumps," 2d ed., McGraw-Hill Book Company, Inc., New York,* 1953.)

Rotary slide-vane, gear, and screw pumps have many important applications, particularly for low flows at high heads or handling liquids of high viscosity.

2. Reciprocating Pumps

a. Capacity.—The capacity of reciprocating pumps is very nearly equal to the displacement. This is readily computed from the piston diameter, stroke, and pump speed. The calculated displacement multiplied by the volumetric efficiency is the capacity of the pump. Volumetric efficiencies of pumps are high (except for heavy viscous liquids) and may be taken as 97 per cent for most purposes since actual values are usually between 94 and 100 per cent. (In very large, efficient pumps, the efficiency may even exceed 100 per cent because of velocity effects.) Below are given formulas for pump capacity.

Single acting:

$$Q = \frac{E_v C N D^2 S}{295}$$

Doubling acting:

$$Q = \frac{E_v C N S (2D^2 - d^2)}{295}$$

where Q = gpm
E_v = volumetric efficiency
C = number of cylinders
N = strokes/min
S = length of stroke, in.
D = diameter of piston, in.
d = diameter of piston rod, in.

Table J-1 gives sizes and capacities of duplex steam pumps.

TABLE J-1.—SIZES AND CAPACITIES OF HORIZONTAL DUPLEX STEAM PUMPS
(Courtesy of Worthington Corp.)

Pump size*	Boiler feed		Maximum recommended capacities, U.S. gpm						
	Gpm	Boiler hp	Cold water	SSU viscosity of cold, viscous liquids					
				250	500	1,000	2,500	5,000	10,000
3 × 2 × 3	7	90	12	12	11	10	9	8	4
3½ × 2¼ × 4	12	160	19	19	18	17	15	13	6
4½ × 2¾ × 4	17	220	29	29	28	26	22	19	9
5¼ × 3½ × 5	32	420	53	53	50	47	41	34	17
6 × 4 × 6	47	610	78	78	75	69	60	51	25
7½ × 5 × 6	73	950	123	123	117	110	95	80	40
7½ × 4½ × 10	74	975	124	124	119	111	96	81	40
9 × 5¼ × 10	101	1,315	169	169	162	151	130	110	55
10 × 6 × 10	132	1,725	220	220	211	196	170	144	72
10 × 7 × 10	180	2,350	300	300	288	268	232	196	98

* Steam cylinder diameter × pump cylinder diameter × length of stroke.

b. Efficiencies.—Figure J-4 gives typical efficiencies of reciprocating pumps handling light liquids. Efficiencies handling viscous liquids will be lower, but unfortunately no specific data can be presented here in this regard, though Table J-1 gives the effect of viscosity on capacity and thereby gives an index of the order of effect to be expected.

c. Control.—Because of the high volumetric efficiencies, the capacity of a reciprocating pump is determined almost entirely on its speed; in fact, the speed of a pump in *good condition* is a fairly accurate measure of the flow. A pump driven by an electric motor will, therefore, deliver a constant flow nearly independent of head; moreover, the flow cannot be readily changed except by returning the excess capacity to the suction or by changing the length of the stroke. This latter expedient is employed in proportioning pumps. Control of flow by regulating the speed of the drive is an efficient method, as there is no loss of power from excess capacity or throttling. This fact accounts for the popularity of the steam pump as it largely offsets the low efficiency of the steam cylinder drive.

3. Centrifugal and Propeller Pumps

a. Performance Characteristics.—As already mentioned under pump types, centrifugals are made to give a wide range of characteris-

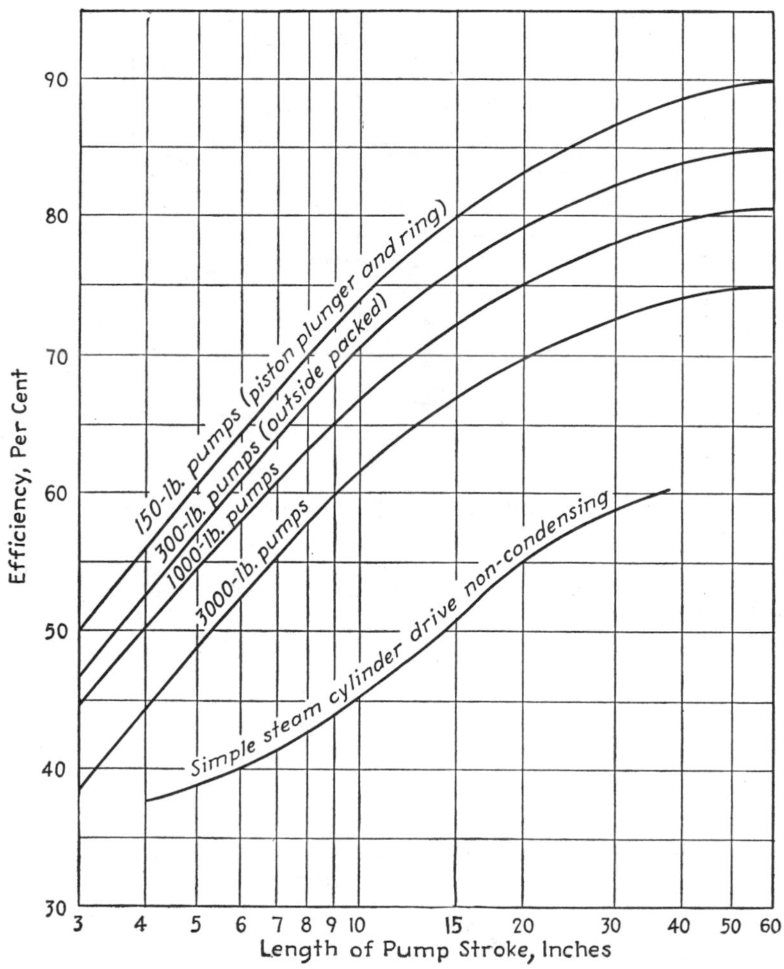

FIG. J-4.—Approximate efficiencies of direct-acting steam pumps. (*Adapted from Nickel, "Direct-acting Steam Pumps," McGraw-Hill Book Company, Inc., New York,* 1915.)

tics by variation of impeller shape, speed, and number of stages. Illustrative performance data of single-stage pumps are given as follows:

Characteristics of a centrifugal pump showing method for correction when speed is changed .... Fig. J-5
Characteristics of pumps handling water compared with those handling viscous oils .... Figs. J-6, J-7, J-16
Comparison of the characteristics of a centrifugal and a propeller pump .... Fig. J-8
Performance on handling paper stock .... Fig. J-9
Performance of an alumina-ceramic pump .... Fig. J-10

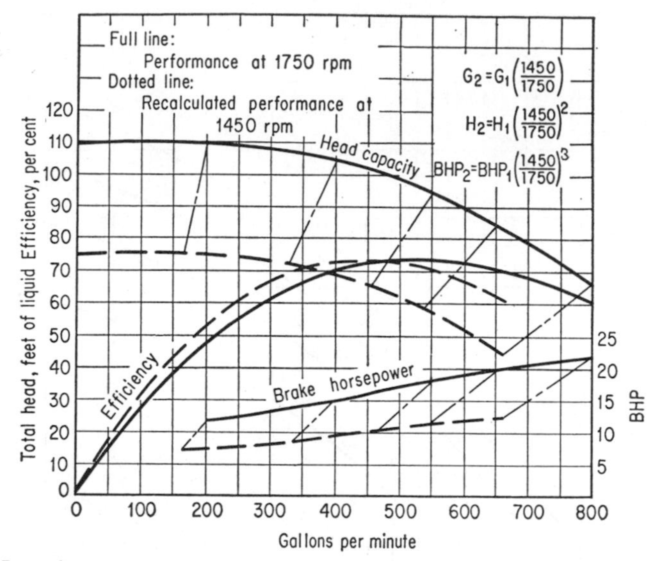

FIG. J-5.—Showing how characteristics at one speed may be converted into those at another. (*Graf, "Gas Engineers' Handbook," McGraw-Hill Book Company, Inc., New York,* 1934.)

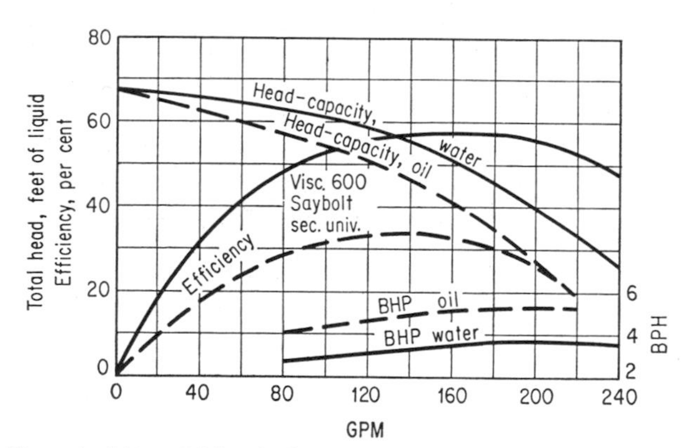

FIG. J-6.—Characteristics of 2-in. single-stage pump for water and 600 SSU oil. (*Graf, "Gas Engineers' Handbook," McGraw-Hill Book Company, Inc., New York,* 1934.)

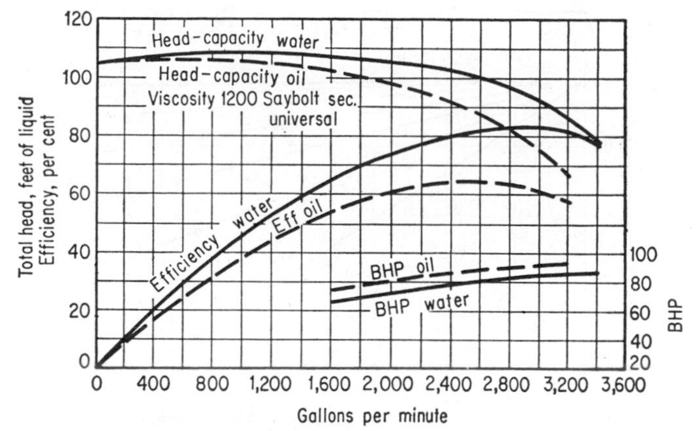

FIG. J-7.—Characteristics of 10-in. double-suction pump for water and 1200 SSU oil. (*Graf, "Gas Engineers' Handbook," McGraw-Hill Book Company, Inc., New York,* 1934.)

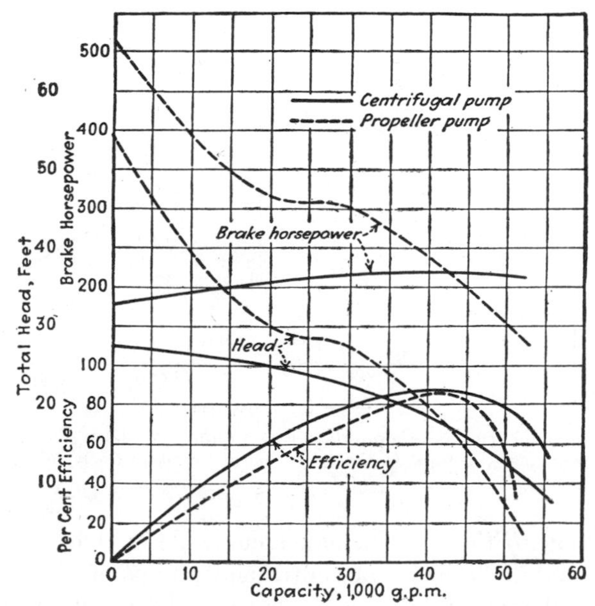

FIG. J-8.—Characteristic curves of a centrifugal- and a propeller-type pump, each rated 42,000 gpm at 17-ft head. (*Kristal and Annett, "Pumps," 2d ed., McGraw-Hill Book Company, Inc., New York,* 1953.)

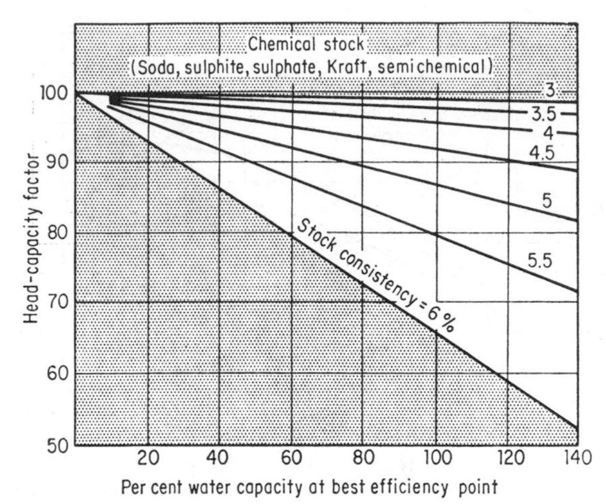

FIG. J-9.—Centrifugal-pump head correction factors for chemical paper stock—soda, sulphite, sulphate, kraft, and semichemical. (*By permission, Goulds Pumps, Incorporated.*)

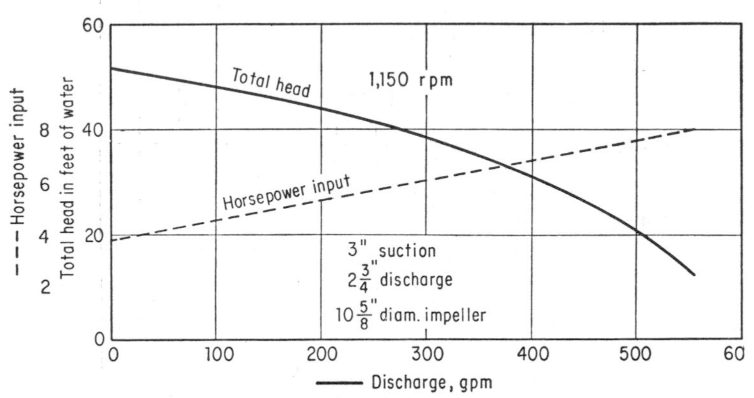

FIG. J-10.—Characteristics of an alumina-ceramic pump (Model LVO-6). (*By permission, the Chemical Equipment Division of General Ceramics Corp., Keasbey, N.J.*)

In addition to these performance curves, Fig. J-11 gives data on the approximate efficiencies of centrifugal pumps of different capacities; Fig. J-12 gives similar data showing the effect of the specific speed. This specific speed is an important criterion for pump design and affords an index as to the number of stages and type of pump best suited for the intended service. The figure indicates the impeller type best suited. For specific speeds below the solid lines in the figure, it is advisable to divide the head into two or more stages so that the specific speed of each stage is in the efficient range; in fact, specific speeds below 800 are seldom recommended.[1] A possible exception is in the case of small pumps, as a very low efficiency may be preferred to the higher cost of multistaging.

The use of this figure is illustrated by Example J-2.

Example J-2. It is desired to pump 150 gpm of water from a main at 35 psi to 220 psi. Find the number of stages required for a centrifugal pump operated at 3,600 rpm.

SOLUTION:

1. Head = 220 − 35 = 185 psi,

or

$$185 \times 2.31 = 428 \text{ ft of water head}$$

2. Specific speed, single stage,

$$N_s = \frac{\text{Rpm}\sqrt{\text{gpm}}}{H^{3/4}} = \frac{3{,}600\sqrt{150}}{(428)^{3/4}} = 470$$

Efficiency from Fig. J-12 = 50 per cent.

3. Specific speed, 2-stage,

$$N_s = \frac{3{,}600\sqrt{150}}{(214)^{3/4}} = 800$$

Efficiency from Fig. J-12 = 60 per cent.
Take 90 per cent of this for 2 stages = 54 per cent.

4. Specific speed, 3-stage,

$$N_s = \frac{3{,}600\sqrt{150}}{(143)^{3/4}} = 1{,}100$$

Efficiency from Fig. J-12 = 63 per cent.
90 per cent of this = 57 per cent.

CONCLUSION: Three stages appear desirable even at this high speed. A single stage, though possible from the standpoint of specific speed, might lead to impractical peripheral speeds. This conclusion is strictly tentative since only one factor of the many entering into pump design has been considered; therefore, it should be used only for preliminary power estimates and tentative choice between centrifugal and other types of pumps.

Centrifugal pumps are widely used for handling thick liquids. Aisenstein[2] states that pumps of 3-in. or smaller suction generally

[1] Recently single-stage hot-oil pumps with high heads have become available. This presumably represents an advance in pump design for low specific speeds.

[2] "Gas Engineers' Handbook," McGraw-Hill Book Company, Inc., New York, 1934.

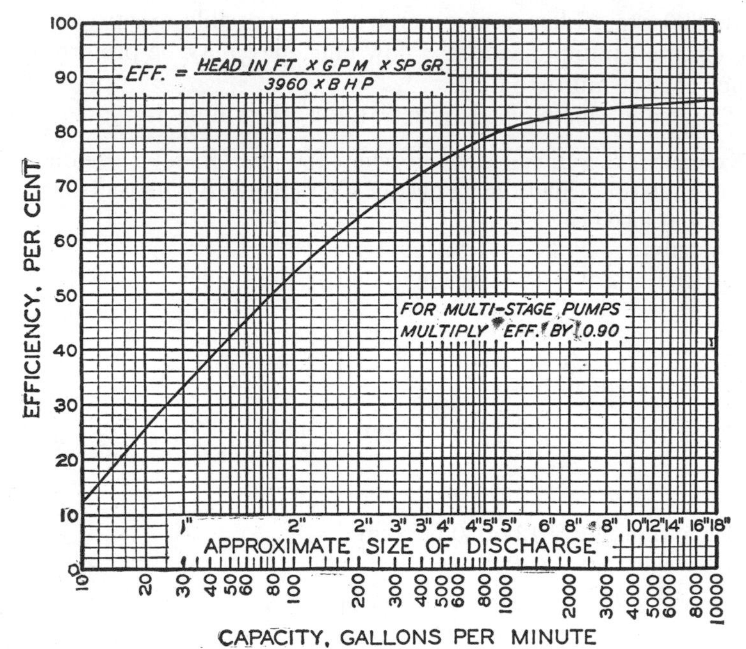

FIG. J-11.—Average efficiencies of double-suction pumps of various capacities. (*Graf, "Gas Engineers' Handbook," McGraw-Hill Book Company, Inc., New York,* 1934.)

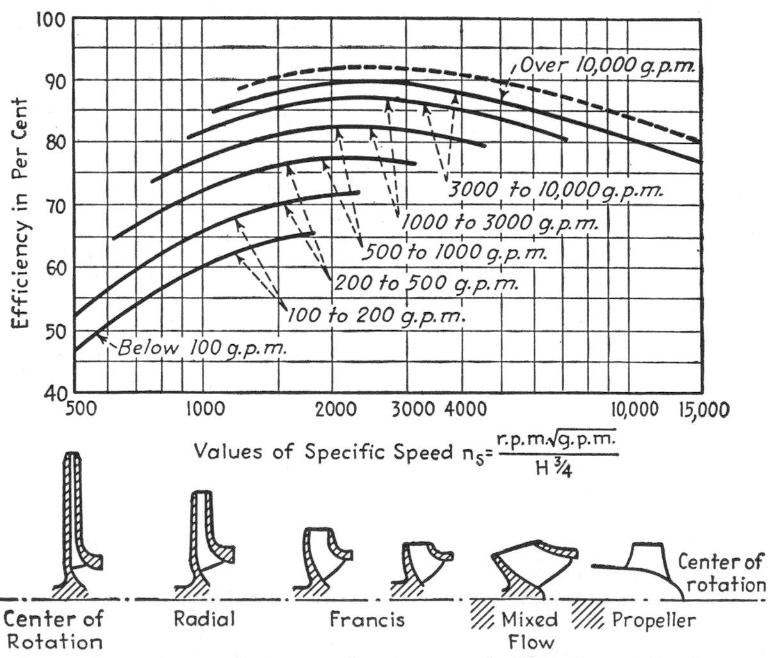

FIG. J-12.—Approximate relative impeller shapes and efficiencies as related to specific speed. (*Courtesy of Worthington Corporation.*)

handle liquids up to 800 SSU and that the larger pumps handle liquids up to 6000 SSU. Figures J-13 to J-15 permit a very approximate calculation of performance of centrifugal pumps handling viscous liquids. *It should be emphasized that these data may fail entirely as applied to some pumps. For example, pumps having water efficiencies of* 70 *per cent may have efficiencies anywhere from* 35 *to* 65 *per cent when handling liquids of* 200 *centistokes* (900 SSU). Inclusion of other

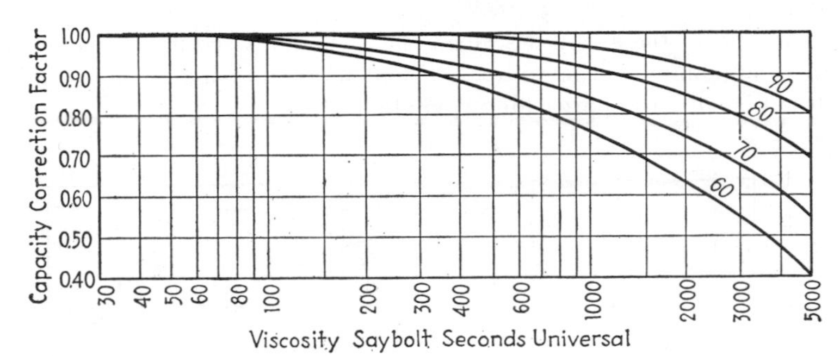

FIG. J-13.—Capacity-correction factor for pumps with water-peak efficiency as marked for capacity at best efficiency point. (*Courtesy of Oil Gas J.*)

factors, such as specific speed and liquid head, might lead to a more nearly adequate relation.

Explanation of Figs. J-13 to J-15 on Effect of Viscosity on Centrifugal Pump Characteristics[1]

Now we take correction factors for the 84 per cent efficient pump for capacity 0.94 and head 0.93 and 84 per cent efficiency will drop to 60 per cent, according to Fig. J-15. Note that Figs. J-13 and J-14 give correction factors for pump capacity and head, but Fig. J-15 gives on scale at left margin corrected efficiencies and not factors. The efficiency at the operating point will be lower in the ratio 81:84, or equals 57.8.

The operating point will come from 638 gpm and 269 ft head water performance as shown in Fig. J-16. Since the shutoff head stays essentially the same irrespective of viscosities, the head capacity curve for 1,000 SSU can be drawn through this point and the specified point using a suitable French curve.

Next calculate the brake horsepower for the specified head and capacity, using 57.8 per cent efficiency:

$$\text{Brake horsepower} = \frac{600 \times 250}{3{,}960 \times 0.578} = 65.5$$

Locate this point in Fig. J-16 and draw a brake-horsepower curve approximately parallel to the water brake horsepower. Now the efficiency for

[1] Quoted from Stepanoff, *Oil Gas J.*, May 6, 1940, p. 78 (with changed figure numbering).

several capacities can be calculated from this brake-horsepower curve and the oil-head-capacity curve as shown in Fig. J-16.

It should be understood that Figs. J-13, J-14, and J-15 are based on limited test data taken on pumps of many types and makes tested at different times by several investigators and therefore will not give results equally accurate for all pumps. We believe, however, it is the best that can be done when taking into account the very complicated nature of pump behavior when handling viscous liquids. It is difficult to estimate the accuracy of these curves. . . .

In addition to the above described effects of the viscosity on the pump performance, pump behavior will be different when pumping oils if the pump operates at conditions approaching cavitation.

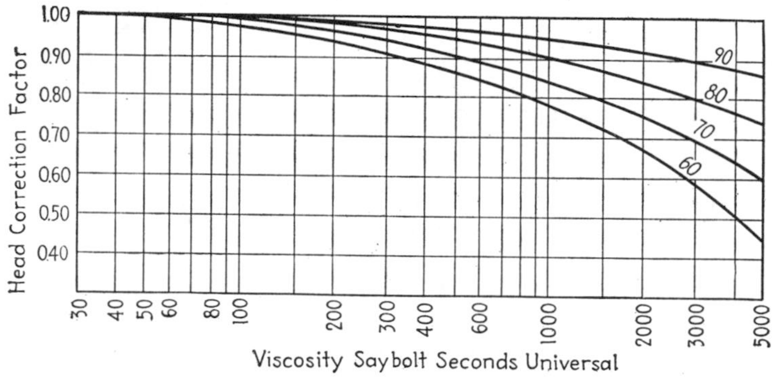

FIG. J-14.—Head-correction factor for pumps with water-peak efficiency as marked for head at best efficiency point. (*Courtesy of Oil Gas J.*)

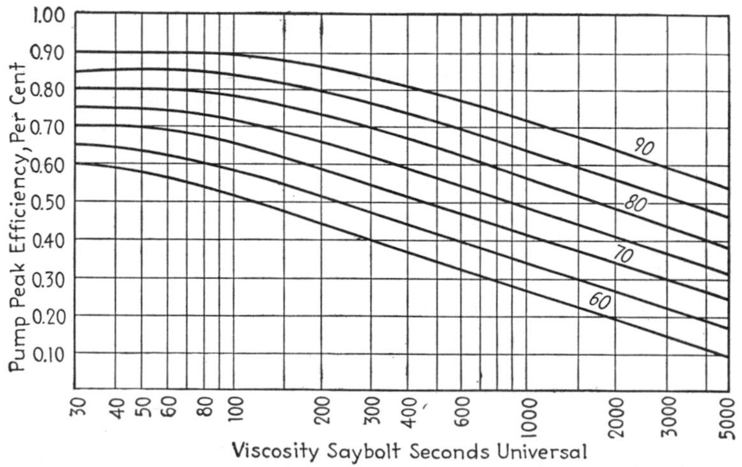

FIG. J-15.—Correction of centrifugal-pump peak efficiency obtained on water for pumping viscous petroleum oil. (*Courtesy of Oil Gas J.*)

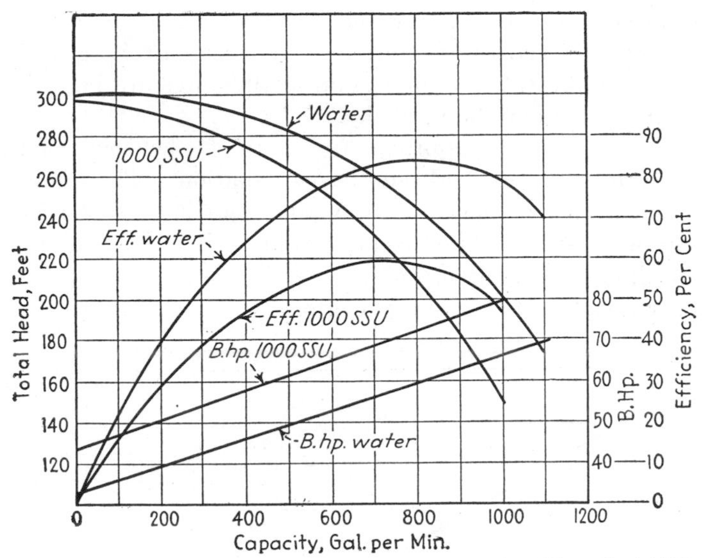

Fig. J-16.—Predicted characteristics for a 4-in. pump handling oil of 1000 Saybolt Universal viscosity. Computed from characteristics on water by factors given in Figs. J-13, J-14, and J-15. (*Courtesy of Oil Gas J.*)

For slurries, clogging characteristics and plasticity must also be taken into account. Figure J-9 illustrates the behavior of pumps handling paper stock. The thickness of this material may be illustrated by the fact that a tumbler nearly full of 6 per cent paper stock must be tipped nearly 45 deg below horizontal before it drops out as a single "slug."

b. Control.—Centrifugal-pump delivery may be controlled by either speed regulation or by throttling. A steep characteristic (*i.e.*, a head that increases rapidly with decreased capacity as the pump in Fig. J-6) is best suited for speed control. A flat characteristic (nearly constant head) is best suited for throttling control.

4. Turbine Pumps

The term "turbine pump" is frequently applied to the diffuser-type centrifugal because of the mechanical similarity of that pump to a steam turbine. However, the term is intended here to refer to turbine pumps differing in type from centrifugals. These pumps, sometimes called "peripheral" or "regenerative" pumps, operate by the impulse principle.

The performance of a typical turbine pump is given in Fig. J-17. It will be noted that the peak of the efficiency curve is sharper than in the case of centrifugals, making the choice of pumps more sensitive. As in the case of centrifugal pumps, capacity may be governed by either speed or throttling. Since shutoff pressures are high, in some cases a relief valve may be needed.

Figure J-18 gives the approximate range of efficiencies of turbine pumps at different heads and requires explanation. The intent is not to give maximum efficiencies obtainable, but rather the probable efficiencies when standard pumps are employed for a particular service. Higher peak efficiencies are obtained, but operation at the peak would afford only a small allowance for overload.

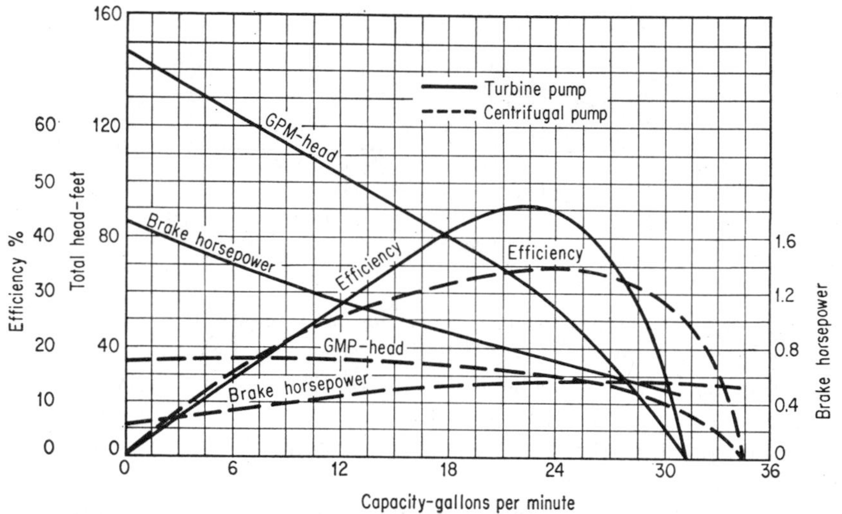

FIG. J-17.—Characteristic curves of centrifugal and turbine pumps. (*Kristal and Annett, "Pumps," 2d ed., McGraw-Hill Book Company, Inc., New York,* 1953.)

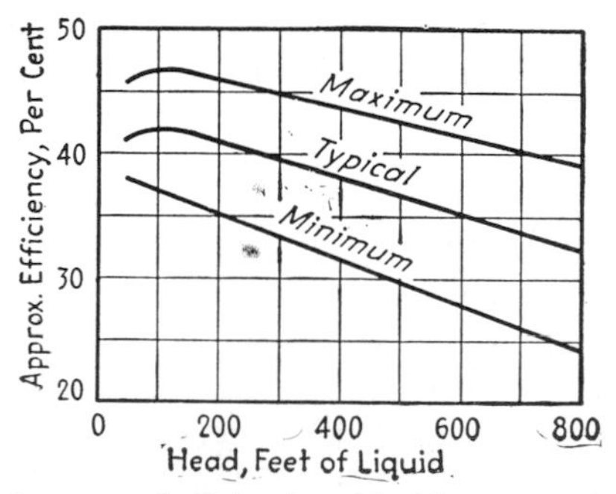

FIG. J-18.—Approximate range of efficiencies of turbine pumps, capacity 20 to 150 gpm.

5. Miscellaneous Types of Pumps

The performance of three pumps is presented as follows:

Rotary pump .... Fig. J-19
Screw pump .... Fig. J-20
Gear pump .... Fig. J-21

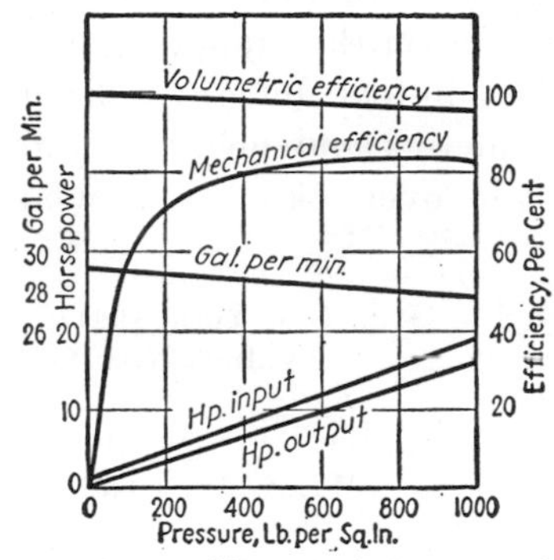

FIG. J-19.—Characteristic curves of a rotary slide-vane pump. (*Kristal and Annett, "Pumps," 2d ed., McGraw-Hill Book Company, Inc., New York,* 1953.)

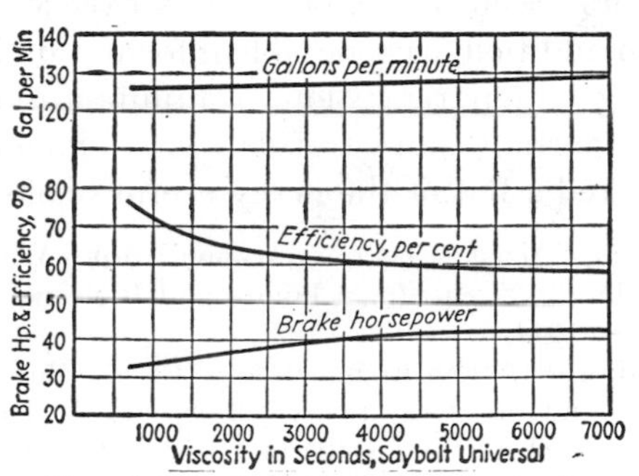

FIG. J-20.—Characteristic curves for a screw pump handling fuel oil. (*Kristal and Annett, "Pumps," 2d ed., McGraw-Hill Book Company, Inc., New York,* 1953.)

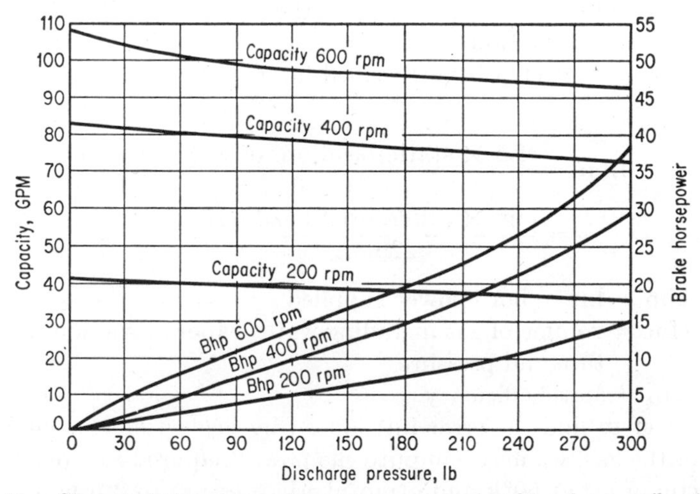

FIG. J-21.—Characteristic curves of spur-gear pump handling heavy oil. (*Kristal and Annett, "Pumps," 2d ed., McGraw-Hill Book Company, Inc., New York,* 1953.)

CHAPTER K

FANS, BLOWERS, AND COMPRESSORS

Wherever air or other gases are handled, fans, blowers, or compressors are encountered. Many excellent texts cover this equipment and, in addition, considerable information is given in texts on air conditioning and refrigeration. Manufacturers' bulletins and quotations are important sources of data and should not be overlooked. Below are listed a few of the many useful books.

Gill, "Air and Gas Compression," John Wiley & Sons, Inc., New York, 1941.
Church, "Centrifugal Pumps and Blowers," John Wiley & Sons, Inc., New York, 1944.
Madison, "Fan Engineering," 5th ed., Buffalo Forge Co., Buffalo, N.Y., 1948.
"Standards, Definitions, Terms, and Test Codes for Centrifugal, Axial and Propellor Fans," Bulletin 110, 2d ed., Air Moving and Conditioning Association, Inc. (Formerly National Association of Fan Manufacturers), Detroit, 1952.

1. Fans and Blowers

a. General Data.—Power required for moving gases may be computed in the same way as for pumping liquids. For last-place accuracy, a correction must be made for the compressibility of gas. In the case of fans and blowers this correction is small and may be neglected. This neglect will be justified later. The resulting equations follow:

$$\text{Bhp} = \frac{\text{cfm} \times \text{static head, in. of water (60°F)}}{6{,}356E_s} = \frac{\text{cfm} \times \text{differential head, psi}}{229.3E_s}$$

where bhp = brake horsepower supplied
cfm = volume of gas including water vapor at suction temperature and pressure
E_s = static efficiency

In order to find the error introduced by neglecting the compressibility of the gas, we may compute the power required for compressing 1,000 cfm of air at 60°F and atmospheric pressure to 2 psig and compare the result with the value computed from the exact equation given under compressors. The results are 12.5 and 12.0 hp, respectively.

Since these figures differ by only 4 per cent, it will be seen that use of the simpler formula is justified for most purposes up to 2 lb discharge. At the low pressures produced by most blowers the difference is entirely negligible.

When fans are used for circulating air against no measurable static head, the power required may be computed from the velocity. This power is the kinetic energy of the gas stream divided by the dynamic efficiency and may be conveniently computed by the equations below.

$$\text{Bhp} = \frac{Qsu^2}{27{,}500{,}000E_d}$$
$$= \frac{Wu^2}{126{,}000{,}000E_d}$$

where bhp = brake horsepower of drive
Q = SCF/min (measured at 60°F, 30 in. Hg)
s = specific gravity of gas; air = 1
W = lb/hr
u = velocity, fps
E_d = dynamic efficiency

Most blowers, even though designed to give static heads, discharge the gas at high velocity, and this velocity head is supplied by the blower. The velocity head may, for most purposes, be considered as a mechanical loss, and the power may be computed directly from the required static head and the static efficiency. If desired, the power may, however, be computed from the total dynamic plus static head and the total efficiency. This may be done either by adding the static plus dynamic power requirements or by first converting the dynamic head to static units and adding to the static head. In the latter case the power should be computed from the static formula and the total efficiency. The velocity head in static units is given by the expressions:

$$\text{Psi} = \frac{\rho u^2}{9{,}250}$$
$$\text{In. water} = \frac{\rho u^2}{326}$$

where ρ = gas density lb/cu ft, discharge conditions
u = gas velocity, fps

b. Performance Characteristics.—Figures K-1 and K-2 give performance characteristics of a propeller and centrifugal fan, respectively. Actually a wide variation of characteristics may be obtained for either type by employing different blade designs. Heads up to

60 in. of water may be obtained by fans, particularly if intended for operation under steady conditions. For heads above 20 in. of water, however, multistage units, *i.e.*, centrifugal and axial compressors, are frequently found desirable. Factors determining such choices are complicated and should be left to the specialist.

For general preliminary estimates the following efficiencies will usually be obtainable for steady services:

	Per Cent
Static efficiency	60
Dynamic efficiency when static head is 0	40
Over-all efficiency	70

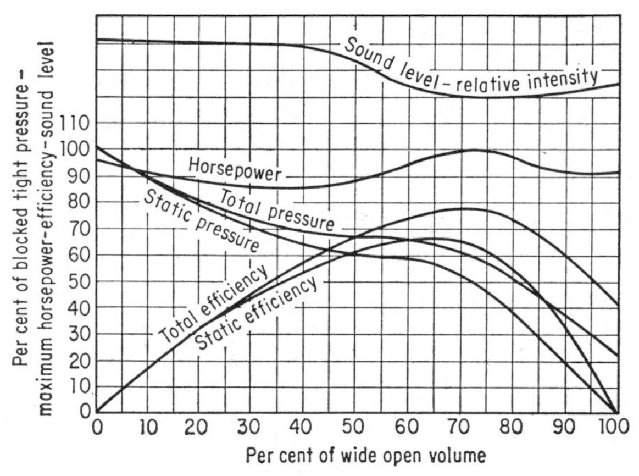

FIG. K-1.—Operating characteristics of axial airfoil type fans. (*Reprinted from "Heating, Ventilating, Air Conditioning Guide," vol.* 36, 1958.)

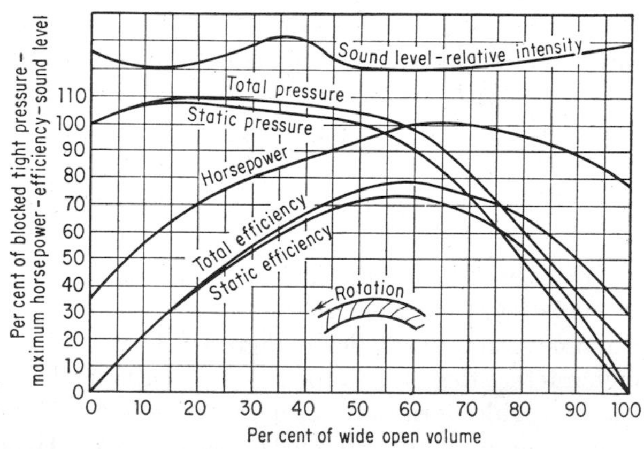

FIG. K-2.—Operating characteristics of centrifugal fans with blades curved backwards (*Reprinted from "Heating, Ventilating, Air Conditioning Guide," vol.* 36, 1958.)

2. General Data for Compressors

a. Power Requirement.—The work of compression of gases can be derived simply from the methods of thermodynamics as outlined in Chapter E. Gases are heated by compression unless the heat of compression is removed. In actual practice a portion, but not all, of this heat is removed. For the sake of definiteness, the calculation of the theoretical power is customarily based on either of two extreme assumptions: either that all of the heat is removed continuously during compression (isothermal), or that none at all is lost (adiabatic). For the isothermal case the work for compressing a perfect gas is [from Eq. (E-36)]

$$w = \int p\,dV = \int \frac{RT}{p}\,dp = RT \ln \frac{p_2}{p_1}$$

where w = work in the same units as RT
p = instantaneous pressure, any unit
p_1, p_2 = suction and discharge pressures
V = instantaneous volume
R = gas-law constant (see Table E-1)
T = absolute temperature, °R or °K, depending on units of R

Converting this equation to specific units of power and introducing the isothermal efficiency, we obtain

$$\text{Bhp} = \frac{MT}{3510E_I} \log \frac{p_2}{p_1}$$

where bhp = brake horsepower
M = SCF/min (measured at 60°F and 30 in. of Hg)
T = suction temperature, °R
p_1, p_2 = absolute pressure at suction, discharge
E_I = isothermal efficiency
= ratio of ideal isothermal power to actual power

The second assumption of no heat loss, *i.e.*, of adiabatic compression, is much more universally used because the results are much closer to actual performance and also because the adiabatic efficiency afforded by a compressor is nearly independent of the properties of the gas being compressed. An additional assumption is also made in the case of perfect gases, namely, that the specific heat is constant. Under these circumstances the *adiabatic* equation of state of a perfect gas becomes

$$pv^{\gamma} = \text{a constant}$$

where p = instantaneous pressure

v = instantaneous volume

γ = adiabatic expansion coefficient

$= \frac{C_p}{C_v} = \frac{C_p}{C_p - R}$

C_p, C_v = specific heats at constant pressure, volume

R = gas-law constant, same units as C_p

The work of ideal adiabatic compression (also called "isentropic" compression) then becomes

$$w = \frac{RT}{n}\left[\left(\frac{p_2}{p_1}\right)^n - 1\right]$$

or

$$\text{Bhp} = \frac{T}{520}\frac{0.0643M}{nE}\left[\left(\frac{p_2}{p_1}\right)^n - 1\right] = \frac{p_1}{p_s}\frac{0.0643M'}{nE}\left[\left(\frac{p_2}{p_1}\right)^n - 1\right]$$

where w = work, same units as RT

R = gas constant

T = absolute temperature at suction, °R

p_1, p_2 = absolute pressures at suction, discharge

p_s = 14.73 psi, 30 in., Hg, etc.

$n = \frac{R}{C_p} = \frac{\gamma - 1}{\gamma}$

$\gamma = \frac{C_p}{C_v}$

C_p, C_v = specific heat at constant pressure, volume

M = SCF/min

M' = cu ft/min measured at suction conditions

E = adiabatic efficiency

Figure K-3 gives a graphical solution of this equation. The accuracy of this figure is about 1 per cent.

Actual gases do not fulfill either of the assumptions used in deriving the adiabatic power equation; the perfect gas law is not obeyed exactly, and specific heats are not constant but increase with both temperature and pressure. (Specific heats of monatomic gases at low pressure are independent of temperature.) Nevertheless the equation does, by good fortune, give reasonably accurate results for actual gases unless the gas-law deviation is very large. For example, the exact theoretical power requirement for two-stage compression of ethylene differs from the value determined by Fig. K-3 by only 2 per cent. It will also be noted by study of Fig. K-3 that the power of compression does not change much for small changes in specific heat or γ. On this account

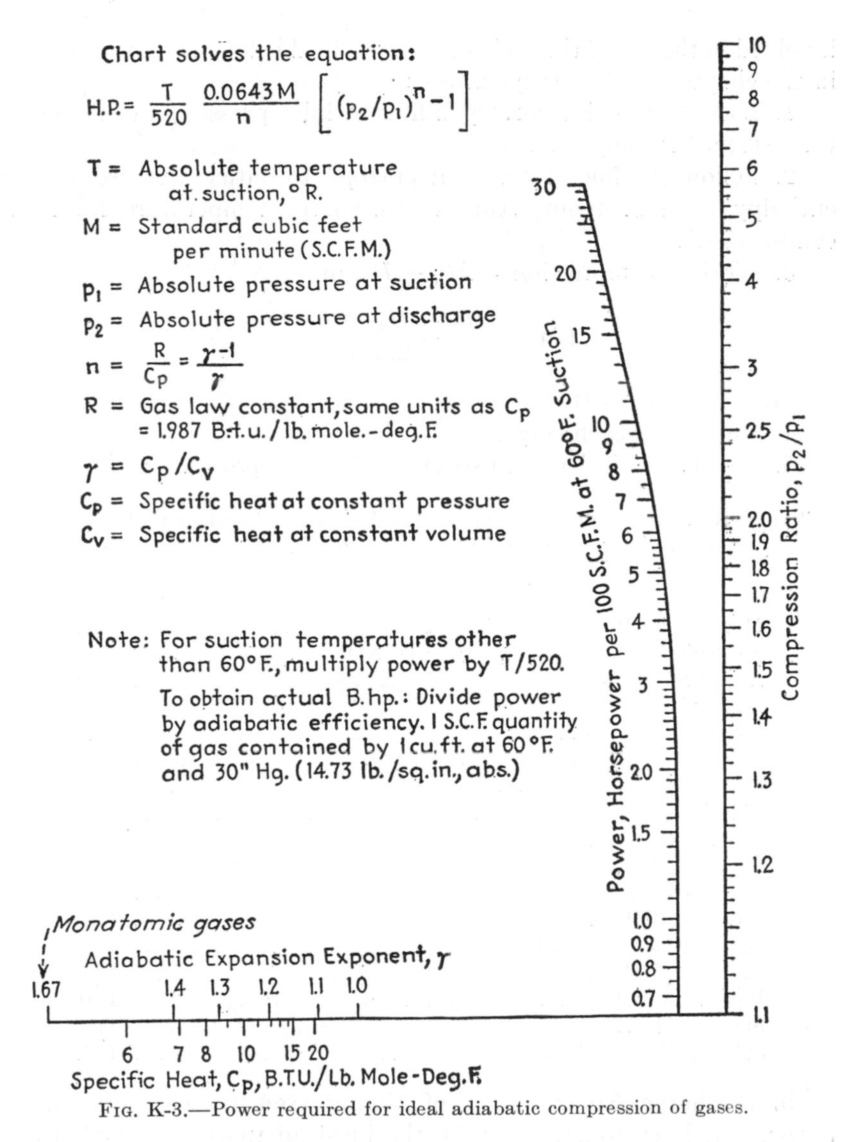

FIG. K-3.—Power required for ideal adiabatic compression of gases.

approximate values of either will suffice. Usually, good accuracy will be obtained by the use of the specific heat (or adiabatic compression exponent) at suction conditions. For gaseous mixtures the average specific heat may be used. Table K-1 gives approximate values of the expansion exponent. Data on specific heats are given in Chapter E.

When imperfect gases are compressed, the use of thermodynamic charts such as Figs. E-20 and E-21 is recommended. The steps involved in the use of these charts are outlined below (see Example E-1 in the chapter on thermodynamics):

1. Locate the point corresponding to inlet pressure and temperature—record enthalpy H_1.
2. Follow the line of constant entropy to outlet pressure, record enthalpy H_2 and temperature t_2 (discharge temperature for ideal compression).
3. Work of compression $= H_2 - H_1$, or

$$\text{Hp} = \frac{(H_2 - H_1) \times \text{lb/hr}}{2{,}544}$$

where H_2, H_1 are in Btu/lb.

4. Repeat for each stage.
5. Divide by efficiency to secure brake horsepower.

TABLE K-1.—APPROXIMATE VALUES OF ADIABATIC EXPANSION EXPONENT OF COMMON GASES AT LOW PRESSURES AND 50 TO 100°F

Gases	Adiabatic Expansion Exponent, γ
Monatomic (He, Ne, Kr, Hg)	1.67
Most diatomic (O_2, N_2, Air, H_2, CO, NO)	1.4
Acetylene (C_2H_2)	1.28
Ammonia (NH_3)	1.31
Butanes (C_4H_{10})	1.1
Butylenes (C_4H_8)	1.1
Carbon dioxide (CO_2)	1.28
Chlorine (Cl_2)	1.33
Ethane (C_2H_6)	1.24
Ethylene (C_2H_4)	1.25
Hydrogen sulphide (H_2S)	1.33
Methane (CH_4)	1.31
Methyl chloride (CH_3Cl)	1.20
Propane (C_3H_8)	1.15
Propylene (C_3H_6)	1.16
Steam (H_2O)	1.32
Sulphur dioxide (SO_2)	1.27

b. Discharge Temperature of Compressed Gases.—For many purposes it is customary to take the ideal adiabatic temperature as close enough. This may be computed by the equation

$$T_2 = T_1 \left(\frac{p_2}{p_1}\right)^{\frac{\gamma - 1}{\gamma}}$$

where T_1, T_2 = absolute temperature at suction, discharge
p_1, p_2 = absolute pressure at suction, discharge
γ = adiabatic expansion exponent

However, it is equally easy to obtain a better value by a heat balance as follows:

$$t_2 = t_1 + \frac{2{,}544P - Q}{60MC_p} = t_1 + \frac{2{,}544P - Q}{Wc_p}$$

where t_1, t_2 = suction, discharge temperature, °F
P = brake horsepower
Q = total heat leaks, Btu/hr
= convection losses + heat removed by jacket water
M = SCF/min
W = lb/hr
C_p = specific heat, Btu/(SCF)(°F)
c_p = specific heat, Btu/(lb)(°F)

In the special case of single-stage reciprocating compressors using normal values of Q, this becomes

$$t_2 = t_1 + \frac{25P}{MC_p} = t_1 + \frac{1{,}500P}{Wc_p}$$

(Example E-1 illustrates the calculation of discharge temperature by use of thermodynamic charts.)

c. Water Vapor in Compressed Gas.—If gases contain water vapor, its volume must be included for computing the power. After compression and cooling, some of this water vapor becomes condensed, and the after- and intercoolers must remove the heat of condensation of the water vapor in addition to the heat of compression. The amount of vapor present may be computed from the vapor pressure of water and Dalton's law:

$$M_w = \frac{M_G p_w}{p - p_w}$$

where M_w = SCF of water vapor (1 lb = 21.02 SCF)
M_G = SCF of fixed gases
p = total pressure, absolute
p_w = vapor pressure water, same units as p (multiply p_w by $\frac{\text{relative humidity}}{100}$ for gases not saturated with water vapor)

The heat of condensation at room temperature is approximately 50 Btu/SCF.

d. Cooling Water.—The work of compression appears quantitatively in the heat content of the gas and must be removed if the compressed gas is desired at the original temperature. This is 2544 Btu/hp-hr. The approximate cooling water requirement for a reciprocating compressor computed on this basis is given in Table K-2. Additional cooling will be required to condense any liquid products, such as water (see previous section), and may also be required because of the Joule-Thomson effect in the case of gases at very high pressures or gases near critical conditions. In this latter case, use of thermodynamic charts is desirable. This use is outlined below (see Example E-1 in the chapter on thermodynamics):

1. Find heat content of inlet gas $= H_1$, Btu/lb.
2. Add power input to obtain heat content of exit gas,

$$H_2 = H_1 + \frac{2{,}544 \text{ bhp}}{\text{lb}} \text{ Btu/lb}$$

(See Paragraph *a* for computing power.)

3. Find heat content of cooled compressed gas $= H_3$, Btu/lb.
4. Heat to be removed by aftercooler (or intercooler) $= H_2 - H_3$ Btu/lb.
5. Cooling water required,

$$\text{Gpm} = \frac{H_2 - H_3}{500 \times \text{temperature rise}}$$

Frequently the gas being compressed is a vapor and is entirely condensed in the aftercooler (condenser). In this case the use of thermodynamic charts, as just described, is identical except that H_3 equals the heat content of the condensed liquid. If thermodynamic charts are not available, the condensing load is simply the latent heat of vaporization. To this must be added about 20 per cent for cooling the compressed vapor to the saturation temperature.

Table K-2.—Approximate Cooling Water Requirement for Reciprocating Compressors Handling Fixed Gases

	Btu/hp-hr	Gpm for each hp		
		10°F rise	20°F rise	40°F rise
Jackets	1000	0.20		
Inter-* and aftercoolers	1500	0.30	0.15	0.04
Total*	2500	0.50	0.25	0.065

* Additional is required to condense any liquid products; also, additional may be required if the Joule-Thomson effect is large.

e. Types of Compressors.—The standard method of compression is still the reciprocating compressor. These compressors are efficient and simple to design for all ranges of pressure and capacity. The reciprocating motion makes them particularly suited to engine drives.

The chief disadvantages are relatively high maintenance and oil consumption.

The centrifugal and axial flow compressors are becoming increasingly popular for the larger capacities. As yet they are not common for the higher pressure ranges, though the only apparent limitation is customer demand. The efficiencies are in general only slightly below the reciprocating type, but frequently this is more than offset by lower costs of purchase and maintenance, together with the almost complete absence of oil spray in the compressed gas. Their rotary motion makes these compressors well suited for motor or turbine drive.

Rotary compressors of various types have important applications. At low heads their efficiencies compare favorably with reciprocating and are well suited to small capacities. The liquid piston type uses no lubrication other than the sealing liquid (usually water).

3. Reciprocating Compressors

a. Power Requirement for Single-stage Compressors.—The approximate efficiencies of reciprocating compressors are given in Fig. K-4. The values given are considered conservative. In fact, considerably higher efficiencies are sometimes quoted; for example, Feller[1] states that "large steam-driven compressors, when in first class condition, show a mechanical efficiency, when fully loaded, of approximately 90%. Smaller steam-driven machines may show lower results, 87 to 90%." However, reciprocating compressors could not be expected to give such high efficiencies without frequent overhaul, and Fig. K-4 is probably about as good as any manufacturer would care to quote for sustained operation with adequate maintenance.

Actually the efficiency probably depends to some extent on the gas being handled. For example, performance on hydrogen would probably be several per cent below that on air or methane.

For most gases the power may be computed directly from Fig. K-5. Where the deviation from the perfect gas law is excessive, use of thermodynamic charts and Fig. K-4 is recommended.

b. Power Requirement for Multistage Compressors.—For compressors developing high pressures it is usually necessary to divide the compression into two or more stages with intercooling between stages, in order to obtain a good mechanical performance and to secure lower power consumption. If a constant adiabatic efficiency is assumed, it can readily be shown that the larger the number of stages, the more the power is reduced. In actual practice, *too* many stages actually increase the power owing to the lower efficiencies obtained and the added power required to supply the pressure drop in the intercoolers. From the

[1] "Air Compressors," McGraw-Hill Book Company, Inc., New York, 1944.

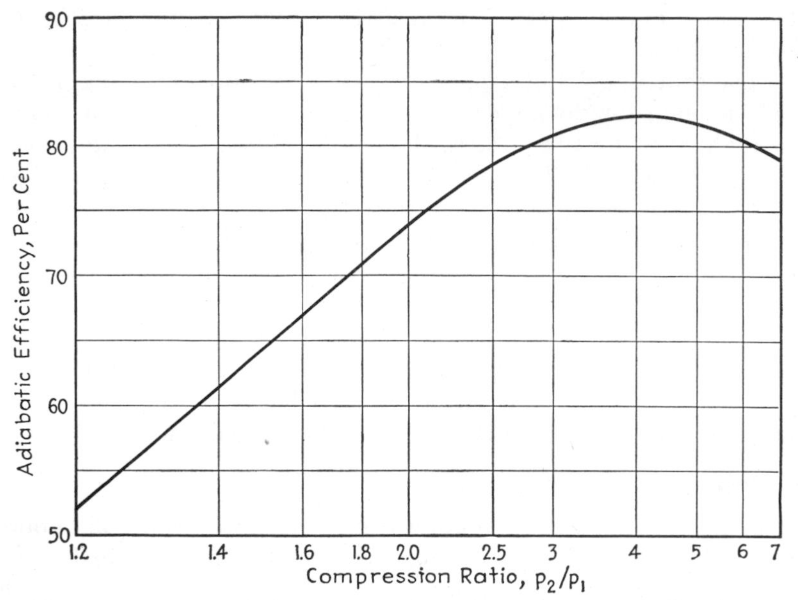

FIG. K-4.—Approximate efficiencies of reciprocating compressors. (*Based on data supplied by the courtesy of Clark Bros. Co., Division of Dresser Industries, Inc.*)

standpoint of lowest power, compression ratios of the individual stages should be, roughly, between 2.5 and 6. Because of higher temperatures developed at high compression ratios it is usually desirable to keep the compression ratio below 5 in order to keep the cylinder walls cool, or for other reasons such as avoiding decomposition of the gas handled or danger of oil ignition.

The power required for multistage compressors may be computed by treating each stage separately as a single-stage machine. For obvious mechanical reasons it is desirable to balance the stages so that each requires the same power. This condition is closely fulfilled when the compression ratios of all stages are identical, that is (neglecting the pressure drop in the intercoolers between stages),

two-stage, $$\frac{p_1}{p_0} = \frac{p_2}{p_1} = \sqrt{\frac{p_2}{p_0}}$$

three-stage, $$\frac{p_1}{p_0} = \frac{p_2}{p_1} = \frac{p_3}{p_2} = \sqrt[3]{\frac{p_3}{p_0}}$$

four-stage, $$\frac{p_1}{p_0} = \frac{p_2}{p_1} = \frac{p_3}{p_2} = \frac{p_4}{p_3} = \sqrt[4]{\frac{p_4}{p_0}}$$

where p_0 = absolute pressure at suction

p_1, p_2, p_3, p_4 = absolute pressure at discharge stages 1, 2, 3, 4

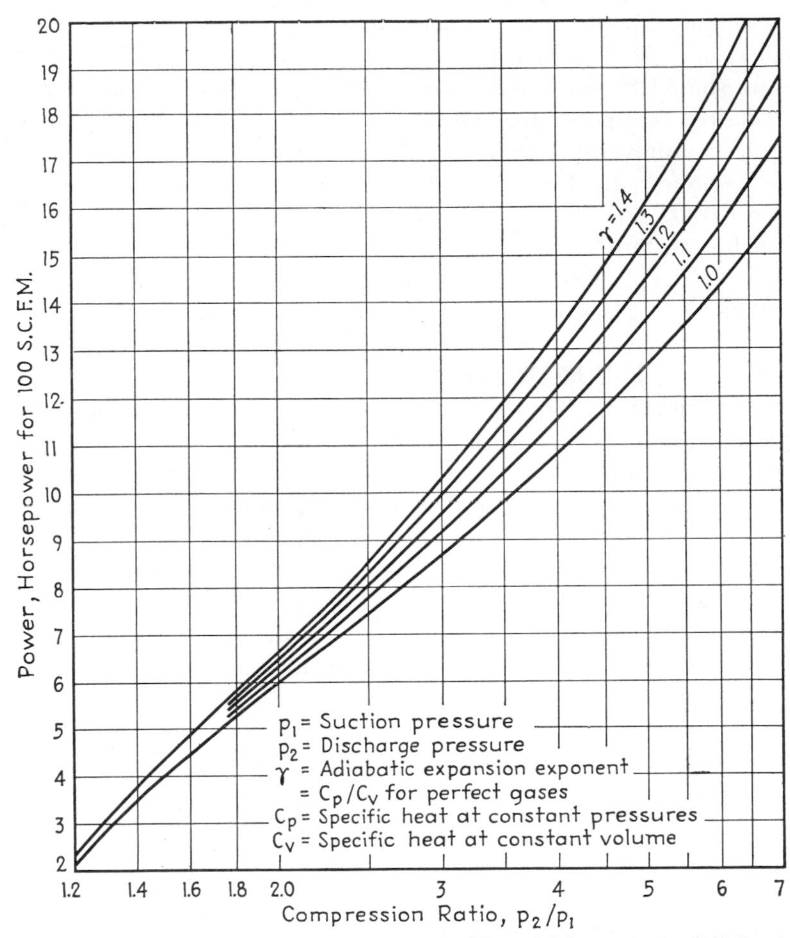

NOTE: For suction temperatures other than 60°F, multiply power by $T/520$, where T = absolute temperature, °R.

FIG. K-5.—Brake horsepower for reciprocating compressors. (*Based on data supplied by the courtesy of Clark Bros. Co., Division of Dresser Industries, Inc.*)

With compression ratios as so calculated, the power for each compression stage may be readily computed, but the total will not include the extra power required because of the pressure drop through the intercoolers. This is most accurately done by estimating the actual pressure drops and adjusting the compression ratios. However, this refinement is seldom necessary, as increasing the power by 2 per cent is usually sufficient allowance, provided the compression ratios of individual stages are above 2.

c. Jacket Water.—The jacket water requirement of compressors depends on mechanical design. In general a high velocity is required so that only about 10°F temperature rise should be assumed for a single pass. If a higher temperature rise is desired, a circulating system should be employed with sufficient fresh water added to hold the desired temperature. Obviously very hard water should be avoided to prevent scale formation. Usually rather warm cooling water is preferred. Feller[1] recommends discharge water temperatures between 100 and 160°F. Approximately 1000 Btu/hp-hr is removed by cooling water. Table K-2 given under general data gives a useful summary of typical rates for cooling water requirements of compressors handling fixed gases. Small compressors are frequently designed for air cooling by fins to obviate the use of cooling water. This expedient is not practiced in the larger sizes.

d. Lubrication.—The lubricating oil actually used by compressors depends largely on the care of supervision by the operator. One gallon per 1,000 hp-hr is a reasonable allowance for preliminary estimates. Nearly all the oil used will leave as spray in the compressed gas and must be removed if an oil-free product is desired.

Reciprocating compressors can be operated with water lubrication by use of leather crimps or carbon rings. In the latter case, lubrication may be eliminated entirely at the expense of increased maintenance. In general, where oil must be excluded, other types of compressors should be considered.

e. Capacity of Reciprocating Compressors.—The volume delivered is the piston displacement divided by the volumetric efficiency. Below are given formulas for capacity:

Single acting:

$$M = \frac{E_v C N D^2 S}{2{,}200}$$

Double acting, no tail rod:

$$M = \frac{E_v C N S(2D^2 - d^2)}{2{,}200}$$

where M = cfm measured at suction conditions
E_v = volumetric efficiency
C = number of cylinders
N = strokes/min
S = length of stroke, in.
D = diameter of piston, in.
d = diameter of piston rod, in.

[1] Feller, "Air Compressors," McGraw-Hill Book Company, Inc., New York, 1944.

The volumetric efficiency depends on a number of factors, of which the piston clearance and the compression ratio are the most important. The clearance is the ratio of minimum volume between piston and exit valves to the displacement. Gill[1] gives a nomograph for computing the volumetric efficiency from the clearance and the compression ratio. Unfortunately, the clearance is usually an unknown factor. However, the equation below based on normal clearance, etc., may be used for estimating volumetric efficiencies directly:

$$E_v = 0.90 - \frac{0.03p_2}{p_1}$$

where E_v = volumetric efficiency
p_1, p_2 = suction, discharge pressures, absolute units
It is perhaps well to point out that very small compressors or those on vacuum service may have somewhat lower volumetric efficiencies.

f. Control.—Since the capacity of compressors changes only about 5 to 15 per cent on doubling the pressure, it is evident that throttling of the discharge is an ineffective method of control. The capacity is changed by different suction pressures, but suction throttling is very seldom advantageous. Consequently, speed control is the best method of regulating output. Where the compressor is electric-driven, the excess capacity may be by-passed back to the suction or "unloading" devices may be employed. These devices usually involve either pockets to increase the clearance, or blocking part of the valves. Unloading changes the compressor capacity by steps, and close regulation for steady delivery at intermediate values must be effected by additional equipment.

4. Centrifugal and Axial Compressors

These machines are really just multistage centrifugal and propeller fans, respectively. The development and application of both machines have expanded rapidly in recent years. Figure K-6 gives performance data for a centrifugal compressor; Fig. K-7 gives performance data of a centrifugal compressor with diffuser vanes for regulation of capacity; Fig. K-8 compares the performance of an axial and centrifugal compressor. The centrifugal machines are quite flexible in their operation. They do exhibit critical speeds, but these are not usually troublesome except for flows very much below the full rating. In the smaller sizes critical speeds may be absent. In the case of axial compressors, critical speeds are an important consideration restricting

[1] "Air and Gas Compression," John Wiley & Sons, Inc., New York, 1941.

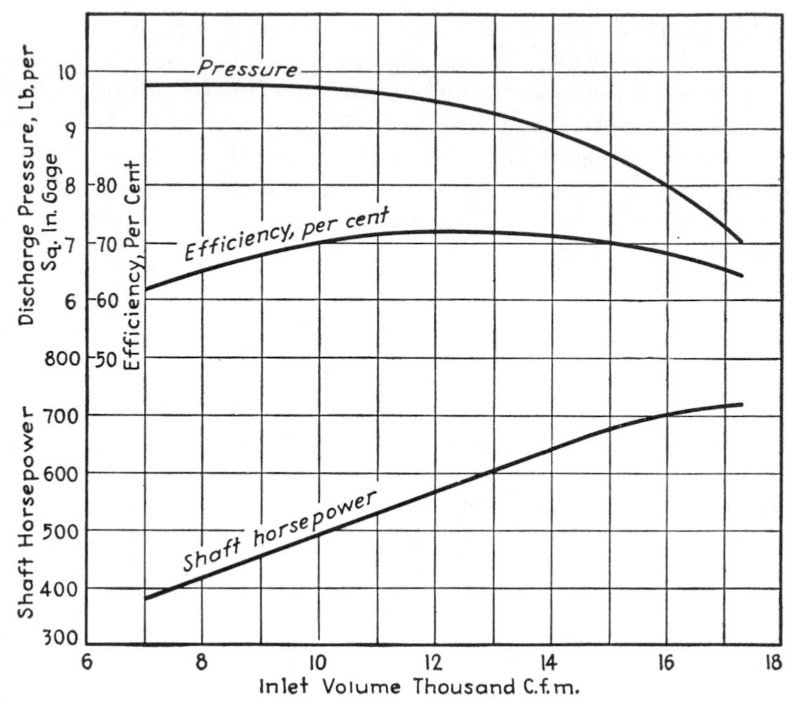

FIG. K-6.—Typical characteristic curve of a constant-speed machine. (*Courtesy of Roots-Connersville Blower, Division of Dresser Industries, Inc.*)

operation to roughly 80 to 120 per cent of the rated capacity. This requirement restricts application of these compressors to uses requiring fairly steady flows against fairly steady heads. Fortunately, however, these are a frequent process requirement so that there are many services in which axials can be used to advantage.

Typical efficiencies of centrifugal and axial compressors are given in Fig. K-9. These values are conservative. For example, Tucker[1] quotes 85 per cent adiabatic efficiencies for axial compressors and still higher values are believed obtainable. Efficiencies are relatively independent of gas density, temperature, and pressure. The number of stages required for high efficiency is, however, dependent on these factors.

Water jackets are not required by these machines but are frequently included to improve the efficiency. For the higher pressure range, intercooling at one or more intermediate pressure is desirable to lower the power and improve the capacity. As in the case of reciprocating

[1] *Chem. & Met. Eng.*, **51**, 96 (March, 1944), also personal communication; cf. Avery, *Allis-Chalmers Power Review*, August, 1939.

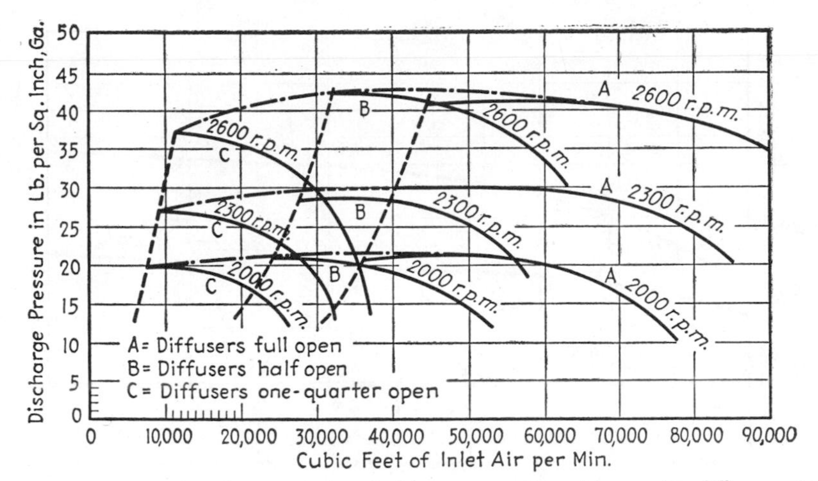

FIG. K-7.—Operating characteristics of a blower operating with movable diffuser vanes. (*Courtesy of Allis-Chalmers Manufacturing Company.*)

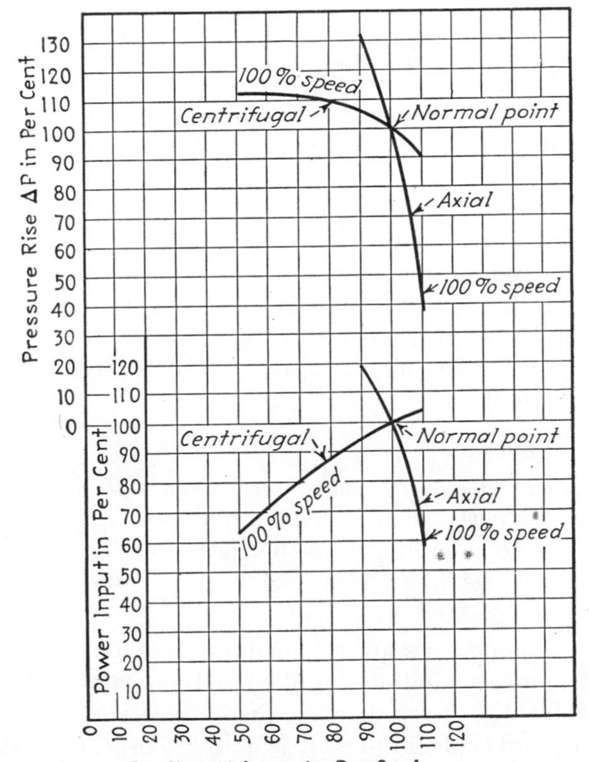

FIG. K-8.—Comparative characteristics of centrifugal and axial compressors at constant speed. (*Courtesy of Allis-Chalmers Manufacturing Company.*)

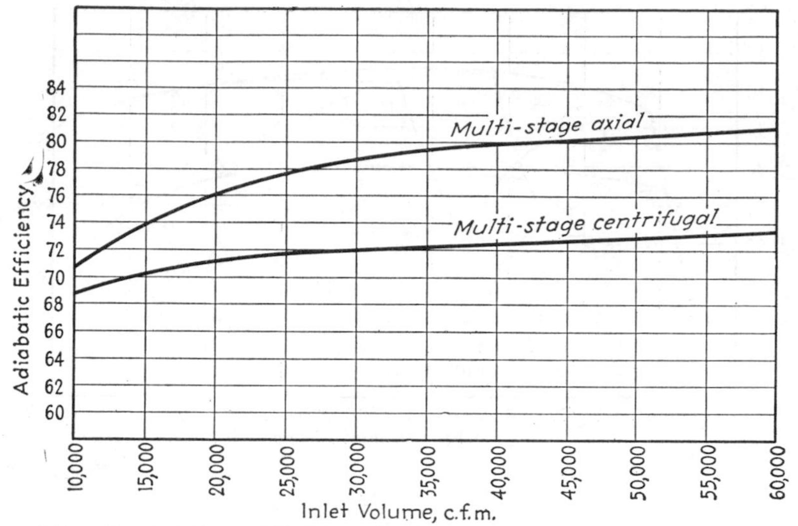

FIG. K-9.—Comparative efficiencies of centrifugal and axial compressors. (*Courtesy of Allis-Chalmers Manufacturing Company.*)

compressors, the power may be computed separately for each stage from suction to first cooler, and from there to second cooler, etc.

Centrifugal compressors are well suited for throttling control or control by diffuser vanes (see Fig. K-7). Speed control is also effective, particularly to obtain a constant flow against a variable head. Axial flow compressors are well suited for speed control when operating against a relatively constant head, but they are not well suited for throttling control.

The bearings of centrifugal and axial compressors are the only parts requiring lubrication. Consequently, the compressed gases will be substantially oil free.

5. Rotary Compressors

The A.S.M.E. test code classifies rotary compressors as *sliding vane,* in which longitudinal vanes slide radially in a rotor mounted eccentrically in a cylinder; *two impeller,* in which two mating lobed impellers revolve within a cylinder; and *liquid piston,* in which a liquid serves to displace the air within a rotating element.[1]

Slide-vane compressors are reasonably priced but have disadvantages of high lubricating-oil consumption and comparatively high maintenance. Table K-3 gives data on single-stage compressors. Two-stage machines are made for pressures up to 125 psig and are not suited to throttling control. Water jackets are required for the larger sizes.

Two impeller compressors are advantageous for pressures between

[1] Feller, *op. cit.*

TABLE K-3.—SINGLE-STAGE SLIDING-VANE COMPRESSOR CAPACITIES
(Feller, "Air Compressors," McGraw-Hill Book Company, Inc., New York, 1944)

60-cycle, rpm	Lb gauge discharge pressure				Lb gauge discharge pressure			
	20	30	40	50	20	30	40	50
	Cfm, actual free air delivery				Hp rating of nearest commercial-size squirrel-cage motor			
1,160	32	31	.....	.....	5	5		
1,160	44	42	.....	.....	5	7½		
1,160	52	50	.....	.....	5	7½		
1,160	76	74	72	70	7½	10	15	15
1,160	83	82	80	78	7½	10	15	15
1,160	112	109	107	105	10	15	20	20
1,160	129	127	124	122	15	15	20	25
1,160	154	152	149	146	15	20	25	25
1,160	197	194	190	186	20	25	30	30
1,160	232	228	224	220	20	25	30	40
870	284	280	275	270	25	30	40	50
870	339	334	329	324	30	40	50	50
870	377	370	365	360	30	40	50	60
870	403	396	390	385	40	50	60	60
870	482	473	467	460	40	50	75	75
690	534	526	519	512	50	60	75	100
690	607	598	592	585	50	75	100	100
690	685	675	665	656	60	75	100	100
690	773	763	754	745	75	100	100	125
575	890	878	866	855	75	100	125	150
575	1,050	1,037	1,023	1,010	100	125	150	150
575	1,410	1,392	1,374	1,355	125	150	200	200
575	1,610	1,592	1,572	1,554	125	200	200	250

the most efficient ranges of blowers and reciprocating compressors. They are reasonably priced, but their oil consumption is appreciable and they are not suited to throttling control.[1] Figure K-10 gives performance data on a large machine.

Liquid piston compressors present the advantages of using no lubrication and developing pressures up to 75 psig by a single stage. They are suited for either speed or throttling control. The performance of a small machine is given in Fig. K-11.

[1] Under proper conditions, two-impeller compressors can be used without lubrication.

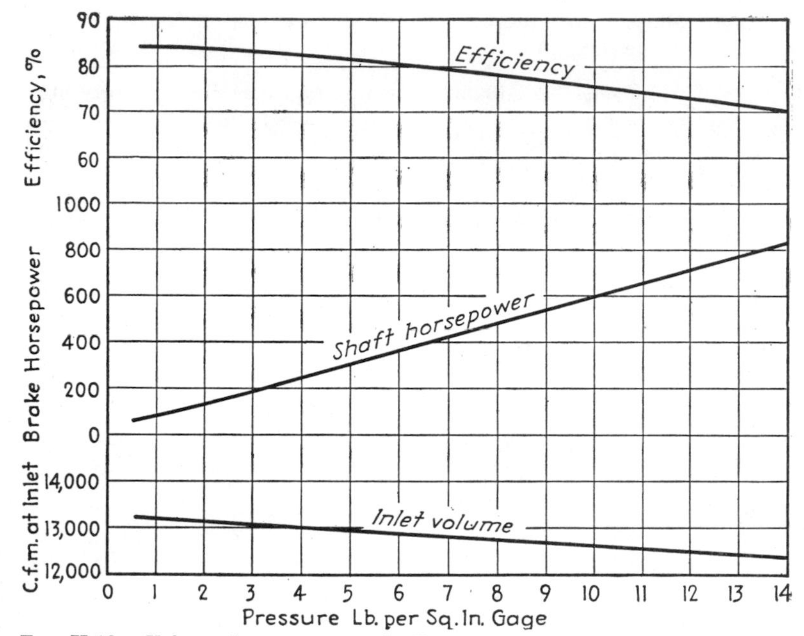

FIG. K-10.—Volume, horsepower, and efficiency curves of a 588-rpm, 12,500-cfm, 12-lb, two-impeller compressor. (*Courtesy of Roots-Connersville Blower, Division of Dresser Industries, Inc.*)

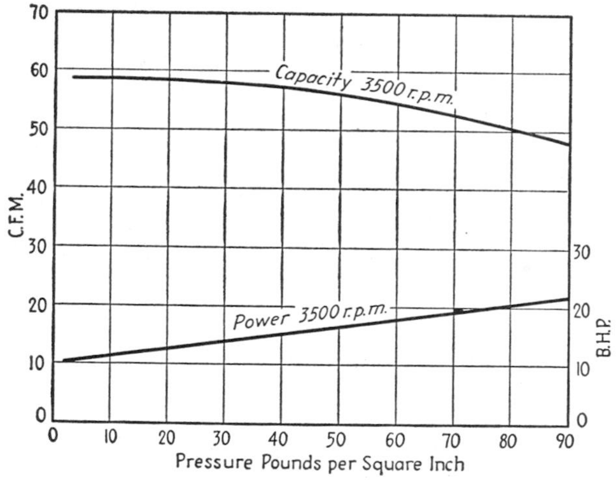

FIG. K-11.—Typical performance curve of a single-stage liquid piston compressor. (*Courtesy of Nash Engineering Company.*)

6. Vacuum Equipment

There is, of course, no magic in the atmospheric pressure of 14.6955 psi, the mean atmospheric pressure at sea level; and compressors of all types will work at pressures well below this (*i.e.*, in the vacuum range). The power requirement is computed in the same manner as the higher pressure equipment. The lower the absolute pressure, however, the larger the displacement required for the same weight of gas so that mechanical vacuum pumps run to size. Further, the volumetric efficiencies of all types of compressors are lower in the vacuum range and at very low pressures approach zero; that is, there is a definite limit to the vacuum any given machine can achieve. In general, rotary machines produce the best vacuums, reciprocating next, and centrifugal machines the poorest.

By far the most popular device for industrial vacuum work is the steam ejector. Steam ejectors are small, inexpensive, and have no moving parts. The steam consumption of these ejectors depends on a large number of factors which are listed below:

1. Absolute pressure of vacuum system.
2. Discharge pressure.
3. Steam pressure.
4. Cooling water temperature.
5. Density and quantity of fixed gases.
6. Quantity of water vapor.
7. Quantity, density, and saturation temperature of other vapors.

Table K-4 gives typical steam requirements for ejection of air and water vapor.

TABLE K-4.—AVERAGE STEAM CONSUMPTION STEAM JET EJECTORS
(Reprinted from "Cameron Hydraulic Data," 12th ed., 1951, by permission of the Ingersoll-Rand Company)
Pounds per hour at 100 psig pressure, steam pressure

Weight mixture, lb/hr	% net dry air by weight	Suction pressure, in. Hg. abs									
		0.5		1.0		1.5	2.0	3.0	4.0		6.0
		3-stage	2-stage	3-stage	2-stage	2-stage	2-stage	2-stage	2-stage	1-stage	1-stage
10	100	73	99	59	70	58	50	42	38	58	36
10	70	59	84	47	60	49	42	35	31	63	39
10	40	45	68	33	47	38	32	26	23	68	41
10	10	24	45	16	28	21	17	14	12	74	42

NOTE: Steam consumption is directly proportional to capacity.

CHAPTER L

PHASE EQUILIBRIA

The difference in composition between two phases affords the basis for common methods of chemical separation. Vapor-liquid equilibria permit distillation and absorption stripping, liquid-liquid equilibria permit liquid-liquid extraction, and solid-liquid equilibria permit crystallization and leaching operations. This chapter gives some of the basic laws of solution together with some experimental data. There are a number of excellent references covering experimental data and rules for extending it. Some of these are listed below:

Hildebrand and Scott, "The Solubility of Non-electrolytes," 3d ed., Reinhold Publishing Corporation, New York, 1955.
Perry, "Chemical Engineers' Handbook," 3d ed., McGraw-Hill Book Company, Inc., New York, 1950.
"International Critical Tables," McGraw-Hill Book Company, Inc., New York.
Chu, "Distillation Equilibrium Data," John Wiley & Sons, Inc., New York, 1950.
Maxwell, "Data Book on Hydrocarbons," 2d ed., D. Van Nostrand Company, Princeton, N.J., 1958.
Seidell (and Linke), "Solubilities, etc.," three companion volumes, each with a slightly different title, D. Van Nostrand Company, Princeton, N.J.

1. Basic Laws and Rules

a. Vapor-Liquid Equilibria.—These obey the general thermodynamic requirement of equal fugacity. This somewhat abstract statement does not of itself yield much information but becomes useful once the fugacities are related to pressure, temperature, and composition. A formal statement of the thermodynamic requirement follows:

$$f_1 = f_1^0 \gamma_1 x_1 \qquad f_2 = f_2^0 \gamma_2 x_2 \tag{L-1}$$

where f_1 = fugacity of component 1 in the vapor
f_1^0 = fugacity of pure liquid component 1 or pure vapor in equilibrium with it
γ_1 = activity coefficient of component 1 in liquid
x_1 = mole fraction of component 1 in liquid

The fugacity is defined by Eqs. (E-46) and (E-47), and the activity coefficient by Eqs. (E-48) to (E-51). However, for this discussion, the fugacity may be regarded as an idealized partial pressure and the activity coefficient as a correction factor. Both may be evaluated by experiment and used to correlate and extend available data.

For ideal liquids, the activity coefficient is unity and drops out, leading to the formal form of Raoult's law:

$$f_1 = f_1^0 x_1 \qquad \text{etc.} \tag{L-2}$$

Moreover, if the fugacity and the pressure are the same (as they are at low pressure), the equation reduces to the familiar form of Raoult's law:

$$p_1 = p_1^0 x_1 \qquad \text{etc.} \tag{L-3}$$

where p_1 = partial pressure of component 1 = py_1
p = total pressure
p_1^0 = vapor pressure of component 1
x_1 = mole fraction of component 1 in liquid
y_1 = mole fraction of component 1 in vapor

This law is directly useful for many purposes including absorption and stripping calculations. It may be used to compute the vapor-liquid equilibria for a binary mixture at constant pressure from the vapor pressures of the two components. To do this, choose a temperature intermediate between the two boiling points and secure the vapor pressures of the pure components:

$$x_1 = \frac{p - p_2^0}{p_1^0 - p_2^0} \tag{L-4}$$

$$y_1 = \frac{p_1^0 x_1}{p} \tag{L-5}$$

where the symbols are defined as for Eq. (L-3).

Raoult's law is usually followed by mixtures of very similar compounds such as benzene-toluene, pentane-heptane, and ethyl-propyl alcohol mixtures. Even for such mixtures, deviations are appreciable at elevated pressures (and corresponding temperatures) owing to deviation of their vapors from the perfect gas law. Correction for this deviation can be made by reference to fugacity correlations. If deviations are not excessively large, the ratio of fugacity to pressure (f/p) for any one component may be considered to be independent of the vapor composition and therefore fixed by its temperature and pressure, either as absolute values or as reduced temperature and

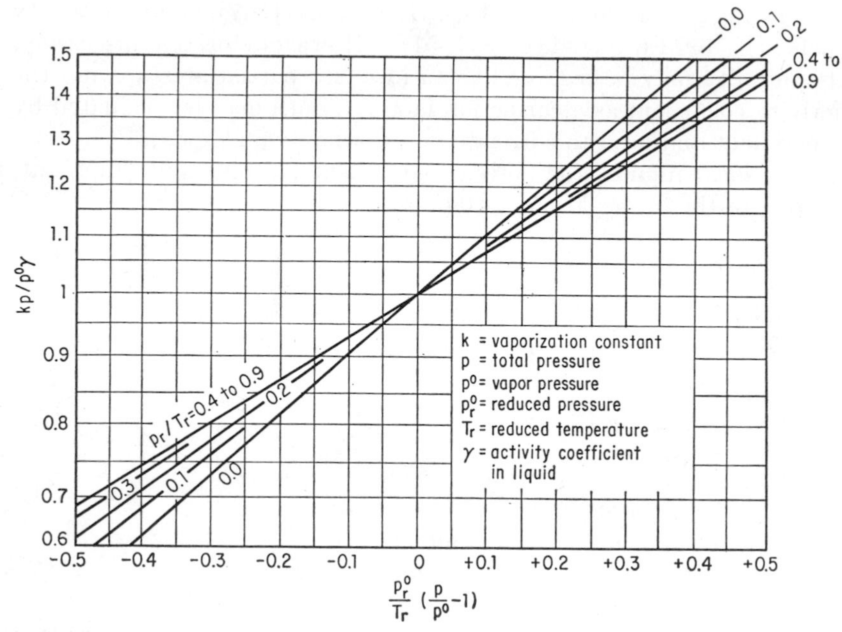

FIG. L-1.—Chart for estimation of vaporization-equilibrium constants—narrow range.

pressure. Use of the latter permits generalized correlations that have been widely used to predict vapor-liquid equilibria.[1] Using the same precept, Eq. (L-6) has been derived:[2]

$$\ln \frac{k_1 p}{\gamma_1 p_1^0} = (1 - z_v + z_l) \frac{p - p_1^0}{RT} \tag{L-6}$$

where $k_1 = y_1/x_1$, equilibrium vaporization constant, component 1
y_1, x_1 = mole fraction of component 1 in vapor, liquid
p = total pressure of system
p_1^0 = vapor pressure of pure component 1 in same units as p
T = absolute temperature
R = gas law constant in units corresponding to p, T, and V
z_v, z_l = compressibility factor for vapor, liquid: pV_v/RT, pV_l/RT
V_v, V_l = volume of component of interest as saturated vapor, liquid
γ_1 = activity coefficient of component 1 in liquid

[1] Method originally proposed by W. K. Lewis. Useful charts are given by Perry, "Chemical Engineers' Handbook," 3d ed., McGraw-Hill Book Company, Inc., New York, 1950.

[2] Scatchard and Raymond, *J. Am. Chem. Soc.*, **68**, 1278 (1944); cf. Perry, *op. cit.*

The difference between the compressibility factors $(z_v - z_l)$ has been correlated[1] [making it possible to generalize Eq. (L-6) still further]. This correlation, though developed for nonpolar compounds, is apparently valid for many polar compounds,[2] but it would fail for substances such as hydrogen fluoride and acetic acid that polymerize to a variable degree in the vapor state. Figures L-1 and L-2 are based

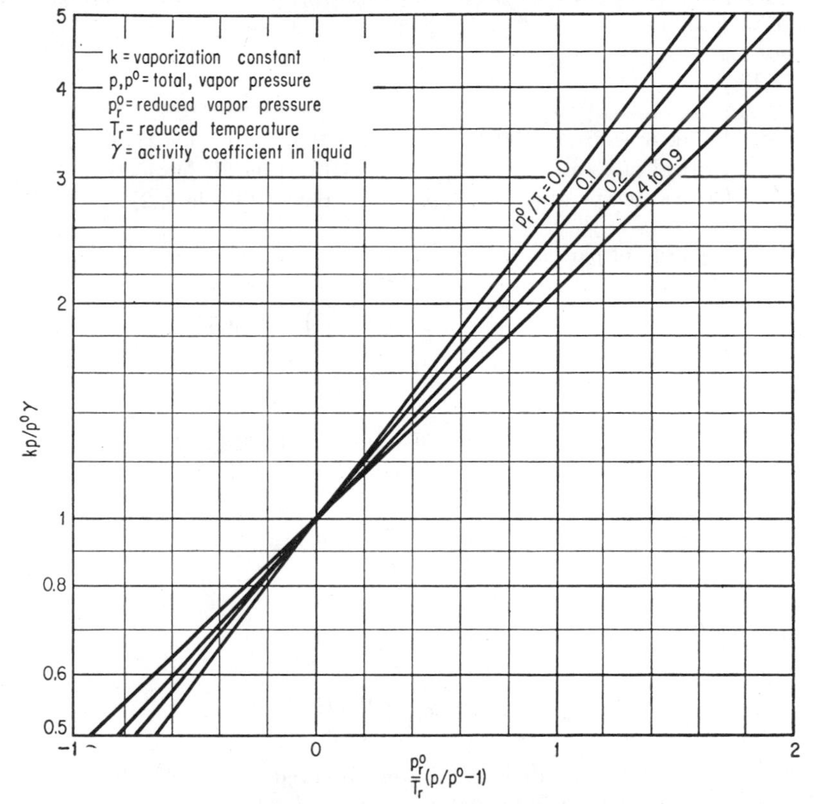

FIG. L-2.—Chart for estimation of vaporization-equilibrium constants—broad range.

on combining correlation and Eq. (L-6). The two figures are mathematically equivalent but cover different ranges. Their use is illustrated in Example L-1.

These charts admittedly suffer in accuracy at extreme conditions owing to inaccuracies in the numerous assumptions made in their

[1] Cope, Lewis, and Weber, *Ind. Eng. Chem.*, **23,** 887 (1931).

[2] Thomson, *Chem. Rev.*, **38,** 36 (1946).

derivations. The most serious of these is neglect of the effect of vapor composition. In general, the charts afford a reliable approximation when the total pressure is neither more than half the critical pressure nor twenty times the vapor pressure. The charts are useful as a first approximation even at appreciably more extreme conditions when experimental data are lacking.

Example L-1. Find the equilibrium vaporization constant for propane at 100°F and 500 psi.

SOLUTION:

1. Physical data:

Vapor pressure at 100°F..........	189 psia (Table E-16)
Critical pressure..................	617 psia (Table E-2)
Critical temperature.............	666°R (Table E-2)

2. Calculation of parameters:

$$\frac{p_r}{T_r} = \left(\frac{189}{617}\right)\left(\frac{666}{560}\right) = 0.365$$

$$\frac{p}{p^\circ} = \frac{500}{189} = 2.65$$

$$\frac{p_r}{T_r}\left(\frac{p}{p^\circ} - 1\right) = 0.365(1.65) = 0.602$$

3. Use of Figure L-2:

$$\frac{kp}{p^\circ\gamma} = 1.56$$

Assume perfect solution, $\gamma = 1$

$$k = \frac{1.56}{2.65} = 0.59$$

4. Discussion: The result is within the observed range of k values for propane. The experimental values spread indicating some dependence on vapor composition, a dependence that becomes more pronounced at still higher pressures.

As already indicated, deviations from Raoult's law due to imperfect liquid mixtures are common and are to be expected for mixtures of liquids that are dissimilar. Nearly all aqueous solutions deviate noticeably from Raoult's law. For many mixtures of two components at any specified temperature, activity coefficients (expressing the deviation) can be correlated by the following equations:

$$\log \gamma_a = [A + 2(B - A - D)x_a + 3Dx_a^2]x_b^2 \tag{L-7}$$

$$\log \gamma_b = [B + 2(A - B - D)x_b + 3Dx_b^2]x_a^2 \tag{L-7a}$$

$$\log \gamma_a = A\left(\frac{Bx_b}{Ax_a + Bx_b}\right)^2 \tag{L-8}$$

$$\log \gamma_b = B\left(\frac{Ax_a}{Ax_a + Bx_b}\right)^2 \tag{L-8a}$$

where γ_a, γ_b = activity coefficient of component a, b
x_a, x_b = mole fraction; in liquid, of component, a, b
A, B = constants for two constant equations
D = added constant for complex systems

Equations (L-7) and (L-7*a*) are the Margules equations in three-constant forms (they reduce to the common two-constant form when $D = 0$), and the equation pair (L-8) and (L-8*a*) are the van Laar equations. Additional constants may be employed,[1] but two constant equations are adequate for most situations. Van Laar and Margules equations prove to be about equally useful. Some mixtures are best fit by one, some by the other. Indeed, for a great many mixtures, either set fits the data equally well. The numerical values computed by Margules become nearly identical with those secured by the van Laar when the two constants A and B are small or do not differ much from each other.

Ternary van Laar equations have been developed, but they are considerably less flexible than the more highly recommended Margules equations[1]. A simplified form of these follows:[2]

$$\begin{aligned}\log \gamma_a = {} & [A_b + 2(B_a - A_b)x_a + (B_c - C_b)x_c]x_b^2 \\ & + [A_c + 2(C_a - A_c)x_a + (C_b - B_c)x_b]x_c^2 \\ & + [\tfrac{1}{2}(A_b + B_b + A_c + C_a - C_b - B_c) \\ & \qquad + (B_a - A_b + C_a - A_c)x_a]x_bx_c\end{aligned} \tag{L-9}$$

$$\begin{aligned}\log \gamma_b = {} & [B_a + 2(A_b - B_a)x_b + (A_c - C_a)x_c]x_a^2 \\ & + [B_c + 2(C_b - B_c)x_b + (C_a - A_c)x_a]x_c^2 \\ & + [\tfrac{1}{2}(A_b + B_a + B_c + C_b - A_c - C_a) \\ & \qquad + (A_b - B_a + C_b - B_c)x_b]x_ax_b\end{aligned} \tag{L-9a}$$

$$\begin{aligned}\log \gamma_c = {} & [C_a + 2(A_c - C_a)x_c + (A_b - B_a)x_b]x_a^2 \\ & + [C_b + 2(B_c - C_b)x_c + (B_a - A_b)x_a]x_b^2 \\ & + [\tfrac{1}{2}(A_c + C_a + B_c + C_b - A_b - B_a) \\ & \qquad + (A_c - C_a + B_c - C_b)x_a]x_ax_b\end{aligned} \tag{L-9b}$$

[1] Wohl, *Trans. Am. Inst. Chem. Engrs.*, **52**, 215 (1946).

[2] Derivation based on two equations given by Perry, "Chemical Engineers' Handbook," 3d ed., McGraw-Hill Book Company, Inc., New York, 1950. The ternary constant C is eliminated from his equation 21, p. 528, by use of his suggested value of C. The results can be rearranged to yield Eq. (L-9) of this text.

where $\gamma_a, \gamma_b, \gamma_c$ = activity coefficient, component a, b, c
x_a, x_b, x_c = mole fraction of component a, b, c in liquid
A_b, B_a = binary Margules constant for mixtures of A and B
A_b = limiting value of $\log \gamma_a$ when $x_b = 1$
B_a = limiting value of $\log \gamma_b$ when $x_a = 1$
A_c, C_a = same for mixtures of A and C
B_c, C_b = same for mixtures of B and C

The original derivation for Margules and van Laar relations was based on constant temperature and constant pressure. The effect of pressures in normal ranges is not significant, but the constants of the equations are usually somewhat temperature dependent. Even so, as an approximation, the temperature dependence may be neglected and the equations applied to data secured at the boiling point of the mixture, *i.e.*, constant pressure, variable temperature. The equations are useful in correlating and extending available data. For example, van Laar or Margules constants can be calculated from the composition of a binary azeotrope and the behavior of mixtures of three components can be estimated from the more readily available data on the three pairs of binary mixtures and even extended to four-component mixtures.[1]

The Margules and van Laar constants are, of course, most reliable when determined by a number of measurements. Two values, such as the activity coefficient of the two components of a binary mixture, are mathematically adequate. The equations applying are as follows:
Margules:

$$A = \frac{1 - 2x_1}{x_2^2} \log \gamma_1 + \frac{2}{x_1} \log \gamma_2 \qquad \text{(L-10)}$$

$$B = \frac{1 - 2x_2}{x_1^2} \log \gamma_2 + \frac{2}{x_2} \log \gamma_1 \qquad \text{(L-10}a\text{)}$$

Van Laar:

$$A = \log \gamma_1 \left(1 + \frac{x_2 \log \gamma_2}{x_1 \log \gamma_1}\right)^2 \qquad \text{(L-11)}$$

$$B = \log \gamma_2 \left(1 + \frac{x_1 \log \gamma_1}{x_2 \log \gamma_2}\right)^2 \qquad \text{(L-11}a\text{)}$$

Numerical values of van Laar constants are given later in this chapter.

For dilute solutions, the derivations from ideality of both liquid

[1] Perry, "Chemical Engineers' Handbook," 3d ed., page 529, McGraw-Hill Book Company, Inc., New York, 1950.

and vapor phases can be lumped together and expressed by Henry's law. Two forms of this follow:

$$p_2 = Hx_2 \tag{L-12}$$
$$C = Bp_2 \tag{L-13}$$

where p_2 = partial pressure of component 2 (solute) in vapor
H = Henry's law constant, same units as p_2
x_2 = mole fraction of component 2 in liquid
C = concentration of solute in solution
B = Bunsen-type coefficient in units corresponding to C and p_2, such as SCF/(cu ft)(atm)

b. Liquid-liquid equilibria refer to the distribution of components between two liquids. This can only occur if miscibility is not complete, and the simpler laws of "ideal" solution, such as Raoult's law, do not apply. The universal thermodynamic requirement for equilibria—that each component have the same fugacity in each phase—does, of course, apply, and Margules equations can be used to correlate and extend liquid-liquid data as for vapor-liquid.

A simple case of great importance is the distribution of a substance (solute) between two liquids immiscible (or partly miscible) in each other. The distribution is conveniently expressed as a ratio:

$$k_3 = x_3z_3 \tag{L-14}$$

where k_3 = distribution coefficient for component 3
x_3 = mole fraction in one liquid phase of component 3
z_3 = mole fraction in the other liquid phase of component 3

For solutions at low concentrations, Henry's law is applicable to both phases and the distribution ratio is constant. The distribution given in the preceding is based on mole per cent. However, weight per cent or other expressions of concentration may be used. In many systems of interest, the "immiscible" liquids have substantial mutual solubility that is affected by addition of the third component (solute) distributed between the two layers.

The complete behavior of any three-component system involving two liquid phases can be described by two charts. A wide variety of chart combinatons can be used. One that is quite convenient is a triangular chart defining the area of miscibility. "Tie lines" may be added to such a chart, but the complete definition is better secured by a separate chart, such as z_3 or k_3 plotted against x_3 [using the symbolism of Eq. (L-14)]. Figures L-4, L-5, and L-6, appearing later in this chapter are so prepared.

2. Experimental Equilibrium Data

A limited number of systems are covered by the data given in this section. It has not been found possible to present a summary of the hydrocarbon vaporization constants due to the comprehensive and voluminous nature of reliable data now available. Figures L-1 and L-2 may be used where applicable, but k values outside this range become increasingly dependent on vapor composition.[1]

The heat of solution of vapors is the sum of the heat of vaporization and the heat of mixing the two liquids at the temperature of interest. Frequently the latter may be neglected and the heat of vaporization alone used. Another rule for estimating the heat of solution is to employ the heat of vaporization at the saturation pressure corresponding to the total pressure at which the column is operating. This latter rule is applicable to gases above the critical pressure. Actually, neither rule is accurate for general application, and the true value frequently lies between the two. In some cases the heat of mixing of the liquids may be quite considerable and specific data must be consulted.

Data appearing in this section are listed in the following:

Van Laar constants for binary mixtures.................. Table L-1
Data on constant-boiling mixtures....................... Table L-2
Vapor-liquid boiling data:
  Acetic acid–water...................................... Table L-3
  Acetone-water.. Table L-4
  Acetone–methyl alcohol................................. Table L-5
  Acetone–ethyl alcohol.................................. Table L-6
  Ethylene dichloride–toluene............................ Table L-7
  Ethyl alcohol–*n*-butyl alcohol (cf. Fig. L-4)........... Table L-8
  Benzene–ethyl alcohol.................................. Table L-9
  Ethyl alcohol–water.................................... Tables L-10 and L-11
  Ethyl acetate–ethyl alcohol............................ Table L-12
  Isopropyl alcohol–water................................ Table L-13
Vapor pressures of ammonia solutions..................... Tables L-14 and L-15
Solubility of gases in water............................. Table L-16
Moisture content of air.................................. Fig. L-3
Liquid-liquid equilibrium: acetic acid in water and isopropyl ether... Figs. L-4, 5, 6
Heats of solution:
  Liquid ammonia in water................................ Table L-17
  Hydrocarbon gases in oil............................... Table L-18

[1] Probably the most accurate values are those secured by the method of Benedict et al., *Chem. Eng. Progr.*, **47,** 571 and 609 (1951). Values so computed have been correlated in chart form: "Vaporization Constants for Hydrocarbons," M. W. Kellogg Co., New York City. Somewhat more convenient charts appear in "Engineering Data Book," 7th ed., Natural Gasoline Supply Men's Association, Tulsa, Oklahoma, 1957.

TABLE L-1.—CONSTANTS OF EQUATIONS FOR ACTIVITY COEFFICIENTS OF BINARY MIXTURES

(Adapted from Perry, "Chemical Engineers' Handbook, 3d ed., McGraw-Hill Book Company, Inc., New York, 1950)

Components of Mixtures		Temp., °F	Van Laar constants		Margules constants*	
1	2		*A*	*B*	*A*	*B*
Acetaldehyde	Ethanol	67–153	−0.10	−0.20	−0.10	−0.20
Acetaldehyde	Water	67–212	0.69	0.78	0.69	0.78
Acetone	Benzene	133–176	0.176	0.176	0.176	0.176
Acetone	Methanol	131–148	0.243	0.243	0.243	0.243
Acetone	Water	133–212	0.89	0.65	0.90	0.65
		77	0.82	0.72	0.82	0.72
Benzene	Isopropanol	162–180	0.591	0.845		
n-Butane	Furfural	100	1.10	1.26	1.09	1.26
		125	1.05	1.17	1.04	1.17
		150	1.00	1.10	0.99	1.10
		200	0.91	0.98	0.91	0.98
n-Butyl alcohol	Butyl acetate	242–259	0.22	0.24	0.22	0.24
n-Butene-2	Furfural	100	0.84	1.03	0.83	1.03
		125	0.80	0.99	0.80	0.99
		150	0.76	0.95	0.76	0.95
		200	0.70	0.90	0.70	0.90
Carbon disulphide	Acetone	95–133	0.556	0.778		
Carbon disulphide	Carbon tetrachloride	115–170	0.10	0.07	0.10	0.07
Carbon tetrachloride	Benzene	170–176	0.052	0.046	0.052	0.046
Carbon tetrachloride	Ethylene dichloride	164–182	0.334	0.258	0.33	0.26
Ethyl alcohol	Benzene	151–176	0.845	0.699		
Ethyl alcohol	Water	77	0.67	0.42	0.65	0.41
Ethyl acetate	Ethyl alcohol	161–171	0.389	0.389	0.389	0.389
Ethyl ether	Acetone	94–133	0.322	0.322	0.322	0.322
Ethyl ether	Ethyl alcohol	94–173	0.42	0.55	0.42	0.55
Isopropyl alcohol	Water	180–212	1.042	0.492		
Isopropyl ether	Isopropyl alcohol	151–180	0.42	0.60		
Methyl alcohol	Benzene	132–148	0.243	0.243	0.243	0.243
Methyl alcohol	Ethyl acetate	138–171	0.505	0.505	0.505	0.505
Methyl alcohol	Water	148–212	0.36	0.22	0.36	0.21
		77	0.25	0.20	0.25	0.20
n-Propyl alcohol	Water	190–212	1.10	0.492		

* Values of Margules constants added by authors.

TABLE L-2.—COMPOSITION AND BOILING POINTS OF SOME CONSTANT BOILING MIXTURES AT ATMOSPHERIC PRESSURE
(Based on data from "International Critical Tables")

System		Mole % of *A*	Temp., °F	Maximum or minimum
Component *A*	Component *B*			
Acetone	Methyl acetate	61	133	Minimum
Acetic acid	Benzene	2.5	176	Minimum
	Toluene	62.7	222	Minimum
n-Butyl alcohol	Toluene	37	222	Minimum
Carbon tetrachloride	Ethyl alcohol	61.3	149	Minimum
	n-Propyl alcohol	75.0	163	Minimum
	Ethyl acetate	43.0	167	Minimum
	Methyl alcohol	44.5	132	Minimum
Carbon disulphide	Acetone	61.0	102	Minimum
	Ethyl alcohol	86.0	108	Minimum
	Methyl acetate	69.5	104	Minimum
	Methyl alcohol	72.0	100	Minimum
Chloroform	Acetone	65.5	148	Maximum
	Ethyl alcohol	84	139	Minimum
	Isopropyl alcohol	92	141	Minimum
Ethyl alcohol	Benzene	44.8	155	Minimum
	Ethyl acetate	46	161	Minimum
	Toluene	81	170	Minimum
Isopropyl alcohol	Benzene	39.3	162	Minimum
	Ethyl acetate	30.5	167	Minimum
	Toluene	77	177	Minimum
Methyl alcohol	Acetone	20	132	Minimum
	Benzene	61.4	137	Minimum
	Ethyl bromide	14	95	Minimum
	Ethyl chloride	62	139	Minimum
	Ethyl acetate	91.7	144	Minimum
	Methyl acetate	35	129	Minimum
Water	Ethyl alcohol	10.57	173	Minimum
	Formic acid	43.3	225	Maximum
	Hydrobromic acid	83.1	258	Maximum
	Hydrochloric acid	88.9	230	Maximum
	Hydrofluoric acid	65.4	248	Maximum
	Isopropyl alcohol	31.46	176	Minimum
	Nitric acid*	62.2*	249*	Maximum

* At 14.23 psia pressure.

TABLE L-3.—VAPOR-LIQUID EQUILIBRIA, ACETIC ACID—WATER SYSTEM AT 1 ATM
[York and Holmes, *Ind. Eng. Chem.*, **34,** 348 (1942). By permission of the American Chemical Society]

Boiling temp., °C	°F	Water in liquid		Water in vapor	
		Weight %	Mole %	Weight %	Mole %
113.7	236.7	2.7	8.5	5.2	15.5
108.7	227.7	9.8	26.7	16.5	39.8
105.7	222.2	18.0	42.3	26.4	54.5
104.2	219.6	27.5	55.9	37.6	66.6
102.5	216.5	36.8	66.0	49.2	76.4
101.8	213.4	48.5	75.8	60.0	83.3
102.2	215.9	58.6	82.5	68.5	87.9
100.8	213.4	66.7	87.0	75.9	91.3

TABLE L-4.—VAPOR-LIQUID EQUILIBRIUM DATA FOR THE SYSTEM ACETONE–WATER AT 1 ATM
[Brunjes and Bogart, *Ind. Eng. Chem.*, **35,** 255 (1943). By permission of the American Chemical Society]

Temp., °F	Acetone in liquid		Acetone in vapor	
	Weight %	Mole %	Weight %	Mole %
133.7	99.50	98.40	99.65	98.44
133.9	99.20	97.47	99.35	97.93
134.2	98.85	96.38	99.10	97.15
134.6	98.00	93.83	98.60	95.62
135.9	97.50	92.37	98.30	94.72
135.7	96.70	90.04	97.90	93.53
135.9	95.40	86.55	97.40	92.08
136.7	93.65	82.08	96.80	90.37
136.9	91.20	76.29	96.20	88.71
137.8	89.50	72.57	95.90	87.90
137.9	87.40	68.29	95.65	87.23
138.9	84.20	62.32	95.20	86.03
139.8	78.60	52.28	94.85	85.12
139.8	77.00	50.96	94.90	85.24
140.5	72.00	44.39	94.50	84.21
141.6	58.20	30.85	94.00	83.09
144.9	40.20	17.27	93.10	80.73
149.4	29.75	11.64	91.85	77.77
161.8	18.00	6.38	87.30	68.09
185.3	6.80	2.21	72.40	44.88
197.6	3.60	1.15	55.50	27.91

TABLE L-5.—VAPOR-LIQUID EQUILIBRIA FOR THE SYSTEM ACETONE–METHYL ALCOHOL AT 1 ATM

(Perry, "Chemical Engineers' Handbook," 3d ed., McGraw-Hill Book Company, Inc., New York, 1950)

Temp., °F	Acetone in liquid, mole %	Acetone in vapor, mole %
148.1	0	0
146.5	5.0	10.2
144.5	10.0	18.6
140.4	20 0	32.2
137.6	30.0	42.8
135.6	40.0	51.3
134.1	50.0	58.6
132.8	60.0	65.6
131.5	70.0	72.5
131.1	80.0	80.0
133.0	100.0	100.0

TABLE L-6.—VAPOR-LIQUID EQUILIBRIA FOR THE SYSTEM ACETONE–ETHYL ALCOHOL AT 1 ATM

(Perry, "Chemical Engineers' Handbook," 3d ed., McGraw-Hill Book Company, Inc., New York, 1950)

Temp., °F	Acetone in liquid, mole %	Acetone in vapor, mole %
172.9	0	0
167.7	5.0	15.5
163.4	10.0	26.2
159.8	15.0	34.8
156.2	20.0	41.7
153.1	25.0	47.8
150.6	30.0	52.4
148.5	35.0	56.6
146.5	40.0	60.5
143.2	50.0	67.4
140.7	60.0	73.9
138.4	70.0	80.2
136.4	80.0	86.5
134.6	90.0	92.9
133.0	100.0	100.0

TABLE L-7.—EXPERIMENTAL VAPOR-LIQUID EQUILIBRIUM DATA FOR ETHYLENE DICHLORIDE–TOLUENE AT 1 ATM

[Jones, Schoenborn, and Colburn, *Ind. Eng. Chem.*, **35**, 666 (1943). By permission of the American Chemical Society]

Temp., °F	Mole fraction ethylene dichloride	
	Liquid x	Vapor y
188.8	0.812	0.909
190.1	0.784	0.893
194.3	0.700	0.844
197.9	0.585	0.765
198.8	0.568	0.750
204.6	0.479	0.674
204.8	0.446	0.645
206.3	0.415	0.614
208.1	0.375	0.578
210.7	0.365	0.565
213.4	0.252	0.424
215.6	0.235	0.401
224.6	0.100	0.197
226.1	0.045	0.095

TABLE L-8.—VAPOR-LIQUID EQUILIBRIUM DATA FOR THE SYSTEM ETHYL ALCOHOL–*n*-BUTYL ALCOHOL AT 1 ATM

[Brunjes and Bogart, *Ind. Eng. Chem.*, **35**, 255 (1943). By permission of the American Chemical Society]

Temp., °F	Ethyl alcohol in liquid		Ethyl alcohol in vapor	
	Weight %	Mole %	Weight %	Mole %
174.4	90.3	93.8	99.6	99.7
176.5	83.0	88.7	98.0	98.7
177.8	77.2	84.5	96.9	98.2
183.0	64.3	74.3	91.4	94.6
186.1	55.5	66.6	88.6	92.7
187.5	48.4	60.1	84.9	90.0
190.9	45.7	57.4	82.7	88.5
190.5	43.7	55.5	82.1	88.1
195.0	36.9	48.8	75.0	82.8
200.3	31.7	42.7	69.3	78.4
205.6	22.9	32.3	60.3	70.8
215.3	17.3	25.2	48.1	59.7
222.5	12.0	17.8	38.0	49.6
232.0	7.3	11.3	22.3	31.6
237.9	4.0	6.2	12.2	18.1

TABLE L-9.—VAPOR-LIQUID EQUILIBRIA FOR THE SYSTEM BENZENE-ETHYL ALCOHOL AT 750 MM

(Perry, "Chemical Engineers' Handbook," 3d ed., McGraw-Hill Book Company, Inc., New York, 1950)

Temp., °F	Benzene in liquid, mole %	Benzene in vapor, mole %
172.6	0	0
165.9	6	20
162.3	11	30
158.2	20	40
154.9	39	50
154.0	57	56
154.9	72	60
159.4	89	70
167.4	96	85
175.5	100	100

TABLE L-10.—VAPOR-LIQUID EQUILIBRIA OF SYSTEM ETHYL ALCOHOL–WATER UNDER PRESSURE

[Griswold, Haney, and Klein, *Ind. Eng. Chem.*, **35**, 701 (1943). By permission of the American Chemical Society]

Temp., °F	Mole % ethyl alcohol in liquid	Mole % ethyl alcohol in vapor	Pressure, psia
302	7.3	33.4	107
	13.8	41.4	119
	26.5	48.7	130
	51.4	61.6	145
	63.9	69.5	147
392	5.8	24.7	300
	11.4	33.8	331
	23.7	43.3	370
	49.7	58.5	415
	63.3	68.1	424
482	6.3	19.2	717
	11.1	26.8	766
	23.5	37.3	869
	50.7	54.9	977
	68.8	69.6	1,010
527	12.6	24.5	1,176
	26.0	34.4	1,341
	40.0	42.5	1,492

TABLE L-11.—EXPERIMENTAL VAPOR-LIQUID EQUILIBRIUM DATA FOR ETHYL ALCOHOL–WATER AT 1 ATM

[Jones, Schoenborn, and Colburn, *Ind. Eng. Chem.*, **35**, 666 (1943). By permission of the American Chemical Society]

Temp., °F	Mole fraction ethyl alcohol	
	Liquid x	Vapor y
212.0	0	0
203.9	0.018	0.179
195.1	0.054	0.3375
185.7	0.124	0.470
182.5	0.176	0.514
181.0	0.230	0.542
179.6	0.288	0.570
177.8	0.385	0.612
176.8	0.440	0.633
175.7	0.514	0.657
174.0	0.673	0.735
172.9	0.840	0.850
173.0	1.0	1.0

TABLE L-12.—VAPOR-LIQUID EQUILIBRIA FOR THE SYSTEM ETHYL ACETATE–ETHYL ALCOHOL AT 1 ATM

(Perry, "Chemical Engineers' Handbook," 3d ed., McGraw-Hill Book Company, Inc., New York, 1950)

Temp., °F	Ethyl acetate in liquid, mole %	Ethyl acetate in vapor, mole %
172.9	0.0	0.0
169.9	5.0	10.2
167.9	10.0	18.7
165.0	20.0	30.5
163.0	30.0	38.9
161.8	40.0	45.7
161.2	50.0	51.6
161.2	54.0	54.0
161.4	60.0	57.6
162.0	70.0	64.4
163.4	80.0	72.6
166.5	90.0	83.7
168.8	95.0	91.4
171.8	100	100

TABLE L-13.—VAPOR-LIQUID EQUILIBRIUM DATA FOR THE SYSTEM ISOPROPYL ALCOHOL–WATER AT 1 ATM

[Brunjes and Bogart, *Ind. Eng. Chem.*, **35**, 255 (1943). By permission of the American Chemical Society]

Temp., °F	Isopropyl alcohol in liquid		Isopropyl alcohol in vapor	
	Weight %	Mole %	Weight %	Mole %
178.2	97.85	93.19	96.82	90.11
177.8	97.29	91.53	96.07	88.01
177.4	96.33	88.72	94.97	85.00
176.9	95.05	85.20	93.54	81.26
176.5	93.42	81.00	91.77	76.98
176.3	91.77	77.02	90.47	74.01
176.1	90.15	73.33	89.28	71.42
176.1	88.48	69.71	88.27	69.31
176.0	88.15	69.05	88.02	68.79
176.0	87.90	68.57	87.74	68.24
176.1	87.82	68.38	87.86	68.46
176.3	87.70	68.13	87.70	68.13
176.1	86.08	64.96	86.91	66.59
176.3	82.39	58.38	85.34	63.58
176.7	73.95	45.97	82.98	59.39
177.4	62.30	33.14	81.27	56.54
178.0	51.15	23.87	80.67	55.59
178.2	45.24	19.86	79.93	54.44
178.4	39.35	16.29	78.99	52.98
178.5	31.90	12.32	79.50	53.78
180.9	23.50	8.43	77.10	50.24
182.8	14.59	4.88	74.42	46.60
192.3	8.01	2.54	63.18	33.99
195.4	6.50	2.04	57.96	23.08
199.7	4.38	1.36	49.00	22.44
203.5	2.72	0.83	36.55	14.73
210.0	0.52	0.16	11.16	3.64

TABLE L-14.—TOTAL VAPOR PRESSURES OF AQUA AMMONIA
(Perry, "Chemical Engineers' Handbook," 3d ed., McGraw-Hill Book Company, Inc., New York, 1950)
Psia

Temp., °F	Molal concentration of ammonia in the solutions in percentages										
	0	5	10	15	20	25	30	35	40	45	50
32	0.09	0.34	0.60	0.97	1.58	2.60	4.20	6.54	9.93	14.18	19.40
40	0.12	0.45	0.77	1.24	2.01	3.25	5.21	8.06	12.05	17.20	23.39
50	0.18	0.64	1.05	1.65	2.67	4.29	6.75	10.35	15.34	21.65	29.26
60	0.26	0.86	1.42	2.21	3.51	5.55	8.65	13.22	19.30	27.05	36.26
70	0.36	1.17	1.84	2.90	4.56	7.13	11.01	16.56	24.05	33.39	44.42
80	0.51	1.52	2.43	3.76	5.85	9.06	13.86	20.61	29.69	40.96	54.08
90	0.70	2.02	3.15	4.83	7.43	11.40	17.23	25.48	36.34	49.82	65.32
100	0.95	2.62	4.05	6.13	9.34	14.22	21.32	31.16	44.12	59.99	78.30
110	1.27	3.34	5.14	7.72	11.64	17.58	26.07	37.81	53.16	71.87	93.19
120	1.69	4.27	6.46	9.63	14.42	21.54	31.69	45.62	63.59	85.33	110.20
130	2.22	5.38	8.07	11.91	17.67	26.20	38.25	54.55	75.55	100.86	129.50
140	2.89	6.70	9.98	14.63	21.49	31.54	45.73	64.78	89.19	118.24	151.30
150	3.72	8.29	12.23	17.81	26.00	37.81	54.43	76.61	104.65	138.10	175.40
160	4.74	10.16	14.92	21.54	31.16	45.02	64.25	89.88	122.10	160.20	202.70
170	5.99	12.41	18.01	25.87	37.11	53.27	75.55	104.84	141.75	185.10	233.20
180	7.51	15.00	21.65	30.86	44.02	62.68	88.17	121.68	163.70	212.60	267.00
190	9.34	18.06	25.87	36.60	51.81	73.32	102.56	140.75	188.10	243.30	304.30
200	11.53	21.60	30.72	43.14	60.62	85.33	118.68	161.81	215.20	277.00	345.50
210	14.12	25.61	36.26	50.58	70.72	98.80	136.42	185.10	245.10	314.50	390.70
220	17.19	30.27	42.47	59.00	81.91	113.81	156.41	211.24	278.20	355.10	439.60
230	20.78	35.59	49.60	68.46	94.43	130.64	178.28	239.70	314.50	400.20	493.40
240	24.97	41.52	57.65	78.91	108.60	149.20	202.74	270.92	354.10	448.90	552.30
250	29.83	48.32	66.67	90.74	124.08	169.48	229.62	305.60	397.60	502.40	

TABLE L-15.—PARTIAL VAPOR PRESSURES OF WATER OVER AQUA AMMONIA
(Perry, "Chemical Engineers' Handbook," 3d ed., McGraw-Hill Book Company, Inc., New York, 1950)
Psia

Temp., °F	Molal concentration of ammonia in the solutions in percentages										
	0	5	10	15	20	25	30	35	40	45	50
32	0.09	0.084	0.079	0.074	0.070	0.065	0.060	0.056	0.051	0.047	0.042
40	0.12	0.115	0.108	0.101	0.095	0.089	0.083	0.076	0.070	0.064	0.058
50	0.18	0.17	0.16	0.15	0.14	0.13	0.12	0.11	0.10	0.094	0.085
60	0.26	0.24	0.23	0.21	0.20	0.19	0.17	0.16	0.15	0.13	0.12
70	0.36	0.34	0.32	0.30	0.28	0.26	0.25	0.23	0.21	0.19	0.17
80	0.51	0.48	0.45	0.42	0.40	0.37	0.34	0.32	0.29	0.27	0.24
90	0.70	0.66	0.63	0.58	0.55	0.51	0.47	0.44	0.40	0.37	0.33
100	0.95	0.90	0.85	0.79	0.74	0.69	0.64	0.59	0.55	0.50	0.45
110	1.27	1.20	1.14	1.07	1.00	0.93	0.86	0.80	0.73	0.67	0.60
120	1.69	1.60	1.51	1.42	1.33	1.24	1.15	1.06	0.97	0.89	0.80
130	2.22	2.10	1.98	1.86	1.74	1.62	1.51	1.39	1.28	1.17	1.05
140	2.89	2.73	2.57	2.42	2.26	2.11	1.96	1.81	1.66	1.52	1.37
150	3.72	3.51	3.31	3.11	2.91	2.72	2.52	2.33	2.14	1.95	1.76
160	4.74	4.48	4.22	3.97	3.71	3.46	3.22	2.97	2.73	2.49	2.25
170	5.99	5.66	5.34	5.02	4.70	4.38	4.07	3.75	3.45	3.15	2.84
180	7.51	7.10	6.69	6.30	5.89	5.49	5.10	4.71	4.33	3.94	3.57
190	9.34	8.83	8.32	7.82	7.32	6.83	6.34	5.86	5.38	4.91	4.44
200	11.53	10.90	10.27	9.65	9.04	8.43	7.83	7.23	6.64	6.06	5.48
210	14.12	13.35	12.58	11.82	11.07	10.32	9.59	8.86	8.13	7.42	6.71
220	17.19	16.25	15.32	14.39	13.48	12.57	11.67	10.78	9.90	9.03	8.17
230	20.78	19.64	18.51	17.40	16.29	15.19	14.11	13.03	11.97	10.91	9.87
240	24.97	23.60	22.25	20.91	19.58	18.26	16.95	15.66	14.38	13.12	11.86
250	29.83	28.20	26.58	25.00	23.39	21.82	20.25	18.71	17.18	15.67	

TABLE L-16.—SOLUBILITY OF GASES IN WATER
(Abstracted from "Gas Engineers' Handbook," McGraw-Hill Book Company, Inc., New York, 1934)

BASIS: Standard cubic feet dissolved in 1 cu ft of water (corrected to 60°F) when saturated at a total presssure of 14.73 psi.

	Temperature, °F							
	32	50	68	86	104	122	140	176
Hydrogen sulphide, H_2S..	4.84	3.52	2.63	2.04	1.62	1.29	1.02	0.514
Sulphur dioxide, SO_2.....	.....	54.8	37.5	26.2	18.5	12.8	9.43	4.10
Carbon dioxide, CO_2.....	1.79	1.24	0.898	0.672	0.519	0.405	0.308	
Carbon monoxide, CO...	0.037	0.030	0.023	0.020	0.018	0.015	0.013	0.008
Methane, CH_4...........	0.058	0.044	0.034	0.028	0.023	0.020	0.017	0.010
Acetylene, C_2H_2.........	1.80	1.35	1.05	0.85				
Ethylene, C_2H_4..........	0.235	0.167	0.125	0.099				
Ethane, C_2H_6...........	0.103	0.068	0.048	0.036	0.028	0.023	0.018	0.010
Hydrogen, H_2...........	0.023	0.020	0.019	0.017	0.016	0.015	0.014	0.009
Atmospheric nitrogen, N_2.	0.025	0.019	0.016	0.014	0.012	0.010	0.009	0.006
Oxygen, O_2.............	0.052	0.040	0.032	0.027	0.023	0.020	0.017	0.010

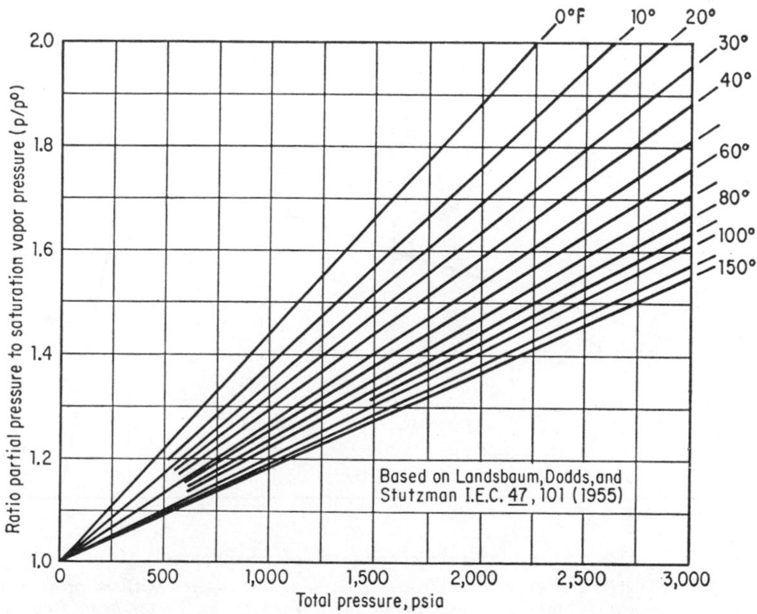

FIG. L-3.—Moisture content of air plotted as partial pressures. [*Based on data of Landsbaum, Dodds, and Stutzman, Ind. Eng. Chem.*, **47**, 101 (1955).]

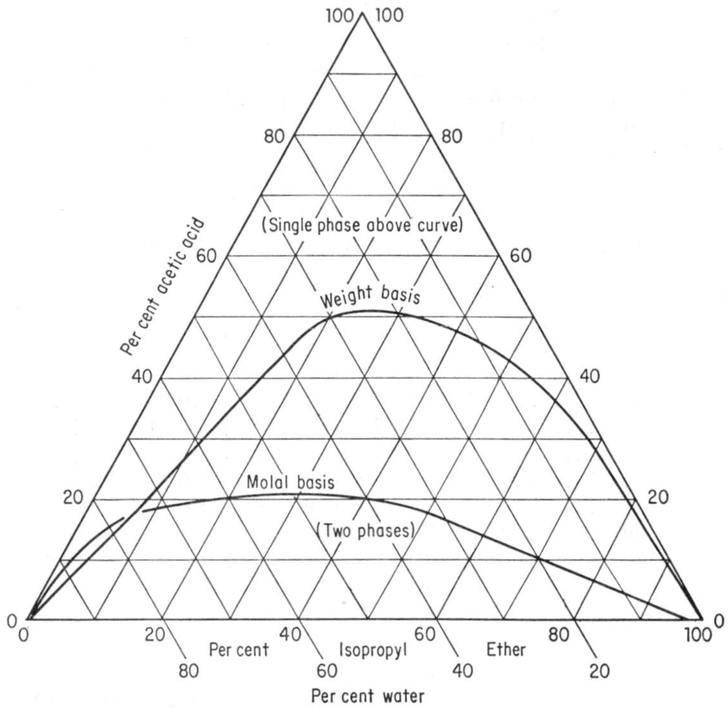

FIG. L-4.—Two-phase boundary: acetic acid–isopropyl ether–water system. [*Based on data reported in Trans. Am. Inst. Chem. Engrs.*, **36**, 601 (1940). See also Othmer, White, and Trueger, *Ind. Eng. Chem.*, **33**, 1245 (1933).]

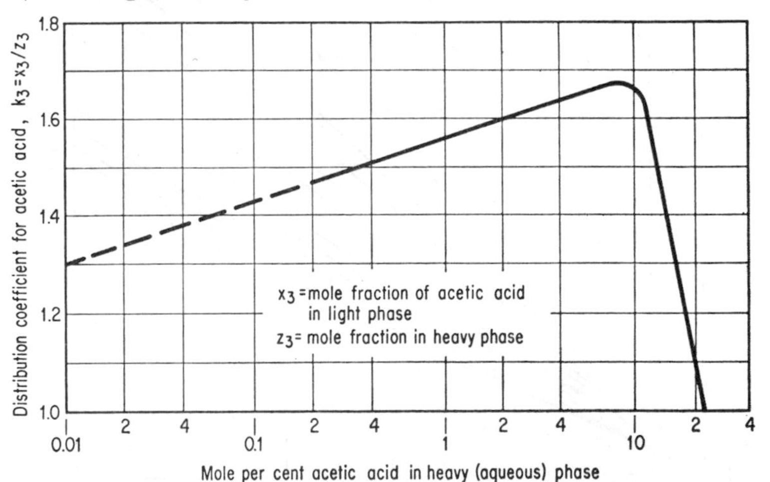

FIG. L-5.—Distribution coefficient for acetic acid in mixtures with isopropyl ether and water at 68°F. [*Computed from data given in Trans. Am. Inst. Chem. Engrs.*, **36**, 601 (1940). Values below 0.2 per cent acid are extrapolated with correction for ionization of acetic acid in water.]

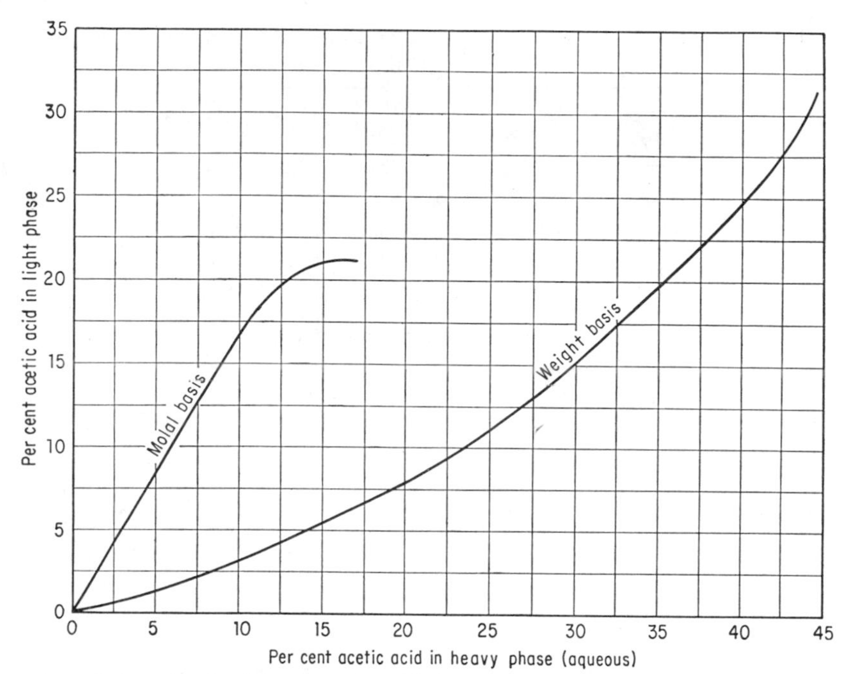

FIG. L-6.—Distribution of acetic acid between phases for acetic acid–isopropyl ether–water system. [*Based on data given in Trans. Am. Inst. Chem. Engrs.*, **36**, 601 (1940).]

TABLE L-17.—INTEGRAL HEATS OF SOLUTION OF LIQUID AMMONIA IN WATER

(Perry, "Chemical Engineers' Handbook," 2d ed., McGraw-Hill Book Company, Inc., New York, 1941)

Per cent of ammonia by weight	Btu/lb of mixture	Btu/lb of ammonia
0	0	358.0
10	34.4	343.8
20	65.7	328.5
30	92.5	308.2
40	108.2	270.0
50	109.4	218.8
60	101.9	169.7
70	84.8	121.1
80	60.7	75.8
90	31.9	35.5
100	0	0

TABLE L-18.—APPROXIMATE HEATS OF SOLUTION OF HYDROCARBON GASES IN 200 MOLECULAR WEIGHT OILS

No. carbon atoms in hydrocarbon	Heat of solution, Btu/lb mole	
	At 80°F	At 250°F
1	1500	−300
2	5500	3500
3	6500	5500
4	8500	6500
5	10,000	8000
6	12,500	9500

CHAPTER M

COUNTERCURRENT SEPARATIONS

The title of this chapter is intended to suggest coverage of absorption, stripping, liquid-liquid extraction, and distillation. All these separation methods involve diffusional mass transfer and are based on differences in composition between a light and a heavy phase in equilibrium. In each case, countercurrent operation is advantageous for all but the simplest applications.

Situations arise in each of the four techniques that can be described by identical mathematical equations. Accordingly, this chapter features a unified treatment of the countercurrent separations. The several arts were developed independently, each with its own terminology. Thus, it has been necessary to employ some rather unfamiliar terminology to secure a unity of written language matching the unity of mathematical language. This type of presentation has been anticipated by numerous procedures and by certain techniques, such as extractive distillation, that fuse the principles of two arts.

The consolidated discussion of the common property of the four procedures is amplified to include the characteristic features of each separation method.

The coverage includes both calculation methods and equipment performance data that are convenient for choice and process design of rectification cycles for separating pure substances from mixtures. The basic theory underlying the methods presented is discussed, but a comprehensive treatment of the theory of mass transfer by diffusion and convection is not justified by the intended scope.

In view of the importance of this group of subjects, it is not surprising that the literature covering them is voluminous. A few reference sources are listed below:

Sherwood and Pigford, "Absorption and Extraction," 2d ed., McGraw-Hill Book Company, Inc., New York, 1952.

Treybal, "Mass-transfer Operations," McGraw-Hill Book Company, Inc., New York, 1955.

Nielson, "Distillation in Practice," Reinhold Publishing Corporation, New York, 1956.

Robinson and Gilliland, "Elements of Fractional Distillation," 4th ed., McGraw-Hill Book Company, Inc., New York, 1950.

Ruhemann, "The Separation of Gases," 2d ed., Oxford University Press, New York, 1932.

Perry "Chemical Engineers' Handbook," 3d ed., McGraw-Hill Book Company, Inc., New York, 1950.

1. General Considerations

a. Techniques of the Separation Processes.—These are illustrated by Figs. M-1 and M-2. The summary given there serves as a reference background for discussion in this section and for the calculation methods to be given later. These illustrations cover only the salient elements (excluding reboilers, condensers, coolers, exchangers, etc.).

Figure M-1 covers single-section columns, the simplest to understand. In absorption, a selective solvent is used to dissolve the more soluble components from a mixture of gases. In a few cases, such as recovery of aqueous ammonia from gas by scrubbing with water, a simple absorber operation produces a useful product directly. Usually a solvent-free product is desired. If the absorption was effected at quite high pressure, its reduction may be adequate to release the dissolved product. Often the rich solvent must be heated, and removal completed by stripping. For example, propane dissolved in a heavy oil can be stripped by steam, which is then condensed releasing propane. The solvent can be stripped by its own vapor. In this case, the condensed vapor must be returned either to the top of the stripper with the liquid feed or to a reflux section above the stripping section (see Fig. M-2). If the dissolved gas stripped out is condensable, the reflux becomes pure product and the stripper is, in reality, a continuous still.

In a continuous still, the feed introduced at the center of the column is directly separated into an overhead and bottom product without need of employing any solvent or other material present. The enrichment section may be likened to an absorber wherein the reflux selectively picks up the heavy component. Even though the enrichment process is an exchange of heavy for light components, it can be described by mathematical expressions used for absorption. The stripping section of a still has a similar relation to a simple stripper (*i.e.*, using relatively nonvolatile solvent). Frequently a mixture of many components is processed so that several distillation cuts are desired and several continuous columns are required. In this case, batch distillation is often used, permitting any number of successive cuts. Although a batch still has no stripping section and cannot, therefore, yield a pure bottom product from a close separation, adequate results are often obtained.

Simple liquid-liquid extraction is completely analogous to absorption and stripping, with one liquid taking the place of the gas. This may be seen by reference to Fig. M-1. As in the case of absorption, a solvent-free product is usually desired. The extract containing the product is usually separated by distillation, and the solvent returned to the extractor.

The above discussion has introduced the simpler cases, and the most complicated rectification cycles can be developed by combining the various component operations represented by these simple cases. The number of ways that the "components" can be appropriately combined is endless. Moreover, in choosing the best combination for a particular problem, the compressors, heat exchangers, pumps, and utility requirements should be considered. A complete discussion of these matters is clearly beyond the scope of a single chapter. Accordingly, a brief description of a few combinations follows in the hope that the reader will find them suggestive.

A simple absorption-stripping cycle produces two products: the "lean" gas leaving the absorber and the "overhead" gas leaving the stripper. The latter represents recovery of all the gases that were absorbed. For multicomponent systems containing more than one appreciably soluble gas, the product from the stripper may not be of adequate purity or more products may be desired. If two gases are dissolved in the absorber, the less soluble one can be removed from the solvent by controlled stripping, yet leave a substantial fraction of the more soluble product in solution. To be sure, some of the more soluble constituent will be stripped, and this must be returned to solution if recovery is to be complete.

The obvious method of doing this is by reabsorption in fresh solvent and combining the fat solvent from the reabsorption with that from the prime absorber. Both the fractional stripping and the reabsorption can be conveniently performed in a single piece of equipment which we may term a partial (or fractional) stripper. The fat liquid from the absorber enters at the middle, liquid flows down through the stripping section (thence to the total stripper), and gases flow upward through the absorbing section (and become the third product of the

Column (plate-packed, spray or bubble)	Absorption Column (or scrub column)	Stripping Column (or lower section of stripping still)	Continuous Distillation Considered as two columns: Enrichment (Upper section)	Continuous Distillation Considered as two columns: Stripping (Lower section)	Batch Distillation	Liquid–liquid Extraction: Case 1: Solvent is heavy phase	Liquid–liquid Extraction: Case 2: Solvent is light phase	Continuous Distillation Involving Three-phase Azeotrope (Feed to decanter): Heavy-product Still	Continuous Distillation Involving Three-phase Azeotrope (Feed to decanter): Light-product Still
Overhead	Lean Gas Product To usage or solvent recovery	Stripped Gas Product (Usually)	Vapor Outlet To condenser for reflux–product split	Vapor Out To enrichment section	Vapor Outlet To condenser for reflux–product split	Raffinate Spent liquor free from extractibles	Extract (or rich solvent) To solvent removal	Vapor Outlets, Both Columns To condenser, thence to decanter	
Top inlet	Inlet Lean Solvent Either fresh or from stripper	Rich Solvent Feed From absorber	Reflux From condenser	Total Liquid Entering From enrichment section plus liquid feed	Reflux From condenser	Inlet Solvent From solvent recovery	Feed Liquid	Reflux Heavy liquid from decanter	Reflux Light liquid from decanter
Bottom inlet	Feed Gas From preceding process steps	Stripping Gas or Vapor From gas supply, steam line, or reboiler	Total Vapor Entering From stripping section plus vapor in feed	Stripping Vapor From reboiler or open steam	Column Bottoms Fall to kettle Vapor Rises from kettle to column	Feed Liquid	Inlet Solvent From solvent recovery	Vapor Inlet From reboiler	Vapor Inlet From reboiler
Bottoms	Rich Solvent To stripper or direct usage	Stripped (Lean) Solvent Recycled to absorber	Total Liquid Out To stripping section	Bottoms Out To supply reboiler needs, balance is product	Bottom Product Remains in kettle at end of run	Extract (or rich solvent) To solvent recovery	Raffinate Spent liquor free from extractibles	Bottoms Out To supply reboiler; balance is heavy product	Bottoms Out To supply reboiler; balance is light product

FIG. M-1.—Descriptive data for single-stage columns.

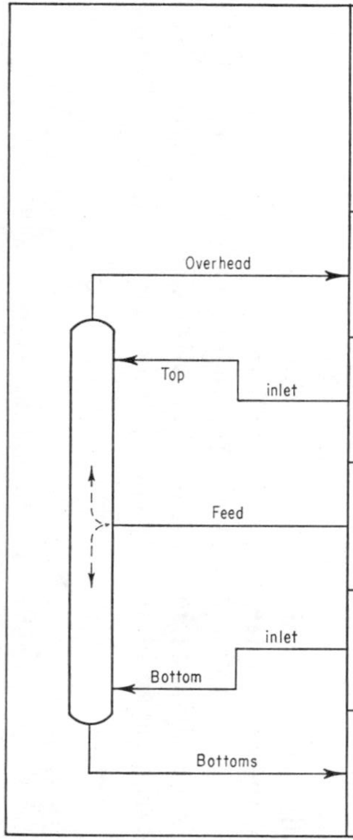

	Continuous Still	Azeotropic Distillation	Upper Two Sections Extractive Distillation	Absorber with Stripping Section	Partial Stripper	Stripper with Reflux	Liquid-Liquid Extractor with Reflux	
							Case 1, Heavy Solvent	Case 2, Light Solvent
	Top section: Enrichment Bottom section: Stripping	Top section: Enrichment Bottom section: Stripping	Top section: Enrichment Bottom section: Extraction	Top section: Absorption Bottom section: Stripping	Top section: Reabsorption Bottom section: Stripping	Top section: Dephlegmating Bottom section: Stripping	Top section: Exhausting Bottom section: Enrichment	Top section: Enrichment Bottom section: Exhausting
Overhead	Vapor Out To condenser for reflux-product split	Vapor Out To condenser, thence to decanter: entrainer is refluxed, other liquid is a product	Vapor Out To condenser for reflux-product split	Lean Gas Product To usage or solvent recovery	Intermediate Product To usage or recycle to absorber	Vapor Out To condenser which condenses solvent but not the gas product	Raffinate Spent liquor, free from extract	Extract To solvent removal; solvent-free material split between product and reflux
Top inlet	Reflux From condenser	Entrainer Feed From decanter plus make-up	Reflux From condenser	Inlet Lean Solvent From stripper	Inlet Solvent From total stripper, or cool rich solvent from absorber	Reflux From condenser	Solvent From solvent recovery	Solvent-free Extract From solvent removal
Feed	Vapor Feed Up to enrichment Liquid Feed Down to stripping	Two-Component Azeotrope Feed From preceding still	Extractive Solvent Recycled from solvent still	Vapor or Gas Feed (May include recycle) upward to absorption	Rich Solvent From absorber; flashed vapor up to reabsorption, liquid down to stripping	Rich Solvent Feed From absorber	Feed Up to exhausting	Feed Down to exhausting
Bottom inlet	Stripping Vapor From reboiler, open steam etc.	Stripping Vapor From reboiler	Vapor Load Out stripping section to absorbing (extracting) section	Stripping Gas or Vapor From reboiler or stripper	Stripping Gas or Vapor From reboiler, etc.	Stripping Vapor From reboiler	Solvent-Free Extract From solvent recovery	Solvent From solvent removal
Bottoms	Total Bottoms Split for reboiling and product draw	Total Bottoms Split for reboiling and product draw	Bottoms Join liquid still feed to stripping section (not shown)	Total Bottoms To reboiler, thence to stripper	Total Bottoms To reboiler, thence to stripper	Total Bottoms To reboiler thence to absorber	Extract To solvent removal; solvent free extract split between product and reflux	Raffinate Spent liquid free from extract

FIG. M-2.—Descriptive data for two-stage columns.

absorption–partial-stripping–total-stripping cycle). If the most soluble gas is the only pure product desired, the prime absorber itself can be used for the reabsorption referred to above and the partial stripping operation so reduced to a stripping section below the feed plate of the absorber. Figure M-2 illustrates qualitative factors in partial strippers and absorbers with stripping sections.

Distillation also has many variants. These arise chiefly because some substances yield mixtures that are azeotropes, *i.e.*, constant-boiling mixtures that boil without change in composition. Such azeotropes may be minimum or maximum boiling (*i.e.*, boil below or above the boiling-point temperatures of the components), but in either case, simple distillation yields the azeotrope as one product and the component in excess as the other. A change in pressure may obviate the azeotrope, but usually some other means of separation must be sought.

If the components are only partially miscible, a minimum-boiling azeotrope is always formed. This is properly called a "three-phase azeotrope," since the azeotrope vapor is in equilibrium with two liquid phases. Such three-phase azeotropes can always be separated by a decanter and two columns, one for each phase decanted from the condensate.

Three-phase azeotropes afford the basis of azeotropic distillation. A third component is added which "entrains" one component of a troublesome azeotrope as a three-phase one. Extractive distillation is another way to separate azeotrope-forming mixtures. In this, a selective solvent is used to absorb one of the azeotropic components, thus representing a fusion of the classical absorption and distillation techniques. Azeotropic and extractive distillations are described briefly in Fig. M-2.

Liquid-liquid extraction is subject to similar variations, notably operation with reflux as shown in Fig. M-2. When aqueous solutions are being extracted, the reflux may be fresh water or a water solution of the solvent-free extract.

It may be appreciated from this brief outline that application of the countercurrent separating techniques to complex mixtures can lead to equally complex equipment combinations, so that a good understanding of their interrelationships is the best basis for choosing separation methods for specific problems. In general, a straightforward distillation is usually most advantageous when it can be employed. Absorption-stripping cycles usually end up to be more complex and to require more control instrumentation, but they are still quite attractive when a moderately selective solvent is available, and they are particularly advantageous for removal of small quantities of soluble gas from large streams. Liquid-liquid extraction is the most feasible method of handling many problems even though mass transfer rates are rather slow, making separation equipment expensive unless the solvent has a high degree of selectivity.

b. The Theoretical Plate.—This is the simplest yardstick for measurement of the difficulty of a separation. A perfect plate is defined as one in which the liquid leaving is in equilibrium with the vapor (or gas) leaving. In liquid-liquid extraction, the light liquid replaces the vapor and the term "stage" is frequently used instead of plate. The number of such plates required for the operation is dependent on the equilibria involved, the ratio of light- to heavy-fluid flows, and the purity and recovery required. The actual number of plates required will, in general, be somewhat greater depending on the plate efficiencies.

In case packed towers are to be used, the number of theoretical plates may still be used as the measure of rectification. The estimation of the height of an actual column equivalent to a theoretical plate (usually referred to as the H.E.T.P.) is not entirely straightforward; moreover, the theoretical treatment of this problem is beyond the scope of the book. Briefly, differential equations are set up for the local rate of diffusional transfer of key components between the liquid and the gas phases. The result may be expressed either in terms of the mass transfer coefficient or as the number of transfer units. This unit is defined mathematically and is similar to the theoretical plate except that it includes allowance for the difference between equilibrium and actual concentrations. In this way rules have been developed for computing the number of transfer units and also the required height for a transfer unit (H.T.U.).

c. Thermodynamic Considerations.—The basic tools in the calculation of countercurrent operations are material balances, heat balances, and phase-equilibria data. If a heat and material balance is made at each theoretical plate of a column consistent with equilibria, the number of theoretical plates, all flows, temperatures, and heating and cooling requirements will be established. Graphical and analytical methods for estimating theoretical plates are based on such iterative treatment. Over-all heat and material balances are necessary to establish consistent over-all flows and utilities.

2. Single-stage Columns

a. Enrichment Factors and Column Efficiencies.—These terms are not in common use but are introduced here to permit a unified discussion of the separation processes. The definition of these terms and their relation to familiar terminology will be developed by a discussion of a single process, absorption, and then generalized.

No matter how high a column may be made, the best it can do is to establish equilibrium between the gas entering and the liquid leaving the base. This equilibrium represents the maximum amount of dissolved gas that can be carried out by a unit amount of liquid and, times the number of units circulated, represents the maximum quantity of gas that may be absorbed. This must be at least equal to the quantity of gas that is to be absorbed. Bare fulfillment of this condition represents the theoretical minimum circulation for complete absorption. *The ratio of liquid actually circulated to this theoretical minimum is the absorption factor.* The above statements are very generalized and may, perhaps, be clarified by repetition in terms of definite units (see nomenclature for Tables M-1 and M-2 on page 413):

Maximum concentration dissolved gas at base of column $= x_2^* = \dfrac{y_1}{k}$

Quantity of gas to be absorbed $= G_1 y_1$, moles/hr

Minimum condition for complete absorption,

$$L_1 x_2^* = \frac{L_1 y_1}{k} = G_1 y_1 \qquad \text{or} \qquad L_1 = kG_1$$

Absorption factor for any liquid rate, $A = \dfrac{L}{kG}$

The absorption efficiency is the fraction of the soluble gas component that is absorbed by a solvent initially free from dissolved gas. In case the inlet liquid contains some dissolved gas, it can then at best reduce the exit gas only to the equilibrium concentration. One may compute this concentration from the equilibrium involved. From this the maximum possible fractional absorption of the gas may be computed. *This fraction times the absorption efficiency E is the net fraction absorbed.* Repeating this in specific units (see nomenclature for Tables M-1 and M-2 on page 413),

Equilibrium concentration exit gas, kx_1

Minimum quantity of gas not absorbed, $G_2 kx_1$

Quantity of gas in inlet stream, $G_1 y_1$

Maximum possible fraction absorbed, $1 - \dfrac{G_2 kx_1}{G_1 y_1}$

Fraction actually absorbed, $E\left(1 - \frac{G_2kx_1}{G_1y_1}\right)$

The above defines absorption efficiency by inference. Directly stated, the efficiency is the fractional recovery of the quantity that is recoverable. In a stripper the gas or vapor stream flowing upward through the column picks up dissolved gases or volatile components. Thus it will be seen that the roles of gas and liquid are exactly reversed from the corresponding roles in an absorber. It should be expected, then, that the same laws should govern the two processes, provided the laws are restated in accordance with this reversal.

In liquid-liquid extraction the role of the solvent is the same as that of the solvent in absorption. The liquid feed stream (whether it be the light or heavy liquid) assumes the role of the gas.

Thus, *stripping factors* and *extraction factors* have analogous definitions to *absorption factors*, and we can use the common term *enrichment factor* for the group and generalize the definition as the ratio of *scavenger* used to the minimum required for complete removal of a stated component from the *feed stream.*

Again efficiencies of stripping, solvent extraction, and absorption are all defined in the same fashion and all can be expressed as the *column efficiency, i.e.*, the fractional recovery of the quantity of the component that is recoverable.

Tables M-1 and M-2 give data for calculating enrichment factors and column efficiencies.

TABLE M-1.—DATA FOR ENRICHMENT FACTORS

Concentration		Equilibrium expression	Absorption factor	Stripping factor
In liquid	In gas			
x	y	Vaporization constant	$\frac{L}{kG}$	$\frac{kG}{L}$
x	y	Raoult's law	$\frac{pL}{p_vG}$	$\frac{p_vG}{pL}$
x	y	Henry's law	$\frac{pL}{HG}$	$\frac{HG}{pL}$
c	y	Bunsen coefficient	$\frac{p\beta L_v}{p_aG_v}$	$\frac{p_aG_v}{p\beta L_v}$
c	y	Ostwald coefficient	$\frac{p - p_s}{p_a - p_s}\frac{519\alpha}{T}\frac{L_v}{G_v}$	$\frac{p_a - p_s}{p - p_s}\frac{T}{519\alpha}\frac{G_v}{L_v}$
In extract	In raffinate		Extraction factor	
		Distribution coefficient:		
x	z	By moles	$\frac{L}{k_d\bar{L}}$	
w	$\bar{w}$	By weight	$\frac{W}{k_w\bar{W}}$	

TABLE M-2.—DATA FOR CALCULATION OF COLUMN EFFICIENCIES

Operation	Simple form for single component transferred		General form single or multi-component transfer†
	E	$1 - E$	$1 - E$
Absorption (end 1 = bottom).......	$\frac{Y_1 - Y_2}{Y_1 - Y_2^*}$	$\frac{Y_2 - Y_2^*}{Y_1 - Y_2^*}$	$\frac{G_2(y_2 - y_2^*)}{G_1y_1 - G_2y_2^*}$
Stripping (end 1 = top)...........	$\frac{X_1 - X_2}{X_1 - X_2^*}$	$\frac{X_2 - X_2^*}{X_1 - X_2^*}$	$\frac{L_2(x_2 - x_2^*)}{L_1x_1 - L_2x_2}$
Liquid-liquid extraction (end 1 = feed end)..............	$\frac{Z_1 - Z_2}{Z_1 - Z_2^*}$	$\frac{Z_2 - Z_2^*}{Z_1 - Z_2^*}$	$\frac{\bar{L}_2(\bar{x}_2 - \bar{x}_2^*)}{\bar{L}_1\bar{x}_1 - \bar{L}_2\bar{x}_2}$

† May be regarded as definition of column efficiency.

NOMENCLATURE FOR TABLES M-1 AND M-2

(Subscript 1 denotes feed end; 2 denotes opposite end of column; asterisk denotes equilibrium value)

E = column efficiency
x = mole fraction (of component transferred) in liquid. Applies to solvent or extract phase in extraction
$\bar{x}$ = mole fraction of solute in raffinate
$X = x/(1 - x)$
y = mole fraction in gas phase
$Y = y/(1 - y)$
z = mole fraction in feed (raffinate) phase
$Z = z/(1 - z)$
c = concentration of dissolved gas, SCF/cu ft of solvent
L = moles liquid per hour. Applies to solvent or extract phase in extraction
$\bar{L}$ = moles of feed (raffinate) phase
L_v = cubic feet of dissolved gas-free liquid per hour
G = moles of total gas per hour
G_v = gas flow, solvent vapor-free basis, SCF/hr
$W, \bar{W}$ = weight flow per hour of solvent (extract) phase, feed (raffinate) phase
$w, \bar{w}$ = weight fraction of solute in solvent (extract) phase, feed (raffinate) phase
p = system pressure, absolute
p_v = vapor pressure of component to be absorbed
p_s = vapor pressure of solvent
p_a = standard atmospheric pressure (1 atm, 14.7 psi, etc.)
k = vaporization constant, the ratio of equilibrium concentrations: y/x
k_d = distribution constant, the ratio of equilibrium concentrations: z/x
k_w = weight-distribution constant, the ratio of equilibrium values: $\bar{w}/w$
H = Henry's law constant, at equilibrium: $H = py/x$ (same units as p)
β = Bunsen coefficient expressed as SCF gas dissolved in 1 cu ft liquid when the partial pressure of the gas is 1 atm
α = Ostwald coefficient, volumes of pure gas dissolved in one volume of liquid when total pressure is 1 atm and volumes are measured at 1 atm and the temperature of interest
T = absolute temperature, °R

b. Analytical Methods for Computing Theoretical Plates.—When the enrichment factor is constant, the number of theoretical plates can be precisely computed from Eq. (M-1), originally derived by Kremser.[1] It is seldom that the condition of constant enrichment factors throughout the column is achieved, and various approximations have been made to correct the Kremser equation. Edmister[2] used an effective factor to average out the variation. His factor is defined by Eq. (M-2), but Eq. (M-3) closely approximates his factors and is more convenient. This method appears to be the most suitable to use in multicomponent calculations for those components that have enrichment factors below unity. Colburn[3] derived relations which can be recast to Eq. (M-4). This equation is an excellent approximation applicable to the great majority of problems. However, the equation pair (M-5) and (M-5*a*) are recommended in preference.[4,5] These represent an empirical variant of Eq. (M-4) and lead to better agreement with results of iteration when values of the enrichment factor are only a little above unity or the column efficiency is low. For the special case of enrichment factors below unity, the Edmister method [*i.e.*, use of Eq. (M-2) or (M-3) and Eq. (M-1)] is recommended, since even the (M-5) equation pair does not yield reliable results for this range.[5]

None of these formulas are very convenient, but graphical methods are adequate for most purposes. Figure M-3 plots the results of the Kremser equation. Eq. (M-5) can be solved by Figs. M-4 and M-5. To determine the number of plates, establish the parameter θ from Fig. M-4 and use this to establish the number of theoretical plates from Fig. M-5. The performance obtainable from a specified number

[1] Kremser, *Natl. Petroleum News*, **22** (21), 42 (May 21, 1930).

[2] Edmister, *Ind. Eng. Chem.*, **35**, 837 (1943).

[3] Colburn, *Ind. Eng. Chem.*, **33**, 489 (1941).

[4] Equation (M-4) has a mathematical discontinuity for either F_2 or $F_1 = 1$. Although this is not true for the (M-5) pair, they predict more theoretical plates than are required when either F_1 or F_2 is below unity. The error may be illustrated by the fact that the equation reduces to $F_2 = \epsilon$ for infinite plates instead of $F_1 = E$, as it should for enrichment factors below unity.

[5] The mathematical equivalence between Eq. (M-4) and the original versions is not, at once, evident from inspection. The original derivation assumes systems that comply with Henry's law for the relatively dilute solutions that prevail at the lean end of the column and also assumes relatively high recovery. The original equation in the mole fraction form for absorption is

$$N_p = \frac{\log [(1 - P)M + P]}{\log (1/P)}$$

where N_p = number of theoretical plates
$P = m_2G_2/L_2$

$$M = \frac{y_1 - mx_2}{y_2 - mx_2} \frac{1 - P}{1 - y^*/y}$$

G, L = gas, liquid rates moles/hr

y, x = mole fraction of component being absorbed in gas, liquid streams

$$m = \frac{dy^*}{dx}$$

(Subscripts 1, 2 denote conditions at the bottom, top plate; * denotes equilibrium value.)

Since assumption of Henry's law at the top is essential to the derivation,

$$m_2 = \left(\frac{dy^*}{dx}\right)_2 = \left(\frac{y^*}{x}\right)_2 = k_2$$

Bearing these identities in mind,

$$P = \frac{m_2 G_2}{L_2} = \frac{1}{A_2} \qquad (A \text{ is defined in Table M-1})$$

$$\frac{y_1 - m_2 x_2}{y_2 - m_2 x_2} = \frac{y_1 - k_2 x_2}{y_2 - k_2 x_2} = \frac{y_1 - y_2^*}{y_2 - y_2^*}$$

This term is identical with the reciprocal of $1 - E$ given in Table M-2 when $G_1 = G_2$. Since the original derivation ignored changes in gas rates, replacement of the above term by $1/(1 - E)$ is justified and, indeed, may be regarded as a corrective extension. This identifies all the terms of the original derivation with the nomenclature of Tables M-1 and M-2 with the exception of y^*/y. For sufficiently high recovery so that y_2 and my_2 can be neglected in comparison with y_1 (assumed in original derivation), a material balance becomes:

$$L_1 x_1 = G_1 y_1$$

$$y_1^* = k_1 x_1$$

$$\frac{y_1^*}{y_1} = \frac{k_1 G_1}{L_1} = \frac{1}{A_1}$$

Completion of the indicated substitutions and algebraic rearrangement lead directly to Eq. (M-4) as applied to absorption ($F = A$).

For the special case of a solvent free of dissolved gas, the approximation in the above material balance can be avoided:

$$L_1 x_1 = E G_1 y_1 \qquad \text{and} \qquad \frac{y_1^*}{y_1} = \frac{E}{A_1}$$

This, of course, represents a deviation from the conditions used in the derivation through introducing a correction beyond the original scope. Substitution of this term leads to an equation resembling the (M-5) pair. On this basis Eq. (M-5*a*) would have been

$$1 - \epsilon = \frac{F_1 - E}{F_1} \frac{F_2}{F_2 - 1} (1 - E)$$

It was found from an empirical study that replacing the $F_2 - 1$ term by $F_2 - \epsilon$ was beneficial and aided in reducing the equation to the graphical form (Figs. M-4 and M-5).

Dr. Colburn himself pointed out that a great many assumptions were required in the derivation and presented a number of examples to show that the resulting errors were quite modest. For this reason the empirical extension offered is justified. In a private communication, the late Dr. Colburn indicated that he was agreeable to the treatment of the subject as offered herein.

of theoretical plates can be done by reversing the process; *i.e.*, find θ from Fig. M-5 and, then, establish the column efficiency from Fig. M-4.

Use of the above methods augmented by heat and material balances is illustrated by Example M-1.

Kremser equation ($F = F_1 = F_2$ or F_e):

$$1 - E = \frac{F - 1}{F^{n+1} - 1} \qquad \text{(M-1)}$$

$$n = \frac{\log [(F - E)/F(1 - E)]}{\log F} \qquad \text{(M-1}a\text{)}$$

$$E = \frac{n}{n + 1} \qquad \text{(when } F = 1\text{)} \qquad \text{(M-1}b\text{)}$$

$$n = \frac{E}{1 - E} \qquad \text{(when } F = 1\text{)} \qquad \text{(M-1}c\text{)}$$

Edmister effective factors:

$$F_e = \sqrt{F_1(F_2 + 1) + 0.25} - 0.50 \qquad \text{(M-2)}$$

Approximate effective factors:

$$F_e = F_1\left(\frac{F_2 + 2}{F_1 + 2}\right) \qquad \text{(M-3)}$$

Colburn equation ($F_2 > F_1 > 1$):

$$n = \frac{\log\left[\frac{F_1}{(1 - E)(F_1 - 1)}\left(\frac{F_2 - 1}{F_2}\right)^2 - \frac{1}{F_2}\right]}{\log F_2} \qquad \text{(M-4)}$$

Colburn-Clarke equation ($F_2 > F_1 \geq 1$):

$$n = \frac{[(F_2 - \epsilon)/F_2(1 - \epsilon)]}{\log F_2} \qquad \text{(M-5)}$$

$$1 - \epsilon = \left(\frac{F_1 - E}{F_1}\right)\left(\frac{F_2}{F_2 - \epsilon}\right)(1 - E) \qquad \text{(M-5}a\text{)}$$

Parameter for Figs. M-4 and M-5:

$$\theta = \left(\frac{F_1 - E}{F_1}\right)(1 - E) = \left(\frac{F_2 - \epsilon}{F_2}\right)(1 - \epsilon) \qquad \text{(M-6)}$$

where E = column efficiency, see Table M-2

F = enrichment factor, *i.e.*, absorption, stripping, or extraction factor, see Table M-1

F_e = effective absorption factor defined by Eq. (M-2) or (M-3)

F_1, F_2 = enrichment factor at feed, other end, *e.g.*, F_1 at bottom of absorber or top of stripper

n = number of theoretical plates

ϵ, θ = parameters as defined by equations

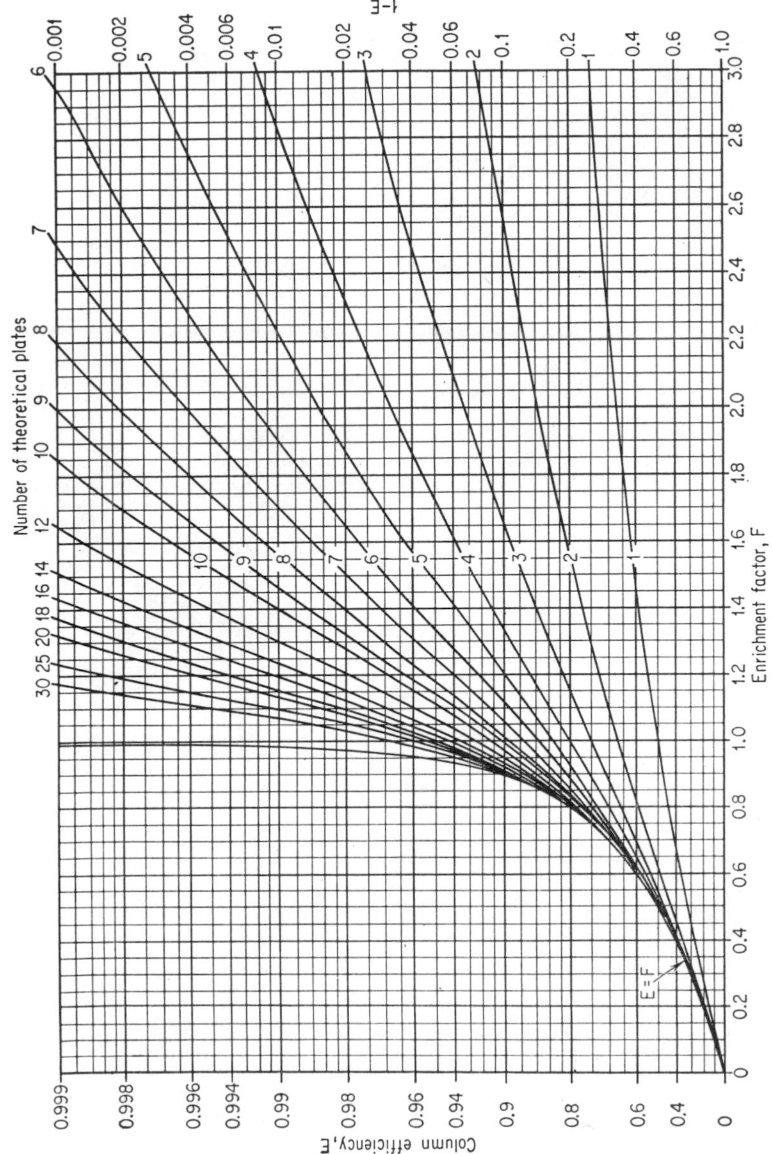

FIG. M-3.—Chart for computing the number of theoretical plates by the Kremser method.

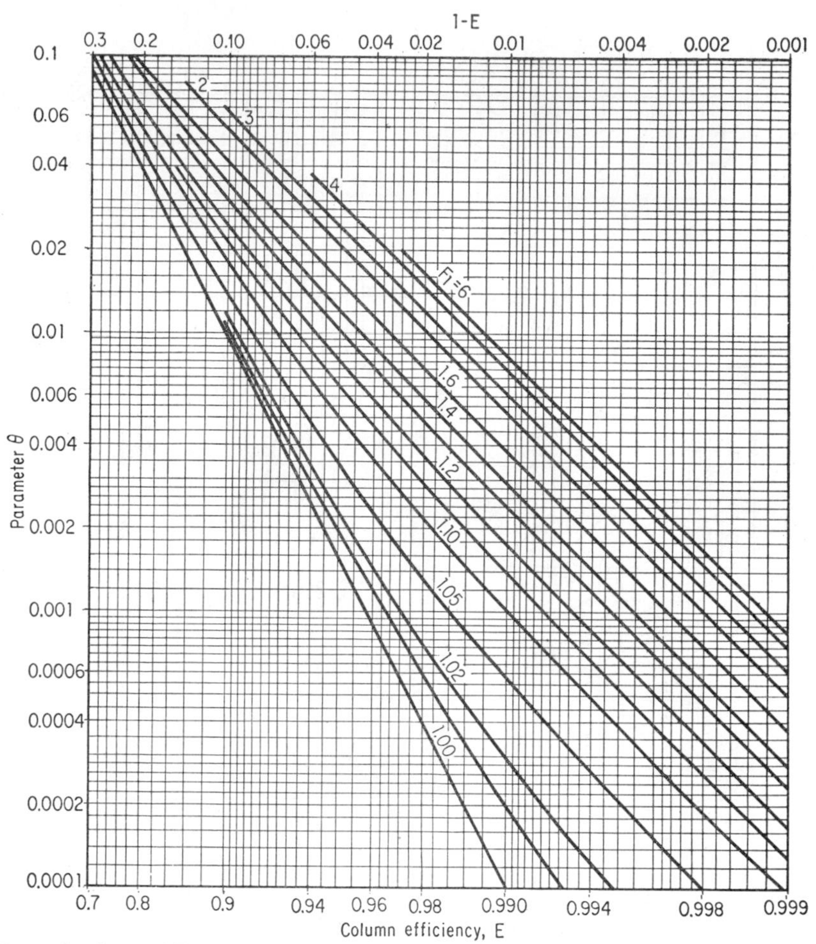

Example of use of Figs. M-4 and M-5:

Given: $F_1 = 1.10$, $E = 0.99$

$\theta = 0.001$ from Fig. M-4

$F_2 = 2$, $\theta = 0.001$

$n = 8$ theoretical plates from Fig. M-5

FIG. M-4.—Parameter for estimating theoretical plates (use parameter θ and Fig. M-5 to determine the number of theoretical plates).

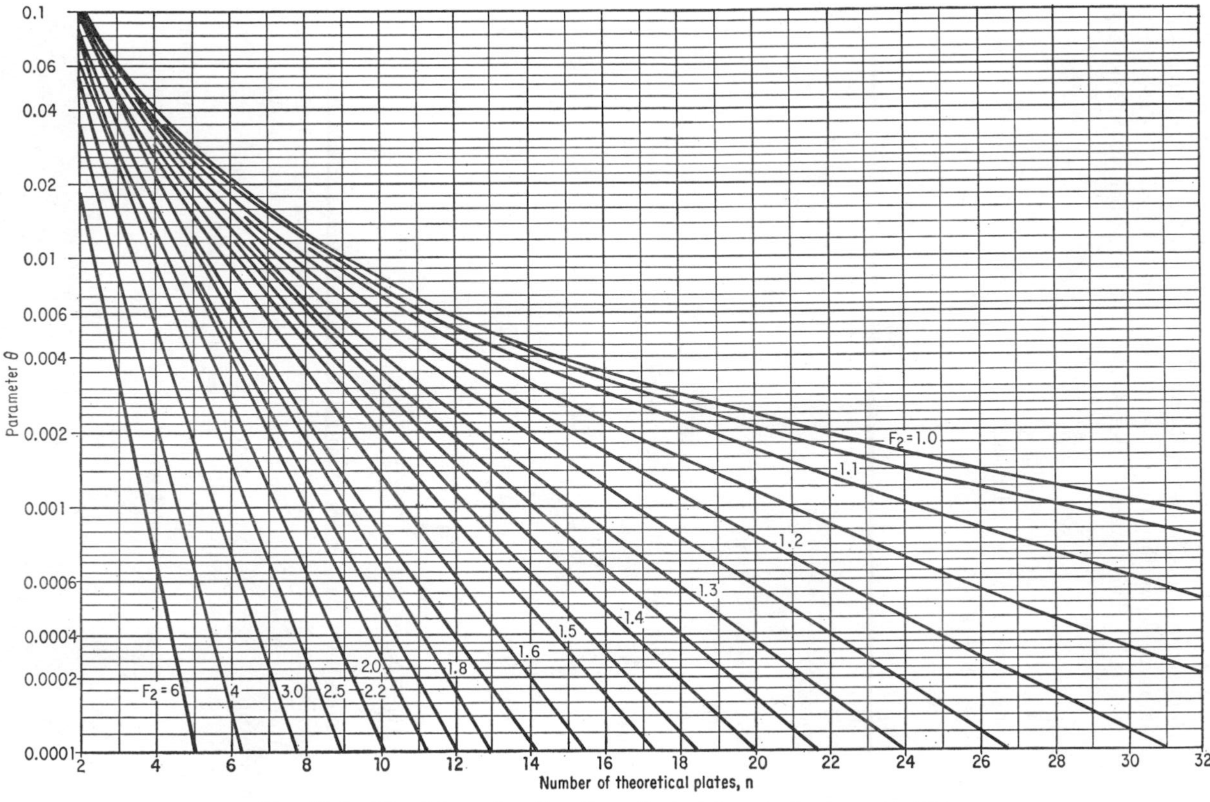

FIG. M-5.—Chart for estimating theoretical plates, single-stage columns.

Example M-1. A study is to be made of the cost of recovery of a crude ethane from natural gas. The cycle to be used is absorption, partial stripping to improve separation for methane, and total stripping to recover the desired product. This example illustrates the absorber calculations for one case.

DATA:

1. Gas available, 500 moles/hr
2. Conditions, 85°F and 410 psia
3. Gas composition:

N_2	2 mole %
CH_4	78 mole %
C_2H_6	16 mole %
C_3H_8	4 mole %
Moisture	−10°F dew point

4. Absorption oil, commercial hexane
   Specific gravity, 60/60°F, 0.73
   Average molecular weight, 130
   Vapor pressure, approximates pure hexane

ASSUMPTIONS FOR CASE 1.

1. Ethane recovery 95%
2. Propane recovery 95%
3. Maximum methane content of product 5%
4. Composition of recycle from partial stripper

C_2H_6	20 mole %
C_3H_8	1 mole %
$CH_4 + N_2$	79 mole %

5. Equilibrium data from Natural Gasoline Supply Men's Association "Engineering Data Book," 1947 (based on 2,000 psi convergence pressure)

6. Absorption conditions for this case: absorption factor at base, 1.10; temperature at base, 20°F

7. Use average pressure of 400 psia for all calculations
8. Assume gas, including recycle, is cooled to 0°F before entering absorber
9. Dissolved gas in absorption oil from stripper

N_2 and CH_4	Negligible
C_2H_6	0.7 mole %
C_3H_8	Develop as part of problem

CALCULATIONS:

1. Material balance over-all cycle:

Component	Natural gas feed		Product			Lean gas	
	Mole %	Moles/hr	Computation	Moles/hr	Mole %	Moles/hr	Mole %
N_2	2	10	Assumed small	0	0	10	2.5
CH_4	78	390	$\frac{5}{95}(76 + 19)$	5	5	385	96.25
C_2H_6	16	80	0.95(80)	76	76	4	1.0
C_3H_8	4	20	0.95(20)	19	19	1	0.25
	100	500		100	100	400	100.00

2. Vaporization constants at base of column, 20°F, 400 psia:
   N_2, $k_c = 23$
   CH_4, $k_1 = 5.2$
   C_2H_6, $k_2 = 0.82$
   C_3H_8, $k_3 = 0.22$
3. Absorption factors (A) at base:

$$C_2H_6,\ A_2 = \frac{L}{k_2G} = \frac{L}{0.82G} \qquad \text{(definition of } A\text{)}$$

$$A_2 = 1.1 \qquad \text{(conditions of case 1)}$$

Hence,

$$\frac{L}{G} = 1.1(0.82) = 0.902$$

$$N_2,\ A_o = \frac{L}{k_oG} = \frac{0.902}{23} = 0.0392$$

$$CH_4,\ A_1 = \frac{0.902}{5.2} = 0.174$$

$$C_3H_8,\ A_3 = \frac{0.902}{0.22} = 4.1$$

4. Fraction of N_2 and CH_4 absorbed:

For even as little as three theoretical plates, the column efficiency is essentially the same as the absorption factor at low absorption factors (see Fig. M-3); hence:

$$\text{For } N_2,\ E_o = A_o = 0.0392$$
$$\text{For } CH_4,\ E_1 = A_1 = 0.174$$

5. N_2 and CH_4 recycle:

	N_2	CH_4
Fraction absorbed, E	0.039	0.174
Fraction not absorbed, $1 - E$	0.961	0.826
Moles not absorbed (lean gas)	10	385 moles/hr
Moles in total feed, $\frac{\text{lean gas}}{1 - E}$	10.04	466 moles/hr
Moles in natural gas feed	10	390 moles/hr
Moles in recycle by difference	0.4	76 moles/hr

6. C_2H_6 and C_3H_8 in recycle:
   $C_2H_6 = 20\%$, $C_3H_8 = 1\%$ of recycle (case condition)
   $N_2 + CH_4 = 76.4$ moles, 79% of total recycle
   $C_2H_6 = {}^{20}\!/_{79}\ 76.4 = 19.2$ moles/hr
   $C_3H_8 = {}^{1}\!/_{79}\ 76.4 = 1.0$ mole/hr
7. Total feed and fractional recoveries:

	Absorber feed, moles/hr			Lean gas, moles/hr (calc. 1)	Gas absorbed, moles/hr	Fraction	
	Natural gas (calc. 1)	recycle (calc. 5,6)	Total			Not absorbed	Absorbed
N_2	10	0.4	10.4	10	0.4	0.961	0.039
CH_4	390	76.0	466.0	385	81.0	0.826	0.174
C_2H_6	80	19.2	99.2	4	95.2	0.040	0.960
C_3H_8	20	1.0	21.0	1	20.0	0.048	0.952
	500	96.6	596.6	400	196.6		

8. Solvent circulation:

$$L/G = 0.902 \text{ (calc. 3)}$$
$$G = 596.6 \text{ moles/hr (calc. 7)}$$
$$L = 0.902(596.6) = 538 \text{ moles/hr}$$
$$\text{Gas absorbed in column} = \underline{197} \text{ (calc. 7)}$$
$$\text{Net oil to top} = 341 \text{ moles/hr}$$
$$341\ (130) = 44{,}330 \text{ lb/hr}$$

$$\frac{44{,}330}{(8.32)60(0.73)} = 121 \text{ gpm (corrected to 60°F)}$$

9. Heat of solution:

	Moles/hr absorbed (calc. 7)	Heat of solution	
		Btu/lb mole	Btu/hr
N_2	0.4		
CH_4	81.0	1500	121,500
C_2H_6	95.2	5500	523,600
C_3H_8	20.0	6500	130,000
	196.6		775,100

10. Approximate temperature inlet oil (as approximation, neglect sensible heat of gas).

$$\text{Heat absorbed by oil} = 44{,}330c_p(20 - t_o) = 774{,}100$$

where c_p = specific heat = 0.47 (Table E-6)
t_o = inlet oil temperature

or
$$20 - t_o = \frac{775{,}100}{44{,}330(0.47)} = 37.2°\text{F}$$

Allow 40°F to provide sensible heat of gas (neglected in calculation).
Try $t_o = -20$°F, for inlet oil.

11. Temperature of top plate basis −20°F oil:

Try −10°F
N_2, $k_o = 19.5$
CH_4, $k_1 = 4.2$
C_2H_6, $k_2 = 0.60$
C_3H_8, $k_3 = 0.15$

$$x = \frac{y}{k} \text{ (equilibrium conditions, top theoretical plate)}$$

$$N_2,\ x_o = \frac{0.025}{19.5} = 0.001 \qquad (y_o = 0.025 \text{ from calc. 1})$$

$$CH_4,\ x_1 = \frac{0.9625}{4.2} = 0.230$$

$$C_2H_6,\ x_2 = \frac{0.01}{0.6} = 0.017$$

$$C_3H_8,\ x_3 = \frac{0.0025}{0.14} = \underline{0.016}$$

0.264

Oil free of dissolved gas $x_8 = 0.736$

Actual feed oil:

$x_2 = 0.007$ (condition of case)
$x_3 = \underline{0.016}$ (value from above, use to be justified later in calc. 15)
0.023

Rates of flow in feed

Total	341 moles/hr (from calc. 8)
C_2H_6	0.007 (341) = 2.4 moles/hr
C_3H_8	0.016 (341) = 5.5 moles/hr

Rate of flow out top plate:

$$\text{Total} = \left(\frac{1 - 0.023}{0.736}\right) 341 = 452 \text{ mole/hr}$$

$$\begin{aligned} N_2,\ 0.001(452) &= 0.5 \text{ mole/hr} \\ CH_4,\ 0.230(452) &= 104.0 \text{ moles/hr} \\ C_2H_6,\ 0.017(452) &= 7.7 \text{ moles/hr} \\ C_3H_8,\ 0.016(452) &= 7.2 \text{ moles/hr} \end{aligned}$$

Net absorbed and heat of solution:

	Moles/hr			Btu/lb mole	Btu/hr
	Total	In feed oil	Net absorbed		
N_2	0.5	0.0	0.5		
CH_4	104.0	0.0	104.0	1500	156,000
C_2H_6	7.7	2.4	5.3	5500	29,100
C_3H_8	7.2	5.5	1.7	6500	11,000
					196,100

$$\text{Temperature rise on oil for this} = \frac{196{,}100}{44{,}330(0.47)} = 9.4°\text{F}$$

This is close enough to 10° rise assumed.

12. Adjusted heat balance:
Lean gas leaving at −10°F is derived from feed at +20°F, the 30° drop being observed by the oil.

Component	Moles/hr	Molal specific heat (from Fig. E-5)	Btu/(hr)(°F)
N_2	10	7.0	70
CH_4	385	8.4	3234
C_2H_6	4	11.3	45
C_3H_8	1	15.4	15
	400		3364

Heat removed from gas 3364(30)	100,920
Heat of solution, absorbed gas (calc 9)	775,100
	876,020

Calculated temperature rise of oil (same method as used in calc. 10):

$$20 - t_o = \frac{876{,}020}{44{,}330(0.47)} = 42^\circ$$
$$t_o = -22^\circ\text{F}$$

This is somewhat cooler than the -20°F assumed but acceptably close for preliminary design.

13. Absorption factors at top of column:

$$L = \text{total flow of liquid out top plate}$$
$$= 452 \text{ moles/hr (calc. 11)}$$
$$G = 400 \text{ moles/hr (lean gas, calc. 1)}$$
$$L/G = 1.13$$

	Vaporization constant k	Absorption factors	
		Top	Base (from calc. 3)
N_2	19.5	0.058	0.0392
CH_4	4.2	0.269	0.174
C_2H_6	0.60	1.88	1.10
C_3H_8	0.15	7.5	4.1

14. Required number of theoretical plates by Figs. M-4 and M-5 for equilibrium of ethane with inlet:

Oil $x_2 = 0.007$

$y_2 = 0.007k_2 = 0.007(0.6) = 0.0042$

Moles lost on this basis $= 400(0.0042) = 1.68$ moles/hr

$$\text{Fractional loss} = \frac{1.68}{99.2} = 0.017$$

Actual recovery desired $= 0.95$

$0.95 = (1 - 0.017)E$ (definition of column efficiency E)

$$E = \frac{0.95}{1 - 0.017} = \frac{0.95}{0.983}$$

$$1 - E = \frac{0.983 - 0.95}{0.983} = \frac{0.033}{0.983} = 0.0335$$

$F_1 = A_b = 1.1$ (by definition of F_1 and calc. 3)

$\theta = 0.0041$ from Fig. M-4

$F_2 = A_T = 1.88$ (by definition of F_2 and calc. 13)

$n = 7$ theoretical plates Fig. M-5

15. Performance on propane:

C_3H_8: For $n = 7$ and $F_2 = 7.5$, θ is much less than 0.0001; and $1 - E$ is much less than 0.0001 for

$$\theta = 0.0001 \qquad F_1 = 4.2$$

Thus absorption of propane with propane-free oil would be virtually complete. Since 95 per cent recovery is all that is required, the propane content of oil used may be allowed to rise to that in equilibrium with the desired lean gas. This justifies the assumption given in calc. 11.

16. Performance on N_2 and CH_4:

Use "Edmister" factors and Fig. M-3 for these.

$$A_E = A_B \frac{A_T + 2}{A_B + 2} \qquad \text{Eq. (M-3)}$$

$$N_2,\ A_E = 0.0392 \frac{2.058}{2.039} = 0.0396$$

$$\text{Fraction absorbed} = E = A_E = 0.0396$$

$$CH_4,\ A_E = 0.174 \frac{2.269}{2.174} = 0.181$$

$$\text{Fraction absorbed} = E = A_E = 0.181$$

These do not differ greatly from values used earlier, and revision is not warranted for intended usage.

17. Solvent vapor loss:

Compute on basis of octane vaporization constant, which is about 0.0003 at $-10°F$ and 400 psi.

$$\text{Mole fraction in vapor phase } y = kx = 0.0003(0.736) = 0.00022$$

$$\text{Moles of vapor} = 0.00022(400) = 0.088$$

$$\text{or } 0.088(130) = 11.4 \text{ lb/hr}$$

18. Composition of rich oil product:

	Moles/hr		Total	Mole %
	Lean oil feed (calc. 11)	Material absorbed (calc. 7)		
N_2	0	0.4	0.4	0.1
CH_4	0	81.0	81.0	15.1
C_2H_6	2.4	95.2	97.6	18.2
C_3H_8	5.5	20.0	25.5	4.7
Oil	333	0	333	61.9
	340.9	196.6	537.5	100.0

CONCLUSIONS

1. 341 moles/hr of absorber oil at −22°F are required. This may contain up to 0.7 and 1.6 mole per cent of ethane and propane respectively. This establishes the principal basis for total stripper design.
2. Seven theoretical plates are required.
3. The rich oil leaving the absorber has the composition as noted in calc. 18. This together with the recycle flow tabulated in calc. 7 becomes the starting point for design of the partial stripper.
4. The above is based on assumption 3, amounting to an assumed performance of the partial stripper.

Discussion:

Calculations can be simplified somewhat if made strictly for first exploratory purposes, but short cuts should be avoided for final calculations. In some cases correction for the plate efficiencies should be considered, as these may differ for the several components.

It will be noted that heat and material balances dominate this type of problem whenever temperature effects of the heat of absorption are significant. The most accurate available thermal data should be sought and used before calculations are accepted for construction.

Assumption 4, in effect, sets the basis of design for the partial stripper and should be easily met, although design calculation on this might show either the need for relaxation of this design basis or the desirability of tighter design to reduce the recycle flow and the oil circulation required by the absorber.

A number of cases would have to be fully evaluated for the three columns and supporting equipment before it could be stated that the assumed operating conditions are the most suitable ones.

In view of the high temperature rise on the oil, one or more intercoolers to reduce the bottom temperature and oil circulation could well be considered. In this case iterative calculations below the intercooler are desirable.

c. McCabe-Thiele Method[1] for Estimating Theoretical Plates.— This graphical method is exact for the absorption of a single component. To apply this method, the concentrations are usually expressed as mole ratios, *i.e.*, moles of component being absorbed to moles of inert gases, and moles dissolved per mole of pure solvent. Weight ratios may also be employed, but percentages cannot be used directly. Two lines are plotted on plain coordinate paper. The *equilibrium line* is merely a plot of the equilibrium concentration in the gas phase plotted against the concentration in the liquid phase at the temperatures prevailing in the column. (If the heat of solution is appreciable, the temperatures must be obtained by heat balances.)

[1] McCabe-Thiele, *Ind. Eng. Chem.*, **17**, 605 (1925).

The *operating line* is a straight line connecting the two points representing the actual operating conditions at the top and the bottom of the column, respectively. The *top* point is the concentration of liquid entering plotted against the concentration of the gas leaving. The *bottom* point is the concentration of liquid leaving the base against the gas entering.

The condition that the proposed operation may be performed is that the operating line be everywhere *above* the equilibrium line (gas concentration plotted upward; liquid, across paper). If this condition is fulfilled, the number of theoretical plates may be stepped off by drawing alternate vertical and horizontal lines between the operating and the equilibrium lines. The number of steps is the number of theoretical plates required. The method is illustrated by Fig. M-6.

In applying the McCabe-Thiele method, it is frequently found that the steps become very small at one end of the figure. This difficulty may be handled by constructing an enlarged chart of the troublesome end and treating it separately.

The above covers only absorption, but the method is equally applicable to stripping and liquid-liquid extraction. In the case of

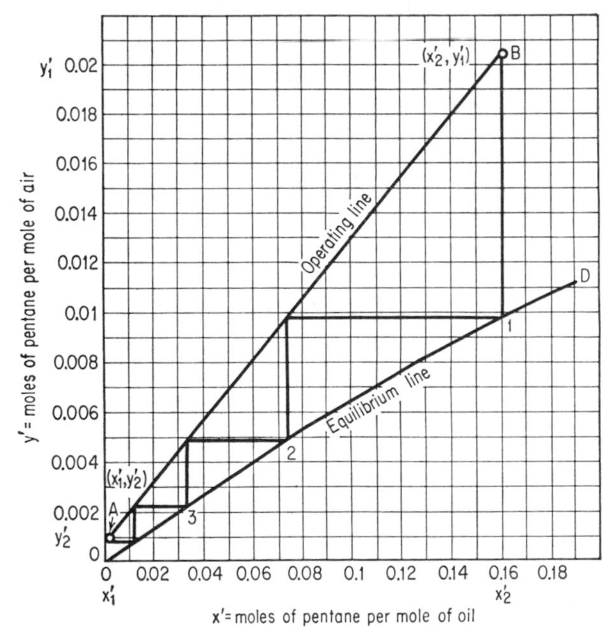

FIG. M-6.—McCabe-Thiele chart for absorption of pentane by oil. (*Nelson, "Petroleum Refinery Engineering," 2d ed., McGraw-Hill Book Company, Inc., New York,* 1941.)

stripping, the condition that the operation is possible is reversed; the operating line must always be below the equilibrium line.

d. Calculation of Transfer Units.—In computing the heights of spray, bubble, or packed columns from mass transfer coefficients, the solution contains two terms, one containing the mass transfer coefficient, mass velocity, etc.; the other a definite integral involving concentrations such as

$$n_T = \int_{y_2}^{y_1} \frac{1}{\ln (y_i/y_i^*)} \frac{dy}{y - y^*} \tag{M-7}$$

where n_T = number of transfer units based on gas phase
y = mole fraction in bulk vapor
y^* = mole fraction in gas in equilibrium with liquid
y_1, y_2 = terminal values of mole fraction
y_i, y_i^* = bulk, equilibrium values of mole fraction of inert (insoluble) gases

The above equation defines the transfer unit as applied to absorption. It should be noted that a similar formula can be used for the liquid phase to secure the number of transfer units based on the liquid. Both the liquid-phase and gas-phase forms are useful. However, many relations based on the gas-phase transfer unit for absorption are valid for the liquid-phase transfer unit in absorption. For simplicity in this presentation advantage will be taken of this fact and we shall here consider only transfer units as based on the phase identified by the feed stream. On this basis, equations given by Colburn[1] can be transposed to Eqs. (M-8) to (M-11) that follow. It should be noted that Eqs. (M-9) and (M-9*a*) are not applicable for small values of enrichment factors. For this case Edmister effective factors may be used in Eq. (M-8) as an approximation. [It is a reasonable inference from its success for theoretical plates that Eq. (M-5*a*) could be used with Eq. (M-9) to improve the accuracy of that equation for transfer units, but the authors have had no opportunity to verify the validity of this proposal by comparison with rigorous results.] Equation (M-12) expresses an approximate relation between transfer units and theoretical plates.[1] This equation is shown graphically by Fig. M-7. Constant enrichment factor:

$$n_T = 2.3 \frac{F}{F - 1} \log \frac{F - E}{F(1 - E)} \tag{M-8}$$

[1] *Ind. Eng. Chem.*, **33**, 459 (1933); see discussion of Eqs. (M-4) and (M-5).

$F_2 > F_1 > 1$:

$$n_T = \phi + 2.3 \frac{F_2}{F_2 - 1} \log \frac{F_2 - \epsilon}{F_2(1 - \epsilon)} \tag{M-9}$$

$$1 - \epsilon = \frac{F_1 - 1}{F_1} \frac{F_2}{F_2 - 1} (1 - E) \tag{M-9a}$$

$$\phi = 1.15 \log \frac{y_2}{y_1} \tag{M-10}$$

Irreversible chemical reaction ($F = \infty$):

$$n_T = \phi + 2.3 \log \frac{1}{1 - E} \tag{M-11}$$

Relation to theoretical plates (approximate):

$$\frac{n_T}{n} = 2.3 \left(\frac{F_2}{F_2 - 1}\right) \log F_2 \tag{M-12}$$

where n_T = number of transfer units (based on feed-stream phase)
n = number of theoretical plates
E = column efficiency (see Table M-2)
F = enrichment factor (see Table M-1)
F_1, F_2 = enrichment factor, at feed, other end
ϕ = transfer units for change in concentration of inerts (frequently negligible)
y_1, y_2 = mole fraction of components *not removed* from feed stream, at feed entry, effluent end

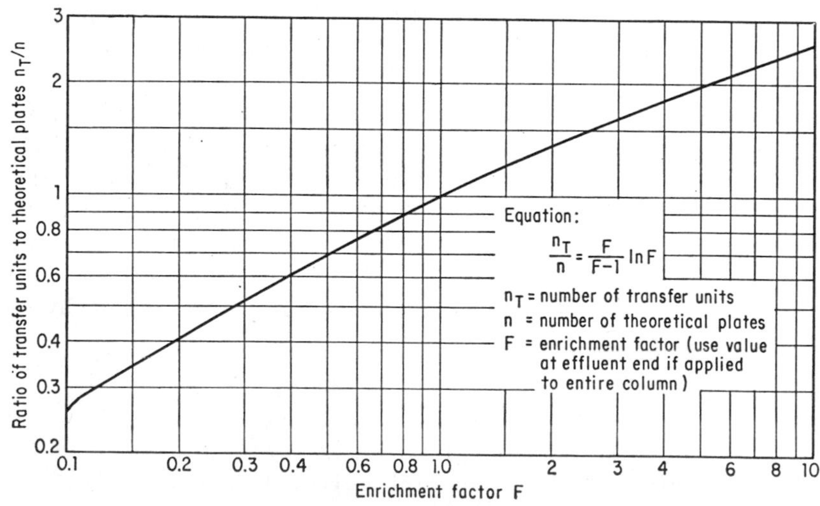

FIG. M-7.—Relation between transfer units and theoretical plates.

3. Flash Separations

When one component is far more volatile then others present or when a rough separation is required, a useful result can be achieved merely by heating (or reducing pressure of warm liquid), leaving the heavy liquid residue and the vapor product. This may be done on a batch or on a continuous basis. The results of continuous flashing can be estimated by trial-and-error calculations based on equations that follow. Table M-3 shows the details of such a calculation.

$$L = \frac{M_1}{1 + k_1V/L} + \frac{M_2}{1 + k_2V/L} + \frac{M_3}{1 + k_3V/L} + \cdots \qquad \text{(M-13)}$$

$$V = M - L \qquad \text{(M-13}a\text{)}$$

$$V = \frac{M_1}{L/(k_1V) + 1} + \frac{M_2}{L/(k_2V) + 1} + \frac{M_3}{L/(k_3V) + 1} + \cdots \qquad \text{(M-14)}$$

$$L = M - V \qquad \text{(M-14}a\text{)}$$

where L = total moles of liquid
V = total moles of vapor
M = total moles of liquid and vapor
M_1, M_2, etc. = total moles of component 1, 2, etc.
k_1, k_2, etc. = vaporization constant for component 1, 2, etc.

For two component systems, the flash equations reduce to

$$L = M_1 \frac{k_2k_2}{k_2 - 1} + M_2 \frac{k_1k_1}{k_1 - 1} \qquad \text{(M-13}b\text{)}$$

$$x_1 = (1 - k_2)/(k_1 - k_2) \qquad \text{(M-13}c\text{)}$$

where x_1 = mole fraction of component 1 in liquid and other terms are as defined above.

4. Continuous Distillation

a. Preliminary Considerations.—The simplest and most frequent application of distillation is the separation of a binary mixture into its two components. Given the desired purity of the products, such a problem can usually be estimated simply from the relative volatility of one component with respect to the other. This is defined by Eq. (M-15).

For a first evaluation it is useful to consider two extreme modes of operation: at *minimum reflux* and *total reflux*. For any case there is a minimum reflux required to effect the desired separation even though the column is made exceedingly high. This is usually (not always)

TABLE M-3.—FLASH CALCULATIONS
(Reprinted from Natural Gasoline Supply Men's Association, "Engineering Data Book," 1957)

Component	Moles m	370 psia k 90°F	Assume $V/L = 0.96$		Assume $V/L = 0.86$		Assume $V/L = 0.8452$		Moles in liquid	Moles in vapor
			$\frac{kV}{L}+1$	$\frac{m}{(kV/L)+1}$	$\frac{kV}{L}+1$	$\frac{m}{(kV/L)+1}$	$\frac{kV}{L}+1$	$\frac{m}{(kV/L)+1}$		
Methane	27.52	7.2	7.912	3.48	7.192	3.83	7.085	3.88	3.88	23.64
Ethane	16.34	1.65	2.584	6.32	2.419	6.75	2.395	6.82	6.82	9.52
Propane	29.18	0.54	1.518	19.22	1.464	19.93	1.456	20.04	20.04	9.14
i-Butane	5.37	0.25	1.240	4.33	1.215	4.42	1.211	4.43	4.43	0.94
n-Butane	17.18	0.185	1.178	14.58	1.159	14.82	1.156	14.86	14.86	2.32
i-Pentane	1.72	0.088	1.084	1.59	1.076	1.60	1.074	1.60	1.60	0.12
n-Pentane	2.18	0.069	1.066	2.05	1.059	2.06	1.058	2.06	2.06	0.12
Hexane	0.47	0.028	1.027	0.46	1.024	0.46	1.024	0.46	0.46	0.01
Heptanes*	0.04	0.0078*	1.0075	0.04	1.0067	0.04	1.0066	0.04	0.04	0.00
Total	100.00			52.07		53.91		54.19	54.19	45.81
Calculated $V/L = \frac{47.93}{52.07} = 0.9205$					$\frac{46.09}{53.91} = 0.8549$		$\frac{45.81}{54.19} = 0.8454$			

* Average of n-heptane and n-octane.

determined by conditions at the feed plate, and its value for this normal case can be computed from Eq. (M-16). For the other extreme, total reflux, no useful product is obtained, but the number of theoretical plates required for the column can be established. This *minimum number of theoretical plates* can be estimated from Eq. (M-17). It is rigorous for mixtures having constant relative volatility. Unfortunately, this condition is seldom fully met. However, if the variation is not large, the numbers computed from the extreme values (top and bottom of column) can be averaged or the McCabe-Thiele[1] method can be used.

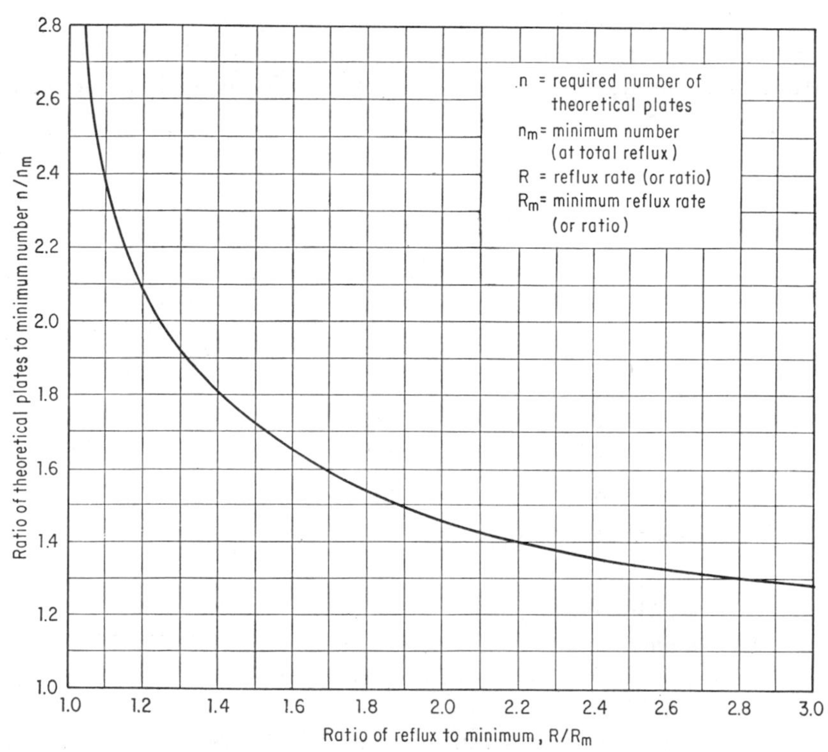

FIG. M-8.—Dimensionless plot showing approximate relation of number of theoretical plates to reflux ratio.

It is obvious that neither of the preceding limiting conditions represents a useful operation, but computation of these conditions is

[1] To do this, draw the diagonal line $x = y$ as the operating line and proceed as in Example M-2. The McCabe-Thiele method is accurate for this application.

useful since proper design conditions are always between the two extremes. This is illustrated by Fig. M-8,[1] a dimensionless chart, which relates useful operation to the two limiting values. This figure is recommended for preliminary rather than final calculations. Under favorable conditions (feed 30 to 70 per cent volatile, constant relative volatility, and comparable latent heats) the correlation is usually within a few per cent of true value, but can be in considerable error for special cases.

Even though high accuracy is not expected from Fig. M-8 under all circumstances, its usage to establish trial design conditions is recommended. To do this, compute the minimum reflux by Eq. (M-16), the minimum theoretical plates by Eq. (M-17), and then use Fig. M-8 to secure the number of theoretical plates at selected reflux ratios. Optimum design usually requires a reflux of 10 to 50 per cent above the minimum reflux.

$$\alpha = k_1/k_2 = \frac{y}{1-y}\,\frac{1-x}{x} \tag{M-15}$$

$$R_m = \frac{C_L F_L + k_f C_v F_v}{\alpha_f - 1}\,C_H \tag{M-16}$$

$$C_L = \frac{x_f - x_b}{x_f}\,\frac{x_p}{x_p - x_b}\left(1 - \frac{1-x_p}{x_p}\,\frac{x_f}{1-x_f}\,\alpha\right) \tag{M-16a}$$

$$C_v = \frac{y_f - x_b}{y_f(x_p - x_b)}\,\frac{x_p - y_f}{1 - y_f} \tag{M-16b}$$

$$C_H = x_f + (1 - x_f)(L_2/L_1) \tag{M-16c}$$

$$n_m = \frac{\log\left(\dfrac{x_p}{1-x_p}\,\dfrac{1-x_b}{x_b}\right)}{\log \alpha} \tag{M-17}$$

where α = relative volatility

x = mole fraction of more volatile component in liquid

y = mole fraction of more volatile component in vapor

k_1, k_2 = vaporization constants of more, less volatile components

k_f = vaporization constant of more volatile component in vapor feed

$= \alpha - (\alpha - 1)y_f$

R_m = minimum reflux for separation (infinite number of theoretical plates), moles/hr

[1] Based on correlation of Brown and Martin, *Trans. Am. Inst. Chem. Engrs.*, **35,** 674 (1938), augmented by authors' calculations.

x_b, x_f, x_p = mole fraction volatile component in bottoms, feed, and top product
y_f = mole fraction volatile component in vapor feed
C_L, C_v = correction factors for product purities applying to liquid, vapor feed. (NOTE: $C_L = C_v = 1$ when $x_p = 1$, $x_b = 0$.)
C_H = approximate correction factor for difference in latent heats
L_2, L_1 = heat of vaporization of more, less volatile component, per mole
n_m = number of theoretical plates at total reflux
F_L, F_v = liquid, vapor feed to column, moles/hr

b. McCabe-Thiele Method for Computing Theoretical Plates.— This graphical method, though not exact, yields moderately accurate results from simple calculations. If the number of moles of vapor flowing up the column is everywhere the same, the method becomes exact. When the molal heat of vaporization of the low-boiling component is only slightly less than that of the higher boiling component, this condition is approximated. If the heats of vaporization differ greatly (such as in acetone-water mixtures), the method will not give entirely accurate results.

To use this method, the mole fraction y of the volatile component in the vapor is plotted against the mole fraction x of the same component in the liquid phase. Three lines are constructed. The equilibrium line is a plot of the equilibrium data. There are two operating lines: the *enrichment line* for the section above the feed plate and the exhausting or *stripping line* for the feed plate and below. The enrichment line represents the composition of the *vapor leaving* and the *liquid entering*. The stripping line represents the *vapor entering* and the *liquid leaving*. To construct these lines it is necessary to assume a reflux. It is frequently convenient to express the amount of reflux as the reflux ratio: moles of liquid condensate returned to moles of product withdrawn. This is the usual way of expressing reflux ratio. A more significant (though seldom used) expression is the ratio of moles reflux to moles of feed. When expressed in this fashion, the optimum reflux is more nearly independent of feed composition. If a very low reflux ratio is chosen, the operating lines may cut the equilibrium line. Such a condition indicates that a larger reflux ratio must be employed in order to accomplish the desired result. Often several reflux ratios must be tried to secure a balance between column height and amount of boiling required. This method is illustrated by Example M-2 and Fig. M-9.

Example M-2. It is desired to distill 100 moles/hr of an ethyl alcohol–*n*-butyl alcohol mixture containing 40 mole per cent ethyl alcohol. 99 mole per cent ethyl alcohol and 98 mole per cent *n*-butyl alcohol are required. For first calculation assume a reflux ratio (reflux to product) of 1.50.

SOLUTION: This problem is solved in Fig. M-9 constructed as follows:

1. Equilibrium line drawn from data presented in Table L-8.

2. Enrichment line AB. Point A represents the composition of liquid entering vs. vapor leaving, *i.e.*, $x = y = 0.99$, the composition of the product.

Point B: $$x = 0, \qquad y = \frac{x_p}{R+1} = \frac{0.99}{2.5} = 0.396$$

where x_p = mole fraction of product = 0.99
R = reflux ratio = 1.5

$\left(\text{As constructed, the slope of the line is } \frac{1}{R+1}\right)$

3. Stripping line CD. Point C represents the mole fraction of vapor entering vs. the liquid leaving. It is customary to take these both as the composition of the bottom product, *i.e.*, $x = y = 0.02$. D is the joint intersection of the two lines at $x = 0.40$, the feed composition.

4. Number of theoretical plates. Beginning with point A, the number of theoretical plates is stepped off by alternate horizontal and vertical lines. 7.7 theoretical plates are required, 3.3 above the feed plate and 4.4 below. This calculation tacitly assumes that the feed is preheated to its bubble point (initial boiling point) and that there are no control fluctuations. To allow for these factors, more theoretical trays should be allowed, say 4 above feed plate and 6 below, making a total of 10. If a 40 per cent plate efficiency is assumed, 25 actual plates will be required.

c. Other Graphical Methods.—Lindsay and Baker[1] and Peters[2] describe modifications of the McCabe-Thiele method that take differences of latent heats into account and are, therefore, capable of higher accuracy. The Peters method employs fictitious molecular weights proportional to the latent heats. All mole fractions and flows are computed on the basis of these pseudo moles and the McCabe-Thiele method used as with real moles. Ponchon[3] and Savarit[4] have devised a method that takes the total enthalpies of vapor and liquid into

[1] *Ind. Eng. Chem.*, **35**, 418 (1943).

[2] *Ind. Eng. Chem.*, **14**, 476 (1922).

[3] *Tech. Modern*, **13**, 20 (1921). Cf. Robinson and Gilliland, "Elements of Fractional Distillation," 4th ed., McGraw-Hill Book Company, Inc., New York, 1950.

[4] *Arts et métiers*, 1922, pp. 55, 142, 178, 241, 266, 307.

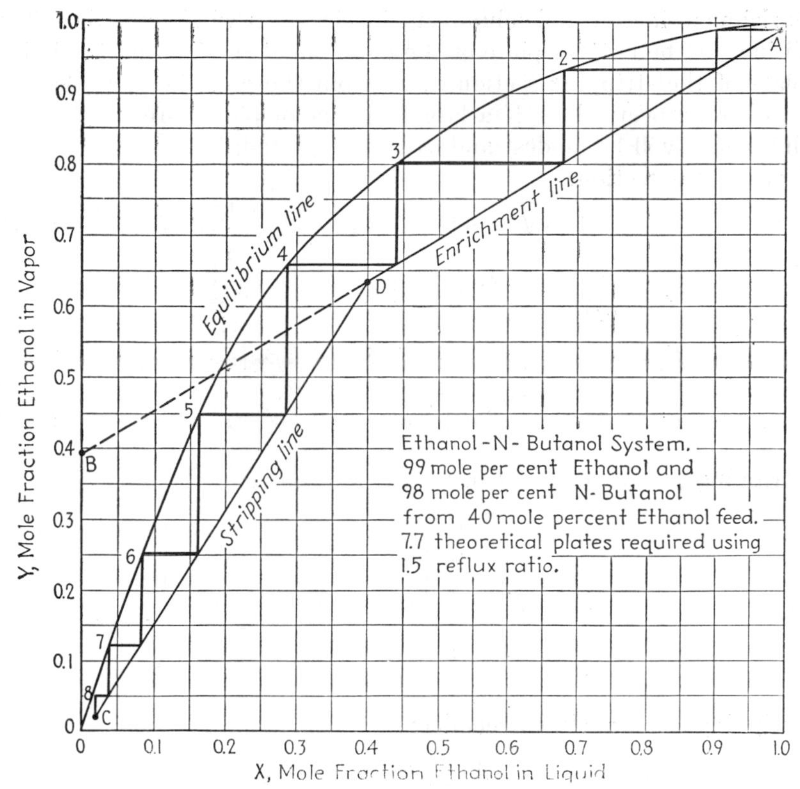

FIG. M-9.—McCabe-Thiele diagram for distillation.

account and is, therefore, free from theoretical error though the graphical construction is tedious when a large number of plates is involved.

d. Modified Colburn Method.—This is the same method as discussed under single-stage columns. It may be applied to distillation by making separate calculations for the enrichment section (treat as an absorber) and the stripping section. This method is, as already mentioned, an approximation but a reasonably good one for usual conditions. It does, however, predict somewhat too many theoretical plates when only a few are required.

Tables M-1 and M-2 may be used to establish the factors for use of Eqs. (M-5) and (M-5*a*) or the graphical equivalent (Figs. M-4 and M-5) in the same manner as already discussed. However, it will be found more convenient to use the approximate equations that follow for the enrichment factors and column efficiencies.[1] It will be noted that conditions at the bottom plate of the enrichment section and the bottom plate of the stripping section differ from those of the feed plate and reboiler, respectively. If a large number of plates are involved, the differences will be modest and can be neglected.

Enrichment section:

$$1 - E = (1 - k_t')r_E = \frac{\alpha_t - 1}{\alpha_t} r_E \tag{M-18}$$

$$r_E = \frac{V_t}{k_f' V_f} \frac{1 - x_t}{1 - x_f} = [1 + (\alpha_n - 1)x_n] \frac{V_t(1 - x_t)}{V_f(1 - x_f)} \tag{M-19}$$

$$F_1 = A_n = \frac{L_n}{k_n' V_n} = [1 + (\alpha_n - 1)x_n] \frac{L_n}{V_n} \tag{M-20}$$

$$F_2 = A_t = \frac{L_t}{k_t' V_t} = \frac{L_t}{V_t} \alpha_t \tag{M-21}$$

Stripping section, open steam:

$$1 - E = r_s = \frac{L_b}{L_f} \frac{x_b}{x_f} \tag{M-22}$$

$$F_1 = S_f = \frac{k_f V_f}{L_f} = \frac{\alpha_f}{1 + (\alpha_f - 1)x_f} \frac{V_f}{L_f} \tag{M-23}$$

$$F_2 = S_b = \frac{k_b V_b}{L_b} = \frac{\alpha_b V_b}{L_b} \tag{M-24}$$

Stripping section, closed reboiler:

$$1 - E = (k_b - 1) \frac{V_b}{L_f} \frac{x_b}{x_f - x_b} = (\alpha_b - 1) \frac{V_b}{V_f} \frac{x_b}{x_f - x_b} \tag{M-25}$$

$$r_s = \frac{k_b V_b + B}{L_f} \frac{x_b}{x_f} = \frac{\alpha_b V_b + B}{L_f} \frac{x_b}{x_f} \tag{M-26}$$

$$F_1 = S_f = \frac{k_f V_f}{L_f} = \frac{\alpha_f}{1 + (\alpha_f - 1)x_f} \frac{V_f}{L_f} \tag{M-27}$$

$$F_2 = S_1 = \frac{k_f V_b}{L_f} = \frac{\alpha_f V_b}{L_b} \tag{M-28}$$

where E = column efficiency
F_1, F_2 = enrichment factors
A = absorption factor
S = stripping factor
r_s = fraction of low boiler not stripped
r_E = fraction of high boiler not retained
k = vaporization constant for low boiler

[1] The approximations are very close, and exact computation does not lead to a significant improvement in accuracy unless either the bottom or top product is less than 95% purity.

k' = vaporization constant for high boiler
x = mole fraction of low boiler
L = liquid flow entering designated plate, moles/hr
V = vapor flow leaving designated plate, moles/hr
B = bottom product withdrawn as liquid from reboiler
α = relative volatility

and subscripts denote

t, top plate
n, bottom plate of enrichment section
f, feed plate
m, bottom plate of stripping section
b, bottom product

The foregoing method is applicable for preliminary calculation on multicomponent distillation wherein the split in products is moderately complete for two key components, since the stripping section can be treated for any component more volatile than the key as the enrichment for the less volatile components and phase equilibria is closely approached for the reverse situations. Applications of this type are probably best limited to developing trial bases for plate-by-plate iterations.

e. Heat and Cooling Requirements for Distillation.—These are computed by standard thermodynamic methods. Heat must be supplied to (1) heat the high-boiling constituent from the inlet temperature to the boiling temperature at the base of the column, (2) heat the volatile component from the feed temperature to its boiling point, (3) vaporize the volatile product and the reflux, and (4) supply the heat of solution, which is the net heat evolved when the two liquids are mixed without vaporization. Because the heat required depends on the temperature of the feed, it is frequently desirable to preheat the feed. This may conveniently be done by heat exchange between the feed and the hot liquid leaving the base, the steam condensate, or even the condensed product. Estimation of steam and cooling water demands is illustrated by Example M-3.

Example M-3. It is desired to find the steam and water requirements for the distillation discussed in Example M-2, *i.e.*, the separation of 100 moles/hr of an *n*-butyl alcohol–ethyl alcohol solution containing 40 mole per cent of ethyl alcohol using a 1.5 reflux ratio. For this calculation the separation may be assumed a perfect one.

SOLUTION:

1. Quantities involved,

Molecular weights:

n-Butyl alcohol = 74
Ethyl alcohol = 46
n-Butyl alcohol: 60 × 74 = 4,440 lb/hr
Ethyl alcohol: 40 × 46 = 1,840 lb/hr

2. Temperature at top of column is 175°F (boiling point of 99 mole per cent ethyl alcohol from Table L-8).

3. Temperature at base,

Allow 1.6 in. of water drop per plate, or 40 in. total.

$$\text{Pressure} = 14.7 + 40\,\frac{0.433}{12} = 16.4 \text{ psi}$$

Temperature of boiling n-butyl alcohol = 250°F (from Fig. E-17)

4. Approximate average column temperature is 201°F at feed plate (boiling point of 40 mole per cent ethyl alcohol at 1 atm from Table L-8).

5. Temperature of feed, assume 185°F, this assumption to be justified later by calculation 12.

6. Sensible heat supplied to n-butyl alcohol, 185 to 250°F, at 212°F the specific heat is 0.78 Btu/(lb)(°F) (from Fig. E-15)

$$\Delta h = 4{,}440 \times 0.78(250 - 185) = 225{,}000 \text{ Btu/hr}$$

7. Sensible heat supplied to ethyl alcohol, 185 to 175°F (a drop in temperature),

Specific heat at 180°F is 0.78 Btu/(lb)(°F)(from Fig. E-15)

$$\Delta h = 1{,}840 \times 0.78(175 - 185) = -14{,}000 \text{ Btu/hr}$$

8. Heat of vaporization of ethyl alcohol,

$$\frac{\Delta H}{T} = 26.2 \text{ Btu/(lb mole)(°R)(from Fig. E-19)}$$

$$\Delta H = \frac{1{,}840}{46} \times (1 + 1.5) \times 26.2 \times (175 + 460)$$

$$= 1{,}664{,}000 \text{ Btu/hr } (1.5 = \text{reflux ratio})$$

9. Heat of solution, neglect as deviation from Raoult's law is small.

10. Total heat supplied,

$$225{,}000 - 14{,}000 + 1{,}664{,}000 = 1{,}875{,}000 \text{ Btu/hr}$$

This is exclusive of heat losses. Making an arbitrary allowance of 5 per cent for these, the total becomes 1,970,000 Btu/hr.

11. Steam required,

Assume steam supplied to the vaporizer at 25 psig.

Temperature = 267°F, latent heat ≈ 934 Btu/lb (Table E-14)

$$\text{Steam} = \frac{1{,}970{,}000}{934} = 2{,}110 \text{ lb/hr}$$

12. Heat exchange,
Assume preheat is supplied to feed by exchange with the n-butyl alcohol product. Heat supplied to heat feed from 80 to 185°F,

$$\begin{aligned}\Delta h &= 4{,}440 \times 0.66 \times (185 - 80) + 1{,}840 \times 0.69 \times (185 - 80) \\ &= 308{,}000 + 133{,}000 = 441{,}000 \text{ Btu/hr}\end{aligned}$$

Heat supplied by n-butyl alcohol,

$$-\Delta h = 4{,}440 \times 0.66 = 2{,}930 \text{ Btu/°F, temp. drop}$$

Temperature drop to supply needed heat,

$$-\Delta t = \frac{441{,}000}{2{,}930} = 150°\text{F}$$

Temperature of product leaving exchanger = 250 − 150 = 100°F
(This makes a temperature difference of 20°F at the cold end of the exchanger.)

13. Cooling requirements,
Condenser removes 1,664,000 Btu/hr (calculation 8).
Product cooler cools alcohol from 175 to 90°F,

$$-\Delta h = 1{,}840 \times 0.69 \times (175 - 90) = 107{,}000 \text{ Btu/hr}$$

Total heat removed,

$$1{,}664{,}000 + 107{,}000 = 1{,}771{,}000 \text{ Btu/hr}$$

The quantity of water required will depend on the temperature rise which depends on cooling water vs. exchanger costs at plant site. Taking a 40°F rise the water required is

$$\frac{1{,}771{,}000}{40 \times 500} = 886 \text{ gpm}$$

DISCUSSION: 2,110 lb/hr of steam are required for the particular process design assumed. It would be desirable to repeat the calculation for different reflux ratios before adopting the final process design—comparing the resulting sizes of the column and exchanger, and the utility requirements. Also the heat loss should be estimated more explicitly.

To check on the accuracy of the calculation made in this example, the number of moles of vapor at the base may be computed. The heat of vaporization of n-butyl alcohol at 250°F is 25.3 × (250 + 460) = 18,000 Btu/lb mole; making the quantity vaporized = 1,970,000/18,000 = 109 moles/hr. This is only slightly more than the 100 moles vapor/hr at the top of the column.

Thus, the condition of a constant molar vapor stream required for accuracy of the McCabe-Thiele method is nearly fulfilled. Furthermore, since the error is in the direction of fewer theoretical plates, the calculation is conservative.

5. Batch Distillation

The simplest case is distillation with direct condensation (no column or rectification) of vapor boiled off. The Rayleigh equations[1] that follow may be used to calculate results for binary mixtures.
From relative volatility:

$$\log \frac{L_o}{L} = \frac{1}{\alpha - 1}\left(\log \frac{x_o}{x} - \alpha \log \frac{1 - x}{1 - x_o}\right) \tag{M-29}$$

From vaporization constant:

$$\log \frac{L_o}{L} = \frac{1}{k - 1} \log \frac{x_o}{x} \tag{M-30}$$

Average composition of overhead product:

$$x_p = \frac{L_o x_o - Lx}{L_o - L} \tag{M-31}$$

where L_o = initial charge, moles
L = residue at any specified time, moles
x_o = initial mole fraction of more volatile component
x = same when residue is L
x_p = average mole fraction more volatile component in accumulated condensate
α = relative volatility, see Eq. (M-15)
k = vaporization constant

A column with reflux is required for any relatively close separation. The minimum reflux required for separation of a batch can be estimated by one of the two equations that follow.[2]
From relative volatility:

$$R = \frac{L_o}{\alpha - 1}\left[\left(\frac{x_o - x_c}{1 - x_c}\right) + (1 - x_o) \ln \left(\frac{1 - x_c}{x_c} \frac{x_o}{1 - x_o}\right)\right] \tag{M-32}$$

From vaporization constant:

$$R = L_o\left[\frac{1 - x_o}{k - 1} \ln \left(\frac{1 - x_c}{x_c} \frac{x_o}{1 - x_o}\right) - \frac{x_o - x_c}{1 - x_c}\right] \tag{M-33}$$

[1] *Phil. Mag.*, **4** (6), 521 (1922).

[2] Based on ideal conditions of constant latent heats, a pure top product, neglect of column capacity, and either constant relative volatility or vaporization constant.

where R = reflux required for initial charge L_o, moles
L_o = initial charge, moles
x_o, x_c = mole fraction volatile component in charge, residue at time of cut
α = relative volatility
k = vaporization constant for most volatile component

In practice, batch distillation requires more than minimum reflux. A separation could be made with only a modest excess reflux over the minimum if the reflux ratio were continuously, and appropriately, increased as the volatile component is stripped from the still pot. This requires a high-quality control system, presumably with differential temperature as one element. Usually simple controls are employed and the reflux ratio changed infrequently, so that *very much more* reflux need be employed. When a number of batches of the same type are to be handled, retention of intermediate (or slop) cuts between the pure product draws is good practice, as these cuts can be added to the next batch and reprocessed without loss.

The number of plates and the reflux ratio can be established for any specified time during the cycle in the same fashion as for continuous distillation. Such calculations are appropriate for conditions at cut points and scheduled changes in reflux ratio. The McCabe-Thiele method is usually adequate for batch distillation since, after all, the technique should be employed only when steam and cooling water requirements are small enough that added capital expenditure for the far more efficient continuous distillation equipment and instrumentation is not warranted.

6. Two-section Columns

Figure M-2 lists typical examples of two-section columns. All these can be treated by dividing the calculations into those for the upper and lower section, respectively. Discussion appearing earlier in the chapter and Fig. M-1 indicate the nature of the division. The special case of continuous distillation has been covered (Section 4).

The principal difficulty in calculations arises from the interrelationships of the two sections that usually necessitate trial-and-error procedure in order to match conditions at the top of the lower section with the base of the upper. In distillation, the *minimum reflux* can be calculated and used to estimate the desired reflux which establishes the basis of calculation for both sections. For any two-section column there is a limiting condition for successful operation that is analogous to minimum reflux, even though the limiting condition is not so easily defined or reduced to a useful universal formula. We may state this both for partial strippers and for absorbers with a stripping section by the equations given below (based on complete separation between two key components).

$$\frac{L_1}{k_1'G_1} = A_b \gtrsim 1 \qquad \text{(M-34)}$$

$$\frac{L_2}{k_2G_2} = \frac{1}{S_t} \lesssim 1 \qquad \text{(M-35)}$$

$$G_1 = G_2 + F_G \qquad \text{(M-36)}$$

$$L_2 = L_1 + F_L \qquad \text{(M-37)}$$

where L_1, L_2 = liquid rate at base of top section, top of bottom section, moles/hr
G_1, G_2 = gas rate at base of top section, top of bottom section, moles/hr
k_1' = vaporization constant of more soluble of two key components at conditions prevailing at base of top section.
k_2 = vaporization constant of less soluble key at top of base section
A_b = absorption factor, base of top section
S_t = stripping factor, top of bottom section
F_G, F_L = moles/hr of gas, liquid, in center feed

These equations can be extended to liquid-liquid extraction by appropriate changes in terminology. It can be noted that if the temperatures of the two sections match and the feed is a single phase, then the flow relationships between the two sections follow at once, either for the minimum condition or for assigned values of enrichment factors (A_b and S_t). The temperature condition is normally met for liquid-liquid extraction but rarely met for absorption-stripping operations owing to heats of solution. The top (absorbing) section is usually considerably cooler than the bottom (stripping) section at their junction, with the net result of *reducing* either the requirement for reflux (partial strippers) or the stripping gas (absorbers with stripping sections). If this situation is kept in mind, Eqs. (M-34) to (M-37) will be found useful in establishing first trial conditions.

7. Phase Separators

a. Basic Factors.—The laws governing the rates of settling of suspended solid particles (or liquid droplets) have been well established.[1]

[1] For an excellent review see Perry "Chemical Engineers' Handbook," 3d ed., McGraw-Hill Book Company, Inc., New York, 1950.

As for all flow problems, the type of settling is dictated by the Reynolds number which is defined for this case as follows:

$$R = \frac{Du_\rho}{\mu} = \frac{du_\rho}{205\mu_c} \tag{M-38}$$

where R = Reynolds number, dimensionless
D, d = particle diameter, ft, microns
u = velocity, fps
ρ = density of continuous phase, lb/cu ft
μ, μ_c = viscosity, lb/(ft sec), centipoises

The flow pattern in the continuous phase resulting from the particle motion is streamline when the Reynolds number is less than 2, and settling rates follow Stokes' law [Eq. (M-39)]. Extremely fine particles are an exception to this rule as they settle at rates below that predicted by Stokes' law due to the effect of individual molecular motion on individual particles. This difficult region begins at particle sizes of around ½ micron for settling in air and around 2 microns for settling in water. For Reynolds numbers between 500 and 200,000, Newton's law, Eq. (M-40), applies.

Stokes' law (R = 0.0001 to 2):

$$u = 0.0285 \left(\frac{d}{1{,}000}\right)^2 \frac{\rho_p - \rho_c}{\mu} \tag{M-39}$$

Newton's law (R = 500 to 200,000):

$$u = 0.178 \sqrt{\frac{d}{100} \frac{\rho_p - \rho_c}{\rho_c}} \tag{M-40}$$

where R = Reynolds number, Eq. (M-38)
u = settling velocity, fps
d = particle diameter, microns (1 mm = 1,000 microns)
ρ_p, ρ_c = density of particle, continuous phase, lb/cu ft
μ = viscosity of continuous phase, centipoises

The intermediate region is covered by Fig. M-10. To use this, compute the "settling velocity" by either Stokes' law, Eq. (M-39), or Newton's law, Eq. (M-40), and then the fictitious Reynolds number. The ratio of the true settling velocity to the calculated value can then be read off the curve and the expected settling velocity computed. In some cases, the fictitious Reynolds number so found will be out of range for the law used, and the calculation must then be repeated using Stokes' law instead of Newton's or vice versa.

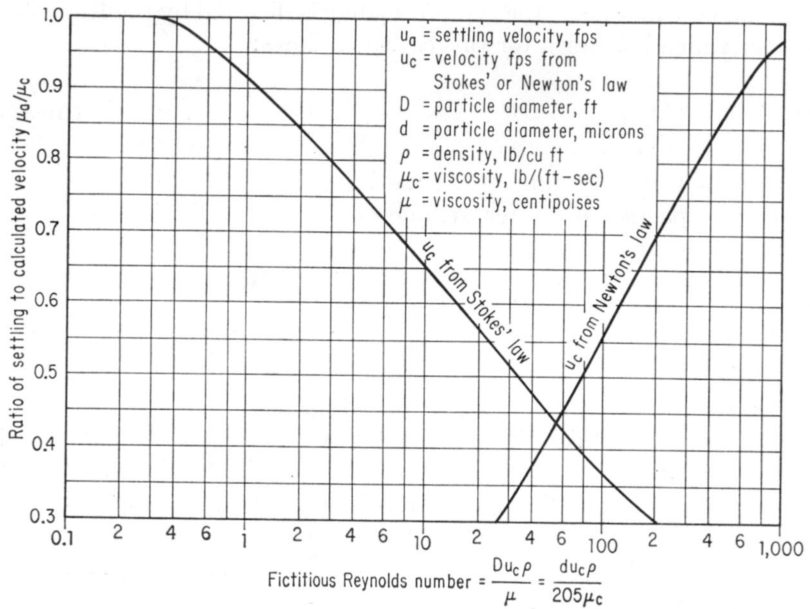

FIG. M-10.—Chart for estimating terminal settling velocities of spherical particles in the intermediate region (between Stokes' and Newton's regions).

In view of the dominance of particle size in settling rates, coalescence of tiny droplets to larger ones greatly favors phase separations involving liquids. High viscosity of either phase impedes coalescence, but high surface tension is favorable. Coalescence is promoted by impingement on surfaces, particularly if the surface is wet by the liquid droplets. This is the basis of the popularity of impingement and centrifugal separators. Electrostatic precipitators are frequently used to promote coalescence of fog or agglomeration of fine dusts.

b. Gravity-type Separators.—These are, basically, just vessels with enough area to keep the velocity below the settling velocity of the finest particle that must be recovered and with enough height to dissipate the entrance velocity effects. In the case of dusts, it is often possible to determine the particle-size distribution and determine the required area for upward flow from Stokes' law, Eq. (M-39), velocity of settling. For coarse solids Newton's law may apply rather than Stokes' law. The minimum height above and below the inlet should be many times the diameter of the inlet nozzle. Addition of baffles, downward pointed inlet nozzle, inlet transitions, etc., all favor good performance.

In the case of separation of liquids, the particle size of the droplets is seldom known and an empirical assumption becomes necessary. For liquid-liquid decanters use of Stokes' law with an assumed particle size in the range of 75 to 150 microns is usually adequate unless the tendency toward stable emulsion formation is strong. If the liquids to be separated are contaminated by fine solid matter, the latter may stabilize the emulsion, but this effect can be eliminated by filtration before decantation.

The situation for separation of liquids from gases is much the same, and the problem is also usually met by rules of thumb or arbitrary formulas. Probably the best of the several arbitrary approaches is assumption of particle size, as many such separations lie in the range between Stokes' and Newton's laws. Design based on removal of 100-micron droplets is conservative and will normally assure a virtually complete removal of entrained liquid. Choice of the design particle in the range of 200 to 400 microns covers the range in difficulty of most separations. In some cases, appreciable entrainment may not be harmful and the design requirement can be relaxed still further. This case is illustrated by the separation required between plates of a distillation column, and the methods of sizing plate columns may be applied to separations of this type.

Separators subject to pulsations, such as separators at the suction or discharge of a reciprocating compressor, should, of course, be based on design velocities of 30 to 50 per cent of nonpulsating equipment. Complete removal of compressor oil is often a severe problem because of the small diameter of the (mechanically produced) droplets and static charge.

Employment of the impingement principle is advantageous for the problems that are difficult because of small droplet size. Frequently "de-misters" are incorporated in gravity separators as a layer of spun wire mesh, Raschig rings, etc., across the flow area. Such aids to coalescence often permit an increase in the design particle size, perhaps up to 1,000 microns.

c. Centrifugal Separators.—A typical centrifugal separator is illustrated in Fig. M-11. The gas enters tangentially, and its spinning section augments the force of gravity by centrifugal force. The higher the velocity, the more effective the separation. Thus high nozzle velocities are desirable. On the other hand, the body of the separator should not be too small lest the separated liquid (or dust) be reentrained by the "turn around." Suitable design velocities are given by the equations that follow:

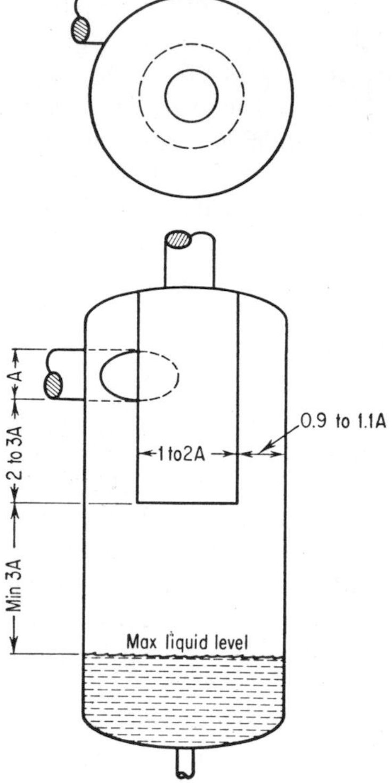

FIG. M-11.—Sketch of typical centrifuge separator for removing liquid entrainment (typical dimension ratios indicated).

Nozzle velocity:

$$u = 3.6\sqrt{\frac{\rho_p - \rho_g}{\rho_g}} \tag{M-41}$$

Velocity in body for normal service:

$$u = 0.4\sqrt{\frac{\rho_p - \rho_g}{\rho_g}} \tag{M-42}$$

For critical services:

$$u = 0.2\sqrt{\frac{\rho_p - \rho_g}{\rho_g}} \tag{M-42a}$$

where u = design velocity, fps

ρ_p, ρ_g = particle, gas density, lb/cu ft

For dust removal Lapple[1] recommends typical dimension ratios based on rectangular inlet nozzles. These correspond to a center outlet tube about twice the diameter of the inlet nozzle and the outlet

[1] Perry, "Chemical Engineers' Handbook," 3d ed., McGraw-Hill Book Company, Inc., New York, 1950.

tube to extend only a short distance below the nozzle. For separation of liquids, the authors prefer a long center tube, extending about three nozzle diameters below the nozzle, and a "turn-around" section between the center tube and base (or maximum liquid level) of three to five nozzle diameters. The greater length gives better opportunity for liquid coalescence on the shell of the separator. Lapple[1] also gives data on minimum particle size that can be removed in centrifugal separation and approximate data on total pressure drop. For first approximation the pressure drop may be taken as three velocity heads computed against velocity in the inlet nozzle plus two velocity heads based on the exit nozzle.

8. Sizing Plate Columns

a. Capacities of Sieve and Bubble Plate Columns.—There are a great many data on the effect of numerous constructional details on capacities of columns for absorption, stripping, and distillation. Such data are useful for establishing detailed design for the intended service. The first design requirement is that the column must accommodate the flow of the liquid countercurrent to that of the gas or vapor. Failure to do so can arise in two manners: The column may *prime* because the vapor carries the liquid upward from one tray to the next; it may *flood* because the liquid cannot flow through the downcomers. The gas bubbling through the liquid produces a seething mass of bubbles or froth that rises well above the liquid level at the overflow weir. If this froth reaches the plate above, priming occurs. Factors leading to flooding are: excessive weir crest, excessive liquid gradient across the plate, too small downcomers, excessive pressure drop of gas across the plate.

A second design requirement arises from consideration of plate efficiency, which can suffer from excessive entrainment of liquid by gas or vapor even though the column exhibits stable mechanical performance.

A convenient procedure for sizing new columns is to set an allowable superficial gas velocity that is permissible from the joint considerations of priming and entrainment and then to set the mechanical details of construction to accommodate the required liquid flow. The recommended procedure for determining the allowable superficial gas velocity is as follows:

1. Set plate spacing and height of clear liquid.
2. Read priming constant from Fig. M-12.

[1] *Loc. cit.*

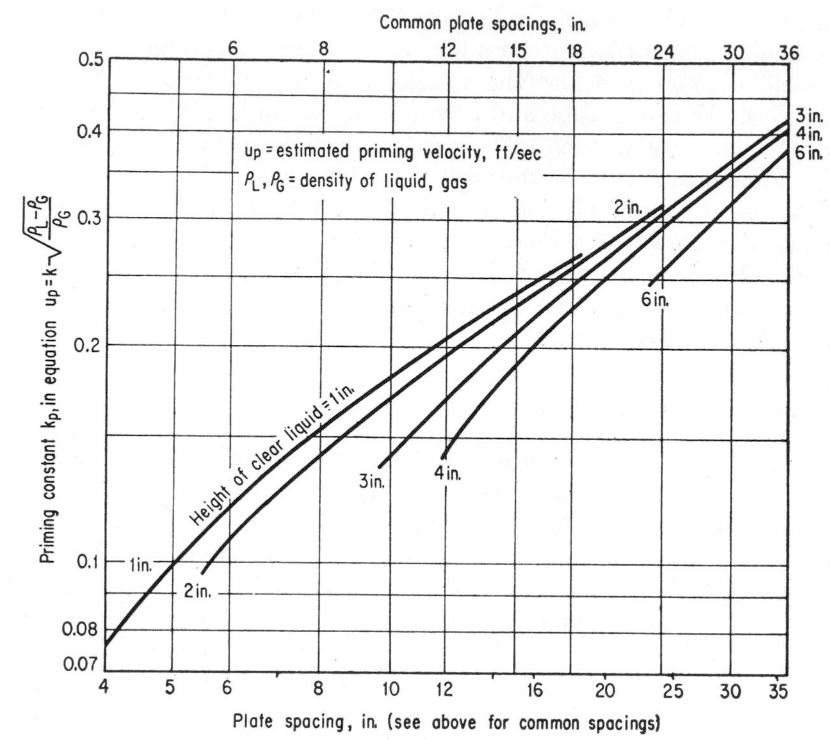

FIG. M-12.—Chart for estimating priming velocities in plate columns. (Does not apply directly to very high liquid-gas ratios. Use with Fig. M-13 to estimate allowable velocity corrected for entrainment and high liquid-gas ratios.)

3. Calculate parameter, $L/G \sqrt{\rho_G/(\rho_L - \rho_G)}$.

4. Set permissible entrainment from consideration of plate efficiency (low entrainment is the normal basis).

5. Calculate allowable velocity from equation below:

$$u = FC_eK_p \sqrt{\frac{\rho_L - \rho_G}{\rho_G}} \tag{M-43}$$

where u = design velocity based on total column area less area of one downcomer
F = factor of safety desired
C_e = correction factor from Fig. M-13
K_p = priming constant from Fig. M-12
ρ_L, ρ_G = liquid, gas density, lb/cu ft

6. Read value of C_e from Fig. M-13.

The figures for sizing plate columns just referred to represent a somewhat different type of correlation than any that have been pub-

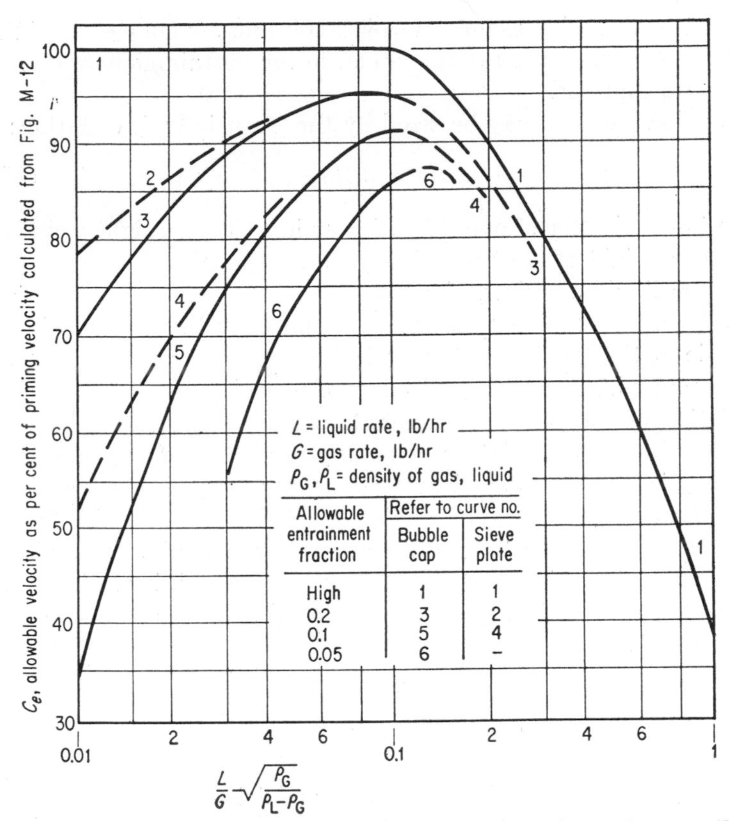

FIG. M-13.—Chart for establishing allowable velocities in plate columns. (Use in conjunction with Fig. M-12.)

lished. Figure M-12 is based on three sources of somewhat divergent nature: the correlation of Fair and Mathews[1] for 18- to 36-in. plate spacings, miscellaneous data for low plate spacings,[2] and the American Institute of Chemical Engineers[3] correlation of froth height. The last correlation provides a convenient means of reconciling data for a wide range of plate spacings. The values of the priming constant by Fig. M-12 range from 60 per cent of that permissible from froth height alone at low spacings up to 80 per cent at high spacings. Figure M-13 gives factors based on the work of Fair and Mathews[1] to correct "priming" velocities for entrainment and for unusually high liquid

[1] Fair and Mathews, *Petroleum Refiner*, April, 1958.

[2] Largely unpublished data. Cf. Wenzel, *Chem. Eng. Progr.*, **53,** 272 (1957), and Sellers and August, *Trans. Am. Inst. Chem. Engrs.*, **34,** 53 (1956).

[3] "Bubble Tray Design Manual," American Institute of Chemical Engineers, New York, 1958.

rates. The dotted lines for sieve plate columns are not well established but take *partial* credit for the much lower entrainment observed in sieve plate columns.[1]

The downcomers may be sized by the Francis Weir equation:

$$m = 0.4lh^{3/2} \tag{M-44}$$

where l = perimeter of overflow pipe or length of straight weir, in.
h = height of crest permissible, in.
m = liquid overflow, cfm

b. Number of Plates Required.—The over-all plate efficiency is the ratio of the number of theoretical plates to the number of actual plates required. Knowledge of this efficiency permits a straightforward calculation of the plate requirement. Because of greater ease in correlation with physical data and column details, the Murphree vapor efficiency is often employed and many of the published data are in terms of the Murphree vapor efficiency. It is, basically, the fractional approach of the vapor to equilibrium with the liquid. Occasionally, the Murphree liquid efficiency, analogously defined, is used.

The over-all efficiency can be estimated from the equation below[2] or Fig. M-14:

$$E_o = \frac{\log \dfrac{A}{A - E_v(A - 1)}}{\log A} \tag{M-45}$$

where E_o, E_v = over-all Murphree vapor efficiency
A = absorption factor[1] (see Table M-1)

Figures M-15 and M-16 are empirical correlations of plate efficiencies for stills and absorbers, respectively. These correlations are, of necessity, only first approximations, since they ignore constructional details and consider only one of the many physical properties involved. The efficiencies are, of course, related to the rates of mass transfer of material in both phases and to the interface area offered by the bubbles in the froth and (to a lesser extent) the droplets above the froth. Numerous investigations have solved the difficult mathematical problems involved in the computation of plate efficiencies, and these efforts

[1] The observed entrainment for sieve plate columns is about 70 per cent that found for bubble plate columns. Cf. Hart, Hansen, and Wilkie, *J. Am. Inst. Chem. Engrs.*, **1**, 441 (1955).

[2] Equation is exact when A is constant. For other conditions many writers recommend use of factor $\frac{L}{G}\Big/\frac{dy}{dx}$ to replace A and reduce errors (see Table M-1 for nomenclature). For derivation of the equation, see Marshall and Pigford, "Application of Differential Equations to Chemical Engineering Problems," University of Delaware, 1947.

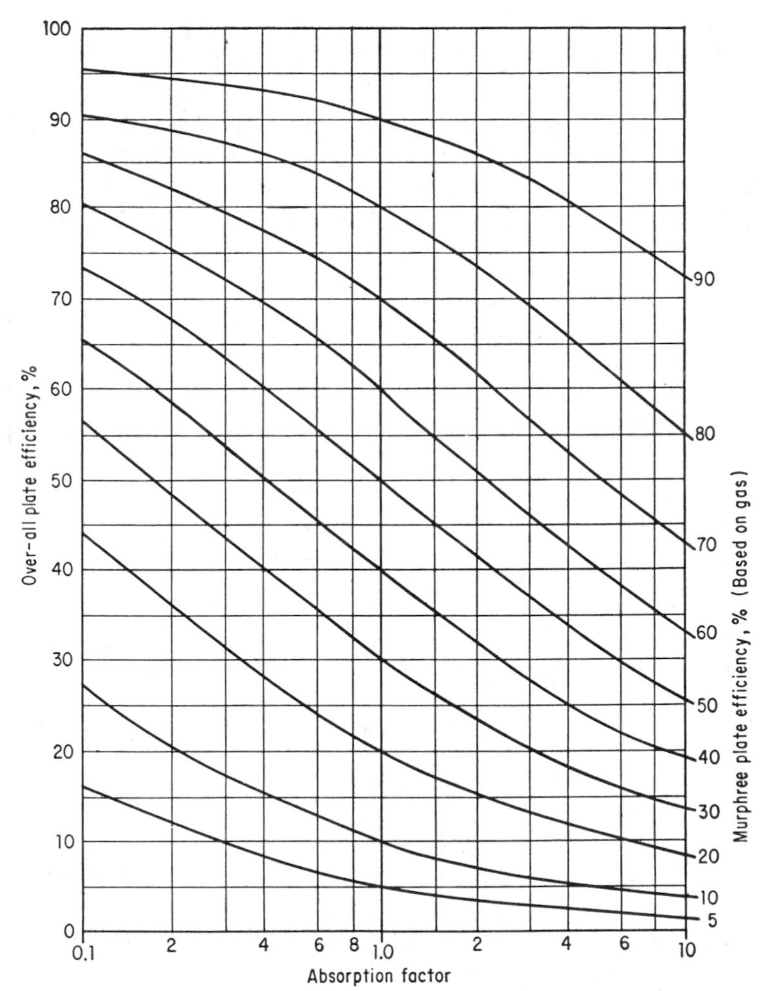

FIG. M-14.—Relation between over-all and Murphree efficiencies, based on Eq. (M-45).

have been summarized in an excellent treatise[1] to which reference is made. To be sure, these calculation procedures are involved and their utility limited by the relatively small amount of good data for diffusivities and accurately measured efficiencies of single plates. The reader may look forward to rapid developments now that the theoretical treatment is beginning to bear fruition. The analysis shows that the following factors are beneficial in improving otherwise low efficiencies:

1. Limit entrainment.

[1] "Bubble Tray Design Manual," American Institute of Chemical Engineers, New York, 1958.

2. Increase of residence time of liquid by raising weir height or increasing liquid path by reducing the number of passes or increasing column diameter.

3. Decreasing cap size or, better, using a perforated plate (sieve plate) is somewhat beneficial.

4. Increasing cap clearances.

5. Using maximum slot (or hole) velocities permitted by pressure drop consideration leads to modest improvement.

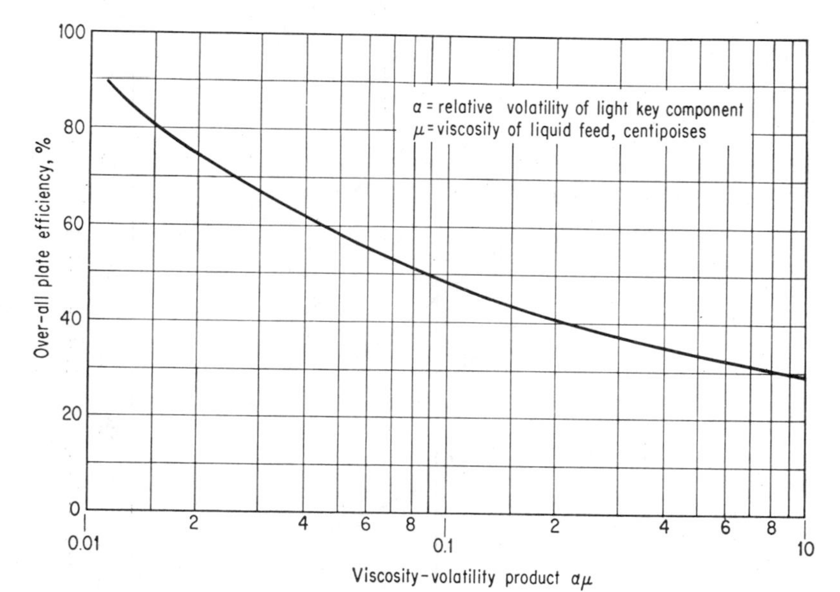

FIG. M-15.—Approximate plate efficiencies of fractionating columns. [*After O'Connel, Trans. Am. Inst. Chem. Engrs.*, **42**, 741 (1946).]

The effect of entrainment is shown by the equation that follows:[1]

$$E_v = \frac{E_d}{1 + \Psi E_d} \tag{M-46}$$

where E_v = Murphree efficiency with entrainment
E_d = Murphree efficiency with no entrainment reaching plate above
Ψ = entrainment impinging on plate above computed as a ratio to *net* liquid downflow

It will be noted that 10 per cent entrainment reduces a 100 per cent Murphree efficiency by almost the same percentage but has little effect on low efficiencies. Thus Eq. (M-46) can be used to establish the

[1] Colburn, *Ind. Eng. Chem.*, **28**, 526 (1936). Note that this equation does not apply to irreversible chemical absorption.

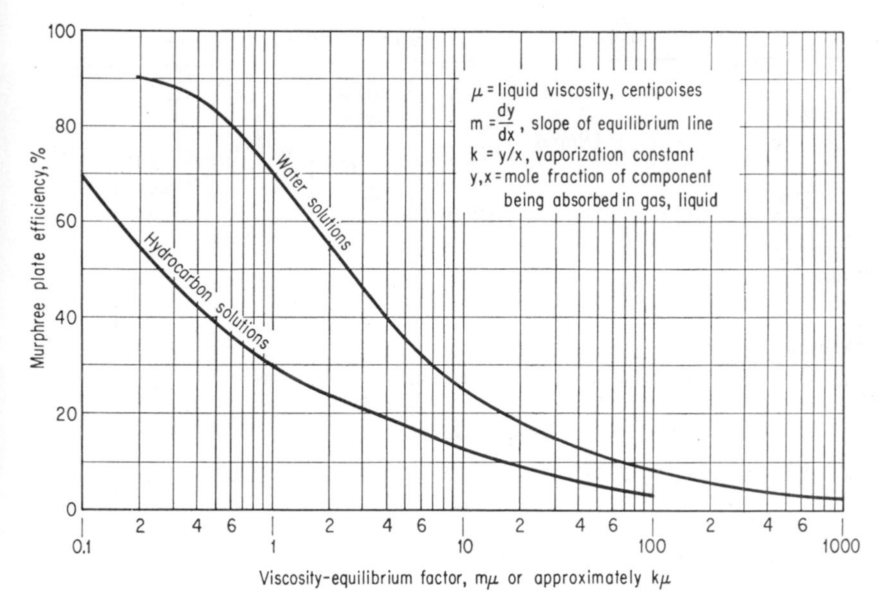

Fig. M-16.—Approximate plate efficiencies of absorbers. (*Adapted from Sherwood and Pigford, "Absorption and Extraction," McGraw-Hill Book Company, Inc., New York, 1952.*)

permissible entrainment ratio in column sizing by methods such as that given in the preceding section.

The special case of irreversible chemical absorption (such as absorption of carbon dioxide in caustic soda) deserves mention. In this case, entrainment has little or no effect on efficiency and the over-all plate efficiency has little meaning, since only one theoretical plate is involved regardless of the performance desired. The performance is directly related to the Murphree efficiency as follows:

$$1 - r = E_v^n \tag{M-47}$$

$$n = \frac{\log (1 - r)}{\log E_v} \tag{M-47a}$$

where r = fraction of component removed by irreversible chemical absorption

E_v = Murphree gas efficiency for component being removed

n = number of plates

9. Sizing of Packed Columns

a. Data on Common Packing Materials.—Tables M-4 and M-5 give surface areas and other data on materials commonly used as filling in packed columns.

b. Capacity of Packed Columns.—Obviously a column cannot be operated if there is danger of liquid being carried over by the gas

TABLE M-4.—PHYSICAL DATA ON DUMPED PACKINGS

Material	Dimensions, in.	Bulk density, lb/cu ft, tower volume	Surface area, sq ft/cu ft, tower volume	% free volume	No. pieces per cu ft, tower volume	Reference
Coke	3	24	12	50	......	1
	6	10	5.5	57	......	1
Ceramic Raschig rings	½ × ½	65	114	53	10,700	3
	1 × 1	45	58	68	1,330	2
	1½ × 1½	45	36	68	380	3
	2 × 2	24	29	83	160	3
Metal Lessing rings	½ × ½	..	130	91	9,000	1
	1 × 1	..	74	93	1,300	1
	2 ×2	..	37	95	160	1
Berl saddles	½	45	141	68	15,000	3
	1	42	79	69	2,300	2
	1½	42	50	70	650	3

1. Sherwood, "Absorption and Extraction," McGraw-Hill Book Company Inc., New York, 1937.
2. Molstad, Abbey, Thompson, McKinney, *Trans. Am. Inst. Chem. Engrs.*, **38,** 387 (1942).
3. Sherwood and Pigford, "Absorption and Extraction," 2d ed., McGraw-Hill Book Company, Inc., New York, 1952.

stream. This danger is present if a column operates at or above the flood point either as defined by the graphical method of White[1] or as visually observed.[2] White further defined a lower limit, the load point, occurring when a further increase of gas flow increases the pressure drop by more than the second power of the superficial gas velocity. (The superficial gas velocity is the average velocity the gas would have if flowing through the cross section of the empty tower.) It is believed that, at flows below the load point, the liquid holdup is relatively independent of gas flow, but that this holdup is increased rapidly by gas flows above the load point. Accordingly, he suggests the load point as the upper limit for operation even though higher flows may be mechanically stable. Consideration of (1) possible control fluctuations, (2) possible future alteration of process design, (3) possible reduction in capacity because of settling and breakage of packing, and (4) difficulty of accurate prediction of load points under

[1] White, *Trans. Am. Inst. Chem. Engrs.*, **31,** 390 (1935).
[2] Sarchet, *Trans. Am. Inst. Chem. Engrs.*, **38,** 283 (1942).

TABLE M-5.—PHYSICAL DATA ON STACKED PACKINGS

Material	Dimensions, in.	Bulk density, lb/cu ft, tower volume	Surface area, sq ft/cu ft, tower volume	% free space	% free transverse area	No. pieces per cu ft tower volume	Reference[1]
Raschig rings	2 × 2	..	32	80	..	....	3
	4 × 3	52	20	64	..	38	1
Partition rings	3 × 3	67	41	44	42	64	2
Single spiral rings	3 × 3	53	36	64	55	64	2
Triple spiral rings	3 × 3	60	42	57	56	64	2
Drip point grids (#6146, General Ceramic Co.)		72	18	48	41	9.2	2

[1] For references, see footnotes for Table M-4 on facing page.

operating conditions, all suggest a more conservative capacity rating: say, 50 to 70 per cent of the load-point capacity. Colburn[1] suggests 50 per cent of the flood-point flows as a desirable operating condition. Distillation columns, however, usually operate between the load and flood point. See discussion of distillation, page 465.

Figures M-17 to M-19 permit the estimation of the load points for gases and nonfoaming liquids. These charts present the irrigation rate at the load point as a function of a term involving the liquid gas ratio, and the ratio of gas and liquid densities. With the exception of the viscosity term, the charts are a direct consequence of the correlation of Sherwood, Shipley, and Holloway.[2] The viscosity correction factors given in Fig. M-19 are based on the findings of later experimenters.[3]

In using these charts, it should be borne in mind that they are most accurate for air-water systems and may suffer somewhat from imperfections in the correlation for systems greatly different. The accuracy will, however, probably be within 20 per cent unless the process fluids are very unusual. *Very* low surface tension or fine suspended solids may possibly promote foaming and lower the tower capacity.

[1] Colburn, Absorption of Gases by Liquids, Collected Papers on Teaching of Chemical Engineering, *Trans. Am. Inst. Chem. Engrs.*, 1940, pp. 269–281.

[2] *Ind. Eng. Chem.* **30**, 765 (1938).

[3] Bain and Hougen, *Trans. Am. Inst. Chem. Engrs.*, **40**, 29 (1944). Schoenborn and Dougherty, *Trans. Am. Inst. Chem. Engrs.*, **40**, 51 (1944).

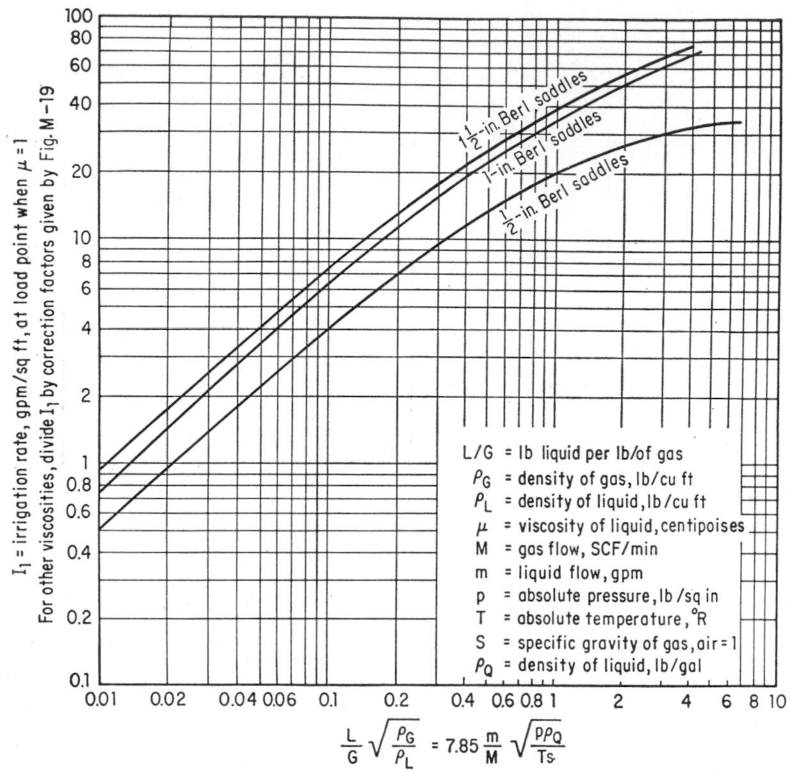

FIG. M-17.—Load points for columns packed with dumped Berl saddles. [*Data of Tillson on Berl saddles reported by Molstad, et al., Trans. Am. Inst. Chem. Engrs.*, **38**, 387 (1942).]

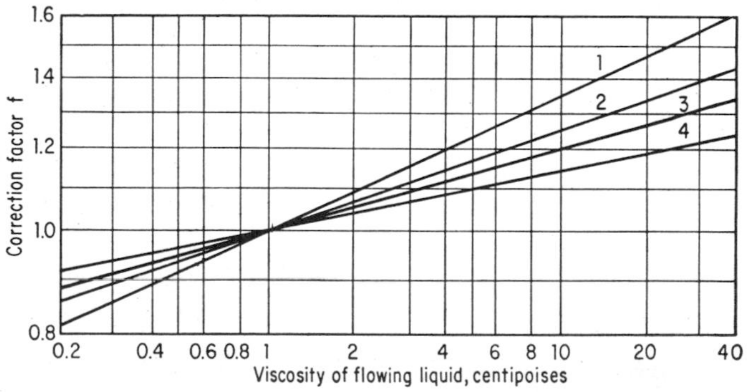

FIG. M-19.—Viscosity correction factors for liquid

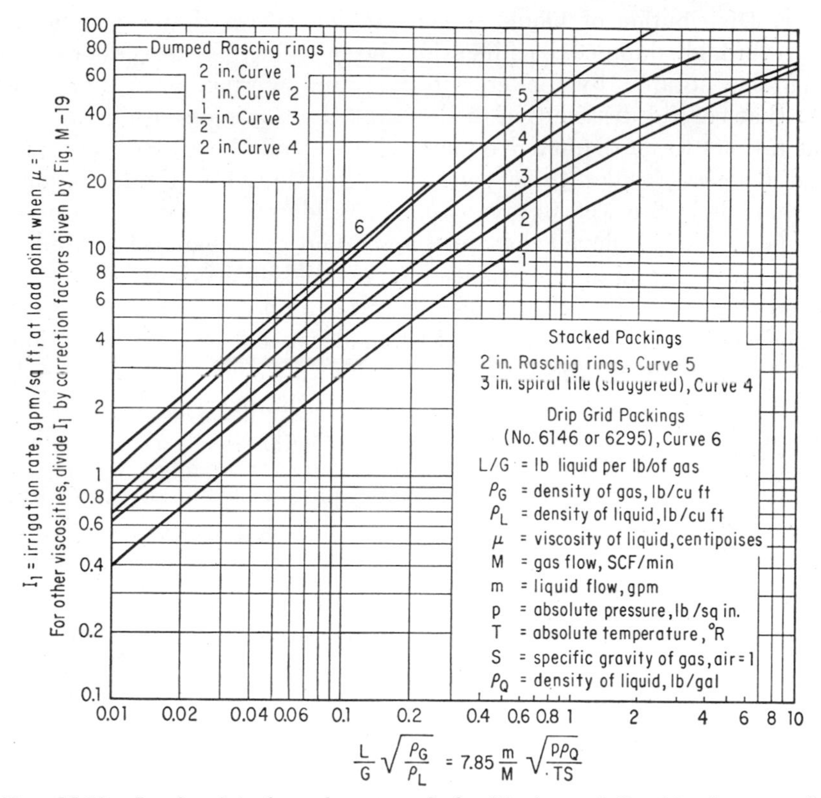

FIG. M-18.—Load points for columns packed with dumped Raschig rings or with stacked packings. [*Data of Tillson and Molstad, et al., Trans. Am. Inst. Chem. Engrs.*, **38**, 387 (1942).]

Curve No.	Suggest for towers filled with:		
	Raschig rings	Berl saddles	Stacked packings
1	$\frac{1}{2}$ in.		
2	1 in.	$\frac{1}{2}$ in.	
3	$1\frac{1}{2}$–2 in.	1 in.	Spiral tile
4		$1\frac{1}{2}$ in.	Grids

$I = f I_1$ = irrigation rate at load point. Where I_1 = load point irrigation based on 1 centipoise. See Figs. M-17 and M-18 for numerical values

flow at load and flood points of packed columns.

c. Distribution of Liquid.—The problem of distribution of liquid flow through dumped packings has been studied by several experimenters, notably by Baker, Chilton, and Vernon[1] who conclude (1) that careful distribution of liquid at the top is essential, (2) that the ratio of tower diameter to packing size should be greater than 8:1, and (3) that, under these conditions, the liquid will not channel. Perhaps it is also well to give careful consideration to distribution of gas flow. It is believed that very low liquid rates should be avoided. Estimates of the minimum irrigation for assured efficiency range between 3 and 5 gpm/sq ft.[1] Examples may be found in the literature of experimental columns operating with moderate efficiency at irrigations of 1 gpm/sq ft.[2] In the absence of specific data, these low rates should be used with caution. Probably liquid distribution is particularly important in this case, and operation near the load point may be desirable. Use of close-packed fillings such as the Stedman type[3] might be helpful.

d. Pressure Drop through Packed Columns.—The pressure drop in packed columns is very small, often smaller than the drop across the rest of the system. Table M-6 gives the range of pressure drops observed at the load point for several types of packing material.

TABLE M-6.—PRESSURE DROPS, PACKED COLUMNS AT LOAD POINT—AIR WATER

[Molstad et al., *Trans. Am. Inst. Chem. Engrs.*, **38**, 387 (1942)]

Packing	Range of Pressure Drop, In. of Liquid (Water)/Ft
½-in. Raschig rings	1.1–1.5
1-in. Raschig rings	0.6–1.8
1½-in. Raschig rings	0.4–0.5
2-in. Raschig rings	0.4–0.9
2-in. Raschig rings, stacked	0.2–0.5
½-in. Berl saddles	0.2–0.8
1-in. Berl saddles	0.7–1.2
1½-in. Berl saddles	0.5–0.6
3 × 3-in. single spiral tile, stacked staggered	0.3–0.6
Drip-point grid #6146, General Ceramic Co., straight flues	0.5–0.6

If a more accurate estimate of the pressure drop is required, it may be obtained by the use of equations of the following type:

$$\Delta p = \frac{U_0^n \rho^{n/2}}{(a - b\sqrt{L})^n} \qquad \text{(equation applies below load point)}$$

[1] *Trans. Am. Inst. Chem. Engrs.*, **31**, 296 (1935).

[2] White and Othmer, *Trans. Am. Inst. Chem. Engrs.*, **38**, 1067 (1942). Dwyer and Dodge, *Ind. Eng. Chem.*, **33**, 485 (1941). Duncan, Koffolt, and Withrow, *Trans. Am. Inst. Chem. Engrs.*, **38**, 259 (1942).

[3] U.S. Patent No. 2,047,444 (1936).

where Δp = inches water head/ft of packed length
U_0 = superficial gas velocity, fpm
L = irrigation rate, gpm/sq ft
ρ = gas density, flowing conditions, lb/cu ft
n = constant (1.8 to 2.0)
a, b = constants (b is somewhat dependent on viscosity)

Below are equations of this type for four packings:

1-in. Berl saddles:

$$\Delta p = \left(\frac{U_0}{105 - 11\sqrt{L}}\right)^2 \rho$$

1½-in. Berl saddles:

$$\Delta p = \left(\frac{U_0}{142 - 14\sqrt{L}}\right)^2 \rho$$

1½-in. Raschig rings:

$$\Delta p = \left(\frac{U_0}{130 - 13\sqrt{L}}\right)^2 \rho$$

No. 6146 drip grids, straight flues:

$$\Delta p = \left(\frac{U_0}{410 - 36.6\sqrt{L}}\right)^2 \rho$$

e. Height Required for Packed Columns.—The mass-transfer rates that determine the height required for a column to perform its design function have been correlated in terms of a two-film theory and consideration of factors that determine the effective available area for mass transfer.

The theory considers the transport of material through one phase to the interface and thence to the bulk of the second phase. Each of these rates is conveniently expressed as a mass-transfer coefficient. Under the assumption that a local equilibrium exists between phases precisely at the interfacial contact, the two may be combined to secure an over-all coefficient of mass transfer. Equations for this follow:

$$\frac{1}{K_G} = \frac{1}{k_G} + \frac{p_H}{k_L} \qquad \text{(M-48)}$$

$$\frac{1}{K_L} = \frac{1}{p_H k_G} + \frac{1}{k_L} \qquad \text{(M-48}a\text{)}$$

where K_G = over-all coefficient of mass transfer based on gas, moles/(hr)(sq ft)(Δp)

K_L = over-all coefficient of mass transfer based on liquid,[1] moles/(hr)(sq ft)(Δx)
k_G = gas film coefficient of transfer, moles/(hr)(sq ft)(Δp)
k_L = liquid film coefficient of transfer, moles/(hr)(sq ft)(Δx)
p_H = Henry's law constant for equation $p_i = p_H x_i$ (see discussion following), atm
$\Delta p = p - p_i$, atm
$\Delta x = x - x_i$
p, p_i = partial pressure of component of interest in bulk of gas phase, at interface, atm
x, x_i = mole fraction of component of interest in bulk of liquid, at interface

Equations (M-48) and (M-48*a*) are accurate for local conditions but can lose accuracy over a section of a column owing to variation of the Henry's law constant. In this case the error can be minimized by replacing p_H by the term

$$p \frac{dy^*}{dx}$$

where p = total pressure, atm
dy^*/dx = slope of equilibrium line
y^* = mole fraction in gas at equilibrium with liquid having mole fraction x

If the effective interface area contained in a cubic foot is known and the coefficients are constant, the height of a column may be computed from either of the equations that follow:

$$Z = \frac{G}{(K_G a)p} n_G \tag{M-49}$$

$$Z = \frac{L}{K_L a} n_L \tag{M-49a}$$

where Z = required height of column, ft
G = gas flow rate, *moles*/(sq ft of column area)(hr)
$K_G a$ = product of interface area a and K_g [see Eq. (M-48)], moles/(hr)(sq ft)(atm)(mole fraction)
$K_L a$ = product of interface area a and K_L [see Eq. (M-48*a*)], moles/(hr)(sq ft)(mole fraction)
p = total pressure of system, atm
L = liquid flow rate, moles/(hr)(sq ft of column area)
n_G, n_L = number of transfer units based on gas, liquid

[1] K_L frequently defined as moles/(hr)(sq ft)(Δc) where Δc is gradient in concentration, moles/cu ft.

a = effective interfacial area, sq ft/cu ft of packed volume (see Fig. M-20)

Tests have shown that the surface of the column packing is not all wetted by the flowing liquid and that the extent of wetting depends on the irrigation rate and properties of the liquid. Moreover, some of the wetting is by stagnant liquid and not effective for mass transfer. Although many useful correlations have been developed before evaluation of the "effective area," such correlations have been short of universality. Recently Shulman[1] and his coworkers have succeeded in making a reasonable approximation of the effective area and developed the correlations that follow:

$$j_D = \frac{k_G p}{G_m}\left(\frac{y_2 \mu_G}{\rho_G D_G}\right)^{2/3} = 1.195\left(\frac{D_e G}{\mu_G}\right)^{-0.36} \quad \text{(M-50)}$$

$$= \frac{k_L}{L}\left(\frac{\mu_L}{\rho_L D_L}\right)^{1/2} = 25.1\left(\frac{L D_p}{\mu_L}\right)^{-0.55} \quad \text{(M-51)}$$

where j_D = modified "j factor" for diffusion

k_G, k_L = mass-transfer coefficients from gas, liquid; see Eqs. (M-48) and (M-48a)

p = total pressure

y_2 = mole fraction of components not absorbed (or stripped)

μ_G, μ_L = viscosity of gas, liquid, lb/(ft)(hr)

ρ_G, ρ_L = density of gas, liquid, lb/cu ft

D_G, D_L = diffusivity of component of interest in gas, liquid, sq ft/hr

D_e = equivalent diameter = $D_p/(1 - \epsilon)$ (see Fig. M-21)

D_p = diameter of sphere having same volume as one piece of packing, ft

= 0.0075, 0.018, 0.033 for ½-, 1-, 1½-in. Raschig rings

= 0.0052, 0.013, 0.024 for ½-, 1-, 1½-in. Berl saddles

$1 - \epsilon$ = space occupied by packing and liquid holdup, cu ft/cu ft

G = mass velocity of gas, lb/(sq ft column area)(hr)

G_m = gas flow rate, *moles*/(sq ft of column area)(hr)

L = liquid flow rate, *moles*/(sq ft of column area)(hr)

To use these equations for calculating column heights it is necessary to evaluate diffusivities in the liquid and in the gas[2] and also to determine liquid holdup[3] and effective areas. Estimation of the latter two

[1] *J. Am. Inst. Chem. Engrs.*, **1**, 247, 253, and 259 (1955); **3**, 157 and 290 (1959).

[2] See Hirschfelder, et al., "Molecular Theory of Gases and Liquids," John Wiley & Sons, Inc., New York, 1954; Sherwood and Pigford, "Absorption and Extraction," 2d ed., McGraw-Hill Book Company, Inc., New York, 1950; and other reference texts.

[3] Shulman et al., *J. Am. Inst. Chem. Engrs.*, **1**, 247, 253, and 259 (1955).

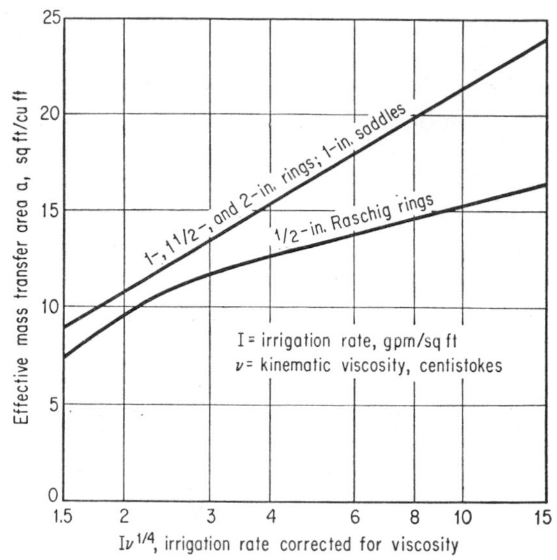

FIG. M-20.—Tentative values of effective mass-transfer areas for packed columns operating at 70 to 90 per cent of load point.

by the best methods is tedious. Figures M-20 and M-21 for direct estimation of effective areas and equivalent diameters may be employed without undue loss in accuracy for most liquids.

The method of employing these data will not be covered in detail, but the order of calculation is as follows:

1. Determine the equivalent diameter d_e and effective area a by Figs. M-20 and M-21 or, more precisely, directly through the Shulman data.

2. Compute k_G and k_L from Eqs. (M-50) and (M-51).

3. For absorbers compute K_G from Eq. (M-48) and the height from Eq. (M-49) based on the number of transfer units computed by Eq. (M-11) (F = absorption factor).

4. For strippers compute K_L from Eq. (M-48a) and the height from Eq. (M-49a) based on the number of transfer units computed by Eq. (M-11) (F = stripping factor).

The above method is admittedly involved, and in many cases, its accuracy is limited by accuracy of both measured and experimental diffusivities. For these reasons the reader should seek specific information on the system of interest in reference texts and in research papers. For many, but not all systems, 1 to 3 ft are equivalent to an actual plate.

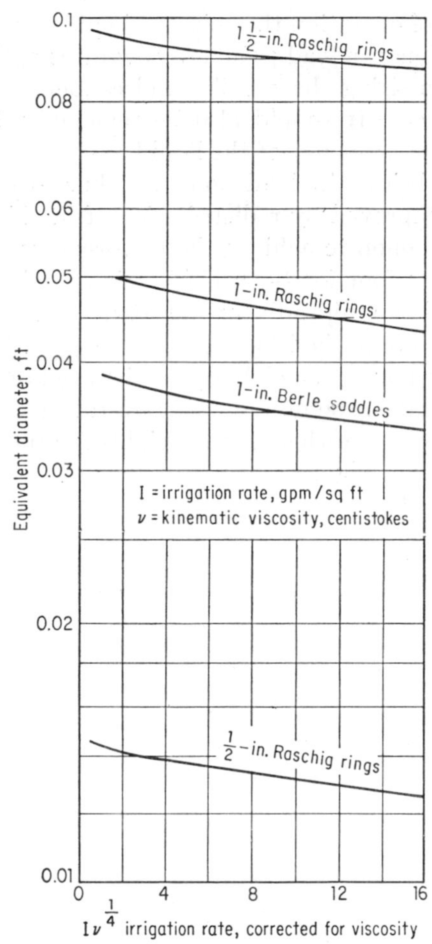

FIG. M-21.—Approximate values of equivalent diameters of column packings (applicable below the load point).

f. Distillation in Packed Columns.—This is usually practiced at rates between the load and flood points for three reasons. First, the volume of liquid flow is normally small compared with the vapor, and the resultant irrigation rate is not enough to wet more than a small fraction of the packing, resulting in loss of efficiency. Second, above the load point the liquid holdup is increased, thereby wetting more packing and increasing efficiency. Last, column stability can be secured by allowing the pressure drop to control boilup, since excess boilup, though wasteful of utilities, does not impair product quality.

For this type of operation the column should be sized for achieving or slightly exceeding the load point by flow rates optimum to the distillation and then sizing the reboiler and condenser for 20 to 30 per cent greater values. It should also be remembered that above the load point there is a tendency of the liquid to channel (a tendency that is largely absent below the load point). Thus, performance of high columns can be improved by redistribution of liquid every 5 to 10 ft.

It is not uncommon to achieve the performance of one theoretical plate in less than a foot for 1- or 1½-in. rings and saddles, though a larger allowance is appropriate for untested systems. In laboratory stills using very close-packed fine fillings, less than 1 in. is frequently required at total reflux. Such laboratory data can readily be obtained, and though they cannot be used directly for full-scale design, they will supply information as to whether the heights for transfer unit may be expected to be low or high.

CHAPTER N

WATER

The water system in a plant is of prime importance. Cooling-water costs may even exceed steam costs. Corrosive water may dictate the use of expensive alloys throughout the plant. The literature on many phases of industrial water is extensive, though cooling water, as such, has not received the attention it deserves. This chapter gives brief mention of the various factors entering into process design and discusses cooling towers, water balance, and contact coolers in modest detail. Below is listed some useful reference material:

Babbitt and Doland, "Water Supply Engineering," 5th ed., McGraw-Hill Book Company, Inc., New York, 1955.

Partridge, "Formation and Properties of Boiler Scale," *Univ. Michigan Eng. Research Bull.* 15, 1930, Ann Arbor, Mich.

Ryan, "Water Treatment and Purification," 2d ed., McGraw-Hill Book Company, Inc., New York, 1946.

Perry, "Chemical Engineers' Handbook," 3d ed., McGraw-Hill Book Company, Inc., New York, 1950.

Keenan and Kaye, "Thermodynamic Properties of Air," John Wiley & Sons, Inc., New York, 1945. (Out of print.)

Nordell, "Water Treatment for Industrial and Other Uses," 2d ed., Reinhold Publishing Corporation, New York, 1958.

1. Water Quality and Treatment

Industrial water is used for a variety of purposes, a few of which are listed below:

1. Boiler feed water.
2. Process water (*i.e.*, water used in contact or to dissolve materials in process of manufacture).
3. Hot-water heaters.
4. Cooling engine and compressor jackets, or other hot surfaces to protect mechanical equipment.
5. General cooling water.

Treated water or condensate is nearly always used for boiler feed water. Very frequently this is also true of process water, particularly if it is used in distillation or for crystallization of high-purity products. Hot-water heaters and jacket cooling water sometimes require treated water. However, water of a total hardness of 50 ppm (expressed as $CaCO_3$) is usually satisfactory. The cost of treating general cooling water is usually excessive, and the whole plant must usually be designed to utilize the available supply even though it may be mildly corrosive and excessively hard.

Because of the many complicated factors in quality of water, the empirical approach is probably the best, *i.e.*, to secure a water survey of quantities and analyses. If, by good fortune, the analysis is typical of waters in the locality, then scale and corrosion problems already encountered by existing plants afford a sound basis for water plant design.

Scale formation results from the precipitation of dissolved materials on the heating of water. Calcium sulphate, calcium carbonate, calcium hydroxide, magnesium hydroxide, and magnesium silicates are examples. Of these, calcium sulphate and carbonate are most troublesome in cooling-water service. Figure N-1 permits the estimation of solubilities of the least soluble forms of these two compounds in the absence of excess carbon dioxide. It is seldom possible to secure water that does not contain more calcium carbonate than represented by these solubilities because of the excess carbon dioxide that forms the soluble bicarbonate. This is not necessarily serious unless conditions favor rapid deposition. Sulphate hardness is usually rather troublesome. Where hard water must be employed, scale formation may be somewhat minimized by high velocity and by avoiding rapid heating. These effects can be more easily achieved by taking small temperature rises in the coolers. Keeping the maximum temperature low is also helpful.

Treatment of water is effected by one or more of five general methods:

1. Precipitation by adding chemicals (lime, soda ash, trisodium phosphate, etc.).
2. Ion exchange by inorganic zeolites.
3. Ion exchange by organic zeolites.
4. Addition of organic materials.
5. Distillation (steam condensate–multiple-effect evaporators).

Chemical precipitation removes most of the scale-forming materials but increases the total dissolved solids. Ion exchange by inorganic zeolites replaces calcium and magnesium ions by sodium. Ion exchange by organic zeolites may merely replace magnesium and calcium ions by sodium as in the case of inorganic zeolites. By a two-stage process, virtually all inorganic material other than silicic acid may be removed by organic zeolites, producing water comparable with distilled.[1] Addition of certain organic matter such as tannic acid to

[1] Nordell, *Chem. & Met. Eng.*, October, 1943, p. 112.

some extent inhibits the formation of scale. Quebracho and similar materials are sometimes added to combat the phenomenon of caustic embrittlement in high-pressure boilers.[1] Steam condensate is nearly always available in plants and, if conserved, affords a supply of high-quality water. Multiple-effect evaporators produce very high-quality water and are used when the less expensive methods are unsuitable.

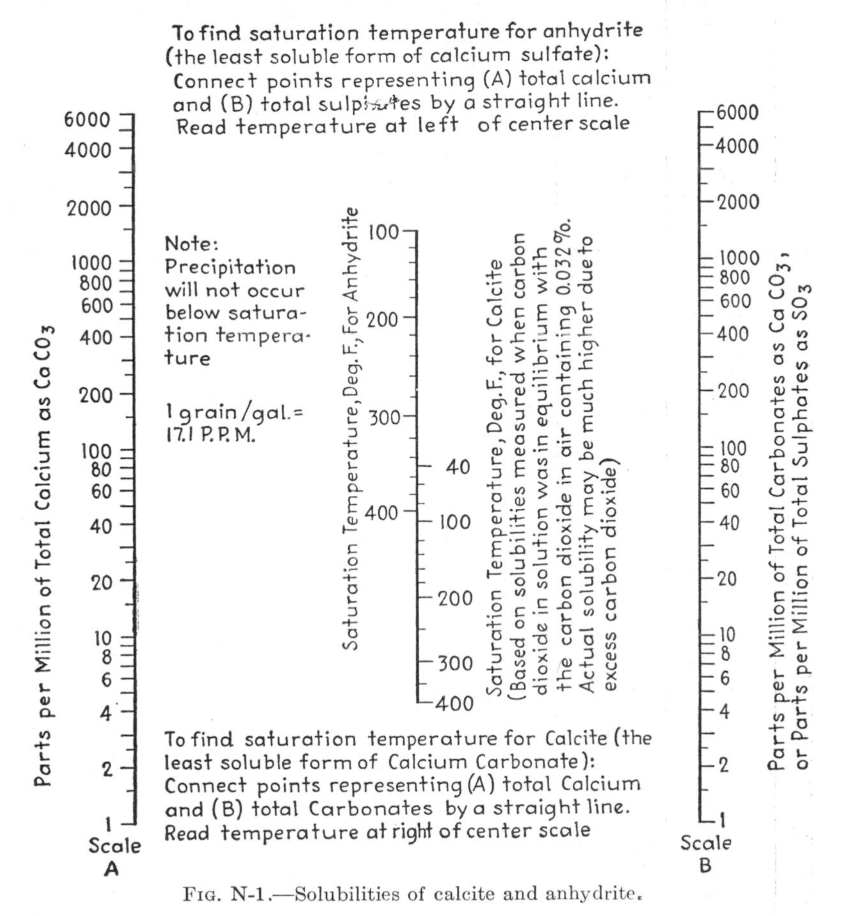

FIG. N-1.—Solubilities of calcite and anhydrite.

As some scale formation is almost inevitable, provision for its periodic removal from equipment should be made. The metaphosphates are suitable for this purpose, provided scale formation has not progressed too far. In some cases the use of corrosive acids or mechanical cleaning is necessary.

[1] Schroeder, Berk, and O'Brien, various papers in *Trans. Am. Soc. Mech. Engrs.* from 1934 on.

TABLE N-1.—PROPERTIES OF AIR MIXED WITH SATURATED WATER VAPOR (Adapted from Perry, "Chemical Engineers' Handbook," 3d ed., McGraw-Hill Book Company, Inc., New York, 1950)

°F	Saturated water vapor*		Volume, cu ft		Total heat† above 0°F, Btu		
	Pressure, psi	Lb/lb dry air	1 lb dry air	1 lb dry air + vapor to saturate it	Of 1 lb dry air above 0°F	Of vapor to saturate 1 lb dry air	Total air + water vapor
0	0.0184	0.000781	11.58	11.59	0.0	0.84	0.84
4	0.0227	0.000963	11.68	11.70	0.96	1.03	1.99
8	0.0279	0.001183	11.78	11.80	1.93	1.27	3.20
12	0.0341	0.001447	11.88	11.91	2.89	1.55	4.44
16	0.0415	0.001764	11.99	12.02	3.86	1.90	5.76
20	0.0504	0.002144	12.09	12.13	4.82	2.30	7.12
24	0.0610	0.002596	12.19	12.24	5.79	2.78	8.57
28	0.0736	0.003134	12.29	12.35	6.75	3.36	10.11
30	0.0809	0.003444	12.34	12.41	7.23	3.69	10.92
32	0.0887	0.003782	12.39	12.47	7.72	4.06	11.78
34	0.0961	0.004100	12.44	12.52	8.20	4.40	12.60
36	0.1041	0.004452	12.49	12.58	8.68	4.77	13.45
38	0.1126	0.004809	12.54	12.64	9.17	5.17	14.34
40	0.1217	0.005202	12.59	12.70	9.65	5.60	15.25
42	0.1315	0.005625	12.64	12.76	10.14	6.07	16.21
44	0.1420	0.006078	12.69	12.82	10.62	6.55	17.17
46	0.1532	0.00656	12.74	12.88	11.10	7.08	18.18
48	0.1652	0.00708	12.79	12.94	11.58	7.65	19.23
50	0.1780	0.00764	12.84	13.00	12.07	8.26	20.33
52	0.1917	0.00823	12.89	13.07	12.55	8.91	21.46
54	0.2063	0.00887	12.95	13.13	13.03	9.61	22.64
56	0.2219	0.00955	13.00	13.20	13.52	10.36	23.88
58	0.2384	0.01028	13.05	13.26	14.00	11.16	25.16
60	0.2561	0.01105	13.10	13.33	14.48	12.00	26.48
62	0.2749	0.01188	13.15	13.40	14.97	12.92	27.89
64	0.2949	0.01276	13.20	13.47	15.45	13.89	29.34
66	0.3162	0.01370	13.25	13.54	15.93	14.93	30.86
68	0.3388	0.01471	13.30	13.61	16.42	16.03	32.45
70	0.3628	0.01578	13.35	13.69	16.90	17.21	34.11
72	0.3883	0.01692	13.40	13.76	17.38	18.47	35.85
74	0.4153	0.01813	13.45	13.84	17.87	19.81	37.68
76	0.4440	0.01942	13.50	13.92	18.35	21.24	39.59
78	0.4744	0.02080	13.55	14.00	18.84	22.76	41.60

TABLE N-1.—PROPERTIES OF AIR MIXED WITH SATURATED WATER VAPOR
(*Concluded*)

°F	Saturated water vapor*		Volume, cu ft		Total heat† above 0°F, Btu		
	Pressure, psi	Lb/lb dry air	1 lb dry air	1 lb dry air + vapor to saturate it	Of 1 lb dry air above 0°F	Of vapor to saturate 1 lb dry air	Total air + water vapor
80	0.5066	0.02226	13.60	14.09	19.32	24.38	43.70
82	0.5406	0.02381	13.65	14.17	19.80	26.11	45.91
84	0.5767	0.02547	13.70	14.26	20.29	27.95	48.24
86	0.6148	0.02723	13.75	14.35	20.77	29.90	50.67
88	0.6551	0.02910	13.80	14.45	21.25	31.98	53.23
90	0.6977	0.03109	13.86	14.55	21.74	34.19	55.93
92	0.7427	0.03320	13.91	14.65	22.22	36.55	58.77
94	0.7901	0.03544	13 96	14.75	22.71	39.06	61.77
96	0.8401	0.03783	14.01	14.86	23.19	41.72	64.91
98	0.8929	0.04036	14.06	14.97	23.67	44.54	68.21
100	0.9486	0.04305	14.11	15.08	24.16	47.56	71.72
105	1.101	0.0505	14.24	15.39	25.37	55.95	81.32
110	1.274	0.0593	14.36	15.73	26.58	65.74	92.32
115	1.470	0.0694	14.49	16.10	27.79	77.16	104.95
120	1.692	0.0813	14.62	16.52	29.00	90.52	119.52
125	1.941	0.0953	14.75	16.99	30.21	106.2	136.41
130	2.221	0.1114	14.88	17.53	31.42	124.6	156.02
135	2.536	0.1305	15.00	18.13	32.63	146.2	178.83
140	2.887	0.1532	15.13	18.84	33.85	171.9	205.75
145	3.280	0.1800	15.26	19.64	35.06	202.4	237.46
150	3.716	0.2122	15.39	20.60	36.27	239.1	275.37
160	4.739	0.2987	15.64	23.09	38.69	337.8	376.49
170	5.990	0.4324	15.90	26.84	41.12	490.8	531.92
180	7.51	0.6577	16.16	33.04	43.55	749.2	792.75
190	9.34	1.0985	16.41	45.00	46.97	1255.7	1302.67
200	11.53	2.2953	16.67	77.24	48.40	2623.0	2671.40

* Below 32°F, the pressure of saturated vapor in contact with ice is given.
† The total air + water vapor added by author.

2. Cooling-tower Theory

Cooling towers remove heat from water largely by evaporation of water that is carried off by air. The amount of water vapor that can be carried off by a given quantity of air depends on its initial water-vapor content and its exit temperature. The latter determines the maximum quantity of water vapor in the exit air. The water-vapor content of air is usually determined by the familiar wet- and dry-bulb temperatures. Figure N-2, page 474, permits the estimation of the dew-point temperature and the relative humidity of air from the wet- and dry-bulb temperatures.

Table N-1, pages 470 and 471, gives thermodynamic properties of air at atmospheric pressure, both as dry air and as air saturated with water vapor; it is also useful for air containing less water vapor than at saturation. Regardless of humidity, the total heat content of air and vapor will be that of saturated air at the *wet-bulb* temperature of the air in question. Similarly, the moisture content of air is specified by the *dew-point* temperature. The specific volume of wet air cannot be obtained from Table N-1 unless the air is saturated. However, the total volume may be obtained by additions of the specific volume of dry air and the volume contributed by the moisture. The latter contribution may be computed by the following equation:

$$V_w = (18.6 + 0.0405t)W$$

where V_w = volume contributed by moisture, cu ft/lb dry air
t = temperature (dry-bulb), °F
W = lb water vapor/lb air

Example N-1 is presented below to illustrate the use of Fig. N-2 and Table N-1 in cooling-tower calculations.

Example N-1. Find the amount of air required to cool 100 gpm of water from 95 to 80°F. The air enters at 78°F dry bulb, 70°F wet bulb, and leaves at 80°F saturated.

SOLUTION:

1. Heat removed from water 95 to 80°F,

$$100 \times 500 \times 15 = 750{,}000 \text{ Btu/hr}$$
$$(1 \text{ gpm} = 500 \text{ lb/hr})$$

2. Heat removed by 1 lb air (heat contents from Table N-1),

Heat content exit, 80°F saturated.......	43.70 Btu/lb
Heat content inlet, 70°F wet bulb.......	34.11
Heat removed from water............	9.59 Btu/lb dry air

3. Pounds of dry air required,

$$\frac{750{,}000}{9.59} = 78{,}200 \text{ lb/hr}$$

4. Moisture content inlet air,

Dew point = 66°F (from Fig. N-2)
Moisture content = 0.0137 lb/lb (from Table N-1)

5. Specific volume of inlet air,

$$\begin{aligned} \text{Dry air at } 78°\text{F} &= 13.55 \text{ cu ft/lb} \quad \text{(Table N-1)} \\ \text{Moisture} &= (18.6 + 0.0405 \times 78) \times 0.0137 = (18.6 + 3.15) \times 0.0137 \\ &= 21.75 \times 0.0137 = 0.30 \text{ cu ft/lb dry air} \\ \text{Total} &= 13.55 + 0.30 = 13.85 \text{ cu ft/lb} \end{aligned}$$

6. Volume inlet air,

$$\frac{78{,}200 \times 13.85}{60} = 18{,}100 \text{ cfm}$$

7. Water evaporated,

Exit air at 80°F saturated carries.......	0.02226 lb/lb dry air
Inlet air at 66°F dew point carries......	0.01370
Net gain...........................	0.00856 lb/lb dry air

or

$$78{,}200 \times 0.00856 = 669 \text{ lb/hr}$$

or

$$\frac{669}{500} = 1.34 \text{ gpm}$$

The water lost by evaporation may be computed as in the example above, but this accuracy is seldom required. The loss is generally between 0.9 and 1 per cent of effluent water for every 10°F through which the water is cooled.

As in the case of absorbing and stripping towers, the height of a cooling tower is related to the number of theoretical plates. However, special methods are used by tower designers.

Because of evaporation in cooling towers, the concentration of dissolved salts is increased; consequently, cooling water may not be circulated indefinitely and a sufficient portion must be discarded and replaced by fresh water to keep the quality up to standard. Usually discard of about 10 per cent of the water leaving the cooling tower is satisfactory; however, the amount discarded will depend on the quality of the fresh supply and on the temperature range of the cooling tower. Example N-2, page 475, illustrates these factors.

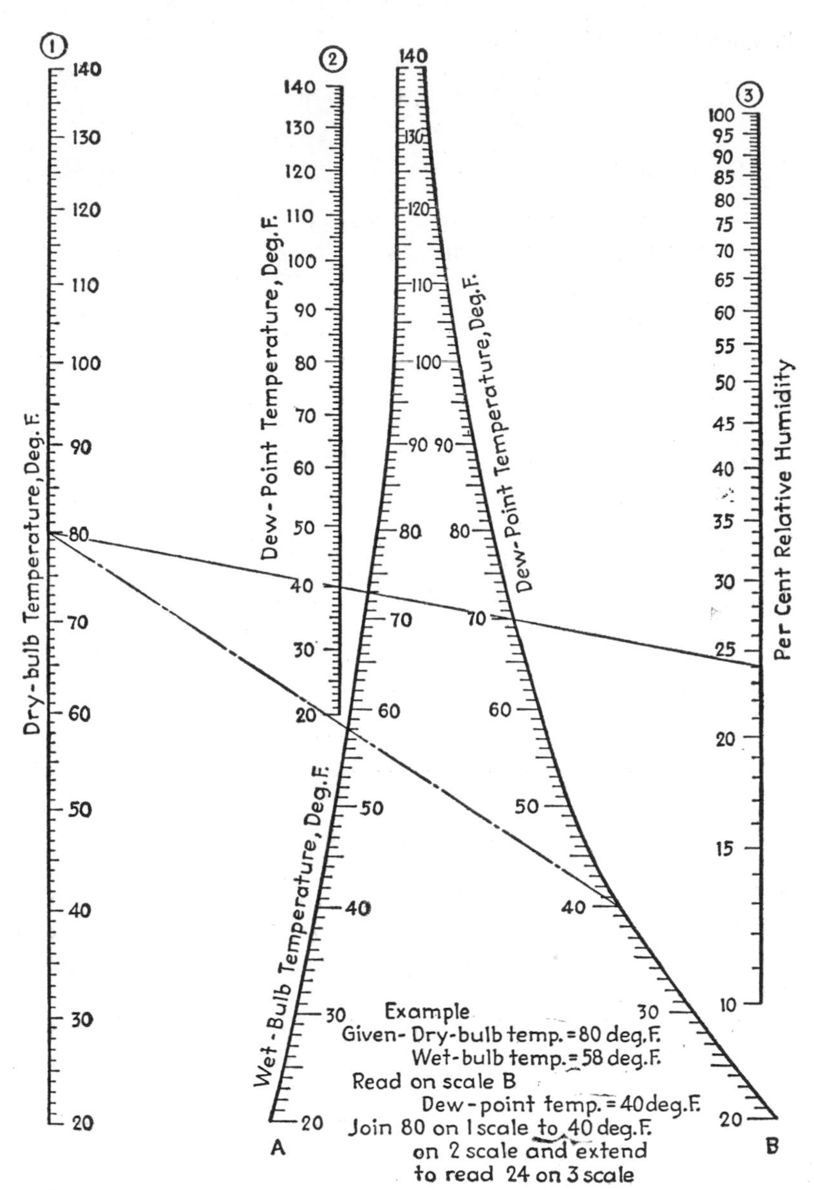

FIG. N-2.—Relation among dry bulb, wet bulb, and dew point and dry bulb, dew point, and relative humidity. (*By E. Cowan, Power, March,* 1934.)

Example N-2. A cooling tower cools 2,000 gpm from 90 to 80°F. 200 gpm of this is discarded after usage. The fresh water used for make-up has the following analysis:

Hardness, 250 ppm as $CaCO_3$
Alkalinity, 180 ppm as $CaCO_3$
Chlorides, 20 ppm as Cl
Sulphates, 25 ppm as SO_3

Find the average analysis of water leaving the cooling tower.

SOLUTION:

1. Evaporation loss,

Assume 1% for 10°F.................... 20 gpm

2. Windage loss,

Assume 1%............................ 20

3. Make-up water,

To replace discard	200
To replace evaporation loss	20
To replace windage loss	20
Total	240 gpm

4. Concentration factor:

Solids contained in 240 gpm must leave with discard and spray, *i.e.*, in 200 + 20 = 220 gpm; therefore, concentration will be increased by factor of 240/220 = 1.09.

5. Average analysis at cooling-tower exit,

Hardness, 1.09 × 250	273 ppm
Alkalinity, 1.09 × 180	196 ppm
Chlorides, 1.09 × 20	22 ppm
Sulphates, 1.09 × 25	27 ppm

6. Scale-forming tendencies. Assume that hardness is all due to calcium. From Fig. N-1, it will be seen that the water is not completely safe with respect to carbonate hardness but is safe for sulphate hardness. As the dissolved salts are only 9 per cent more than the make-up water, the quality is only slightly inferior to the fresh-water supply.

3. Weather Data

The performance of cooling towers is dependent on the local weather. Figure N-3, giving the design of wet-bulb temperatures for the United States, is based on an extensive survey and is reliable for open country. In very hilly or mountainous country, local variations may have to be taken into account. In any event a comparison of the local temperature with near-by weather stations will (in conjunction

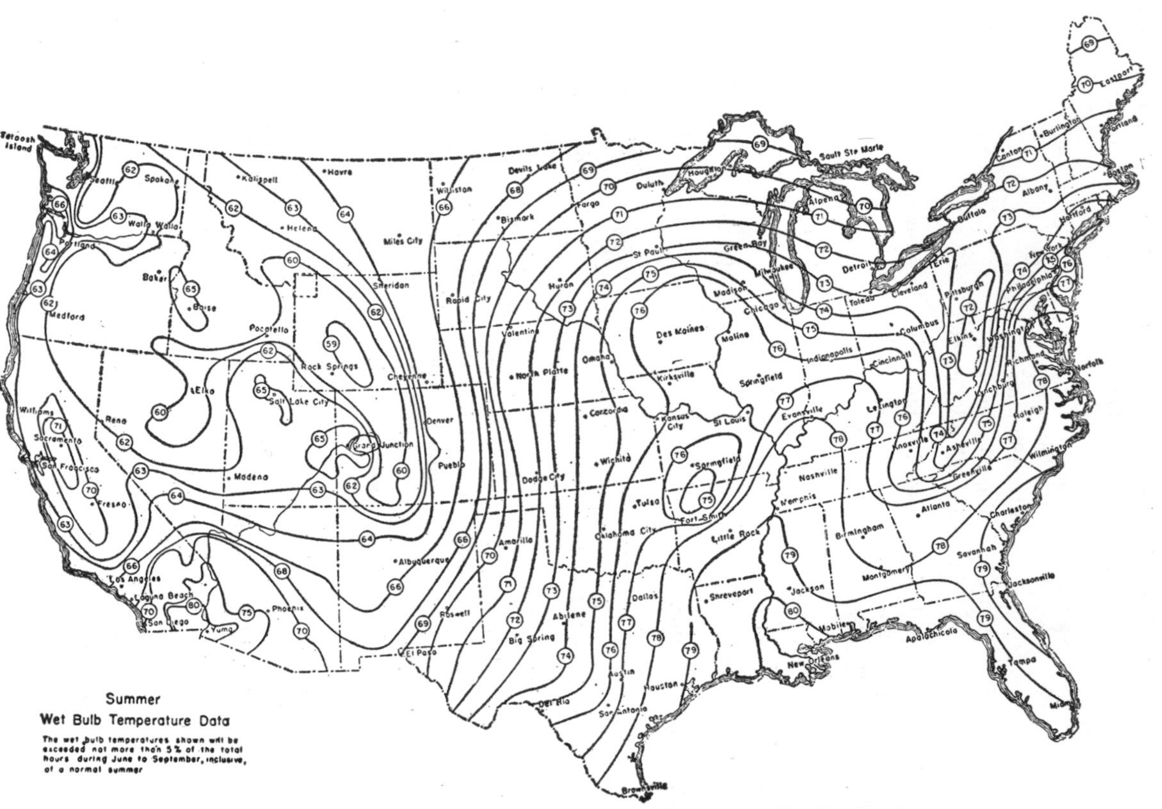

FIG. N-3.—Summer wet-bulb-temperature data. (*Courtesy of The Marley Company.*)

with Fig. N-3) permit the choice of a wet-bulb temperature adequate for design.

Wind velocities are subject to wide local variations. A 3-mph velocity is usually safe for design factors based on the minimum. Maximum wind velocities determine the mechanical strength necessary in the construction of cooling towers and, for that matter, of all equipment and buildings. To a small extent, cooling-tower performance depends on the atmospheric pressure. Altitude largely determines the barometric pressures (see Fig. C-1). Winter temperatures are important in regard to freeze-up problems.

4. Cooling Towers and Spray-pond Performance

Cooling towers may secure contact of water and air either by natural draft or by mechanical draft. Natural-draft towers are sometimes constructed with chimneys to increase the air flow, but they usually depend largely on the velocity of the prevailing wind. Because of variations in direction and velocity of the wind, their performance is somewhat changeable. However, if the tower design is adequate for the most unfavorable conditions, the resulting variation in water temperature is seldom serious enough to produce difficulty in process control. The use of fans to produce mechanical draft gives somewhat steadier performance and reduces the sizes required where space is limited. The fans may be placed either at the base or the top of the tower. Spray ponds are frequently used where spray loss is not serious and where their performance meets the service demands. Performance data are given as follows:

An atmospheric cooling tower.................. Fig. N-4
An induced-draft cooling tower................ Figs. N-5 and N-6

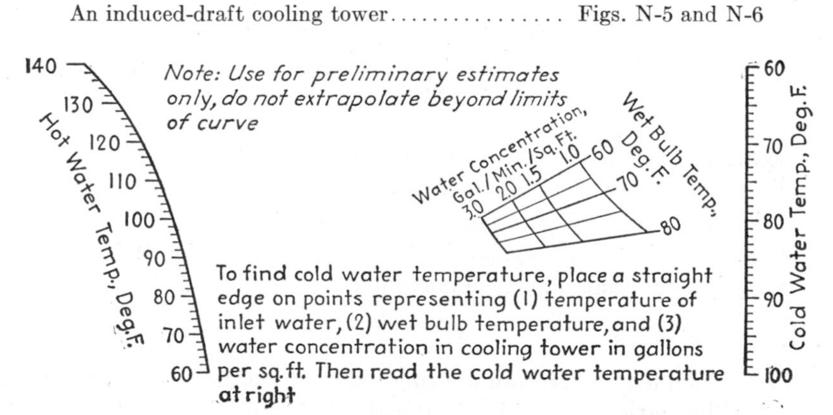

FIG. N-4.—Performance curve of a 12-deck, 35-ft-high atmospheric cooling tower at a wind velocity of 3 mph. (*By permission, The Fluor Corp., Ltd.*)

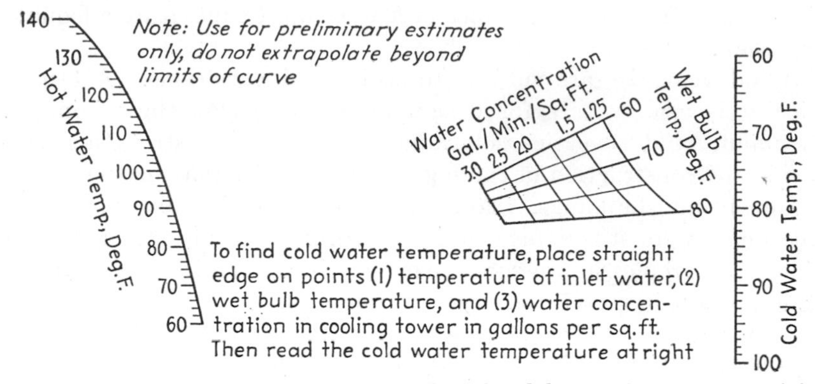

FIG. N-5.—Performance curve of a counterflow, induced-draft cooling tower containing 24 ft of filling. (*By permission, The Fluor Corp., Ltd.*)

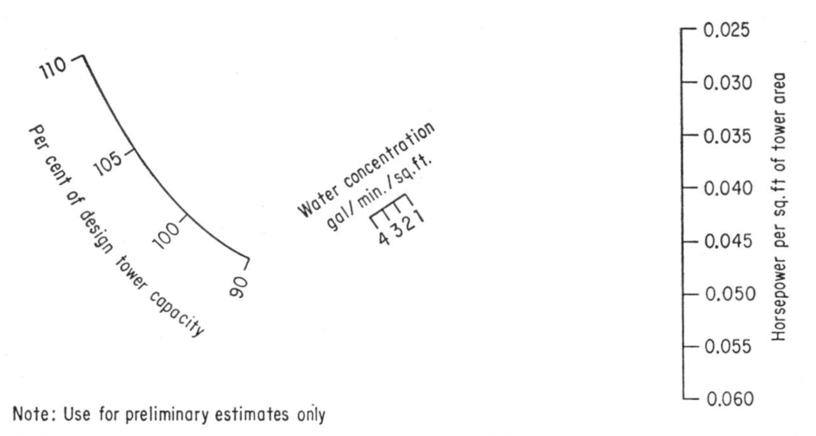

Note: Use for preliminary estimates only

To find fan horsepower requirements, place straight edge at 100% on design tower capacity curve and at water concentration used in Fig. N-5. Fan horsepower requirement shown multiplied by total tower area (as found in Fig. N-5) will give total horsepower requirements.

FIG. N-6.—Fan-horsepower curve for induced-draft cooling towers with 24 ft of filling. (*By permission, The Fluor Corp., Ltd.*)

These are not adequate for mechanical design but are suitable for an intelligent choice of temperatures for preliminary process design and for estimating space requirements prior to quotations from the manufacturer. Cooling towers should be kept reasonably distant from other equipment that might be corroded by the spray.

5. Water Balances

The principles involved in making balances on plant water are best given by illustration. An idealized water balance, given in Table N-2, is based on the following assumptions:

1. Treated water or condensate is used for all boiler feed and process water except contact coolers.
2. Treated water is used in engine jackets, with closed recirculation and cooling by exchange with cooling-tower water. Only pump leakage and similar losses to be made up by treated water. These losses are assumed to be 2 per cent of the amount circulated.
3. 1,000 gpm of fresh water is available at 72°F. Some of this will be treated for boiler use, etc., the remainder to be used in services requiring the coldest available water. Finally, this water will go to the cooling tower.
4. Additional cooling-water requirements to be supplied by cooling-tower water. The cooling tower is assumed to supply 85°F water when the feed is 100°F.

TABLE N-2.—AN IDEALIZED WATER BALANCE
All figures in gpm
Fresh-water Balance

Treated water used:		
Boiler feed	72	
Process water	112	
Engine jackets	8	
Total	192	
Less condensate returned	52	
Total water to treatment plant		140
Fresh water for cooling		770
Total fresh water required		910
Installed capacity		1,000
Overage		90

Cooling-tower Balance

	Service		
	Supplied	Returned	Lost or discarded
85 to 90°F	1,200	1,200	0
85 to 100°F	4,300	4,210	90
85 to 105°F	1,500	1,300	200
85 to 110°F and above	300	0	300
Subtotals	7,300	6,710	590
Evaporation loss	......	.....	108
Spray loss	......	.....	72
Totals	7,300	.....	770
Installed capacity	10,000		
Fresh water	......	.....	770
Overage	2,700	.....	None

The water balance is, of course, affected by the temperature rises taken throughout the plant. Only small rises can be taken for some purposes, such as use in ammonia condensers. Large ones may be taken in other cases such as in stills, where the product cooler, vent cooler, and condenser may be placed in series. Consequently, no general rule can be given. The average rise should be consistent with good cooling-tower design. If very hard water is employed, small temperature rises are recommended.

6. Contact Coolers

Because heat-transfer coefficients of gases are frequently low, it is sometimes desirable to cool gases in a tower by direct contact with water or other liquid. Although either tray or packed towers may be employed, the latter are generally preferred for this service. Barometric condensers are essentially contact coolers, and the same construction is suitable for quenching hot gases. Contact coolers may be employed for any pressure service, provided the gas being cooled does not have too high a solubility in the cooling liquid. The principles used in estimating these towers are the same as for cooling water towers.

Published information on the performance of contact coolers is very meager. Parekh,[1] however, has studied the cooling of air at atmospheric pressure by water in a column packed with Raschig rings. Using an empirical correlation of his results together with the definition of the transfer unit, the following equation has been developed. To what extent these data can be extrapolated to gases of different specific gravity, and to pressures other than atmospheric, cannot be stated. The heights required, however, are usually low so that large factors of safety may frequently be employed without incurring unduly high costs.

$$Z = a(H_T - H_B)\left[\frac{\ln(\Delta H_B/\Delta H_T)}{\Delta H_B - \Delta H_T}\right]\sqrt{\frac{G}{L}} = a\frac{H_T - H_B}{\Delta H_m}\sqrt{\frac{G}{L}}$$

where Z = height of packing required, ft

H_T = heat content of exit gas and water vapor, Btu/lb of dry gas

H_B = heat content of inlet gas and water vapor, Btu/lb of dry gas

$\Delta H_T = H_T^* - H_T$

$\Delta H_B = H_B^* - H_B$

[1] Quoted by McAdams, "Heat Transmission," 3d ed., McGraw-Hill Book Company, Inc., New York, 1954. Some other data are also presented, and the definition of transfer units based on heat transferred is given.

H_T^* = heat content of gas saturated with water vapor at the inlet water temperature, Btu/lb of dry gas

H_B^* = heat content of gas saturated with water vapor at the exit-water temperature, Btu/lb of dry gas

ΔH_m = logarithmic mean of ΔH_T, ΔH_B

G/L = lb of dry gas per lb of water circulated

a = 1.4 for 1-in. Raschig rings

= 1.7 for 1½-in. Raschig rings

= 2.1 for 2-in. Raschig rings

INDEX[1]

[1] Page references in boldface are more important than the ones preceding.